Principles and Applications of Quinoproteins

Principles and Applications of Quinoproteins

edited by

Victor L. Davidson
The University of Mississippi Medical Center
Jackson, Mississippi

Marcel Dekker, Inc. New York • Basel • Hong Kong

Library of Congress Cataloging-in-Publication Data

Principles and applications of quinoproteins / edited by Victor L.
Davidson.
p. cm.
Includes bibliographical references and index.
ISBN 0-8247-8764-1
1. Quinoproteins. 2. PQQ (Biochemistry) I. Davidson, Victor L.
QP552.Q55P75 1992
574.19'258--dc20 92-26045
CIP

This book is printed on acid-free paper.

MARCEL DEKKER, INC.
270 Madison Avenue, New York, New York 10016

Current printing (last digit):
10 9 8 7 6 5 4 3 2 1

PRINTED IN THE UNITED STATES OF AMERICA

Preface

The study of quinoproteins and their quinonoid cofactors is a relatively young and rapidly emerging field. The first major symposium on this topic was held in 1988, and aside from the proceedings of that symposium, this is the first book devoted entirely to basic and applied research on quinoproteins and quinonoid cofactors such as PQQ. It is rare that a relatively new field touches upon as wide a range of disciplines as does quinoprotein research. Recent meetings on this topic have drawn experts from such diverse fields as enzymology, microbiology, molecular biology, nutrition, pharmacology, analytical chemistry, physical biochemistry, medicine, and organic chemistry. Not surprisingly, the literature on quinoproteins can be found among a wide range of journals that relate to these disciplines. Because of this many people have heard of quinoproteins, but very few, including most workers in the field, are aware of the full body of knowledge that has accumulated. The intention of this book is to provide a single, comprehensive source of information on all aspects of basic and applied research that relate to quinoproteins and quinonoid cofactors. It will provide information to investigators on recent developments in their own and related areas, and also educate the general reader on the scope, direction, and potential value of research in this multi-disciplinary field.

Victor L. Davidson

Contents

Contributors

Osao Adachi Department of Biological Chemistry, Yamaguchi University, Yamaguchi, Japan

Christopher Anthony Biochemistry Department, University of Southampton, Southampton, England

Peter R. Bergethon Department of Biochemistry, Boston University School of Medicine, Boston, Massachusetts

David Bui Department of Nutrition, University of California, Davis, California

Doreen E. Brown Department of Chemistry, Amherst College, Amherst, Massachusetts

Victor L. Davidson Department of Biochemistry, The University of Mississippi Medical Center, Jackson, Mississippi

David M. Dooley Department of Chemistry, Amherst College, Amherst, Massachusetts

Rudolf Flückiger Laboratory of Human Biochemistry, Children's Hospital Medical Center, Harvard Schools of Medicine and Dental Medicine, Boston, Massachusetts

Paul M. Gallop Laboratory of Human Biochemistry, Children's Hospital Medical Center, Harvard Schools of Medicine and Dental Medicine, Boston, Massachusetts

Christa Hartmann* Department of Pharmaceutical Chemistry, University of California, San Francisco, California

Edward Henson Laboratory of Human Biochemistry, Children's Hospital Medical Center, Harvard Schools of Medicine and Dental Medicine, Boston, Massachusetts

Norio Hobara Department of Hospital Pharmacy, Okayama University Medical School, Okayama, Japan

Wim G. J. Hol BIOSON Research Institute, University of Groningen, Groningen, The Netherlands

Shinobu Itoh Department of Applied Chemistry, Osaka University, Osaka, Japan

Herbert M. Kagan Department of Biochemistry, Boston University School of Medicine, Boston, Massachusetts

Isao Karube Research Center for Advanced Science and Technology, University of Tokyo, Tokyo, Japan

Jack Kilgore Department of Nutrition, University of California, Davis, California

Yasushi Kitagawa Applied Chemistry Research Department, Asahi Breweries, Ltd., Tokyo, Japan

Kerstin Laufer Keck Laboratories, California Institute of Technology, Pasadena, California

*Current affiliation: Department of Chemistry, Temple University, Philadelphia, Pennsylvania.

Mary E. Lidstrom Keck Laboratories, California Institute of Technology, Pasadena, California

F. Scott Mathews Department of Cell Biology and Physiology, Washington University School of Medicine, St. Louis, Missouri

Kazunobu Matsushita Department of Biological Chemistry, Yamaguchi University, Yamaguchi, Japan

William S. McIntire Molecular Biology Division, Department of Veterans Affairs Medical Center and Departments of Biochemistry and Biophysics, and Anesthesia, University of California, San Francisco, California

A. Netrusov Microbiology Department, Moscow University, Moscow, Russia

Hideo Nishigori Faculty of Pharmaceutical Sciences, Teikyo University, Kanagawa, Japan

Yoshiki Ohshiro Department of Applied Chemistry, Osaka University, Osaka, Japan

Mercedes A. Paz Laboratory of Human Biochemistry, Children's Hospital Medical Center, Harvard Schools of Medicine and Dental Medicine, Boston, Massachusetts

Nadia Romero-Chapman Department of Nutrition, University of California, Davis, California

Robert B. Rucker Department of Nutrition, University of California, Davis, California

Carsten Smidt Department of Nutrition, University of California, Davis, California

Francene M. Steinberg Department of Nutrition, University of California, Davis, California

Philip C. Trackman Department of Biochemistry, Boston University School of Medicine, Boston, Massachusetts

David Tran Department of Nutrition, University of California, Davis, California

Toshihiro Tsuchida Third Department of Internal Medicine, Toyama Medical and Pharmaceutical University, Toyama, Japan

Clifford J. Unkefer Isotope and Nuclear Chemistry Division, Los Alamos National Laboratory, Los Alamos, New Mexico

Teizi Urakami Biochemical Division, Mitsubishi Gas Chemical Company, Tokyo, Japan

Akiharu Watanabe Third Department of Internal Medicine, Toyama Medical and Pharmaceutical University, Toyama, Japan

Kenji Yokoyama Research Center for Advanced Science and Technology, University of Tokyo, Tokyo, Japan

I

Introduction

1

A Brief History of Quinoproteins

Victor L. Davidson

The University of Mississippi Medical Center, Jackson, Mississippi

I. THE DISCOVERY OF PYRROLOQUINOLINE QUINONE (PQQ)

The history of quinoproteins began in the 1960s. During that decade the investigations of two very different phenomena resulted in what is now recognized as the birth of quinoproteins. Research in the area of nonphosphorylative bacterial glucose metabolism revealed that a pyridine nucleotide-independent enzyme was responsible for the primary oxidation of the sugar substrate. Hauge [1] reported in 1964 the characterization of a bacterial glucose dehydrogenase with a dissociable cofactor that was neither a pyridine nucleotide nor a flavin. This cofactor had unprecedented properties, and it was suggested that it could be a substituted napthoquinone. At approximately the same time, several researchers began to study in detail the process of bacterial methanol metabolism. The realization that several bacterial species were capable of growth on either methane or methanol as a sole source of carbon had stimulated interest not only among microbial physiologists but also among commercial interests, which hoped to develop economically viable fermentation processes based on methylotrophic bacteria. A key enzyme in bacterial methanol metabolism is methanol dehydrogenase. Studies of this enzyme revealed that, like the above-mentioned glucose dehydrogenase, it also possessed a dissociable organic cofactor that was neither a pyridine nucleotide nor a flavin. Based upon its fluorescence properties, Anthony and Zatman proposed in 1967 [2] that this cofactor might be an unusual pteridine. For the next decade the

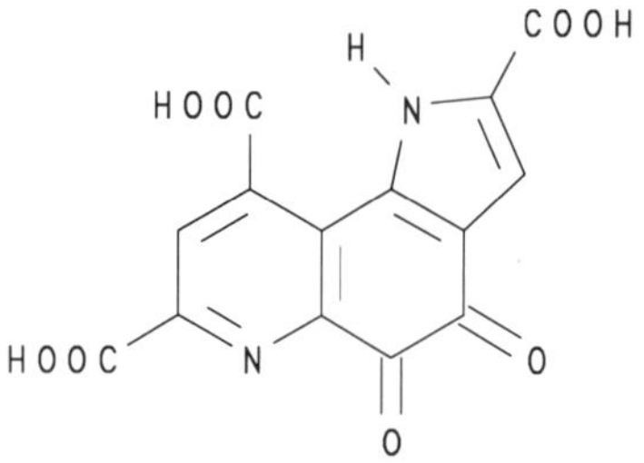

Figure 1 The structure of pyrroloquinoline quinone (PQQ).

status of the identity of these unusual cofactors remained unchanged. In the late 1970s Duine and coworkers reinvestigated the question of the structure of the cofactor of methanol dehydrogenase. Studies that applied electron spin resonance, nuclear magnetic resonance, and mass spectroscopy to this problem led to the proposal in 1979 that the cofactor was not a pteridine but a quinone structure that contained two nitrogen atoms [3]. At approximately the same time, the precise structure of this cofactor was deduced by Salisbury et al. [4] from X-ray diffraction studies of a crystalline acetone adduct of the cofactor. The structure was that of 4,5-dihydro-4,5-dioxo-1-H-pyrrolo[2,3-f]quinoline-2,7,9-tricarboxylic acid (Fig. 1). For a brief time this cofactor was referred to by the trivial name methoxatin. It is now most commonly called pyrroloquinoline quinone, or PQQ.

II. THE CHARACTERIZATION OF PQQ-BEARING QUINOPROTEINS

Once the structure and certain properties of PQQ were known, it became possible to design analytical methods to detect PQQ in proteins and biological fluids. Early methods were based upon the comparison of certain physical and chemical properties, such as fluorescence [5] of the dissociable cofactors or derivatives of such molecules, with authentic PQQ and derivatives of it. Furthermore, it was demonstrated that PQQ could reconstitute the biological activity of an apoenzyme form of glucose dehydrogenase [6]. Reconstitution was possible with either the purified apoenzyme or with crude membrane preparations that contained the apoenzyme. This finding was the basis for development of a biological assay to identify and quantitate the presence of free PQQ in biological fluids and in extracts of proteins or cells. Given the knowledge of PQQ and the availability of analytical techniques to screen for its presence, a large number of bacterial methanol and glucose dehydrogenases were shown to possess noncovalently bound PQQ as a cofactor. Using the above-mentioned techniques, a bacterial pyridine nucleotide–independent aldehyde dehydrogenase [7] was also identified. Subsequently, several similar PQQ-dependent bacterial enzymes were identified that catalyzed transformations of the following compounds: long-chain alcohols and polyethylene glycol [8,9], secondary and polyvinyl alcohols [10], polyhydroxy alcohols [11], hydroaromatics [12], lactate [13], and nitriles [14]. Thus, by the early 1980s it became clear that PQQ was a widely distributed and important cofactor in prokaryotic systems.

III. THE CHARACTERIZATION OF COVALENTLY BOUND QUINONE PROSTHETIC GROUPS

Whereas the characterization of PQQ from enzymes in which it is freely dissociable had become a reasonably straightforward process, the question of whether PQQ functioned as a covalently bound prosthetic group in other enzymes proved to be a far more difficult and controversial topic to address. An inherent problem in efforts to isolate covalently bound PQQ from enzymes lies in the *o*-quinone structure, which exhibits a high level of reactivity with amino acid side chains of proteins. As such it has never been possible to isolate native PQQ from a protein after it has been subjected to proteolysis and the subsequent treatments necessary to isolate those peptides that possess the covalently bound prosthetic group. The first suggestion of an enzyme with a novel covalently bound quinone was presented in 1980. Duine and coworkers proposed [15], based upon its redox behavior and electron spin resonance and electron-nuclear double resonance spectra, that bacterial methylamine dehydrogenase possessed such a prosthetic group. The absorption spectra of methylamine dehydrogenases were, however, significantly different from those exhibited by methanol and glucose dehydrogenases [16,17], leaving open to question the precise nature of this cofactor. In 1987, van der Meer et al. [18] reported that this cofactor was in fact PQQ based upon analysis of a phenylhydrazine derivative of the isolated cofactor. As discussed later, this conclusion proved to be incorrect.

Given the demonstrated presence of PQQ in bacterial enzymes, it was only natural that investigators would begin to search for this cofactor in eukaryotic enzymes. The foci of these studies were known enzymes for which the precise identity of an organic cofactor was uncertain and enzymes for which convincing reaction mechanisms were difficult to postulate given the presence of the redox center known to be present. The first class of eukaryotic enzymes proposed to contain such a previously unrecognized quinone prosthetic group was the copper-containing amine oxidases. These enzymes had long been thought to possess pyridoxal phosphate in addition to copper. In 1984 two independent studies implicated covalently bound PQQ as a redox cofactor of amine oxidases. Lobenstein-Verbeek et al. [19] based this claim on HPLC analysis of a dinitrophenylhydrazine derivative of the cofactor, and Ameyama et al. [20] based their proposal on fluorescence properties and the biological activity of acid hydrolysates of the enzyme. Later, Dooley and coworkers employing resonance Raman spectroscopy [21] and Klinman and coworkers utilizing a chemical reductive trapping technique [22] provided additional evidence that pyridoxal phosphate was not a cofactor for the amine oxidases and that a quinone similar to PQQ was more likely present as a redox cofactor. In 1986, two independent studies based upon the HPLC analysis of dinitrophenylhydrazine derivatives [23] and resonance Raman spectroscopy [24] suggested that mammalian lysyl oxidase also contained a PQQ-like species rather than pyridoxal phosphate as a cofactor. In the following years several reports, primarily from Duine and coworkers, indicated that PQQ was a redox cofactor for a variety of eukaryotic oxidoreductases. These enzymes included dopamine β-hydroxylase [25], DOPA decarboxylase [26], galactose oxidase [27], soybean lipoxygenase [28], and laccase [29]. In each of these studies, identification of PQQ was based upon the analysis of derivatized cofactors that were isolated from enzymes treated with phenylhydrazines or hexanol as a part of the derivatization protocol. As a result of these studies with eukaryotic systems, it had become acceptable by the late 1980s to believe that PQQ was a universally distributed cofactor relevant to both prokaryotes and to higher organisms.

Concomitant with the growing acceptance of PQQ, however, was the accumulation of data that cast serious doubt on the presence of PQQ in mammalian enzymes. The first convincing report to the contrary was that of James et al. [30] in 1990, which provided strong evidence that the organic redox prosthetic group of bovine serum amine oxidase was not PQQ, but a modified tyrosine residue, 6-hydroxydopa or topaquinone (Fig. 2). This report of a quinone structure other than PQQ at the active site of this enzyme was soon followed by reports that the bacterial methylamine dehydrogenase also possessed a modified aromatic amino acid rather than covalently bound PQQ at its active site [31–33]. In this case the prosthetic group was derived from two gene-encoded tryptophan residues that had been covalently linked and to which two oxygens had been inserted to generate an indole quinone structure, which was designated tryptophan tryptophylquinone or TTQ (Fig. 3). At this point it became clear that the reference to quinoproteins as

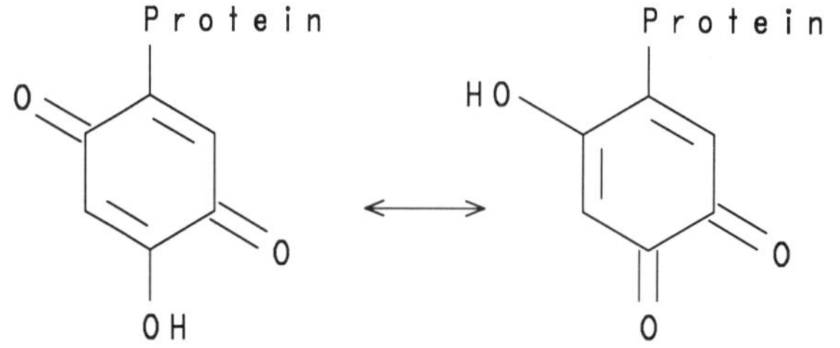

Protein O O OH ⟷ Protein HO O O

Figure 2 The structure of topaquinone. The two possible tautomeric forms of the prosthetic group are shown.

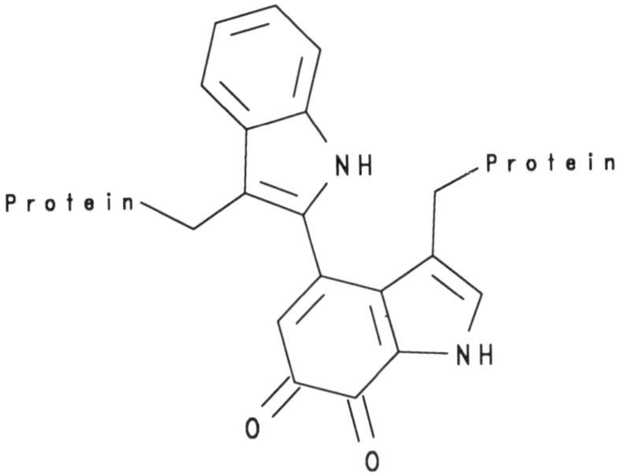

Figure 3 The structure of tryptophan tryptophylquinone (TTQ).

enzymes that possessed PQQ would have to be reevaluated. It was apparent that while many enzymes classified as quinoproteins did possess PQQ, some of these enzymes, particularly those with covalently bound prosthetic groups, possessed instead other previously unrecognized and novel quinone species at their active sites.

For certain other enzymes that had been purported to contain PQQ, strong evidence was subsequently presented that no external redox cofactor was present at all. For galactose oxidase, it was demonstrated that the redox role that had been attributed to PQQ was in fact performed by a tyrosyl radical [34]. X-ray crystallographic studies [35] confirmed this finding and further indicated that a covalent bond existed between a cysteine residue and the tyrosine residue in question. Extensive analysis of dopamine β-hydroxylase [36] and soybean lipoxygenase [37] also revealed no evidence for the presence of PQQ in these enzymes. The reasons for the apparent artifactual identification of PQQ as the cofactor of these enzymes has never been fully explained. It is now quite clear, however, that the protocols used in those studies, the so-called hydrazine [18] and hexanol extraction [38] methods, are inappropriate for the analysis of covalently bound cofactors. Given this history of conflicting reports regarding the presence of PQQ in proteins, it is clear that refinement of the analytical techniques used in the identification and quantitation of PQQ and related quinonoid cofactors is critical to the continued development of this field. Furthermore, only by direct structural analysis, preferably of the underivatized protein-bound prosthetic group, as has been done with methylamine dehydrogenase [33] and galactonse oxidase [35], will one be able to conclude with absolute certainty its precise nature.

As a consequence of the above-mentioned studies, it now appears possible that the original PQQ cofactor described in 1979 may only be found noncovalently bound in prokaryotic enzymes. Novel covalently bound quinone prosthetic groups are, however, clearly functioning in a number of important eukaryotic enzymes. It is also evident that aromatic amino acids, as a result of previously unrecognized posttranslational modifications or peculiarities of their immediate environment within the enzyme molecule, may exhibit novel catalytic and redox roles that had not before been thought to exist. These observations have certainly revolutionized the fields of enzymology and bioenergetics and placed an important focus on the continued characterization of the precise structures and functions of quinoproteins.

IV. THE PHYSIOLOGICAL RELEVANCE OF PQQ

The purported presence of PQQ in mammalian enzymes led to a great deal of speculation that this cofactor could be a previously unrecognized vitamin and provided the impetus for several experimental studies related to the nutritional and biomedical relevance of PQQ to mammals. To place these studies in their proper perspective, it is first necessary to review a number of significant observations that relate to the physiological roles of PQQ in prokaryotic systems.

A peculiar and very interesting observation regarding prokaryotic quinoproteins is that in many bacteria the synthesis of the apoenzyme and PQQ is not coordinate [39]. It was also demonstrated that certain bacteria synthesized apoglucose dehydrogenase but did not synthesize PQQ [40]. These apoenzymes became activated only when the bacterium was provided with an external source of PQQ. Furthermore, it was demonstrated that many methylotrophic bacteria, which depend heavily for survival upon the action of the periplasmic quinoprotein methanol dehydrogenase, excreted large amounts of free PQQ into their growth medium [41]. From these findings, it could be argued that for certain bacteria PQQ was a vitamin that was synthesized and excreted into the environment by other bacteria. This phenomenon was well characterized by Ameyama and coworkers [42,43], who clearly demonstrated the effects of growth stimulation of certain microorganisms by PQQ.

The demonstration that PQQ could function as a growth factor for microorganisms, and the belief at that time that PQQ was present as the cofactor for certain critical mammalian enzymes, led to the speculation that PQQ could perhaps be a previously unrecognized vitamin for mammals as well. This hypothesis was tested by Rucker and coworkers, who examined the nutritional importance of PQQ to mice and reported in 1989 that, indeed, mice fed a diet deficient in PQQ failed to develop properly [44]. Many of the manifestations of PQQ deficiency appeared to be related to defects in the activity of lysyl oxidase, a quinoprotein. Thus, at the same time that evidence was accumulating against the presence of PQQ in mammalian enzymes, nutritional studies suggested that PQQ was an essential factor for mammals.

Another question raised by the suggestions that PQQ was present in mammalian enzymes was the possibility that free PQQ could exist in mammalian tissues and fluids as well. If true, this would pose significant biomedical ramifications. PQQ, as other quinones, has the capability of undergoing a process called redox cycling. This involves a reaction with molecular oxygen that generates the superoxide radical species. It was speculated that if free PQQ were present in mammalian tissues and allowed to react in such a manner, then it could contribute significantly to the condition known as oxidative stress. Conversely, the structure of PQQ and related quinones is such that under appropriate circumstances it could also act as a scavenger of free radicals and as such be of protective or therapeutic value. The potential for therapeutic roles of PQQ was highlighted by reports of the preventative effects of PQQ against experimental liver injury in rats [45,46] and against the glucocorticoid-induced formation of cataracts in developing chick embryos [47]. A key issue in addressing the relevance of PQQ in mammalian systems is the methodology for identifying and quantitating PQQ and related quinonoid species. As mentioned above, the appropriateness of the techniques used in the identification of PQQ in proteins has been a controversial point. This controversy has extended to the analysis of free PQQ. In 1989, Paz et al. [48] described a highly sensitive colorimetric assay referred to as the redox cycling assay. Using this assay, the presence of PQQ could be demonstrated in a number of biological fluids. The validity of the redox cycling assay has been vigorously disputed by Duine and coworkers [49–51], but nonetheless has gained wide acceptance. An amperometric electrochemical method for the detection of PQQ has also been developed [52] recently, and data obtained with this technique appear to provide support for the validity of the redox cycling assay.

Regarding the physiological relevance of PQQ to mammals and other eukaryotes, the sum of the evidence accumulated in the past few years clearly indicates that PQQ and related compounds are capable of exerting a variety of effects in mammalian systems. Using the bioassay for PQQ, Xiong et al. [53] also demonstrated the occurrence of PQQ in plants and reported its ability to stimulate pollen germination. The presence of PQQ in animals and plants would seem to be at variance with the revelations that PQQ is perhaps not a cofactor in eukaryotic enzymes. Further study is required to fully understand the precise functions of PQQ and of quinoproteins in higher organisms.

V. QUINOPROTEINS AND BIOTECHNOLOGY

While applied biomedical research on PQQ and related quinonoid compounds has generated a great deal of excitement, there has been relatively little commercial exploitation to date of quinoproteins themselves. This is somewhat ironic in that the discovery of quinoproteins resulted, in large part, from commercial interest in methanol dehydrogenase and the methylotrophic bacteria that synthesized the enzyme. This lack of progress has been due, in part, to a lack of understanding of the physiology and genetics of these bacteria. Recent advances in the molecular biology of methylotrophs may change this situation [54]. However, at least two promising quinoprotein-based biocatalytic processes have been described thus far [55,56], and the potential for using quinoproteins, either isolated or in their host cells, to perform catalytic reactions of commercial interest remains strong.

Another area in which researchers have attempted to exploit quinoproteins for commercial purposes is that of biosensor technology. The quinoproteins characterized thus far utilize a wide range of substrates including sugars, alcohols, aldehydes, and amines. Devices capable of detecting such compounds have potential applications in the areas of medicine, fermentation, food processing, and environmental studies. Enzyme-based biosensors could theoretically be used for this purpose. As efficient coupling of the enzyme and sensor is critical to the overall performance of such a device, it would be desirable to employ an enzyme that could transfer electrons directly to an electrode. Given that quinoprotein dehydrogenases do not require additional soluble cofactors and normally donate electrons to a redox center within the complex macromolecular matrix of another protein, they have been implicated as possible candidates for use in the development of bioanalytical devices [57–59]. The construction of quinoprotein-based biosensors that utilize glucose dehydrogenase [60], alcohol dehydrogenase [61], and methanol dehydrogenase [62] have been reported thus far.

VI. QUINOPROTEINS AND PQQ—WHAT LIES AHEAD?

The field of quinoprotein research is relatively young, and the body of information on quinoproteins and quinone cofactors and prosthetic groups, particularly PQQ, is growing at an increasingly rapid rate. The further characterization of the structures, reaction mechanisms, and electron transfer properties of quinoproteins should remain an active and productive area. The knowledge to be gained from these studies will certainly further our understanding of fundamental processes such as enzymatic catalysis, long-range biological electron transfer, and protein biosynthesis. Much of this knowledge should also be transferable to applied quinoprotein-based technologies. Further characterization of the biological roles of PQQ and related compounds in higher organisms will determine the potential for application of these compounds to problems of pharmacological and biomedical relevance. Thus, future research in this field promises not only the clarification of past controversies, but also the continued discovery of new and exciting properties and applications of quinoproteins and PQQ.

REFERENCES

1. Hauge, J. G. (1964). Glucose dehydrogenase from *Bacterium antiratum*: An enzyme with a novel prosthetic group. *J. Biol. Chem.*, *239*: 3630.
2. Anthony C., and Zatman, L. J. (1967). The microbial oxidation of methanol. The

prosthetic group of the alcohol dehydrogenase of *Pseudomonas* sp. M27. *Biochem. J.*, *104*: 960.
3. Westerling, J., Frank, J., and Duine, J. A. (1979). The prosthetic group of methanol dehydrogenase from *Hyphomicrobium* X: Electron spin resonance evidence for a quinone structure. *Biochem. Biophys. Res. Commun.*, *87*: 719.
4. Salisbury, S. A., Forrest, H. A., Cruse, W. B. T., and Kennard, O. (1979). A novel coenzyme from bacterial primary alcohol dehydrogenases. *Nature*, *280*: 843.
5. Duine, J. A., Frank, J., and Jongejan, J. A. (1983). Detection and determination of pyrroloquinoline quinone, the coenzyme of quinoproteins. *Anal. Biochem.*, *133*: 239.
6. Ameyama, M., Nonobe, M., Shinagawa, E., Matsushita, K., and Adachi, O. (1985). Method of enzymatic determination of pyrroloquinoline quinone. *Anal. Biochem.*, *151*: 263.
7. Ameyama, M., and Adachi, O. (1982). Aldehyde dehydrogenase from acetic acid bacteria, membrane-bound. *Methods Enzymol.*, *89*: 491.
8. Kawai, F., Yamanaka, H., Ameyama, M., Shinagawa, E., Matsushita, K., and Adachi, O. (1985). Identification of the prosthetic group and further characterization of a novel enzyme, polyethylene glycol dehydrogenase. *Agri. Biol. Chem.*, *49*: 1071.
9. Duine, J. A., Frank, J. Jzn., and Jongejan, J. K. (1987). Enzymology of quinoproteins. *Adv. Enzymol.*, *59*: 169.
10. Shimoa, M., Ninomiya, K., Kuno, O., Kato, N., and Sakazawa, C. (1986). Pyrroloquinoline quinone as an essential growth factor for a polyvinyl alcohol-degrading symbiont, *Pseudomonas* sp. VM15C. *Appl. Environ. Microbiol.*, *51*: 268.
11. Ameyama, M., and Shinagawa, E., Matsushita, K., and Adachi, O. (1985). Solubilization, purification, and properties of a membrane-bound glycerol dehydrogenase from *Gluconobacter industrius*. *Agri. Biol. Chem.*, *49*: 1001.
12. van Kleef, M. A. G., and Duine, J. A. (1988). Bacterial NAD(P)-independent quinate dehydrogenase is a quinoprotein. *Arch. Microbiol.*, *150*: 132.
13. Duine, J. A., and Frank, J. (1981). Quinoproteins: A novel class of dehydrogenases. *Trends Biochem. Sci.*, *6*: 278.
14. Nagasawa, T., and Yamada, H. (1987). Nitrile hydratase is a quinoprotein. *Biochim. Biophys. Res. Commun.*, *147*: 701.
15. De Beer, R., Duine, J. A., Frank, J., and Large, P. J. (1980). The prosthetic group of methylamine dehydrogenase from *Pseudomonas* AM1. Evidence for quinone structure. *Biochim. Biophys. Acta*, *622*: 370.
16. Kenny, W. C., and McIntire, W. (1983). Characterization of methylamine dehydrogenase from bacterium W3A1. Interaction with reductants and amino-containing compounds. *Biochemistry*, *22*: 3858.
17. Husain, M., Davidson, V. L., Gray, K. A., and Knaff, D. B. (1987). Redox properties of the quinoprotein methylamine dehydrogenase from *Paracoccus denitrificans*. *Biochemistry*, *26*, 4139.
18. van der Meer, R. A., Jongejan, J. A., and Duine, J. A. (1987). Phenylhydrazine as probe for cofactor identification in amine oxidoreductases. *FEBS Lett.*, *221*: 299.
19. Lobenstein-Verbeek, C. L., Jongejan, J. A., Frank, J. J., and Duine, J. A. (1984). Bovine serum amine oxidase: A mammalian enzyme having covalently-bound PQQ as a prosthetic group. *FEBS Lett.*, *170*: 305.

20. Ameyama, M., Hayashi, M., Matsushita, K., Shinagawa, E., and Adachi, O. (1984). Microbial production of PQQ. *Agri. Biol. Chem.*, *48*: 561.
21. Moog, R. S., McGuirl, M. A., Cote, C. E., and Dooley, D. M. (1986). Evidence for methoxatin (pyrroloquinoline quinone) as the cofactor in bovine plasma amine oxidase from resonance Raman spectroscopy. *Proc. Natl. Sci. USA*, *83*.: 8435.
22. Hartman, C., and Klinman, J. P. (1988). Pyrroloquinoline quinone: a new redox cofactor in eukaryotic enzymes. *Biofactors*, *1*: 41.
23. van der Meer, R. A., and Duine, J. A. (1986). Covalently bound pyrroloquinoline quinone is the organic prosthetic group in human placental lysyl oxidase. *Biochem. J.*, *239*: 789.
24. Williamson, P. R., Moog, R. S., Dooley, D. M., and Kagan, H. M. (1986). Evidence for pyrroloquinoline quinone as the carbonyl cofactor in lysyl oxidase by absorption and resonance Raman spectroscopy. *J. Biol. Chem.*, *261*: 16302.
25. van der Meer, R. A., Jongejan, J. A., and Duine, J. A. (1988). Dopamine β-hydroxylase from bovine adrenal medulla contains covalently-bound pyrroloquinoline quinone. *FEBS Lett.*, *231*: 303.
26. Groen, B. W., van der Meer, R.A., and Duine, J. A. (1988). Evidence for PQQ as cofactor in 3,4-dihydroxyphenylalanine (dopa) decarboxylase of pig kidney. *FEBS Lett.*, *237*: 98.
27. van der Meer, R. A., Jongejan, J. A., and Duine, J. A. (1989). Pyrroloquinoline quinone as cofactor in galactose oxidase. *J. Biol. Chem.*, *264*: 7792.
28. van der Meer, R. A., and Duine, J. A. (1988). Pyrroloquinoline quinone (PQQ) is the organic cofactor in soybean lipoxygenase-1. *FEBS Lett.*, *235*: 194.
29. Karhunen, E., Niku-Paavola, M. L., Viikari, L., Haltia, T., van der Meer, R. A., and Duine, J. A. (1990). A novel combination of prosthetic groups in a fungal laccase; PQQ and two copper atoms. *FEBS Lett.*, *267*: 6.
30. James, S. M., Mu, D., Wemmer, D., Smith, A. J., Kaur, S., Burlingame, A. L., and Klinman, J. P. (1990). A new redox cofactor in eukaryotic enzymes: 6-Hydroxydopa at the active site of bovine serum amine oxidase. *Science*, *248*.: 981.
31. Christoserdov, A. Y., Tsygankov, Y. D., and Lidstrom, M. E. (1990). Cloning and sequencing of the structural gene for the small subunit of methylamine dehydrogenase from *Methylobacterium extorquens* AM1: Evidence for two tryptophan residues involved in the active center. *Biochem. Biophys. Res. Comm.*, *172*: 211.
32. McIntire, W. S., Wemmer, D. E., Christoserdov, A. Y., Lidstrom, M. E. (1991). A new cofactor in a prokaryotic enzyme: Tryptophan tryptophylquinone as the redox prosthetic group in methylamine dehydrogenase. *Science*, *252*: 817.
33. Chen, L., Mathews, F. S., Davidson, V. L., Huizinga, E., Vellieux, F. M. D., Duine, J. A., and Hol, W. G. J. (1991). Crystallographic investigations of the tryptophan-derived cofactor in quinoprotein methylamine dehydrogenase. *FEBS Lett.*, *287*: 163.
34. Whittaker, M. M., and Whittaker, J. W. (1990). A tyrosine-derived free radical in apogalactose oxidase. *J. Biol. Chem.*, *265*: 9610.
35. Ito, Nobutoshi, Phillips, S. E. V., Stevens, C., Ogel, Z. B., McPherson, M. J., Keen, J. N., Yadav, K. D. S., and Knowles, P. F. (1991). Novel thioether bond revealed by a 1.7 Å crystal structure of galactose oxidase. *Nature*, *350*: 87.
36. Robertson, J. G., Kumar, A., Mancewicz, J. A., and Villafranca, J J. (1989). Spectral

studies of bovine dopamine β-hydroxylase. Absence of covalently bound pyrroloquinoline quinone *J. Biol. Chem.*, *264*: 19916.
37. Michaud-Soret, I., Daniel, R., Chopard, C., Mansuy, D., Cucurou, C., Ullrich, V., and Chottard, J. C. (1990). Soybean lipoxygenases-1, −2a, −2b, and −2c do not contain PQQ. *Biochem. Biophys. Res. Comm.*, *172*: 1122.
38. van der Meer, R. A., Mulder, A. C., Jongejan, J. A., and Duine, J. A. (1989). Determination of PQQ in quinoproteins with covalently bound cofactor and in PQQ-derivatives. *FEBS Lett.*, *254*: 99.
39. van Schie, B. J., van Dijken, J. P., and Kuenen, J. G. (1984). Non-coordinated synthesis of glucose dehydrogenase and its prosthetic group in *Acinetobacter* and *Pseudomonas* species. *FEMS Microbiol. Lett.*, *24*: 133.
40. van Schie, B. J., Hellingwerf, K. J., van Dijken, J. P., Elferink, M. G. L., van Dijl, J. M., Kuenen, J. G., and Konings, W. N. (1985). Energy transduction by electron transfer via a pyrroloquinoline quinone-dependent glucose dehydrogenase in *Escherichia coli*, *Pseudomonas aeruginosa*, and *Acinetobacter calcoaceticus* (var. *Iwoffi.*). *J. Bacteriol.*, *163*: 493.
41. McIntire, W. S., and Weyler, W. (1987). Factors affecting the production of pyrroloquinoline quinone (PQQ) by the methylotrophic bacterium W3A1. *Appl. Environ. Microbiol.*, *53*: 2187.
42. Ameyama, M., Shinagawa, E., Matsushita, K., and Adachi, O. (1984). Growth stimulation of microorganisms by pyrroloquinoline quinone. *Agri. Biol. Chem.*, *48*: 2909.
43. Ameyama, M., Shinagawa, E., Matsushita, K., and Adachi, O. (1985). Growth stimulating activity for microorganisms in naturally occurring substances and partial characterization of the substance for the activity as pyrroloquinoline quinone. *Agri. Biol. Chem.*, *49*: 699.
44. Killgore, J., Smidt, C., Duich. L., Romero-Chapman, N., Tinker, D., Reiser, K., Melko, M., Hyde, D., and Rucker, R. B. (1989). Nutritional importance of pyrroloquinoline quinone. *Science*, *245*: 850.
45. Watanabe, A., Hobara, N., and Tsuji, T. (1988). Protective effect of pyrroloquinoline quinone against liver injury in rats. *Curr. Ther. Res.*, *44*: 896.
46. Hobara, N., Watanabe, A., Kobayashi, M., Tsuji, T., Gomita, Y., and Araki, Y. (1988). Quinone derivatives lower blood and liver acetaldehyde but not ethanol concentrations following ethanol loading to rats. *Pharmacology*, *37*: 264.
47. Nishagori, H., Yasunaga, M., Mizumura, M., Lee, J. W., and Iwatsuru, M. (1989). Preventative effects of pyrroloquinoline quinone on formation of cataract and decline of lenticular and hepatic glutathione of developing chick embryo after glucocorticoid treatment. *Life Sci.*, *45*: 593.
48. Paz, M. A., Gallop, P. M., Torrelio, B. M., and Fluckiger, R. (1988). The amplified detection of free and bound methoxatin (PQQ) with nitroblue tetrazolium redox reactions: Insights into the PQQ-locus. *Biochem. Biophys. Res. Comm.*, *154*: 1330.
49. van der Meer, R. A., Groen, B. W., Jongejan, J. A., and Duine, J. A. (1990). The redox-cycling assay is not suited for the detection of pyrroloquinoline quinone in biological samples. *FEBS Lett.*, *261*: 131.
50. Paz, M. A., Fluckiger, R., and Gallop, P. M. (1990). Comment: Redox-cycling is a property of PQQ but not of ascorbate. *FEBS Lett.*, *264*: 283.

51. van der Meer, R. A., Groen, B. W., Jongejan, J. A., and Duine, J. A. (1990). Reply: The redox-cycling assay and PQQ. *FEBS Lett.*, *264*: 284.
52. Bergethon, P. R. (1990). Amperometric electrochemical detection of pyrroloquinoline quinone in high-performance liquid chromatography. *Anal. Biochem.*, *186*: 324.
53. Xiong, L. B., Sekiya, J., and Shimose, N. (1990). Occurrence of pyrroloquinoline quinone (PQQ) in pistils and pollen grains of higher plants. *Agri. Biol. Chem.*, *54*: 249.
54. Lidstrom, M. E., and Stirling, D. I. (1990). Methylotrophs: Genetics and commercial applications. *Ann. Rev. Microbiol.*, *44*: 27.
55. van Huynh, N., Decleire, M., Motte, J. C., and Monseur, X. (1989). Production of gluconic and galactonic acids from whey. In *PQQ and Quinoproteins* (Jongejan, J. A., and Duine, J. A., eds.), Kluwer Academic Publishers, Dordrecht, The Netherlands, p. 97.
56. Dinarieva, T., and Netrusov, A. (1991). Lactic acid formation by free and immobilized cells of an obligate methylotroph. *Biotechnol. Prog.*, *7*: 234.
57. Turner, A. P. F., D'Costa, E. J., and Higgins, I. J. (1987). Enzymatic analysis using quinoprotein dehydrogenases. *Ann. NY Acad. Sci.*, *501*: 283.
58. Davidson, V. L. (1990). Quinoproteins: A new class of enzymes with potential use as biosensors. *Amer. Biotech. Lab*, *8* (2): 32.
59. Davidson, V. L., and Jones, L. H. (1991). Intermolecular electron transfer from quinoproteins and its relevance to biosensor technology. *Anal. Chim. Acta*, *249*: 235.
60. D'Costa, E. J., Higgins, I. J., and Turner, A. P. F. (1986). Quinoprotein glucose dehydrogenase and its application in an amperometric glucose sensor. *Biosensors*, *2*: 71.
61. Kitagawa, Y., Kitabatake, K., Kubo, I., Tamiya, E., and Karube, I. (1989). Alcohol sensor based on membrane-bound alcohol dehydrogenase. *Anal. Chim. Acta*, *218*: 61.
62. Zhao, S., and Lennox, R. B. (1991). Pyrroloquinoline quinone enzyme electrode based on the coupling of methanol dehydrogenase to a tetrathiafulvalene-tetracyanoquinodimethane electrode. *Anal. Chem.*, *63*: 1174.

II

Selected Quinoproteins

2

Methanol Dehydrogenase in Gram-Negative Bacteria

Christopher Anthony

University of Southampton, Southampton, England

I. INTRODUCTION

Methanol oxidation in gram-negative bacteria is always catalyzed by the NAD^+-independent alcohol dehydrogenase, which was described originally in a pink facultative methylotroph, now called *Methylobacterium extorquens* [1,2]. Although not specific for methanol, its usual function is to catalyze methanol oxidation, and so it is usually referred to as methanol dehydrogenase (MDH; EC 1.1.99.8). Possession of this enzyme is one feature that appears to be common to all gram-negative methane- and methanol-oxidizing bacteria. Complete descriptions of all MDHs described up to 1985 have been published in an extensive review [3]. The first quinoproteins to be described and shown to have an unusual prosthetic group were this dehydrogenase [2,4,5] and glucose dehydrogenase [6], and the prosthetic group of MDH was the first PQQ to be purified and characterized [5]. This first description showed that the absorption spectrum of MDH was clearly different from that of flavoproteins, having an absorption band due to the prosthetic group with a peak at 345 nm and a shoulder at about 400 nm (Fig. 1); denaturation released a reddish-brown, small, acidic molecule with a characteristic green fluorescence having an excitation maximum at 365 nm and emission maximum at 470 nm. After purification and characterization of this novel prosthetic group, it was concluded that it was either a completely novel compound or an unusual pteridine [5]. The first of these suggestions proved to be correct when it was shown, more than 10 years later, to be PQQ [7–9].

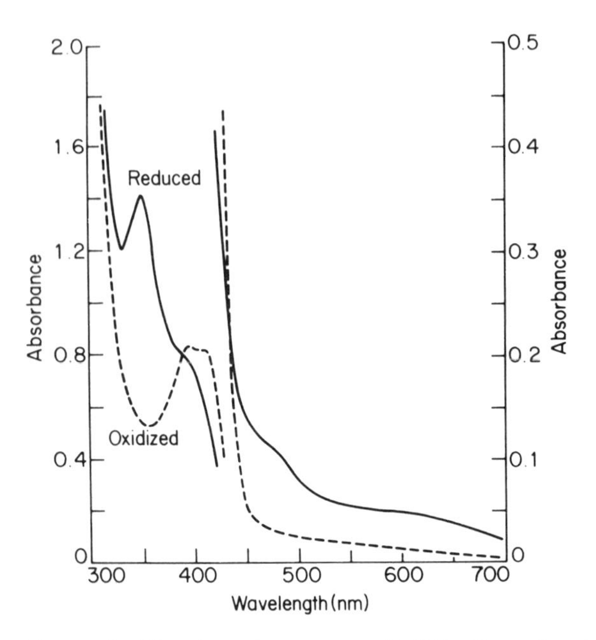

Figure 1 The absorption spectra of MDH. The reduced form (——) corresponds to MDH_{red} in Figure 2 and contains $PQQH_2$ (Fig. 3). The oxidized form (– –) corresponds to MDH_{ox} in Figure 2 and contains PQQ (Fig. 3); this spectrum is very similar to that of the semiquinone form (MDH_{sem}) in which the enzyme is isolated [4,5]. (From Ref. 38 with permission.)

MDH has a molecular mass of about 149 kDa and has an $\alpha_2\beta_2$ structure, each tetrameric molecule containing two molecules of prosthetic group. It is assayed with an artificial electron acceptor (usually phenazine ethosulfate) at its pH optimum (about pH 9) in the presence of its usual activator, NH_4Cl. Its natural electron acceptor is a specific cytochrome *c* called cytochrome c_L. MDH is a soluble protein that constitutes 10–15% of the soluble bacterial protein. It is located, together with cytochrome c_L, exclusively in the periplasm, where its concentration is about 0.5 mM [3,10]. If it is assumed that these proteins are globular in shape, then it is possible that they could form a monolayer one molecule deep in the periplasm. Indeed, if the distance between the periplasmic membrane and the outer membrane is about 70 Å, then there is hardly space for the large MDH to move in more than two dimensions.

This chapter will cover these features in more detail. This will be followed by a discussion of cytochrome c_L, an analysis of possible reaction mechanisms, and a discussion of recent work on structural aspects of the enzyme.

II. GENERAL CHARACTERISTICS

A. The Primary Electron Acceptor

The physiological electron acceptor is cytochrome c_L, and the reaction with this cytochrome can be measured directly (see below for problems with this) or with a second (terminal electron) acceptor, which is usually a typical Class I cytochrome *c* [11]. A more recent method, which has the advantage of using the physiological electron acceptor but does not have the disadvantage of having two cytochromes in the assay system, uses a dye, 2,6-dichlorophenolindophenol (DCIP), as terminal electron acceptor [12]. For the assay of the extracted enzyme it is more convenient to use the cationic dye phenazine methosulfate (PMS) or phenazine ethosulfate (PES) at pH 9. It has been suggested [13] that the free radicals of the dyes produced at high pH form the actual electron acceptors. Reduction of PES is measured by oxidizing it with DCIP (measured spectrophotometrically) or with oxygen (measured in an oxygen electrode) [11,14]. An alternative to PES is Wurster's blue, the perchlorate salt of the cationic free radical of *N,N,N′,N′*-tetramethyl-*p*-phenylenediamine [13,14]. In the absence of substrate or competitive inhibitor such as cyanide, the enzyme is rapidly destroyed by PES [3].

B. Substrate Specificity

MDH has a wide but well-defined specificity; only primary alcohols are oxidized, and the most important determinant of activity is the steric configuration of substituents [3]. Using the enzyme from *M. extorquens* it was shown that a second substituent on the C-2 atom prevents binding, the general formula for an oxidizable substrate being $R{\cdot}CH_2OH$, where R may be H, OH (as in hydrated aldehydes), $R'CH_2$—, or $R''R'C{=}C$—; the rate of oxidation of these substrates is usually at least 30% of that with the best substrate, which is methanol (K_m value about 20 μM) [3,15]. Analysis of the reaction with ^{13}C deuterated benzyl alcohols has demonstrated that MDH specifically removes the pro-*S* hydrogen at the C-1 carbon atom, the pro-*R* hydrogen being retained in the aldehyde product [16].

A common characteristic of MDHs is their ability to oxidize formaldehyde (probably as the gem-diol hydrated form) at a rate similar to that measured with methanol, although mutants lacking MDH are unaltered in their ability to oxidize formaldehyde. It thus appears that there must be some mechanism by which MDH is prevented from oxidizing formaldehyde. This may be effected by a periplasmic modifier protein (M-protein), which decreases the affinity of MDH for formaldehyde [17–19]. This protein has a second effect on the activity of MDH; it increases its affinity for some substrates that, in its absence, have very low affinities for MDH (e.g., 1,2-butanediol, 1,3-butanediol, 4-hydroxybutyrate). In practice this leads to an observed oxidation of these substrates by pure MDH, which would not otherwise occur because they would be unable to protect MDH from inactivation by PES [3,17,19,20,21].

A final point to make about the substrate specificity of MDH is that, even after purification and extensive dialysis, the enzyme catalyzes a rapid reduction of electron acceptor in the absence of any added substrate [3]. In a study of MDH of *Methylophilus methylotrophus* it was shown that this endogenous substrate is not bound methanol; each molecule of tetrameric MDH was shown to have two molecules of bound methanol (or formaldehyde) plus a further 90 molecules of unidentified reductant [22]. Addition of PMS plus activator (ammonia) leads to oxidation of the endogenous substrate, but the enzyme becomes inactivated (by the PMS) unless the enzyme is protected by competitive inhibitors such as cyanide [23].

C. Activators and Inhibitors of MDH

1. Activators

When prepared aerobically and assayed with dye electron acceptors, MDH has an absolute requirement for ammonium salts or methylamine. The relatively higher concentrations required at lower pH values suggest that the free base is the active species [2]. Whatever the mechanism of ammonia activation, it appears that high pH values and ammonia are only essential for reoxidation of the reduced enzyme, and not for the reduction of enzyme by methanol [23]. Ammonia can be replaced by esters of glycine or β-alanine, the K_m values for these activators being much lower than for ammonia [3]. When the enzyme is prepared anaerobically, ammonia is not always required for activity of crude extracts, and it appears likely that ammonia is replacing a "natural" activator that is destroyed by oxygen [3,24]. When assayed with cytochrome c_L, at the physiological pH optimum, ammonia usually has little effect [25–27]. The significance of the role of ammonia in the reaction mechanism is discussed further in Sections III.A and IV.C.

2. Inhibitors

Remarkably few inhibitors of MDH are known. The oxidation of methanol by whole bacteria is inhibited by EDTA, *p*-nitrophenylhydrazine, and high phosphate concentrations, but these compounds do not inhibit the isolated enzyme when assayed with PES [1,2,28]. In this assay PES and PMS are potent irreversible inhibitors of MDH, complete inactivation occurring within a few seconds [3]; substrates or competitive inhibitors (e.g., cyanide) are able to protect the enzyme. As discussed in Section III, cyclopropanol irreversibly inhibits MDH by direct enzyme-catalyzed reaction with the prosthetic group [29–33].

Cyanide and hydroxylamine are competitive inhibitors for some MDHs, usually competing for the substrate-binding site but sometimes competing for the activator site [3,23,25,34]. This property of cyanide makes it a useful protecting agent when purification of enzyme in the absence of substrate is required [23,25].

Cyclopropane-derived inhibitors such as cyclopropanol act by an irreversible reaction with the PQQ, catalyzed by the enzyme, and these compounds have been particularly useful in investigations of the mechanism of MDH [29–33]. Inhibitors of whole cell oxidation also inhibit the reaction with the physiological electron acceptor cytochrome c_L. Phosphate inhibits reduction of cytochrome c_L by diminishing the ionic bonding required for electron transfer [12,35]. EDTA is, however, a much more potent inhibitor of the reaction, preventing binding of the two proteins in a competitive manner [25,35,36]. Phenylhydrazine presumably acts by direct reaction with PQQ as shown for other hydrazine derivatives [37].

III. THE MECHANISM OF MDH

A. The Catalytic Cycle

A major reason for the 20-year delay between the first description of the spectrum of MDH and its prosthetic group and insight into its mechanism is that addition of substrate to the enzyme has no effect on its absorption spectrum, and it becomes inactivated when incubated with artificial electron acceptors [4,5,23]. A major step forward was taken when it was found that cyanide prevents inactivation, facilitating isolation of the oxidized form of the enzyme, containing PQQ, which on addition of substrate gives the reduced form, containing $PQQH_2$ [23]. This work was later extended in an investigation of the kinetics of the reaction by stopped-flow spectrophotometry, which led to the proposed reaction cycle summarized in Figure 2 [32]. It was concluded that the high pH optimum of the reaction is related to the pH dependency of the oxidation by the dye electron acceptor of the reduced (MDH_{red}) and semiquinone (MDH_{sem}) forms of MDH (reactions 1 and 4). The rate-limiting step is the conversion of the complex containing the oxidized form of the enzyme ($MDH_{ox}S$) to MDH_{red} plus product (P) (reaction 3) and is the only step affected by the activator, ammonia. Making use of the large deuterium isotope effect for this step of the reaction (in the absence of ammonia), the transient $MDH_{ox}\cdot C^2H_3OH$ complex has been isolated and its decomposition to MDH_{red} plus formaldehyde product observed [33].

B. The Prosthetic Group of Methanol Dehydrogenase

We first isolated, purified, and characterized the prosthetic group of MDH from *M. extorquens* [5], its X-ray structure was determined by Kennard and her colleagues [8], and an extensive chemical characterization was achieved by Frank, Duine, and coworkers (summarized and reviewed in Refs. 3 and 37). The fact that the PQQ remains (noncovalently) bound throughout the reaction cycle classifies it as a prosthetic group rather than a coenzyme that is necessarily released during the cycle or a cofactor that can be added to an apoenzyme to produce a functional enzyme.

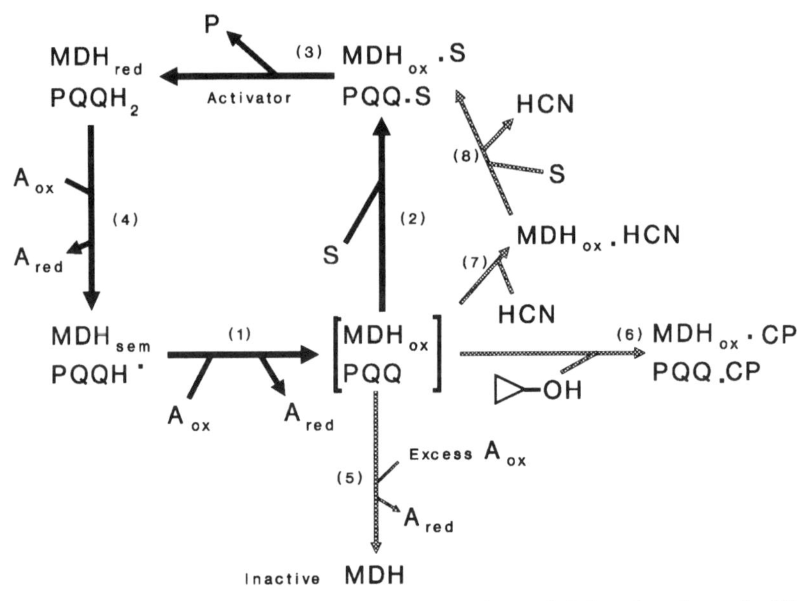

Figure 2 The reaction cycle of MDH. This reaction cycle is based on the work of Frank et al. [27,32]. The normal reaction cycle is outlined using dark arrows. The starting point is the semiquinone form of the enzyme as it is usually isolated. Reactions 1, 2, and 4 are reversible; reaction 3 is not. A is the one-electron acceptor, which is either PES, PMS, Wurster's blue, or cytochrome c_L. MDH_{red} is the fully reduced form; MDH_{ox} is the fully oxidized form; MDH_{sem} is the semiquinone form. The reaction with cyclopropanol (reaction 6) is described in Figure 4, and the mechanisms of the reactions catalyzed in steps 2 and 3 are discussed in Figures 3, 5, 6, and 7.

Each $\alpha_2\beta_2$ tetramer of the enzyme has two molecules of tightly bound PQQ [38] present in the semiquinone free radical form (PQQH$^{\cdot}$) [37]. This explains why the enzyme cannot be reduced by addition of substrate, because the free radical form of PQQ is unable to accept two electrons from methanol. ENDOR experiments have suggested that the ring structures of the PQQ may be in a hydrophobic environment in the protein [39] and that the 9-carboxyl is probably essential for binding PQQ in the active site [37]. Of particular relevance in proposing reaction mechanisms are the extensive studies with isolated PQQ, which have demonstrated that the C-5 carbonyl is very reactive towards nucleophilic reagents [40] in compounds forming adducts with PQQ including aldehydes, ketones, methanol, urea, cyanide, ammonia, and amines [3,37,40]. It is this feature of the chemistry of PQQ that encourages the assumption that a PQQ-substrate complex may be important in the reaction mechanism.

C. The Reaction Mechanism

Work on the catalytic cycle (above) suggests that MDH oxidizes its substrate and releases the product prior to reaction with its electron acceptor, and the two types of mechanism considered here are those that are consistent with this conclusion. In the first type, reducing equivalents are transferred directly from substrate to PQQ, generating $PQQH_2$ plus the oxidized product. Figure 3 illustrates the most likely reaction if the first type of mechanism is involved. In this mechanism there is an acid-base–catalyzed hydride transfer to the C-5 carbonyl that is entirely analogous to the mechanism of typical alcohol:NAD^+ oxidoreductases [41]. In the second type of reaction mechanism, a PQQ-substrate complex is formed at the active site prior to reduction of PQQ to the quinol. However, at present there is little direct evidence that a PQQ-substrate product is formed on the enzyme, and this is clearly a key point in future work on the mechanism. The direct evidence relates to the change in spectrum seen on reaction of MDH with methanol, cyanide, or cyclopropanol. This reaction must be done in the presence of electron acceptor, which oxidizes the semiquinone prosthetic group to the oxidized form containing PQQ [32]. In order to "see" any intermediates, the reaction was carried out with deuterated methanol as substrate, which diminished the rate of reaction of substrate with enzyme. Rapid gel filtration after mixing MDH with 1 mM C^2H_3OH plus excess oxidant yielded a preparation with a characteristic absorption spectrum (not shown). This product had a half-life of only 2 minutes, and no evidence was presented to indicate whether its spectrum was due to a mixture of reduced plus oxidized MDH or due to a mixture of reduced MDH plus MDH containing a covalent PQQ-methanol complex [32].

The most important piece of evidence relating to the possibility of PQQ reacting directly with substrate is the work with cyclopropanol; this reacts at the

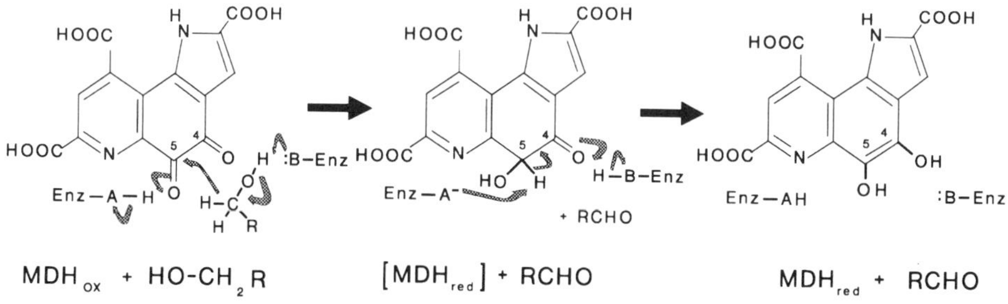

Figure 3 A simple hydrogen transfer mechanism for MDH, involving hydride transfer to C-5. This is analogous to the mechanism of typical NAD^+-linked dehydrogenases and involves acid/base-catalyzed hydride transfer to the C-5 carbonyl.

active site to form a nonreactive form of oxidized PQQ, which has been isolated from the enzyme and identified as a ring-closed form of a C-5 3-propanal adduct (see Fig. 4) [33]. The most direct way for ring opening of cyclopropanol to occur is by way of deprotonation by a base, and this has led to the suggestion that a basic group must be present at the active site. It is suggested that the mechanism of reaction with cyclopropanol consists of a concerted proton abstraction followed by rearrangement of the cyclopropoxy anion to a ring-opened carbanion and attack of the latter on the electrophilic C-5 of PQQ [33,42] (Fig. 4). It is suggested that this is aberrant behavior during the reaction mechanism, such than an unbreakable carbon-carbon bond is formed instead of a carbon-oxygen bond, which might be formed with a typical alcohol substrate as in Figures 5, 6, and 7. It should be noted, however, that although the observed reaction with cyclopropanol is consistent with these three mechanisms, the straightforward mechanism in Figure 3 is perhaps even more consistent with the cyclopropanol result because it also requires the presence of the basic group but does not require different types of bonds being formed with the two types of substrate.

Figure 5 presents the mechanism proposed by Frank and Duine and their colleagues [33,37,42]. Their proposed mechanism involves a base-catalyzed proton abstraction concerted with attack of the oxyanion on the C-5 of PQQ. I have slightly modified this in Figure 6 to a more conventional mechanism in which the addition of the oxyanion is aided by protonation of the C-5 carbonyl by an acid group on the enzyme (AH) leading eventually to formation of the hydroxyl group on C-5 instead of the postulated hydroxide. Another variation on this theme is the suggested mechanism of Houck and Unkefer [16] (Fig. 7), in which they propose that the C-4 carbonyl accepts a hydride from the pro-*S* position of the alcohol C-1 after formation of the same hemiketal intermediate shown in Figures 5 and 6.

The mechanisms discussed here all involve direct hydrogen (proton plus hy-

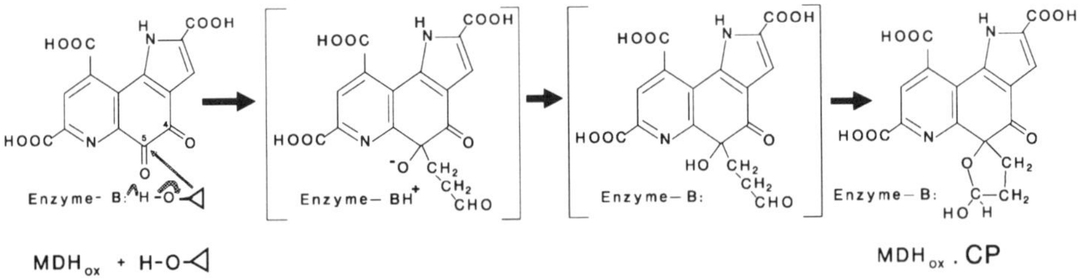

Figure 4 The reaction of MDH with cyclopropanol. The reaction mechanism is based on the work of Frank et al. [33]. Identification of the product of the reaction demonstrated the necessity for a base (–B:) at the active site This demonstration is the basis of the other mechanisms described for alcohol oxidation in Figures 3, 5, 6, and 7.

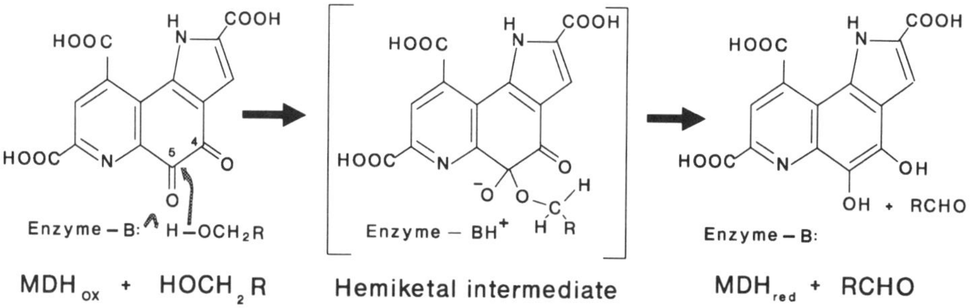

MDH_{ox} + $HOCH_2R$ Hemiketal intermediate MDH_{red} + RCHO

Figure 5 MDH mechanism involving a hemiketal intermediate and a base at the active site. This mechanism is taken from the proposal of Frank et al.; it involves "a base-catalyzed proton abstraction concerted with attack of the oxyanion on the C-5 of PQQ" [33].

dride) transfer or reaction of substrate with the C-5 carbonyl. It has been pointed out, however, by Duine et al. [37] that the scarce evidence that is available does not rule out even more unusual mechanisms that might involve, for example, the PQQ acting as a *p*-quinone, or in which the pyrrole ring is open on the enzyme and only closed during isolation in a manner analogous to that previously thought for the prosthetic group of methylamine dehydrogenase.

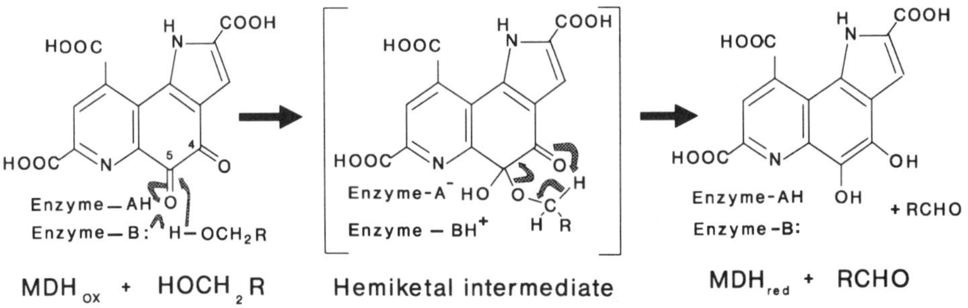

MDH_{ox} + $HOCH_2R$ Hemiketal intermediate MDH_{red} + RCHO

Figure 6 MDH mechanism involving a hemiketal intermediate and both acid and basic groups at the active site. This differs from the mechanism in Figure 5 in that the addition of the oxyanion is facilitated by the protonation of the oxygen of the C-5 carbonyl by an acid group on the enzyme (AH) leading eventually to formation of a hydroxyl group on C-5 instead of the hydroxide postulated in Figure 5. It should be noted that the electrocyclic mechanism proposed for the fragmentation of the hemiketal intermediate could be (more probably) replaced by a second set of reactions involving acidic and basic groups in the active site.

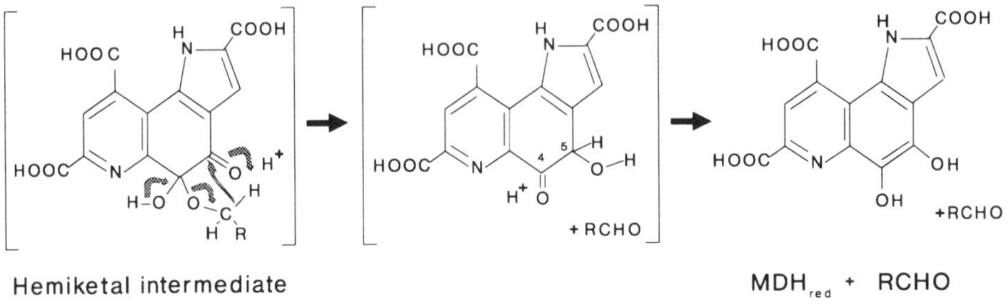

Hemiketal intermediate

MDH_{red} + RCHO

Figure 7 Mechanism involving a hemiketal intermediate and hydride transfer to the C-4 of PQQ. This mechanism has been modified (for convenience) from the mechanism of Houck and Unkefer [16]; in the original proposal there was a direct proton transfer from the C-5 hydroxyl to the C-4 carbonyl.

D. Role of Ammonia in the Reaction Mechanism

It has been demonstrated that activation by ammonia is confined to the substrate oxidation step (see above), but it is not clear how it is involved in the mechanism, although addition of ammonia to the C-5 position of PQQ in solution is known to occur [40].

An early example of a mechanism in which a PQQ-substrate complex is an intermediate was that of Forrest et al. [43], who proposed that amines play a key role by forming an iminoquinone with a covalent carbon-nitrogen bond at the C-4 position. The alcohol becomes bonded to the same carbon atom and is then released as the aldehyde, the hydrogen being transferred to the PQQ. However, this scheme is not consistent with the observation that the reactivity of isolated PQQ resides in the C-5 carbonyl center nor with the demonstration that ammonia or amines are not always essential for activity and that they may be readily removed by dialysis. This mechanism and others (see Ref. 37) share the assumption that an iminoquinone structure is active in catalysis. However, because addition of ammonia does not induce spectral changes in MDH, it is possible that it reacts with some other part of the enzyme or that the ammonia is replaced by an amine group on the enzyme.

E. The Reaction Mechanism of Alcohol Dehydrogenase

It has been suggested that studies of the reaction mechanism of the closely related quinohemoprotein alcohol dehydrogenase (ADH) of *Pseudomonas testosteroni*, which involved the reconstitution of active enzyme with analogs of PQQ, may provide a clue to the mechanism of MDH [42,44]. It is postulated that there is a basic residue at the active site that is supposed to favor hydride transfer from the alcohol to PQQ [44]. It was suggested that a probable course of alcohol dehydrogenation by ADH involves formation of an intermediate alkylidene-bridged amine-PQQ adduct, from which aldehyde is released by subsequent hydrolysis, a Schiff base forming an intermediate [44]. It should be noted, however, that there are no precedents for such reactions where the nitrogen of a basic group attacks the C-1 of an alcohol releasing a hydride.

IV. REACTION OF MDH WITH ITS CYTOCHROME ELECTRON ACCEPTOR

A. Introduction

It was known for some years that MDH interacts with the electron transport chain at the level of the *c*-type cytochromes, but a direct, methanol-dependent interaction between MDH and cytochrome *c* was extremely difficult to demonstrate. The earlier indications included the high concentrations of cytochrome *c* in methylotrophs, the reduction of cytochrome *c* by methanol in whole bacteria, the measurements of proton translocation in whole bacteria and ATP synthesis in membrane vesicles, and the demonstration that mutants lacking these cytochromes were able to oxidize other substrates but were unable to oxidize methanol (for an extensive review see Ref. 3). Methanol-dependent reduction of *c*-type cytochromes was eventually demonstrated using anaerobically prepared crude extracts of *Hyphomicrobium* X [24] and of *M. extorquens* [45] and by using pure proteins from *M. extorquens* AM1, *Methylophilus methylotrophus*, and *Paracoccus denitrificans* [25,36]. The reasons for the difficulties involved in these investigations are discussed at length in Ref. 3 and are summarized below.

B. Cytochrome c_L, the Physiological Electron Acceptor for MDH

1. Introduction

All methylotrophic bacteria that have been investigated contain at least two soluble *c*-type cytochromes [3,26,46–50,80]. One of these, originally called cytochrome c_H [46], is a typical Class I *c*-type cytochrome. The other is distinguished from this basic cytochrome by its larger size and lower isoelectric point and is called cytochrome c_L [46]. It is the only cytochrome *c* able to act as electron acceptor for MDH in *M. extorquens* [11,25,35,51], *M. methylotrophus* [25,35], *P. denitrificans* (called cytochrome c_{551i} or c_{552} [25,50]), *Hyphomicrobium* [14,49], and *Acetobacter methanolicus* [12,26]. Its specific role in the oxidation of methanol has been confirmed by characterizing mutants lacking it [52–54]. Its properties are discussed at some length below because of their relevance to the mechanism of bonding and electron transfer between MDH and this cytochrome.

2. General Properties

In many respects cytochrome c_L is a typical Class I cytochrome c [55]. These soluble, high potential cytochromes are usually small, basic cytochromes that have a single, low spin, heme prosthetic group, bonded covalently to cysteine residues in the protein, with histidine as the 5th ligand and the sulfur atom of a methionine residue as the 6th ligand. They have typical absorption spectra in the reduced form (α maximum at 549–552 nm and 25°C) and in the oxidized form (about 410 nm in the Soret region and 695 nm in the near infrared). In most of these features cytochrome c_L is similar to other cytochromes in this class; it differs from typical Class I cytochromes c in being larger (17–21 kDa) and in having a low isoelectric point (3.5–4.5); these properties reflect the fact that the primary sequence of the cytochrome is completely different from any other c-type cytochrome (see below) [3,46,56].

3. Primary Structure of Cytochrome c_L

We have confirmed that cytochrome c_L constitutes a novel class of c-type cytochromes by determining its primary sequence (deduced from its gene sequence [57]. Except for the typical heme-binding site (Cys-Ser-Gly-Cys-His), the sequence of cytochrome c_L shows no homology with cytochromes in any databases; in particular, none of the conserved features of c-type cytochromes are seen in the sequence of cytochrome c_L. One key feature of special importance in a typical Class I cytochrome c_L is the sixth ligand methionine, which is usually 60 residues (or more) towards the C-terminal from the heme-binding histidine. In cytochrome c_L the three methionines are all within 50 residues of this histidine, and the sequences around the methionines bear no relation to those around the methionines of other c-type cytochromes. The position of many aromatic and lysine residues in the polypeptide chains of c-type cytochromes are highly conserved. The lysine residues arranged around the pocket are of particular importance in binding to its electron donors and acceptors (e.g., oxidases). That there is no homologous arrangement of lysines in cytochrome c_L is not surprising, because this cytochrome is not a substrate for the oxidase; its function is to mediate electron transfer between MDH and the typical Class I cytochrome c_H [10,58,80].

4. The Heme Environment of Cytochrome c_L and Its Reaction with CO

The cytochrome c_L of many methylotrophs is able to react with CO. Such a reaction has been used to indicate an oxidase function [59], but there is no evidence that this cytochrome functions as an oxidase [3]. The slow, incomplete reaction with CO probably reflects the structure around the heme pocket that allows a more readily dissociable iron-methionine bond [46,60]. That the heme environment is slightly unusual is indicated by the unusual response of the midpoint redox potential to changing pH values; two ionizing groups affect redox potentials, the pK values being 3.5 and 5.5 in the oxidized form and 4.5 and 6.5 in the reduced form [46]. If these dissociations arise from the heme, then the higher of the pK values is likely to be due to the inner heme propionate in the hydrophobic environment of the heme cleft and the lower pK values due to the outer propionate in its more hydrophilic environment. The suggestion that the heme environment may be unusual is supported by the demonstration that the axial methionine ligand has a novel configuration as directly observed in NMR studies [61].

5. Autoreduction of Cytochrome c_L

The cytochrome c_L from some bacteria undergoes rapid autoreduction of the heme iron in the ferricytochrome when the pH is raised in the absence of any reducing agent [3,26,45,49,56]. The process is, by definition, a first-order intramolecular reaction that occurs at high pH values; it occurs in some typical *c*-type cytochromes, but the rate is very much higher with cytochrome c_L. In bacteria that grow at pH 7, the pK for autoreduction is about pH 10, but in an acidophilic methylotroph, *A. methanolicus*, the pK is about pH 7 [26]. It has been proposed that the mechanism involves dissociation of a weakly acidic group, which dissociates on raising the pH to give a negatively charged species able to donate an electron to the heme, the free radical produced by this process being stabilized by sharing an electron with the heme iron [3,45]. As discussed below, this mechanism may be involved in the MDH/cytochrome interaction. Whether or not this is the case, the phenomenon of autoreduction in this cytochrome is an intriguing characteristic.

C. The Methanol-Dependent Reduction of Cytochrome c_L by MDH

1. Introduction

This was remarkably difficult to demonstrate. Whenever MDH and cytochrome *c* of *M. extorquens* or *M. methylotrophus* are mixed, the cytochrome is immediately reduced even in the absence of methanol. There are two possible explanations for this. The first is that the enigmatic endogenous reductant on MDH provides electrons for reduction of the cytochrome in the absence of methanol [3]. The second possibility is that MDH stimulates autoreduction of cytochrome c_L to occur at lower pH values than usual by stimulating a change in pK of the acidic group involved in the autoreduction phenomenon [45]. If this is so, then no electron transfer between the proteins need occur, the presence of methanol will clearly make no difference and, because autoreduction is, by definition, a first-order, intramolecular reaction, the kinetics will be first order with respect to oxidized cytochrome c_L. This was indeed shown to be the case (for review, see Ref. 3). Support for this idea also came from work with the acidophilic *A. methanolicus*, in which autoreduction occurs at pH 7 (instead of pH 10), and MDH stimulates it to occur at pH 4, the growth pH of the organism [26]. Against this explanation, however, is the fact that first-order kinetics are difficult unequivocally to demonstrate, and that MDH-stimulated reduction of cytochrome c_L can be demonstrated with proteins of *Hyphomicrobium* at a rate that is considerably greater than the rate of autoreduction of the cytochrome c_L of this organism [49]. On balance, therefore, it appears more probable that MDH-catalyzed reduction of cytochrome c in the absence of methanol may be due to the endogenous reductant.

Because of the problems outlined above, it was important to find an alternative way to demonstrate electron transfer from methanol to cytochrome c_L catalyzed by MDH; for this to be unequivocal it was necessary to show a rate stimulation by methanol, together with production of formaldehyde. This was eventually achieved by coupling the system to a second electron acceptor (horse heart cytochrome c) [25,36]. These experiments demonstrated that, after oxidation of endogenous substrate, further reduction of cytochrome depends on methanol. The system was specific for cytochrome c_L, had the same affinity for methanol, the same substrate specificity, and the same sensitivity to EDTA. It has been used to demonstrate unequivocally that MDH catalyzes electron transfer directly to cytochrome c_L from methanol with concomitant production of formaldehyde in five completely different types of methylotroph: *Methylomonas* [48], *Methylobacterium*, *Methylophilus*, *Paracoccus* [25,36], and *Acetobacter* [12,26]. Despite this unequivocal evidence of the specific, direct, reaction of cytochrome c_L with MDH, considerable discussion of the significance of these results has occurred because we calculated that the rates measured were too slow to account for the rate of respiration in bacteria [25]. This conclusion has been confirmed for the MDH and cytochrome c_L from *Hyphomicrobium* [27]. It will be of great importance to see if addition of factor X, described by Dijkstra et al. [62], is sufficient to increase these rates to those occurring in whole bacteria.

2. The Reaction Cycle of MDH with Ferricytochrome c_L as Electron Acceptor

The experiments described in this section are all described in a paper by Dijkstra et al. [27] that analyzes the separate steps in the methanol:cytochrome c_L oxidoreductase reaction using steady-state and stopped-flow kinetic techniques and studies of isotope effects using deuterated methanol. The substrates used in this work were fully reduced MDH (MDH_{red}, containing fully reduced quinol, $PQQH_2$) and MDH as normally isolated (MDH_{sem}, containing half-reduced PQQ, $PQQH^{\cdot}$). Rates were measured at the pH optimum for reaction with cytochrome c_L (pH 7.0) and the pH optimum for reaction with the artificial electron acceptor (pH 9.0); the effect of ammonia as activator was also considered. The spectra of the MDH intermediates during the oxidation of MDH with cytochrome were the same as found when Wurster's blue was used for their oxidation (see Section III.A), so the overall scheme (Fig. 2) is essentially the same as that proposed for the dye-linked reaction.

Two conditions are required for efficient turnover: rapid intramolecular oxidation of methanol (S) in the enzyme substrate complex, $MDH_{ox}{\cdot}S$, and rapid oxidation of the reduced enzyme by the ferricytochrome. It was shown that which of these steps becomes rate-limiting depends on the nature of the oxidant, the pH, the ionic strength, and presence or absence of activator.

From the stopped-flow experiments it was concluded that ferricytochrome c_L is an excellent oxidant of both forms of reduced MDH at pH 7, but a poor oxidant at pH 9. The opposite was true when Wurster's blue replaced cytochrome c_L as oxidant; the reaction was much faster at pH 9 than at pH 7. EDTA was a potent inhibitor of oxidation of MDH by the cytochrome (see Section II.C). The rate of oxidation of MDH by cytochrome c_L was shown to be sufficiently rapid to account for the observed in vivo rate of methanol oxidation.

Under steady-state conditions, the overall turnover of the enzyme was slower at pH 9 than at pH 7 because the rate of oxidation of MDH (MDH_{sem} and MDH_{red}) becomes rate-limiting at pH 9. The lack of a deuterium isotope effect (using C^2H_3OH) at this pH indicates that substrate conversion is not rate-limiting, and ammonia is not an effective activator. By contrast, at pH 7 the rate-limiting step is substrate conversion, and ammonia is an effective activator.

In summary, ferricytochrome c_L is an excellent oxidant of reduced MDH at pH 7, but the substrate oxidation step is very slow and activation by ammonia very poor. By contrast, at pH 9 the cytochrome is such a poor oxidant that the activation of the substrate oxidation step that does now occur becomes irrelevant. It was concluded that the relatively slow overall rate is because of the need for an activator for the substrate oxidation step. It was suggested that this might be the low-M_r oxygen-labile component described previously [62], but this was not tested in this system.

D. Analysis of the MDH/Cytochrome c_L Interaction by Chemical Modification of the Proteins

It has been suggested that the interactions of MDH with cytochrome c_L is likely to be electrostatic in nature, involving ionic interactions between lysine or arginine residues on MDH and carboxyl residues on the cytochrome [63,64]. It was speculated that the lysine residues that form such a well-defined pattern in predictions of the secondary structure of the β-subunit of the MDH of *M extorquens* (Section V.B) might be involved in "docking" with carboxyls of cytochrome c_L, which might also be involved in interaction with lysines of the binding domain of the typical Class I cytochrome, cytochrome c_H [63,64]. That ionic interactions are involved has now been demonstrated using the two proteins from *M. extorquens*, *M. methylotrophus* and *Acetobacter methanolicus*; the reaction was strongly inhibited by low concentrations of salts; the extent of inhibition was directly related to the square root of the ionic strength of the medium, NaCl inhibiting the reaction by decreasing the affinity of the cytochrome for MDH [12,35].

That lysine residues on MDH are involved in docking with carboxyl groups on the cytochrome was indicated by chemically modifying MDH. The modified MDH retained activity in the dye-linked assay systems showing that the active site for reaction with substrate had not been altered. It was shown that reagents that change the charge on lysines led to inactive MDH, whereas those that modified MDH with retention of charge had relatively little affect. The inhibition by reagents specific for arginine residues suggests that thcsc may also be involved. When cytochrome c_L was modified with lysine-modifying reagents, its activity was retained, but those reagents that modified carboxyl groups led to greatly diminished activity [12,35].

E. Analysis of the MDH/Cytochrome c_L Interaction by Cross-Linking Studies

Although initially our attention was drawn to the potential importance of lysines in the interaction by the high proportion of lysine residues in the β subunit of the MDH of *M. extorquens* [63], it appears that these lysines are unlikely to be involved in docking with cytochrome c_L. The small β subunit from the acidophilic methylotroph *A. methanolicus* does not have those lysines thought to be important in the docking process [12] (see Fig. 11). It now appears that the lysines on the larger α subunit are involved in this process. The evidence for this comes from cross-linking the two proteins using zero-length cross-linking agents recently described by Grabarek and Gergely [65]. When carboxyl groups on the cytochrome are modified and then attacked with the unmodified MDH, lysine residues on the MDH displace the reagent from the cytochrome and form isopeptide bonds with its carboxyl groups. The groups responsible for the docking and subsequent cross-linking can then be identified by hydrolysis of the cross-linked product followed by protein sequencing. Using proteins from three different methylotrophs, these experiments demonstrated in every case that carboxyl groups on cytochrome c_L interact only with lysyl residues on the larger α subunit of MDH [12,35].

As described previously [10,58], MDH, cytochrome c_L, cytochrome c_H, and the membrane oxidase form a complete electron transport chain, catalyzing the oxidation of methanol by molecular oxygen. The three periplasmic proteins of this chain may either operate as separate entities, forming short-lived bimolecular complexes during which electron transfer occurs, or they might form stable complexes of more than two proteins in a "wire" system. In such a system electrons would flow from MDH through cytochrome c_L to cytochrome c_H, the site of entry of electrons into cytochrome c_L being different from the site of exit. If cytochrome c_L has only a single site for electron transfer, dissociation from MDH must occur prior to reaction with the typical Class I cytochrome *c*, and it should thus be possible to demonstrate cross-linking between the two cytochromes. This has indeed been demonstrated in conditions where the carboxyl groups of cytochrome c_L are modified and lysine residues on the second cytochrome are involved in forming the isopeptide bond [35]. By contrast, no cross-linking occurred when the carboxyls of the Class I cytochrome were modified for the reaction. Furthermore, it proved impossible to cross-link the three proteins in a ternary complex together; MDH and cytochrome c_H competed for the cytochrome c_L. These results all suggest that cytochrome c_L has a single site by which carboxyl groups are involved in docking with lysyl groups on MDH and cytochrome c_H and by which electron transfer occurs [35].

V. THE STRUCTURE OF METHANOL DEHYDROGENASE

A. The Subunit Composition of MDH and the Binding of PQQ

Although thought for many years to be a dimer of identical subunits (M_r about 60 kDa), the MDH of *M. extorquens* has recently been shown to be a tetramer arranged in an $\alpha_2\beta_2$ configuration, having α and β subunits of about 66 and 8.5 kDa, respectively [63]. This configuration has now been confirmed for MDH of the obligate methylotrophs *Methylobacillus glycogenes* [66] and *M. methylotrophus* [35] and in the acidophilic *A. methanolicus* [12]. A gene coding for the small subunit has also been identified in *P. denitrificans* [54]. The molecular masses for the subunits of MDH from *M. extorquens* are about 66 and 8.5 kDa, giving a total molecular mass of about 149 kDa. This is remarkably high when it is considered that the value as estimated by gel filtration is only about 120 kDa; the reason for this discrepancy is not known; it was first observed in one of the first studies of this enzyme when it was shown that the molecular mass by analytical ultracentrifugation is about 146 kDa [4].

Each tetramer has two molecules of PQQ [38]. It is not known whether or not the complete tetramer can oxidize two molecules of methanol at the same time or if there is any cooperativity in the reaction mechanism between two α/β dimers, each containing one PQQ molecule. It has proved impossible to separate the subunits in any way that does not also lead to denaturation and loss of PQQ; similarly, all methods for removal of PQQ lead to denaturation of the enzyme. The subunits are not dissociated by high salt concentrations but require strong detergents such as SDS, 8 M urea, 2.5 M guanidinium hydrochloride, or treatment at high pH (pH 12) for dissociation [12,63]. These observations suggest that the prosthetic group is held between the tightly bound subunits or that it is held so tightly to one of the subunits that the conditions required to remove it are sufficiently harsh as to also dissociate the enzyme. Clearly the bonding between subunits, and perhaps also between PQQ and the enzyme, involves hydrophobic bonds rather than simple ionic bonds. ENDOR spectroscopy [39] indicated that PQQ is held in a hydrophobic environment but the bonding must be more complex than this; other constraints include the requirement to neutralize the highly polar carboxyl groups that render the isolated PQQ extremely hydrophilic. A second consideration is the requirement for the C-5 carbonyl of PQQ to be exposed for interaction with the wide range of hydrophobic and hydrophilic substrates that are oxidized by MDH.

Earlier work indicated that MDH does not contain any metal atoms [4], but recent work suggests that calcium is important in the structure of the MDH of some bacteria [66,67,81].

B. The Primary Sequence of MDH and Related Proteins

The brief outline above makes it obvious that when consideration is given to the structure of MDH, the following regions and interactions are of particular interest: the substrate/MDH-binding site, the site of PQQ binding, the region involved in docking and electron transfer between MDH and cytochrome c_L, binding of Ca^{++} and binding of α and β subunits to each other.

Ideas about the regions of MDH likely to be involved in these interactions can obviously be developed from a consideration of the primary structures of MDHs from different sources and also from comparison with primary structures of the quinoproteins alcohol dehydrogenase (ADH) and glucose dehydrogenase (GDH). This is likely to be instructive because their basic structure will have some differences related to their different locations in the cell (membrane or periplasm), some difference related to different strength of PQQ binding ("tight" in MDH, "loose" in GDH), and some differences related to their different substrates (ADH being similar, GDH being completely different). This is discussed further in a comprehensive review from which much of this material is taken [68].

The regions of similarity between the different types of quinoprotein are summarized in Figure 8. The only region of similarity between all four types of quinoprotein has been suggested to be the PQQ-binding site. In addition to this, the MDHs and ADH show considerable similarity, with the ADH having an extra "cytochrome domain." GDH-A has an extra N-terminal membrane-anchoring region followed by a region with some similarity to MDH and ADH. The only region of similarity to the periplasmic GDH-B is the putative PQQ-binding site.

1. *The Putative PQQ-Binding Site*

This sequence of about 65 amino acids was independently proposed as the PQQ-binding site by Cleton-Jansen et al. [69] and Inoue et al. [70], and my proposal [68] is derived from these (Fig. 9). At the outset it is important to consider what is

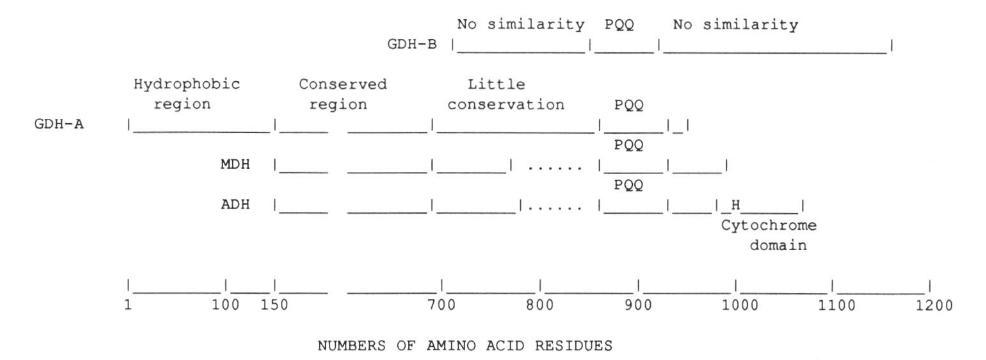

Figure 8 Alignment of the Amino Acid Sequences of the Quinoproteins. The lengths of sequence are approximately to scale (but note the break in the scale). The sequences of MDH and ADH are very similar throughout the proteins; the comments on degree of conservation in the diagram refer to similarities between these two proteins and the GDH-A. The only region of the two types of GDH showing similarity with each other is the postulated PQQ site. (From Ref. 68 with permission.)

meant by "binding site." Binding of ligands in proteins usually involves relatively few amino acids, and these are not necessarily adjacent or even on the same part of the primary sequence. In the present case this length of 65 amino acids probably includes amino acids that are involved in specific interactions with PQQ, and it will also constitute essential secondary structural features and perhaps substrate-binding regions. Of the 65 amino acids included in this possible site, 19 are identical in all the quinoproteins except for the B-type GDH, which has only six amino acids that are identical to all the others. This might suggest at first that GDH-B does not have a similar PQQ-binding site. However, the only part of GDH-B that has any similarity to other quinoproteins (and therefore is a candidate for a conserved PQQ-binding site) is within this same region. If the sequences for the two types of GDH are compared, they have 15 out of 65 identical amino acids. There are thus two obvious possibilities. The first is that the GDH-B is exceptional in its PQQ-binding site. The second possibility is that the MDHs and ADH have one site (of 29 identical amino acids), whereas the GDHs have a second site (of 15 identical amino acids). The two potential sites have only six amino acids in common. If the region described here is also involved in binding substrates and electron acceptors, then the second suggestion is perhaps more likely because these ligands are completely different for the two types of enzyme; for glucose dehydrogenase they are glucose and ubiquinone, and for the other enzymes they are alcohols and (probably always) a cytochrome *c*.

```
MDH M.ext  477- GGTMATAGDL VFYGTLDGY L-KARDSDTG DL-LWKFKIP SGAIGYPMT YTHKGTQYVA IYYGVGG
MDH M.org  477- GGTLATAGDL VFYGTLDGY L-KARDSDTG DL-LWKFKIP SGAIGYPMT YTHKGTQYVA IYYGVGG
MDH P.den  476- GGTMATAGGL TFYVTLDGF I-KARDSDTG EL-LWKFKLP SGVIGHPMT YKHDGRQYVA IMYGVGG
ADH A.ace  486- GGILATGGDL LFQGLANGE F-HAYDATNG SD-LYKFDAQ SGIIAPPMT YSVNGKQYVA VEVGWGG
                **••**•*•*  *    •*  • •* *•••* •  *•**• • **•*• *** *• •* **** •  * ** -Identity in MDHs and ADH
                !! ••!•!••  !    ••    •! • ••! •  !•••  • •!  • !!! !• •! !!!• •  !  ! -Identity in MDHs, ADH & GDHA
GDHA E.col 713- GGPISTAGNV LFIAATADN YLRAYNMSNG EK-LWQGRLP AGGQATPMT YEVNGKQYVV ISAGGHG
GDHA A.cal 719- GGSISTAGNV MFVGATQDN YLRAFNVTNG KK-LWEARLP AGGQATPMT YEINGKQYVV IMAGGHG
                * • *•• •    •  ••*•  •     ••*  • *    ** • •* **·   *•***•* •   •     -Identity in all GDHs
GDHB A.cal 137- GLPSSKDHQS GRLVIGPDQ KIYYTIGDQG RNQLAYLFLP NQAQHTPTQ QELNGKDYHT YMGKVLR
                *   ••• •         •         • *    *     • •     *    • •* •*   •       -Identity in all proteins
```

Figure 9 The putative PQQ-binding site on quinoproteins. The residue numbers refer to the mature proteins with signal peptides excluded. The methanol dehydrogenases (α subunits) (MDH) are from *M. extorquens* [68,71], *M. organophilum* [72], and *P. denitrificans* [73]. The glucose dehydrogenases (GDHA and GDHB) are from *Acinetobacter calcoaceticus* [74,75] and *E. coli* [69], and the alcohol dehydrogenase is from *Acetobacter aceti* [70,76]. (!), Identity in all quinoproteins except for the periplasmic B-type glucose dehydrogenase of *A. calcoaceticus*; (*), identity; (•), a conserved residue. It should be noted that although there is considerable identity between all the GDHs (15 amino acids out of 65), only 6 of these are the same as the 19 amino acids that are identical in all the other 6 quinoproteins. (From Ref. 68 with permission.)

2. Methanol Dehydrogenase: Large (α) Subunit

The three α-subunits of MDH shown in Figure 10 are identical in the number of amino acids present, and they show an extremely close similarity to each other. The proteins from the two *Methylobacterium* species are, as may be expected from such similar bacteria, almost identical. The relatively greater differences between the *Methylobacterium* species and the *Paracoccus* are reflected in the failure of antibodies to the *Paracoccus* enzyme to cross-react with MDH from *Methylobacterium* [73]. The greater ratio of basic to acidic amino acids in the *Paracoccus* enzyme is reflected in its lower acidic isoelectric point (3.7 compared with 8.8 for *Methylobacterium*) [3]. As expected for a periplasmic protein, there are no predicted membrane-spanning helices. It has been suggested that the more hydrophobic region towards the C-terminal might be involved in linking the MDH to the membrane bilayer providing the enzyme with greater accessibility to cytochrome c_L [72]. This is very unlikely. When proteins bind to membranes, they do so either by way of membrane-spanning hydrophobic regions or by interactions with charged membrane proteins or phospholipids. The observation that MDH is sometimes associated with membranes is probably spurious. When care is taken, MDH is located exclusively in the periplasmic fraction [3]. Any hydrophobic regions of the MDH are more likely to be involved in binding PQQ or in binding the α and β subunits together.

3. Methanol Dehydrogenase: Small (β) Subunit

Figure 11 shows the structure of the β subunit of MDH. When this small subunit was first described, its very high concentration of lysine residues (15 out of 74) drew attention to the possibility that they might have some special role in the interaction of MDH with the acidic cytochrome c_L [63,64]. In particular, consideration of secondary structure predictions of the C-terminal half of the protein suggested that the five lysines in this region (residues 53, 59, 60, 63, and 70) might be of special importance because they form a well-defined positively charged region on one side of a predicted amphipathic alpha helix. The β subunit from *Paracoccus* has four of these five lysine residues (all except 53), apparently strengthening the argument. However, the β subunit from the acidophilic methylotroph *A. methanolicus* has no lysines in this region and indeed has no more than three of the nine lysines that are common to the other two MDHs, and these are in the N-terminal region [12]. If positive charges on the β subunits are especially important, it appears that the two conserved arginines are the more likely candidates. The N-terminal half of the β subunit is relatively more hydrophobic, and if hydrophobic reactions are involved in binding the two types of subunit together, then this region is more likely to be involved. No sequences having similarity to the β subunit are found in databases, and there are no regions of similarity to the MDH β subunit in the other quinoproteins. This suggests either that it is not involved in binding PQQ or that it has some special function with respect to this in MDH. This might indeed be the case because MDH differs from the GDHs and ADH in its tight (noncovalent) binding of PQQ; it is neces-

```
                   **     **** *     **** *  *    * *     ** *********  *** * **** ****    *  *  ***
MDH M.ext     1-  NDK-----LVELSKS-DDNWVMPGKNYDSNNFSDLKQINKGNVKQLRPAWTFSTGLLNGHEGAPLVVDGKMYIHTSFPNN
MDH M.org     1-  NDK-----LVELSKS-DDNWVMPGKNYDSNNYSELKQVNKSNVKQLRPAWTFSTGLLNGHEGAPLVVDGKMYVHTSFPNN
MDH P.den     1-  NDQ-----LVELAKD-PANWVMTGRDYNAQNYSEMTDINKENVKQLRPAWSFSTGVLHGHEGTPLVVGDRMFIHWPFPNT
ADH A.ace     1-  DGQGNTGEAIIHADDHPENWLSYGRTYSEQRYSPLDQINRSNVGDLKLLGYYTLDTNRGQEATPLVVDGIMYATTNWSKM
                                 !!   !  !       !       !  !!                 ! !  !!!!   !

                   **** *  ** **** ****** *  ****************      **  ****      * ******* *  **  ***
MDH M.ext    75-  TFALGLDDPGTILWQDKPKQNPAARAVACCDLVNRGLAYWPGDGKTPALILKTQLDGNVAALNAETGETVWKVEN--SDI
MDH M.org    75-  TFALDLDDPGHILWQDKPKQNPAARAVACCDLVNRGLAYWPGDGKTPALILKTQLDRHVVALNAETGETVWKVEN--SDI
MDH P.den    75-  TFALDLNEPGKILWQNKPKQNPWARWVACCDVVNRGLAYWPGDDQVKPLIFRTQLDGHIVAMDAETGETRWIMEN--SDI
ADH A.ace    81-  E-ALD-AATGKLLWQYDPKVPGNIADKGCCDTVNRGAGYWNG------KVFWGTFDGRLVAADAKTGKKVWAVNTIPADA
                    !     !  !!!  !!             !!! !!!!  !! !          !     !  ! !!   !      !

                  ***     ********* ** *  ************ ****** *   *** *****  ****   **   * ***** **
MDH M.ext   153-  KVG----STLTIAPYVVKDKVIIGSSGAELGVRGYLTAYDVKTGEQVWRAYATGPDKDLLLASDFNIKNPHYGQKGLGTG
MDH M.org   153-  KVG----STLTIAPYVVKDKVIIGSSGAELGVRGYLTAYDVKTGGQVWRAYATGPDKDLLLADDFNVKNAHYGQKGLGTA
MDH P.den   153-  KVG----STLTIAPYVIKDLVLVGSSGAELGVRGYVTAYDVKSGEMRWRAFATGPDEELLLAEDFNAPNPHYGQKNLGLE
ADH A.ace   153-  SLGKQRSYTVDGAVRVAKGLVLIGNGGAEFGARGFVSAFDAETGKLKWR-FYTVPNNK----NEPDHAASDNILMNKAYK
                   !                 !  !  !  !  !!! ! !!   !     !   !!   ! !

                  ***  ***  ***************     * * * ************     ******* * ** * *****  *********
MDH M.ext   229-  TWEG-DAW--KIGGGTNWGWYAYDPGTNLIYFGTGNPAPWNETMRP---GDNKWTMTIFGRDADTGEAKFGYQKTPHDEW
MDH M.org   229-  TWEG-DAW--KIGGGTNWGWYAYDPGTNLIYFGTGNPAPWNETMRP---GDNKWTMTIFGRDADTGEAKFGYQKTPHDEW
MDH P.den   229-  TWEG-DAW--KIGGGTNWGWYAYDPEVDLPYYGSGNPAPWNETMRP---GDNKWTMAIWGREATTGEAKPAYQKTPHDEW
ADH A.ace   228-  TWGPKGAWVRQGGGGTVWDSLVYDPVSDLIYLAVGNGSPWNYKYRSEGIGSNLFLGSIVALKPETGEYVWHFQATPMDQW
                  !!    !!     !!!! !     !!!    ! !    !!  !!!    !      ! !      !       !!!     ! !! ! !

                  ********* *** ** *    ******************* * *  ** * *****   * * * ** ****** ****** *
MDH M.ext   303-  DYAGVNVMMLSEQKDKDGKARKLLTHPDRNGIVYTLDRTDGALVSANKLDDTVNVFKSVDLKTGQPVRDPEYGTRMDHLA
MDH M.org   303-  DYAGVNVMMPSEQKDKDGKTRKLLTHPDRNGIVYTLDRTDGALVSANKLDDTVNVFKTVDLKTGQPVRDPEYGTRMDHLA
MDH P.den   303-  DYAGVNVMMLSEQEDKQGQMRKLLTHPDRNGIVYTLDRTNGDLISADKMDDTVNWVKEVQLDTGLPVRDPEFGTRMDHKA
ADH A.ace   308-  DYTSVQQIMTLDMPVK-GEMRHVIVHAPKNGFFYVLDAKTGEFLS-GKNYVYQNWANGLDPLTGRPMYNPDGLYTLNGKF
                  !!  !   !         ! !  !     !    !!  ! !!    !    !    !        !         !! !   !

                   * **************** *   *  *****************  ********** ******        *      ****
MDH M.ext   383-  KDICPSAMGYHNQGHDSYDPKRELFFMGINHICMDWEPFMLPYR--AGQFFVGATLNMYPGPKGDRQNYEG----LGQIK
MDH M.org   383-  KDVCPSAMGYHNQGHDSYDPKRELFFMGINHICMDWEPFMLPYR--AGQFFVGATLNMYPGPKGDRQNYEG----LGQIK
MDH P.den   383-  RDICPSAMGYHNQGHDSYDPERKVFMLGINHICMDWEPFMLPYR--AGQFFVGATLTMYPGPKATAER-AG----AGQIK
ADH A.ace   386-  WYGIPGPLGAHNFMAMAYSPKTHLVYIPAHQIPFGYKNQVGGFKPHADSWNVGLDMTKNGLPDTPEARTAYIKDLHGWLL
                      !   ! !!       ! !               !                       !    !!         !

                  ***** *  ******** ** *** ****** ** ****  ******** ****** *** ******* * * ***** *
MDH M.ext   457-  AYNAITGDYKWEKMERFAVW-GGTMATAGDLVFYGTLDGYLKARDSDTGDLLWKFKIPSGAIGYPMTYTHKGTQYVAIYY
MDH M.org   457-  AYNAITGSYKWEKMERFAVW-GGTLATAGDLVFYGTLDGYLKARDSDTGDLLWKFKIPSGAIGYPMTYTHKGTQYVAIYY
MDH P.den   456-  AYDAISGEMKWEKMERFSVW-GGTMATAGGLTFYVTLDGFIKARDSDTGELLWKFKLPSGVIGHPMTYKHDGRQYVAIMY
ADH A.ace   466-  AWDPVKMETVW-KIDHKGPWNGGILATGGDLLFQGLANGEFHAYDATNGSDLYKFDAQSGIIAPPMTYSVNGKQYVAVEV
                  !         !          ! !!  !!  ! !  !       !     !   !    !  !!    !! !  !!!!   !  !!!!

                  *** ****************** ***** *    ** **** ****** ** ******
MDH M.ext   536-  GVGGWPGVGLVFDLADPTAGLGAVGAFKKLANYTQMGGGVVVFSLDGKGPYDDPNVGEWKSAAK    -599 (end)
MDH M.org   536-  GVGGWPGVGLVFDLADPTAGLGAVGAFKKLANYTQQGGGVIVFSLDGKGPYDDPNVGEWKSASK    -599 (end)
MDH P.den   535-  GVGGWPGVGLVFDLADPTAGLGSVGAFKRLQEFTQMGGGVMVFSLDGESPYSDPNVGEYAPGEPR   -599 (end)
ADH A.ace   545-  GWGGI-----------YPISMGGVG---RTSGWTVNHSYIAAFSLDGKAKLPALNNRGFLPVKPPA-596
                  ! !!                   ! !!          !           !!!!!         !

CYT C-555     1-  APVDQATYNGFKIYKQQRCETCHGATGEGSAAFPNLLNSLKNLSKDQFKEVVLKGRNAMP
                               **    *  *   ** ***     * *                *  *   **  **  *
ADH A.ace   597-  QYDQKVVDNGYFQY-QTYCQTCHGDNGEGAGMLPDLRWAGAIRHQDAFYNVV--GRGALT
                      *       **       *   **     **       * *
CYT CL       48-  IDDKSCLRNG-ESLFATSCSGCHGHLAEGK-LGPGLNDNYWTYPSNTTDVGLFATIFGGA

CYT C-555    61-  PFEANK----KVAEGIDDLYTYIKGRSDGTVPAGELEKPQ -96
                               *      *     *     *
ADH A.ace   654-  AYGMDRFDTSMTPDEIEAIRQYLIKRANDTYQREVDARKNDKNIPENPTLGINP -707
                     *        ****                  *    *        *          *
CYT CL      106-  NGMMGPHNENLTPDEMLQTIAWIRHLYTGPKQDAVWLNDEQKKAYTPYKQGEVI -159
```

sary to dissociate the MDH into its subunits in order to release its prosthetic group.

4. The Alcohol Dehydrogenase of Acetic Acid Bacteria

The sequence of this quinohemoprotein dehydrogenase is shown in Figure 10 together with those of the α subunits of methanol dehydrogenase. ADH is isolated from membranes using the detergent Triton X-100, and this is included in all subsequent purification steps; it appears, therefore to be a typical membrane protein [77,79]. However, except for an N-terminal signal peptide, there are no strongly hydrophobic regions in the structure and certainly no membrane-spanning helices. There is considerable similarity between the first 596 amino acids of the ADH and the 599 amino acids of the three MDHs shown here, with the putative PQQ-binding site located near the N-terminus. No region of the primary sequence of ADH show any similarity to the β subunit of MDH.

The remaining 110 residues of ADH constitute a separate domain that was suggested to be a cytochrome domain [70]. The isolated ADH contains heme *c*, as indicated by its spectrum, and there is a typical heme-binding site (CXXCH; residues 614–618). The cytochrome *c* that shows most identity with this region of the ADH is a small Class I cytochrome *c* (cytochrome *c*-555) of no known function from the methanotroph *Methylococcus capsulatus* [78]. It should be noted that the similarity to this particular cytochrome *c* is not necessarily especially significant because other small cytochromes are also fairly similar at this relatively low level. Furthermore, the sixth ligand to the iron (the fifth being the histidine residue) in *c*-type cytochromes is a methionine, and there is no methionine residue on the ADH corresponding to the single methionine residue [59] in the cytochrome *c*-555. An alternative possibility is suggested in Figure 10, in which the electron acceptor for MDH, cytochrome c_L, is also aligned with the C-terminal region of ADH. This does not show as many identical residues as does

---

Figure 10 Sequences of three methanol dehydrogenases, alcohol dehydrogenase, cytochrome c-555 and cytochrome c_L. Methanol dehydrogenases (α subunits) (MDHs) are from *Methylobacterium extorquens* (*M. ext*) [68,71], *Methylobacterium organophilum* (*M. org*) [72], and *Paracoccus denitrificans* (*P. den*) [73]. The alcohol dehydrogenase (ADH) is from *Acetobacter aceti* [76], which is very similar to that of *A. polyoxygenes* [77]. The cytochrome that shows most similarity to the C-terminal region of ADH is cytochrome *c*-555 from *Methylococcus capsulatus* [78]. The sequence of cytochrome c_L is that of the electron acceptor for MDH in *M. extorquens* [57]. A possible PQQ-binding site is placed in a box and so is a potential heme-binding site. The cytochrome c_L has a methionine (in bold print), which may form the 6th ligand with the heme iron, which corresponds to a methionine residue in ADH. Numbers refer to number of residues in the mature protein excluding signal peptides. (!), Identical amino acids in all MDHs and ADH; (*), identical amino acids in the 3 MDHs, and, in the C-terminal region, the similarity of the ADH to the two *c*-type cytochromes. (From Ref. 68 with permission)

```
M.extorquens      YDGTKCKAAGNCWEPKPGFPEKIAGSKYDPKHDPKELNKQADSIKQMEERNKKRVENFKKTGKFEYDVAKISAN
                  ***••***•*•******••*•*••******•*** **•**••*•  *••* • ** *•********** **••
P.denitrificans   YDGANCKAPGTCWEPKPDYPAKVEGSKYDPQHDPAELSKQGESLAVMDARYEWRVWNMKKTGKFEYDVKKIDGYGDETKAPPAD
                  ====•== ==•=====•======•••= =====• ======• = ==   =•===== =• = ==••= == •=
A.methanolicus   AYDGTHCKKPGVCWEPQPGYPEQLVGVKYDPHFDPAELGVQ-ES---MNARNEARTAYFILTGTWEEDVNQIPK
                  !!!••!! •!•!!!!•!••!••••! !!!!• !! !!  ! •!   !••! • !• • !!••! !! •!
```

Figure 11 The complete sequences of the β-subunit of MDH (MDHB). The sequences are of the β-subunits of methanol dehydrogenase from *M. extorquens* [63], *P. denitrificans* [68], and *Acetobacter methanolicus* [12]. The sequence from *A. methanolicus* was obtained by sequencing the isolated protein [12]. Signal peptides are not included. (*), Identity between MDHs of *M. extorquens* and *P. denitrificans*; (=), indentity between the MDH of *A. methanolicus* and either of the other two MDHs; (!), identity between all three proteins; (•), a conservative replacement. (From Ref. 68 with permission.)

cytochrome *c*-555 (20 by contrast with 25), but the methionine residue that is conserved in the two examples of cytochrome c_L that have been sequenced is also present on ADH very near to a region with four other identical amino acids. Because there is so much similarity between the sequences of MDH and ADH, it is perhaps to be expected that the electron acceptor for ADH will be similar to that for MDH, which is the cytochrome c_L drawn here.

REFERENCES

1. Anthony, C., and Zatman, L. J. (1964). Isolation and properties of *Pseudomonas* sp. M27. *Biochem. J. 92*: 609.
2. Anthony, C., and Zatman, L. J. (1964). The methanol-oxidizing enzyme of *Pseudomonas* sp. M27. *Biochem. J. 92*: 614.
3. Anthony, C. (1986). The bacterial oxidation of methane and methanol. *Adv. Microbial Physiol. 27*: 113.
4. Anthony, C., and Zatman, L. J. (1967). The microbial oxidation of methanol: Purification and properties of the alcohol dehydrogenase of *Pseudomonas* sp. M27. *Biochem. J. 104*: 953.
5. Anthony, C., and Zatman, L. J. (1967). The microbial oxidation of methanol: The prosthetic group of alcohol dehydrogenase of *Pseudomonas* sp. M27; a new oxidoreductase prosthetic group. *Biochem. J. 104*: 960.
6. Hauge, J. G. (1964). Glucose dehydrogenase of *Bacterium antitratum*: An enzyme with a novel prosthetic group. *J. Biol. Chem. 239*: 3630.
7. Duine, J. A., and Frank, J. (1980). The prosthetic group of methanol dehydrogenase: Purification and some of its properties. *Biochem. J. 187*: 221.
8. Salisbury, S. A., Forrest, H. S., Cruse, W. B. T., and Kennard, O. (1979). A novel coenzyme from bacterial primary alcohol dehydrogenases. *Nature 280*: 843.
9. Duine, J. A., Frank, J., and Verwiel, P. E. J. (1980). Structure and activity of the prosthetic group of methanol dehydrogenase. *Eur. J. Biochem. 108*: 187.
10. Anthony, C. (1992). The role of quinoproteins in bacterial energy transduction. In *Quinoproteins and Their Applications* (V. L. Davidson, ed.), Marcel Dekker, New York.
11. Day, D. J., and Anthony, C. (1990). Methanol dehydrogenase from *Methylobacterium extorquens* AM1. *Methods Enzymol. 188*: 210.
12. Chan, H. T. C., and Anthony, C. (1991). The interaction of methanol dehydrogenase and cytochrome c_L in an acidophilic methylotroph *Acetobacter methanolicus*. *Biochem. J. 280*: 139.
13. Duine, J. A., Frank, J., and Westerling, J. (1978). Purification and properties of methanol dehydrogenase from *Hyphomicrobium* X. *Biochim. Biophys. Acta 524*: 277.
14. Frank, J., and Duine, J. A. (1990). Methanol dehydrogenase from *Hyphomicrobium* X. *Methods Enzymol. 188*: 202.
15. Anthony, C., and Zatman, L. J. (1965). The alcohol dehydrogenase of *Pseudomonas* sp. M27. *Biochem. J. 96*: 808.
16. Houck, D. R., and Unkefer, C. J. (1991). Stereospecificity of the PQQ-dependent methanol dehydrogenase. Personal communication.
17. Page, M. D., and Anthony, C. (1986). Regulation of formaldehyde oxidation by the

methanol dehydrogenase modifier proteins of *Methylophilus methylotrophus* and *Pseudomonas* AM1. *J. Gen. Microbiol. 132*: 1553.
18. Long, A. R., and Anthony, C. (1990). Modifier protein for the methanol dehydrogenase of methylotrophs. *Methods Enzymol. 188*: 216.
19. Long, A. R., and Anthony, C. (1991). The periplasmic modifier protein for methanol dehydrogenase in the methylotrophs, *Methylophilus methylotrophus* and *Paracoccus denitrificans. J. Gen. Microbiol. 137*: 2353.
20. Bolbot, J. A., and Anthony, C. (1980). The metabolism of 1,2-propanediol by the facultative methylotroph, *Pseudomonas* AM1. *J. Gen. Microbiol. 120*: 245.
21. Ford, S., Page, M. D., and Anthony, C. (1985). The role of methanol dehydrogenase modifier protein and aldehyde dehydrogenase in the growth of *Pseudomonas* AM1 on 1,2-propanediol. *J. Gen. Microbiol. 131*: 2173.
22. Ghosh, R., and Quayle, J. R. (1981). Purification and properties of the methanol dehydrogenase from *Methylophilus methylotrophus. Biochem. J. 199*: 245.
23. Duine, J. A., and Frank, J. (1980). Studies on methanol dehydrogenase from *Hyphomicrobium* X: Isolation of an oxidised form of the enzyme. *Biochem. J. 187*: 213.
24. Duine, J. A., Frank, J., and Ruiter, L. G. (1979). Isolation of a methanol dehydrogenase with a functional coupling to cytochrome *c*. *J. Gen. Microbiol. 115*: 523.
25. Beardmore-Gray, M., O'Keeffe, D. T., and Anthony, C. (1983). The methanol: cytochrome *c* oxidoreductase activity of methylotrophs. *J. Gen. Microbiol. 129*: 923.
26. Elliott, E. J., and Anthony, C. (1988). The interaction between methanol dehydrogenase and cytochrome *c* in the acidophilic methylotroph *Acetobacter methanolicus. J. Gen. Microbiol. 134*: 369.
27. Dijkstra, M., Frank, J., and Duine, J. A. (1989). Studies on electron transfer from methanol dehydrogenase to cytochrome c_L, both purified from *Hyphomicrobium* X. *Biochem. J. 257*: 87.
28. Anthony, C. (1975). The microbial metabolism of C1 compounds: The cytochromes of *Pseudomonas* AM1. *Biochem. J. 146*: 289.
29. Mincey, T., Bell, J. A., Mildvan, A. S., and Abeles, R. H. (1981). Mechanism of action of methoxatin-dependent alcohol dehydrogenase. *Biochemistry 20*: 7502.
30. Parkes, C., and Abeles, R. H. (1984). Studies on the mechanism of action of methoxatin-requiring methanol dehydrogenase: reaction of enzyme with electron-acceptor dye. *Biochemistry 23*: 6355.
31. Dijkstra, M., Frank, J., Jongejan, J. A., and Duine, J. A. (1984). Inactivation of quinoprotein alcohol dehydrogenases with cyclopropane-derived suicide substrates. *Eur. J. Biochem. 140*: 369.
32. Frank, J., Dijkstra, M., Duine, A. J., and Balny, C. (1988). Kinetic and spectral studies on the redox forms of methanol dehydrogenase from *Hyphomicrobium*. X. *Eur. J. Biochem. 174*: 331.
33. Frank, J., van Krimpen, S. H., Verwiel, P. E. J., Jongejan, J. A., Mulder, A. C., and Duine, J. A. (1989). On the mechanism of inhibition of methanol dehydrogenase by cyclopropane-derived inhibitors. *Eur. J. Biochem. 184*: 187.
34. Bamforth, C. W., and Quayle, J. R. (1978). The dye-linked alcohol dehydrogenase of *Rhodopseudomonas acidophila*: Comparison with dye-linked methanol dehydrogenase. *Biochem. J. 169*: 677.
35. Cox, J. M., Day, D. J., and Anthony, C. (1992). The interaction of methanol

dehydrogenase and its electron acceptor, cytochrome c_L, in methylotrophic bacteria *Biochim. Biophys. Acta 1119*: 97.

36. Beardmore-Gray, M., and Anthony, C. (1984). Methanol: cytochrome *c* activity of methylotrophs. In *Microbial Growth on C-1 Compounds* (R. L. Crawford and R. S. Hanson, eds.), American Society for Microbiology, Washington, DC, pp. 97–105.
37. Duine, J. A., Frank, J., and Jongejan, J. A. (1987). Enzymology of quinoproteins. In *Advances in Enzymology*, John Wiley & Sons, New York, pp. 169–212.
38. Duine, J. A., Frank, J., and Verwiel, P. E. J. (1981). Characterization of the second prosthetic group in methanol dehydrogenase from *Hyphomicrobium* X. *Eur. J. Biochem. 118*: 395.
39. Duine, J. A., Frank, J., and de Beer, R. (1984). An electron-nuclear double-resonance study of methanol dehydrogenase and its coenzyme radical. *Arch. Biochem. Biophys. 233*: 708.
40. Dekker, R. H., Duine, J. A., Frank, J., Verwiel, P., and Westerling, J. (1982). Covalent addition of water, enzyme substrates and activators to pyrrolo-quinoline quinone, the coenzyme of quinoproteins. *Eur. J. Biochem. 125*: 69.
41. Akhtar, M., and Wilton, D. C. (1973). Enzyme mechanisms. *Ann. Rep. Chem. Soc. B 70*: 98.
42. Frank, J., Dijkstra, M., Balny, C., Verwiel, P. E. J., and Duine, J. A. (1989). Methanol dehydrogenase: Mechanism of action. In *PQQ and Quinoproteins*, (J. A. Jongejan and J. A. Duine, eds.), Kluwer Academic Publishers, Dordrecht, pp. 13–22.
43. Forrest, H. S., Salisbury, S. A., and Kilty, C. G. (1980). A mechanism for the enzymic oxidation of methanol involving methoxatin. *Biochem. Biophys. Res. Commun. 97*: 248.
44. Jongejan, J. A., Groen, B. W., and Duine, J. A. (1989). Structural properties of PQQ involved in the activity of quinohaemoprotein alcohol dehydrogenase from *Pseudomonas testosteroni*. In *PQQ and Quinoproteins* (J. A. Jongejan and J. A. Duine, eds.), Kluwer Academic Publishers, Dordrecht, pp. 205–216.
45. O'Keeffe, D. T., and Anthony, C. (1980). The interaction between methanol dehydrogenase and the autoreducible cytochromes *c* of the facultative methylotroph *Pseudomonas* AM1. *Biochem. J. 190*: 481.
46. O'Keeffe, D. T., and Anthony, C. (1980). The two cytochromes *c* in the facultative methylotroph *Pseudomonas* AM1. *Biochem. J. 192*: 411.
47. Cross, A. R., and Anthony, C. (1980). The purification and properties of the soluble cytochromes *c* of the obligate methylotroph *Methylophilus methylotrophus*. *Biochem. J. 192*: 421.
48. Ohta, S., and Tobari, J. (1981). Two cytochromes *c* of *Methylomonas*. *J. Biochem. 90*: 215.
49. Dijkstra, M., Frank, J., van Wielink, J. E., and Duine, J. A. (1988). The soluble cytochromes *c* of methanol-grown *Hyphomicrobium* X: Evidence against the involvement of autoreduction in electron-acceptor functioning of cytochrome c_L. *Biochem. J. 251*: 467.
50. Long, A. R., and Anthony, C. (1991). Characterization of the periplasmic cytochromes *c* of *Paracocus denitrificans*: Identification of the electron acceptor for methanol dehydrogenase, and description of a novel cytochrome *c* heterodimer. *J. Gen. Microbiol. 137*: 415.

51. Day, D. J., and Anthony, C. (1990). Soluble cytochromes *c* of methanol-utilising bacteria. *Methods Enzymol. 188*: 298.
52. Nunn, D. N., and Lidstrom, M. E. (1986). Isolation and complementation analysis of 10 methanol oxidation mutant classes and identification of the methanol dehydrogenase structural gene of *Methylobacterium* sp. strain AM1. *J. Bacteriol. 166*: 581.
53. Nunn, D. N., and Lidstrom, M. E. (1986). Phenotypic characterisation of 10 methanol oxidation mutant classes of *Methylobacterium* sp. strain AM1. *J. Bacteriol. 166*: 591.
54. Harms, N., and van Spanning, R. J. M. (1991). C-1 metabolism in *Paracoccus denitrificans*: Genetics of *Paracoccus denitrificans*. *J. Bioenerg. Biomembr. 23*: 187.
55. Pettigrew, G. W., and Moore, G. R. (1987). *Cytochrome* c: *Biological Aspects*. Springer-Verlag, New York.
56. Beardmore-Gray, M., O'Keeffe, D. T., and Anthony, C. (1982). The autoreducible cytochromes *c* of the methylotrophs, *Methylophilus methylotrophus* and *Pseudomonas* AM1. *Biochem. J. 207*: 161.
57. Nunn, D. N., and Anthony, C. (1988). The nucleotide sequence and deduced amino acid sequence of the cytochrome c_L gene of *Methylobacterium extorquens* AM1: A novel class of *c*-type cytochromes. *Biochem. J. 256*: 673.
58. Anthony, C. (1988). Quinoproteins and energy transduction. In *Bacterial Energy Transduction* (C. Anthony, ed.), Academic Press, London, pp. 293–316.
59. Poole, R. K. (1988). Bacterial cytochrome oxidases. In *Bacterial Energy Transduction* (C. Anthony, ed.), Academic Press, London, pp. 231–291.
60. Widdowson, D., and Anthony, C. (1975). The microbial metabolism of C1 compounds: The electron transport chain of *Pseudomonas* AM1. *Biochem. J. 152*: 349.
61. Santos, H., and Turner, D. L. (1988). Characterization and NMR studies of a novel cytochrome *c* isolated from *Methylophilus methylotrophus* which shows a redox-linked change of spin state. *Biochim. Biophys. 954*: 277.
62. Dijkstra, M., Frank, J., and Duine, J. A. (1988). Methanol oxidation under physiological conditions using methanol dehydrogenase and a factor isolated from *Hyphomicrobium* X. *FEBS Letts. 227*: 198.
63. Nunn, D. N., Day, D. J., and Anthony, C. (1989). The second subunit of methanol dehydrogenase of *Methylobacterium extorquens* AM1. *Biochem. J. 260*: 857.
64. Anthony, C. (1990). The oxidation of methanol in gram-negative bacteria. *FEMS Microbiol. Rev. 87*: 209.
65. Grabarek, Z., and Gergely, J. (1990). Zero-length crosslinking procedure with the use of active esters. *Anal. Biochem. 185*: 131.
66. Adachi, O., Matsushita, K., Shinagawa, E., and Ameyama, M. (1990). Purification and properties of methanol dehydrogenase and aldehyde dehydrogenase from *Methylobacillus glycogenes*. *Agric. Biol. Chem. 54*: 3123.
67. Adachi, O., Matsushita, K., Shinagawa, E., and Ameyama, M. (1990). Calcium in quinoprotein methanol dehydrogenase can be replaced by strontium. *Agric. Biol. Chem. 54*: 2833.
68. Anthony, C. (1992). The structure of bacterial quinoprotein dehydrogenases. *Int. J. Biochem. 24*: 29.
69. Cleton-Jansen, A. M., Goosen, N., Fayet, O., and van de Putte, P. (1990). Cloning, mapping, and sequencing of the gene encoding *Escherichia coli* quinoprotein glucose dehydrogenase. *J. Bacteriol. 172*: 6308.

70. Inoue, T., Sunagawa, M., Mori, A., Imai, C., Fukuda, M., Takagi, M., and Yano, K. (1990). Possible functional domains in a quinoprotein alcohol dehydrogenase from *Acetobacter aceti*. *J. Ferment. Bioeng. 70*: 58.
71. Anderson, D. J., Morris, C. J., Nunn, D. N., Anthony, C., and Lidstrom, M. E. (1990). Nucleotide sequence of the *Methylobacterium extorquens* AM1 moxF and moxJ genes involved in methanol oxidation. *Gene 90*: 173.
72. Machlin, S. M., and Hanson, R. S. (1988). Nucleotide sequence and transcriptional start site of the *Methylobacterium organophilum* XX methanol dehydrogenase structural gene. *J. Bacteriol. 170*: 4739.
73. Harms, N., de Vries, G. E., Maurer, K., Hoogendijk, J., and Stouthamer, A. H. (1987). Isolation and nucleotide sequence of the methanol dehydrogenase structural gene from *Paracoccus denitrificans*. *J. Bacteriol. 169*: 3969.
74. Cleton-Jansen, A-M., Goosen, N., Odle, G., and van de Putte, P. (1988). Nucleotide sequence of the gene coding for quinoprotein glucose dehydrogenase from *Acinetobacter calcoaceticus*. *Nucleic Acids Res. 16*: 6228.
75. Cleton-Jansen, A-M., Goosen, N., Vink, K., and van de Putte, P. (1989). Cloning, characterization and DNA sequencing of the gene encoding the Mr 50 000 quinoprotein glucose dehydrogenase from *Acinetobacter calcoaceticus*. *Mol. Gen. Genet. 217*: 430.
76. Inoue, T., Sunagawa, M., Mori, A., Imai, C., Fukuda, M., Takagi, M., and Yano, K. (1989). Cloning and sequencing of the gene encoding the 72-kilodalton dehydrogenase subunit of alcohol dehydrogenase from *Acetobacter aceti*. *J. Bacteriol. 171*: 3115.
77. Takami, T., Fukaya, M., Takemura, H., Tayama, K., Okumura, H., Kawamura, Y., Nishiyama, M., Horinouchi, S., and Beppu, T. (1991). Cloning and sequencing of the gene cluster encoding two subunits of membrane-bound alcohol dehydrogenase from *Acetobacter polyoxygenes*. *Biochim. Biophys. Acta 1088*: 292.
78. Ambler, R. P., Dalton, H., Meyer, T. E., Bartsch, R. G., and Kamen, M. D. (1986). The amino acid sequence of cytochrome *c*-555 form the methane-oxidising bacterium *Methylococcus capsulatus*. *Biochem. J. 233*: 333.
79. Ameyama, M., and Adachi, O. (1982). Alcohol dehydrogenase from acetic acid bacteria, membrane-bound. *Methods Enzymol. 89*: 450.
80. Anthony, C. (1992). The *c*-type cytochromes of methylotrophic bacteria. *Biochim. Biophys. Acta 1099*: 1.
81. Richardson, I. W., and Anthony, C. (1992). Characterization of mutant forms of the quinoprotein methanol dehydrogenase lacking an essential calcium ion. *Biochem. J.* (in press).

3

Bacterial Quinoproteins Glucose Dehydrogenase and Alcohol Dehydrogenase

Kazunobu Matsushita and Osao Adachi

Yamaguchi University, Yamaguchi, Japan

I. INTRODUCTION

Quinoprotein dehydrogenases function as primary dehydrogenases that transfer reducing equivalents directly or indirectly to a bacterial aerobic respiratory chain. The quinoprotein dehydrogenases have been shown to be located exclusively in the periplasm of Gram-negative bacteria. This characteristic is in contrast with NAD(P)-dependent dehydrogenases, which are located in the cytoplasm, and with flavoprotein dehydrogenases, which are mainly located at the inner surface of the cytoplasmic membrane. Furthermore, since some of the quinoprotein dehydrogenases are isolated in a soluble form and others as membrane-bound species, they are considered to be located either in the periplasmic space or on the outer surface of the cytoplasmic membrane. Since quinoprotein dehydrogenases donate reducing equivalents that drive respiration, it is very important to know how they are linked to the respiratory chain. To date, there remains some ambiguity as to the precise identities of the natural electron acceptors for quinoprotein dehydrogenases and how these proteins are coupled to the respiratory chain. This is partly due to the relatively short history of quinoprotein research and partly due to an incomplete understanding of bacterial respiratory chains.

Glucose and alcohol dehydrogenases are typical quinoproteins, which are distributed in a relatively wide variety of Gram-negative bacteria. Different soluble and membrane-bound forms of each have been found. Thus, both of these enzymes provide good systems for comparative studies of their structures and the mechanisms by which they are coupled to the respiratory chain. This chapter describes the structures and functions of the two bacterial quinoproteins, glucose dehydrogenase and alcohol dehydrogenase.

II. STRUCTURE AND FUNCTION OF GLUCOSE DEHYDROGENASE

Glucose dehydrogenase is a quinoprotein that catalyzes the direct oxidation of D-glucose to D-gluconate at the outer surface of the cytoplasmic membrane, and it is known as an alternative pathway to the phosphotransferase system of bacteria to catalyze D-glucose assimilation [1]. The enzyme is found in a wide variety of bacteria including Gram-negative facultative anaerobes such as enteric bacteria and *Zymomonas*, as well as aerobic bacteria such as pseudomonads and acetic acid bacteria.

Glucose dehydrogenase was initially investigated in the early 1960s by Hauge [2–6] and subsequently in the late 1970s by Duine et al. [7], who showed the enzyme to be a quinoprotein. Furthermore, in the mid-1980s, the same enzyme was purified and investigated again by two groups [8,9]. The enzymes used in this research are the same glucose dehydrogenases purified from the soluble fraction of *Acinetobacter calcoaceticus*. This bacterium contains a soluble form of glucose dehydrogenase in addition to a membrane-bound form, which had been believed for a long time to be the same enzyme or interconvertible forms [6,10]. However, recent evidence has shown that this soluble enzyme is not a typical glucose dehydrogenase, and it should not be confused with the typical membrane-bound glucose dehydrogenase. The membranes of several strains such as *Escherichia coli*, *Klebsiella aerogenes*, *Pseudomonas aeruginosa*, *Gluconobacter suboxydans*, *Acetobacter aceti*, and *A. calcoaceticus* contain antigens cross-reactive with an antibody of glucose dehydrogenase purified from *Pseudomonas fluorescens* [11], while glucose dehydrogenase purified from the soluble fraction of *A. calcoaceticus* does not cross-react with the antibody [12]. Subsequently, soluble glucose dehydrogenase (s-GDH) and membrane-bound glucose dehydrogenase (m-GDH) have been purified separately from *A. calcoaceticus* and shown to be distinctive, as summarized in Table 1 in all aspects, i.e., optimum pH, kinetics, substrate specificity, ubiquinone reactivity, molecular size, and immunoreactivity [13]. The immunochemical evidence also denies the possibility that s-GDH is a degradation product or a precursor of the membrane-bound enzyme. The notion that s-GDH and m-GDH are independent molecular species is consistent with the genetic evidence obtained by Cleton-Jansen et al. [14,15], who showed that *A. calcoaceticus* contains two different *gdh* genes having no sequence homology. Furthermore, no antigens cross-react with the antibody specific for soluble dehydrogenase in both soluble and membrane fractions of any bacterial strains tested so far except for *A. calcoaceticus*, of which all nine strains tested contain both s-GDH

Table 1 Properties of s-GDH and m-GDH

Properties	s-GDH	m-GDH
Localization	Soluble	Membrane-bound
Subunit structure	α_2	α
Molecular weight	50 kDa	83–87 kDa
Prosthetic group	2 PQQ	1 PQQ
Metal ion	2 Ca^{2+}	1(?) Ca^{2+}
Metal ion replaced	$Cd^{2+} > Mn^{2+} > Mg^{2+}$	$Mg^{2+} > Sr^{2+}$
Isoelectric point	9.5	8.7
Electron acceptor(s)	PMS, DCIP, WB, Q_1, Q_2	PMS, DCIP, WB, Q_1, Q_2, Q_6
Natural electron acceptor	Cytochrome b_{562} (?)	Ubiquinone
Substrate specificity	Glucose > Maltose > Lactose	Glucose > Xylose > Galactose
K_m for glucose	3.3 and 24.5 mM (negative cooperativity)	4.2 mM at pH 8.5, 0.5 mM at pH 6.0
Turnover number	5870 e/s/mol	1700 e/s/mol
Occurrence	*A. calcoaceticus*	A wide variety of bacteria
Purified from [Ref.]	*A. calcoaceticus* [5,8,9,13]	*A. calcoaceticus* [12,13], *P. fluorescens* [16,21], *G. suboxydans* [17,35], *E. coli* [18]

and m-GDH (Matsushita et al., unpublished). Thus, s-GDH could be used as a taxonomic marker for *A. calcoaceticus*.

s-GDH of *A. calcoaceticus* is a dimer consisting of identical subunits of 48–55 kDa, and each subunit contains one molecule of pyrroloquinoline quinone (PQQ) [8,9]. A gene (*gdh*B) for the enzyme has been cloned and sequenced [15]. The gene product is estimated to be a polypeptide of 52.8 kDa that contains a 24-amino-acid signal sequence at its N-terminus and thus is expected to become the mature protein of 50.2 kDa having no hydrophobic regions. Thus, s-GDH seems to be translocated through the cytoplasmic membrane into the periplasmic space. s-GDH is capable of catalyzing the oxidation of disaccharides, lactose or maltose, as well as D-glucose. The enzyme is able to donate electrons to several artificial dyes, including phenazine methosulfate (PMS), dichlorophenolindophenol (DCIP), and Wurster's blue (WB) [8,9], as well as short-chain ubiquinone homologs, Q_1 and Q_2, but it is unable to react with long-chain ubiquinones, Q_6 and Q_9 [13].

On the other hand, m-GDH has been solubilized and purified, in the presence of detergent, from the membranes of *P. fluorescens* [16], *G. suboxydans* [17], *E. coli* [18], and *A. calcoaceticus* [13]. The membrane-bound enzymes of all bacterial strains seem to be closely related structurally to each other since all the enzymes of seven strains tested cross-reacted immunologically [11]. The purified enzyme is a single polypeptide of 83–87 kDa at least in the presence of detergent; probably it exists as a monomer in detergent micelles or in phospholipid membranes. The gene for m-GDH has been cloned and sequenced from both *A. calcoaceticus* [19] and *E. coli* [20]. The amino acid sequences deduced from the nucleotide sequences of both genes (*gdh*A for *A. calcoaceticus* m-GDH and *gcd* for *E. coli* m-GDH) show that both m-GDHs are very homologous proteins of 87 kDa containing five hydrophobic domains in the N-terminal part, which might serve as membrane-spanning segments. Surprisingly, however, there is no significant homology in the amino acid sequence between s-GDH and m-GDH, which have only small homologous regions in a putative C-terminal PQQ-binding domain. m-GDH exhibits a relatively broad substrate specificity for monosaccharides including D-glucose, D-fucose, D-galactose, or D-xylose, but the enzyme is incapable of oxidizing lactose, unlike s-GDH. The enzyme, like s-GDH, can reduce several artificial electron acceptors such as PMS, DCIP, WB, Q_1, and Q_2. In contrast to the soluble enzyme, however, m-GDH is also able to reduce long-chain ubiquinones, Q_6 and Q_9 [12,13,21].

Glucose dehydrogenase binds PQQ noncovalently, but the strength of the binding seems to be somewhat different from enzyme to enzyme. PQQ can be readily removed from m-GDH of *P. fluorescens* [22], *P. aeruginosa* [23], or *E. coli* [24] by dialysis with EDTA, in which the apoenzyme thus inactivated can be successfully reconstituted with PQQ in the presence of calcium or magnesium ion. On the other hand, the enzymes of *A. calcoaceticus* and *G. suboxydans* bind PQQ so tightly that PQQ can not be removed by EDTA-dialysis [7,17]. Instead, PQQ of *A. calcoaceticus* s-GDH was removed by acidification [5], treatment with potassium bromide at high concentration [7], or thermal treatment [25]. The apo-form of s-GDH was reactivated with PQQ in the presence of calcium or cadmium ion but not of magnesium ion [25]. Recently, the apo-forms of s-GDH and m-GDH have been purified from a PQQ-deficient mutant of *A. calcoaceticus*, and there has been shown to be a critical difference in the metal requirement between s-GDH and m-GDH (Matsushita et al., unpublished). s-GDH can bind PQQ in the presence of calcium or cadmium ion, while magnesium or strontium ion can replace calcium ion in m-GDH. Furthermore, studies on the binding of PQQ analogs show that the tricarboxyl group of PQQ is essential for its binding to the apoenzyme [26]. Interestingly, glucose dehydrogenase has been found in a wide variety of enterobacterial strains, of which many species, including *Escherichia* or *Salmonella*, contain the apo-form of the enzyme, but some other strains such as *Klebsiella pneumoniae* and *Serratia marcescens* have the holoenzyme [27,28]. The physiological significance of the presence of apo-glucose dehydrogenase in such enterobacterial strains remains unclear at this moment, though several approaches have been taken to address this question [27,29,30]. Since the activity of such an apoenzyme can be reconstituted by the addition of PQQ in the presence of magnesium or calcium ion [24], the apoenzyme of *E. coli* has been widely used for enzymatic determination of PQQ [31,32].

III. HOW GLUCOSE DEHYDROGENASE IS COUPLED TO THE RESPIRATORY CHAIN

There is long-standing controversy with respect to the natural electron acceptor for glucose dehydrogenase. The notion that cytochrome *b* is an electron acceptor of glucose dehydrogenase from *A. calcoaceticus* was given originally by Hauge [3,5] and more recently shown by Dokter et al. [33]. The latter authors described that s-GDH of *A. calcoaceticus* donates electrons to the respiratory chain via a cytochrome b_{562}. On the other hand, Beardmore-Gray and Anthony [34] suggested that glucose dehydrogenase is linked to the respiratory chain via ubiquinone in *A. calcoaceticus*. The former authors described s-GDH that was confused with m-GDH at that moment, as mentioned previously. It has been shown extensively that only m-GDH, and not s-GDH, purified from *A. calcoaceticus* is able to react with more native ubiquinones such as Q_6 or Q_9 [13]. Likewise, m-GDHs purified from *P. fluorescens* and *G. suboxydans* have been shown to react with the longer-chain ubiquinone homolog, Q_6, as well as short-chain homologs, Q_1 and Q_2, in an aqueous detergent solution [21,35]. Since such a hydrophobic electron acceptor is very hard to work with properly in aqueous solution, however, ubiquinone reduction by the enzyme should be studied ideally in a phospholipid environment. This can be performed by reconstitution of glucose dehydrogenase into a phospholipid bilayer containing ubiquinone, where a terminal oxidase capable of oxidizing the resultant ubiquinol must be reconstituted at the same time, in order to know whether or not a proper electron transfer from the enzyme to ubiquinone occurs. Thus, m-GDHs purified from the membranes of *E. coli* and *G. suboxydans* have been reconstituted into proteoliposomes together with the respective ubiquinone, Q_8 for *E. coli* and Q_{10} for *G. suboxydans*, and cytochrome *o* terminal oxidase purified, respectively, from *E. coli* and *G. suboxydans* [35,36]. The proteoliposomes thus reconstituted are able to catalyze the oxidation of glucose with a reasonable rate, and the subsequent electron transfer reaction also generates an electrochemical proton gradient at the site of the terminal oxidase. The apparent turnover of glucose oxidase with respect to glucose dehydrogenase in the proteoliposomes is comparable to that of the native membrane. Since cytochrome *o* has been shown to oxidize ubiquinol [37], such a reconstitution of glucose oxidase activity indicates that glucose dehydrogenase is able to reduce Q_8 or Q_{10} to produce Q_8H_2 or $Q_{10}H_2$, which is in turn oxidized rapidly by the cytochrome *o* in the proteoliposomes. Thus, ubiquinone has been extensively demonstrated to be the natural electron acceptor for m-GDH, which is further supported by the notion that m-GDH might have a ubiquinone-binding domain motif similar to that of mitochondrial NADH dehydrogenases [38]. The same reference suggests that the ubiquinone-binding site could be located inside the bacterial cytoplasmic membrane. If this is the case, the electron transfer from PQQ to ubiquinone in the enzyme molecule would generate a membrane potential, which is in contrast to the notion that only a terminal oxidase is responsible for the generation of an electrochemical proton gradient. Thus, although the coupling mechanism in the glucose oxidase respiratory chain has not yet been conclusively proven, it is reasonable to conclude that m-GDH, present in a wide variety of bacteria, is linked to the respiratory chain via ubiquinone.

On the other hand, the situation of s-GDH is much more ambiguous. s-GDH exhibits no glucose oxidase activity in spite of binding to proteoliposomes containing Q_9 and cytochrome *o*, while *A. calcoaceticus* m-GDH reconstituted into the proteoliposomes reproduces glucose oxidase activity as in the case of *E. coli* or *G. suboxydans*, suggesting that s-GDH does not react with ubiquinone (Matsushita et al., unpublished). Cytochrome *b* may be required for such a reconstitution, but it still remains unsolved whether or not cytochrome *b* is the natural electron acceptor for s-GDH. Furthermore, more surprisingly, it has been questioned whether or not s-GDH really functions in vivo, since a mutant strain containing only s-GDH but not m-GDH does not shown any in vivo activity for glucose oxidation [14,15].

IV. STRUCTURE AND FUNCTION OF ALCOHOL DEHYDROGENASE

Several different types of PQQ-dependent alcohol dehydrogenase have been characterized in oxidative bacteria such as pseudomonads and acetic acid bacteria. The enzymes catalyze the oxidation of a wide variety of alcohols, except for methanol, which is not oxidized at all or with very low affinity, and thus differs from methanol dehydrogenase. These alcohol dehydrogenases could be divided into three subgroups according to their structure: (1) quinoprotein alcohol dehydrogenase, (2) quinohemoprotein alcohol dehydrogenase, and (3) quinohemoprotein-cytochrome *c* alcohol dehydrogenase complex (Table 2).

Quinoprotein alcohol dehydrogenase has been purified from the soluble fractions of *P. aeruginosa* and *P. putida* [39,40]. The enzyme was thought to be a dimer of identical subunits of 60 kDa. As in the case of methanol dehydrogenase, however, an additional small subunit has been found in the enzyme of *P. aeruginosa*, suggesting that it may have an $\alpha_2\beta_2$ structure (Schrover et al., unpublished). Thus, the enzyme complex contains two PQQ per mole. With respect to catalytic properties, the alcohol dehydrogenase is similar to methanol dehydrogenase; it has a high optimum pH of 9.0–9.5 and exhibits the enzyme activity using an artificial electron acceptor such as PMS or WB in the presence of ammonia or amine. The enzyme is capable of oxidizing a great number of primary and secondary alcohols and aldehydes. Though the enzyme is able to oxidize methanol, the K_m value for methanol is 1000 times higher than that for ethanol (13–18 μM).

Quinohemoprotein alcohol dehydrogenase has been purified as an apo-form of a single polypeptide of 67 kDa from the soluble fraction of *Comamonas*

Table 2 Alcohol Dehydrogenases Having PQQ as the Prosthetic Group

Properties	Quinoprotein alcohol dehydrogenase	Quinohemoprotein alcohol dehydrogenase	Quinohemoprotein-cytochrome *c* alcohol dehydrogenase complex
Soluble or membrane bound	Soluble	Soluble	Membrane-bound
Subunit structure	$\alpha_2\beta_2$	α	$\alpha\beta\gamma$
(molecular weight)	(60, 9 kDa)	(67 kDa)	(78, 48, 12 kDa)
Prosthetic group(s)	PQQ	PQQ (apo) and heme *c*	PQQ (α) and 3 hemes *c* (α and β)
Electron acceptor(s)	PMS, WB, cytochrome c_{551}	PMS, WB, ferricyanide	PMS, DCIP, ferricyanide, Q_1, Q_2, ubiquinone
Optimum pH	9.0	7.7	4.5 or 6.0
Activator (NH_4^+)	Required for PMS, WB; not required for cytochrome c_{551}	Not required	Not required
Turnover number	33 e/s/mol	12 e/s/mol	700 e/s/mol
Substrate (K_m)			
Methanol	2 or 94 mM	ND	ND
Ethanol	14 or 18 μM	5 mM	1.6 mM
Strains	*P. aeruginosa* and *P. putida*	*C. testosteroni*	*A. aceti* and *G. suboxydans*
Ref.	39, 40, unpublished	41	42, 43, 45, unpublished

ND, not detected.

testosteroni [41]. This enzyme is quite different from the quinoprotein alcohol dehydrogenase described above in several respects. It has a neutral optimum pH of 7.7, does not require any activator, and is able to use ferricyanide as well as PMS or WB as the electron acceptor. The alcohol dehydrogenase is unable to oxidize methanol, though it is capable of oxidizing a variety of primary and secondary alcohols and aldehydes like the quinoprotein alcohol dehydrogenase. Furthermore, it is characteristic that the enzyme contains one molecule of heme *c*/mol and no PQQ, and that the full enzyme activity is reconstituted with one molecule of PQQ/mol in the presence of calcium ion. Thus, this enzyme is called *quinohemoprotein alcohol dehydrogenase*. Recently, a quinohemoprotein alcohol dehydrogenase has been purified from *P. putida*, which is isolated as holo-enzyme, unlike the enzyme of *C. testosteroni* (Adachi et al., unpublished).

The alcohol dehydrogenase of acetic acid bacteria functions in vinegar fermentation to oxidize ethanol to acetic acid together with aldehyde dehydrogenase. This alcohol dehydrogenase is totally different from the quinoprotein alcohol dehydrogenase, while it is similar to quinohemoprotein alcohol dehydrogenase in some aspects but different in several other aspects. The alcohol dehydrogenase is tightly bound to the cytoplasmic membranes of the organisms, which distinguishes it from other alcohol dehydrogenases, which exist in a soluble form. Alcohol dehydrogenase of acetic acid bacteria has been solubilized and purified in the presence of detergent from the membranes of *G. suboxydans* and *A. aceti* [42,43]. Subsequently, the enzyme purified from *G. suboxydans* has been crystallized as large red crystals [44]. The purified enzyme is composed of three subunits: dehydrogenase peptide of 72–80 kDa, cytochrome c_{553} of 48–53 kDa, and a 14–17 kDa peptide whose function is unknown. The enzyme complex (about 140 kDa) contains one molecule of PQQ and three molecules of heme *c* as the prosthetic group [45]. Heme *c* is bound to both the first and second subunits of the enzyme, and PQQ is probably bound to the first dehydrogenase subunit (Matsushita et al., unpublished). Therefore, the dehydrogenase subunit is considered a quinohemoprotein, and alcohol dehydrogenase of acetic acid bacteria should be referred to as a quinohemoprotein-cytochrome *c* alcohol dehydrogenase complex. The gene for the dehydrogenase subunit of alcohol dehydrogenase of *A. aceti* has been cloned and sequenced [46,47]. The gene encodes a polypeptide of 78 kDa and contains a preceding signal sequence consisting of 35 amino acid residues, which is consistent with the notion that alcohol dehydrogenase of acetic acid bacteria is located on the periplasmic side of the cytoplasmic membrane. Furthermore, of the total amino acid sequence of the 78-kDa dehydrogenase subunit, about 600 amino acid residues from the N-terminal end exhibit close similarity to the amino acid sequence of quinoprotein methanol dehydrogenase, and the following residues at the C-terminal end contain a heme *c*–binding sequence (Cys-X-X-Cys-His). Thus, the quinohemoprotein dehydrogenase subunit of alcohol dehydrogenase of acetic acid bacteria is considered a conjugated product of a quinoprotein part and a cytochrome *c* part. Furthermore, the gene for the second subunit cytochrome *c* has also been detected downstream of the *adh* gene and sequenced (Inoue et al., unpublished). This gene encodes a polypeptide having a signal sequence in the N-terminal region and a subsequent mature protein portion of 48 kDa. This mature polypeptide has three heme *c*–binding motifs, which is inconsistent with the notion that the purified enzyme contains three heme *c* moieties in the whole enzyme complex. The gene encoding the third subunit has not been found.

The quinohemoprotein-cytochrome *c* alcohol dehydrogenase complex does not require any activator for activity and is able to react with ferricyanide as well as PMS, DCIP, or WB, as is also true for the quinohemoprotein alcohol dehydrogenase. However, the enzyme shows a lower optimum pH of 4–6 and relatively restricted substrate specificity compared with quinoprotein or quinohemoprotein alcohol dehydrogenase; it oxidizes only primary alcohols from ethanol to hexanol, but not methanol or secondary or tertiary alcohols, and shows some activity for formaldehyde and acetaldehyde. In contrast to the other types of alcohol dehydrogenase, the enzyme of acetic acid bacteria is able to reduce short-chain ubiquinone homologs, Q_1 and Q_2, and native ubiquinones, Q_9 and Q_{10}, under specialized conditions. Another characteristic of the quinohemoprotein-cytochrome *c* alcohol dehydrogenase complex is that the enzyme activity (70–150 μmol of ferricyanide reduced/min/mg protein) is about 10 times higher than that of the quinohemoprotein alcohol dehydrogenase ($\approx$10 μmol of ferricyanide reduced/min/mg of protein). There are variant species of *G. suboxydans* incapable of oxidizing ethanol, of which the membranes do not exhibit any alcohol dehydrogenase activity and contain the quinohemoprotein subunit of alcohol dehydrogenase but not the cytochrome *c* subunit [48]. The reconstitution of the cytochrome *c* subunit to the membranes is able to restore the enzyme activity as well as ethanol oxidase respiratory activity. More directly, dissociation of cytochrome *c* from the quinohemoprotein-cytochrome *c* complex has also been shown to largely decrease the enzyme activity, which can be reactivated by the reconstitution of the isolated cytochrome *c* to the residual components (Matsushita et al., unpublished). Thus, the cytochrome *c* confers higher enzyme activity to the quinohemoprotein dehydrogenase and thus seems to be essential for formation of a complex functional in the respiratory chain. Although no function of the third small subunit is known, since it has never been dissociated from the complex, it may confer additional properties such as membrane binding and/or ubiquinone binding on the enzyme.

There are several other enzymes related to the quinoprotein alcohol dehydrogenase described above, which include aromatic alcohol [49], polyethylene glycol [50,51], and polyvinyl alcohol [52] dehydrogenases. These alcohol dehydrogenases do not react with methanol but exhibit different substrate specificities for a wide variety of alcohols. Aromatic alcohol dehydrogenase oxidizes aromatic alcohols such as vanillyl, benzyl, or cinnamyl alcohol in addition to primary alcohols, secondary alcohols, and aldehydes. The enzyme has recently been shown to also oxidize ethylene glycols and polyethylene glycols [53]. The aromatic alcohol dehydrogenase is a soluble single protein of 72 kDa and contains heme *c* in the molecule, and thus is similar to quinohemoprotein alcohol dehydrogenase. Polyethylene glycol dehydrogenase is distinguished by the ability to oxidize long-chain primary alcohols up to C_{18} in addition to polyethylene glycol. The enzyme is membrane bound and has been purified in the presence of detergent. The purified enzyme is a tetramer of identical subunits of 60 kDa in the detergent solution, and it appears to be similar to quinoprotein alcohol dehydrogenase because of the lack of heme *c* in the molecule. The specificity of polyvinyl alcohol dehydrogenase is to oxidize secondary alcohols, but not primary alcohols, in addition to polyvinyl alcohol. This enzyme is membrane bound and able to reduce cytochrome *c* in the membranes [54], but the structure is not known since it has not been purified.

V. HOW ALCOHOL DEHYDROGENASE IS COUPLED TO THE RESPIRATORY CHAIN

There is little information on the coupling of alcohol dehydrogenases to the respiratory chain. Since quinoprotein alcohol dehydrogenases of *P. aeruginosa* and *P. putida* are analogous to methanol dehydrogenase in terms of the structure and kinetics, the natural electron acceptor for the alcohol dehydrogenase is expected to be a soluble cytochrome *c*. Recently, such a cytochrome *c* was purified from the periplasmic fraction of ethanol-grown *P. aeruginosa* and shown to react with the quinoprotein alcohol dehydrogenase (Schrover et al., unpublished). Since *P. aeruginosa* has been shown to have cytochrome *c* oxidase as the terminal oxidase [55], the ethanol oxidase respiratory chain would work just like a methanol oxidase system of methylotrophs. On the other hand, the natural electron acceptor for the quinohemoprotein alcohol dehydrogenase of *C. testosteroni* has never been investigated and thus is unknown. Since the enzyme is different from the quinoprotein alcohol dehydrogenase in several aspects, quinohemoprotein alcohol dehydrogenase may be coupled to the respiratory chain in a manner different from that of the quinoprotein alcohol dehydrogenase. However, it should be noted that some alternative way to oxidize ethanol is also present, because this strain, which has an apo-alcohol dehydrogenase, can grown on ethanol in the absence of PQQ. In fact, the intact cells are able to oxidize ethanol even in the absence of PQQ, the activity of which is increased threefold by the addition of PQQ (Matsushita et al., unpublished). The oxidation of ethanol can be performed by an NAD-dependent alcohol dehydrogenase located in the cytoplasm and coupled with the NADH oxidase respiratory chain in the cytoplasmic membrane. The same system may also work in the cytoplasm of *P. aeruginosa*.

The coupling mechanism to the respiratory chain of alcohol dehydrogenase of acetic acid bacteria has been investigated [56]. As described already, this alcohol dehydrogenase is distinctive in several aspects from quinoprotein or quinohemoprotein alcohol dehydrogenase, especially in that it forms a complex with cytochrome *c* and is tightly bound to the membrane. Furthermore, the enzyme is able to react with ubiquinone or ubiquinone homologs as mentioned previously, and acetic acid bacteria, both *Acetobacter* and *Gluconobacter* genera, have a highly active respiratory chain, which is terminated by either cytochrome *o* or cytochrome a_1 [57,58,59], each of which is known to be a ubiquinol oxidase. The ethanol oxidase respiratory chain has been reconstituted in a manner similar to that of the glucose oxidase system; alcohol dehydrogenase purified from *G. suboxydans* or *A. aceti* was reconstituted together with ubiquinone, Q_{10} or Q_9, and a terminal oxidase, cytochrome *o* or a_1 [56]. In contrast to the case of glucose dehydrogenase, alcohol dehydrogenase was incorporated successfully into preformed proteoliposomes, a phenomenon that may be explained by hydrophobicity differences between the two dehydrogenases. The proteoliposomes thus prepared exhibit an ethanol oxidase activity, with 10 times higher specific activity than the intact membranes, which is about 50% of the turnover number, with respect to alcohol dehydrogenase, observed in the intact membranes. Thus, alcohol dehydrogenase of acetic acid bacteria seems to be linked to the respiratory chain via ubiquinone, which is just like glucose dehydrogenase. However, this alcohol dehydrogenase consists of three subunits—quinohemoprotein, cytochrome *c*, and a small polypeptide of which function is unknown—and the details of the intramolecular electron transfers that occur between these subunits is not known. Therefore, it remains unclear which components of the alcohol dehydrogenase donate electrons to ubiquinone.

VI. CONCLUDING REMARKS

Bacterial quinoprotein dehydrogenases function as oxidoreductases in the periplasm of gram-negative bacteria. The periplasm and the periplasmic surface of the cytoplasmic membrane have only recently been recognized as important sites for metabolism, especially for electron transport, of gram-negative bacteria; more than 20 types of electron transport proteins have been assigned to work in this region [60]. Such periplasmic electron transport proteins include quinoprotein dehydrogenases, such as methanol dehydrogenase occurring only in methylotrophs, alcohol dehydrogenase detected in *Pseudomonas* species, and glucose dehydrogenase found in a wide variety of gram-negative bacteria including facultative anaerobes. Additional quinoprotein dehydrogenases are found in acetic acid bacteria, which contain alcohol, aldehyde, glucose, glycerol, and fructose dehydrogenases [61]. The latter quinoprotein dehydrogenases function in a highly active respiratory chain, which is coupled to specific sugar oxidation reactions [59,62].

Of the quinoproteins described above, the amino acid sequences of methanol, alcohol, and glucose dehydrogenases have been established from the nucleotide sequences, which enable us to compare those quinoproteins [20,47]. These quinoproteins have a relatively high sequence homology—about 30% homology towards each other. In particular, the C-terminal region of these quinoproteins is highly conserved and has been suggested to be a PQQ-binding domain. One exception is s-GDH, which does not have such a sequence homology to other quinoproteins.

Quinoprotein dehydrogenases function in the aerobic respiratory chain as the primary dehydrogenase and couple substrate oxidation to the remaining part of the respiratory chain. In this respect, there have been shown to be two different ways in which these quinoprotein dehydrogenases work in the bacterial aerobic respiratory chain (Fig. 1). Glucose dehydrogenase and the quinohemoprotein-cytochrome *c* alcohol dehydrogenase complex are bound tightly to the periplasmic side of the cytoplasmic membrane and donate electrons to ubiquinone, which is embedded in the membrane phospholipids. The resulting ubiquinol is subsequently oxidized by a terminal oxidase in the membrane. Thus, the terminal oxidase involved in the glucose and alcohol oxidase systems must be a ubiquinol oxidase, the activity of which has been found in cytochrome *o*, cytochrome *d*, and cytochrome a_1. Quinoprotein alcohol dehydrogenase may be freely soluble in the periplasmic space or loosely bound to the periplasmic side of the cytoplasmic membrane. And

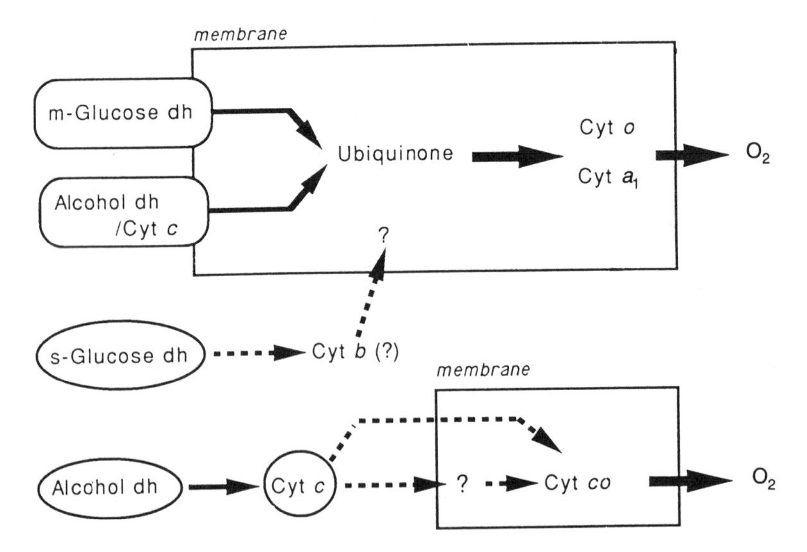

Figure 1 Possible coupling mechanism of glucose dehydrogenase and alcohol dehydrogenase in the aerobic bacterial respiratory chain. Broken lines show the coupling route that has not been demonstrated experimentally. dh, Dehydrogenase; Cyt, cytochrome.

thus, quinoprotein alcohol dehydrogenase may couple with the respiratory chain like methanol dehydrogenase, which donates electrons to a terminal oxidase via cytochrome *c*, a typical periplasmic electron transport protein [63]. In the alcohol oxidase system of *Pseudomonas* as well as the methanol oxidase system, thus, the terminal oxidase must be a cytochrome *c* oxidase, which is reasonable because *Pseudomonas* species contain a cytochrome *co* terminal oxidase capable of oxidizing cytochrome *c* [55].

As a whole, respiratory chains in which quinoprotein dehydrogenases are involved as a primary dehydrogenase are simple, and only the terminal oxidase may be involved in generation of an electrochemical proton gradient. However, such a truncated respiratory chain may be favorable for the rapid oxidation of a large amount of substrate. The notion may also be supported by the periplasmic location of quinoprotein dehydrogenase, since there is no requirement for an energy-consuming transport of the substrates into the cells.

REFERENCES

1. Midgley, M., and Dawes, E. A. (1973). The regulation of transport of glucose and methyl α-glucoside in *Pseudomonas aeruginosa*. *Biochem. J. 132*: 141.
2. Hauge, J. G. (1959). Purification and properties of a bacterial glucose dehydrogenase. *Acta Chem. Scand. 13*: 2125.
3. Hauge, J. G. (1960). Purification and properties of glucose dehydrogenase and cytochrome *b* from *Bacterium anitratum*. *Biochim. Biophys. Acta 45*: 250.
4. Hauge, J. G. (1961). Mode of action of glucose dehydrogenase from *Bacterium anitratum*. *Arch. Biochem. Biophys. 94*: 308.
5. Hauge, J. G. (1964). Glucose dehydrogenase of *Bacterium anitratum*: An enzyme with a novel prosthetic group. *J. Biol. Chem. 239*: 3630.
6. Hauge, J. G., and Hallberg, P. A. (1964). Solubilization and properties of the structurally-bound glucose dehydrogenase of *Bacterium anitratum*. *Biochim. Biophys. Acta 81*: 251.
7. Duine, J. A., Frank, J., and van Zeeland, J. K. (1979). Glucose dehydrogenase from *Acinetobacter calcoaceticus*. A "quinoprotein." *FEBS Lett. 108*: 443.
8. Geiger, O., and Gorisch, H. (1986). Crystalline quinoprotein glucose dehydrogenase from *Acinetobacter calcoaceticus*. *Biochemistry 25*: 6043.
9. Dokter, P., Frank, J., and Duine, J. A. (1986). Purification and characterization of quinoprotein glucose dehydrogenase from *Acinetobacter calcoaceticus* L.M.D. 79.41. *Biochem. J. 239*: 163.
10. Dokter, P., Pronk, J. T., van Schie, B. J., van Dijken, J. P., and Duine, J. A. (1987). The *in vivo* and *in vitro* substrate specificity of quinoprotein glucose dehydrogenase of *Acinetobacter calcoaceticus* LMD 79.41. *FEMS Microbiol. Lett. 43*: 195.
11. Matsushita, K., Shinagawa, E., Inoue, T., Adachi, O., and Ameyama, M. (1986). Immunological evidence for two types of PQQ-dependent D-glucose dehydrogenase in bacterial membranes and the location of the enzyme in *Escherichia coli*. *FEMS Microbiol. Lett. 37*: 141.

12. Matsushita, K., Shinagawa, E., Adachi, O., and Ameyama, M. (1988). Quinoprotein D-glucose dehydrogenase in *Acinetobacter calcoaceticus* LMD 79.41: The membrane-bound enzyme is distinct from the soluble enzyme. *FEMS Microbiol. Lett. 55*: 53.
13. Matsushita, K., Shinagawa, E., Adachi, O., and Ameyama, M. (1989). Quinoprotein D-glucose dehydrogenase of the *Acinetobacter calcoaceticus* respiratory chain: Membrane-bound and soluble forms are different molecular species. *Biochemistry 28*: 6276.
14. Cleton-Jansen, A-M., Goosen, N., Wenzel, T. J., and van de Putte, P. (1988). Cloning of the gene encoding quinoprotein glucose dehydrogenase from *Acinetobacter calcoaceticus*: Evidence for the presence of a second enzyme. *J. Bacteriol. 170*: 2121.
15. Cleton-Jansen, A-M., Goosen, N., Vink, K., and van de Putte, P. (1988). Cloning of the genes encoding the two different glucose dehydrogenases from *Acinetobacter calcoaceticus*. In *PQQ and Quinoproteins* (Jongejan, J. A., and Duine, J. A., eds.), Kluwer Academic Publishers, Dordrecht, pp. 79–85.
16. Matsushita, K., Ohno, Y., Shinagawa, E., Adachi, O., and Ameyama, M. (1980). Membrane-bound D-glucose dehydrogenase from *Pseudomonas* sp.: Solubilization, purification and characterization. *Agric. Biol. Chem. 44*: 1505.
17. Ameyama, M., Shinagawa, E., Matsushita, K., and Adachi, O. (1981). D-Glucose dehydrogenase of *Gluconobacter suboxydans*. Purification and characterization. *Agric. Biol. Chem. 45*: 851.
18. Ameyama, M., Nonobe, M., Shinagawa, E., Matsushita, K., Takimoto, K., and Adachi, O. (1986). Purification and characterization of the quinoprotein D-glucose dehydrogenase apoenzyme from *Escherichia coli*. *Agric. Biol. Chem. 50*: 49.
19. Cleton-Jansen, A-M., Goosen, N., Odle, G., and van de Putte, P. (1988). Nucleotide sequence of the gene coding for quinoprotein glucose dehydrogenase from *Acinetobacter calcoaceticus*. *Nucleic Acids Res. 16*: 6228.
20. Cleton-Jansen, A-M., Goosen, N., Fayet, O., and van de Putte, P. (1990). Cloning, mapping, and sequencing of the gene encoding *Escherichia coli* quinoprotein glucose dehydrogenase. *J. Bacteriol. 172*: 6308.
21. Matsushita, K., Ohno, Y., Shinagawa, E., Adachi, O., and Ameyama, M. (1982). Membrane-bound, electron transport-linked, D-glucose dehydrogenase of *Pseudomonas fluorescens*. Interaction of the purified enzyme with ubiquinone or phospholipid. *Agric. Biol. Chem. 46*: 1007.
22. Imanaga, Y., Hirano-Sawatake, Y., Arita-Hashimoto, Y., Itou-Shibouta, Y., Katoh-Semba, R. (1979). On the cofactor of glucose dehydrogenase of *Pseudomonas fluorescens*. *Proc. Japan Acad. 55*: 264.
23. Duine, J. A., Frank, J., and Jongejan, J. A. (1983). Detection and determination of pyrroloquinoline quinone, the coenzyme of quinoproteins. *Anal. Biochem. 133*: 239.
24. Ameyama, M., Nonobe, M., Hayashi, M., Shinagawa, E., Matsushita, K., and Adachi, O. (1985). Mode of binding of pyrroloquinoline quinone to apo-glucose dehydrogenase. *Agric. Biol. Chem. 49*: 1227.
25. Geiger, O., and Gorisch, H. (1989). Reversible thermal inactivation of the quinoprotein glucose dehydrogenase from *Acinetobacter calcoaceticus*. *Biochem. J. 261*: 415.
26. Shinagawa, E., Matsushita, K., Nonobe, M., Adachi, O., Ameyama, M., Ohshiro, Y., Itoh, S., and Kitamura, Y. (1986). The 9-carboxyl group of pyrroloquinoline

quinone, a novel prosthetic group, is essential in the formation of holoenzyme of D-glucose dehydrogenase. *Biochem. Biophys. Res. Commun. 139*: 1279.
27. Homes, R. J. W., Postma, P. W., Neijssel, O. M., Tempest, D. W., Dokter, P., and Duine, J. A. (1984). Evidence of a quinoprotein glucose dehydrogenase apoenzyme in several strains of *Escherichia coli. FEMS Microbiol. Lett. 24*: 329.
28. Bouvet, O. M. M., Lenormand, P., and Grimont, P. A. D. (1989). Taxonomic diversity of the D-glucose oxidation pathway in the *Enterobacteriaceae. Int. J. Syst. Bacteriol. 39*: 61.
29. Homes, R. J. W., van Hell, B., Postma, P. W., Neijssel, O. M., and Tempest, D. W. (1985). The functional significance of glucose dehydrogenase in *Klebsiella aerogenes. Arch. Microbiol. 143*: 163.
30. Biville, F., Turlin, E., and Gasser, F. (1991). Mutants of *Escherichia coli* producing pyrroloquinoline quinone. *J. Gen. Microbiol. 137*: 1775.
31. Ameyama, M., Nonobe, M., Shinagawa, E., Matsushita, K., and Adachi, O. (1985). Methods of enzymatic determination of pyrroloquinoline quinone. *Anal. Biochem. 151*: 263.
32. Geiger, O., and Gorisch, H. (1987). Enzymatic determination of pyrroloquinoline quinone using crude membranes from *Escherichia coli. Anal. Biochem. 164*: 418.
33. Dokter, P., van Wielink, J. E., van Kleef, M. A. G., and Duine, J. A. (1988). Cytochrome b-562 from *Acinetobacter calcoaceticus* L.M.D. 79.41. *Biochem. J. 254*: 131.
34. Beardmore-Gray, M., and Anthony, C. (1986). The oxidation of glucose by *Acinetobacter calcoaceticus*: Interaction of the quinoprotein glucose dehydrogenase with the electron transport chain. *J. Gen. Microbiol. 132*: 1257.
35. Matsushita, K., Shinagawa, E., Adachi, O., and Ameyama, M. (1989). Reactivity with ubiquinone of quinoprotein D-glucose dehydrogenase from *Gluconobacter suboxydans. J. Biochem. 105*: 633.
36. Matsushita, K., Nonobe, M., Shinagawa, E., Adachi, O., and Ameyama, M. (1987). Reconstitution of pyrroloquinoline quinone-dependent D-glucose oxidase respiratory chain of *Escherichia coli* with cytochrome *o* oxidase. *J. Bacteriol. 169*: 205.
37. Matsushita, K., Patel, L., and Kaback, H. R. (1984). Cytochrome *o* type oxidase from *Escherichia coli*. Characterization of the enzyme and mechanism of electrochemical proton gradient generation. *Biochemistry 23*: 4703.
38. Friedrich, T., Strohdeicher, M., Hofhaus, G., Preis, D., Sahm, H., and Weiss, H. (1990). The same domain motif for ubiquinone reduction in mitochondrial or chloroplast NADH dehydrogenase and bacterial glucose dehydrogenase. *FEBS Lett. 265*: 37.
39. Groen, B W., Frank, J., and Duine, J. A. (1984). Quinoprotein alcohol dehydrogenase from ethanol-grown *Pseudomonas aeruginosa. Biochem. J. 223*: 921.
40. Rupp, M., and Gorisch, H. (1988). Purification, crystallisation and characterization of quinoprotein ethanol dehydrogenase from *Pseudomonas aeruginosa. Biol. Chem. Hoppe-Seyler 369*: 431.
41. Groen, B. W., van kleef, D. A. M., and Duine, J. A. (1986). Quinoprotein alcohol dehydrogenase apoenzyme from *Pseudomonas testosteroni. Biochem. J. 234*: 611.
42. Adachi, O., Tayama, K., Shinagawa, E., Matsushita, K., and Ameyama, M. (1978).

Purification and characterization of particulate alcohol dehydrogenase from *Gluconobacter suboxydans*. *Agric. Biol. Chem.* *42*: 2045.

43. Adachi, O., Miyagawa, E., Shinagawa, E., Matsushita, K., and Ameyama, M. (1978). Purification and properties of particulate alcohol dehydrogenase from *Acetobacter aceti*. *Agric. Biol. Chem.* *42*: 2332.
44. Adachi, O., Shinagawa, E., Matsushita, K., and Ameyama, M. (1982). Crystallization of membrane-bound alcohol dehydrogenase of acetic acid bacteria. *Agric. Biol. Chem.* *46*: 2859.
45. Ameyama, M., and Adachi, O. (1982). Alcohol dehydrogenase from acetic acid bacteria, membrane-bound. *Methods Enzymol.* *89*: 450.
46. Inoue, T., Sunagawa, M., Mori, A., Imai, C., Fukuda, M., Takagi, M., and Yano, K. (1989). Cloning and sequencing of the gene encoding the 72-kilodalton dehydrogenase subunit of alcohol dehydrogenase from *Acetobacter aceti*. *J. Bacteriol.* *171*: 3115.
47. Inoue, T., Sunagawa, M., Mori, A., Imai, C., Fukuda, M., Takagi, M., and Yano, K. (1989). Possible functional domains in a quinoprotein alcohol dehydrogenase from *Acetobacter aceti*, *J. Ferm. Bioeng.* *70*: 58.
48. Matsushita, K., Nagatani, Y., Shinagawa, E., Adachi, O., and Ameyama, M. (1991). Reconstitution of the ethanol oxidase respiratory chain in membranes of quinoprotein alcohol dehydrogenase-deficient *Gluconobacter suboxydans* subsp. α strains. *J. Bacteriol.* *173*: 3440.
49. Yamanaka, K., and Tsuyuki, Y. (1983). A new dye-linked alcohol dehydrogenase (vanillyl alcohol dehydrogenase) from *Rhodopseudomonas acidophila* M402. Purification, identification of reaction product and substrate specificity. *Agric. Biol. Chem.* *47*: 2173.
50. Kawai, F., Kimura, T., Tani, Y., Yamada, H., and Kurachi, M. (1980). Purification and characterization of polyethylene glycol dehydrogenase involved in the bacterial metabolism of polyethylene glycol. *Appl. Environ. Microbiol.* *40*: 701.
51. Kawai, F., Yamanaka, H., Ameyama, M., Shinagawa, E., Matsushita, K., and Adachi, O. (1985). Identification of the prosthetic group and further characterization of a novel enzyme, polyethylene glycol dehydrogenase. *Agric. Biol. Chem.* *49*: 1071.
52. Shimao, M., Ninomiya, K., Kuno, O., Kato, N., and Sakazawa, C. (1986). Existence of a novel enzyme, pyrroloquinoline quinone dependent polyvinyl alcohol dehydrogenase, in a bacterial symbiont, *Pseudomonas* sp. strain VM15C. *Appl. Environ. Microbiol.* *51*: 268.
53. Yamanaka, K. (1991). Polyethylene glycol dehydrogenating activity demonstrated by dye-linked alcohol dehydrogenase of *Rhodopseudomonas acidophila* M402. *Agric. Biol. Chem.* *55*: 837.
54. Shimao, M., Onishi, S., Kato, N., and Sakazawa, C. (1989). Pyrroloquinoline quinone-dependent cytochrome reduction in polyvinyl alcohol-degrading *Pseudomonas* sp. strain VM15C. *Appl. Environ. Microbiol.* *55*: 275.
55. Matsushita, K., Shinagawa, E., Adachi, O., and Ameyama, M. (1990). *o*-Type cytochrome oxidase in the membrane of *Pseudomonas aeruginosa*. *FEBS Lett.* *139*: 255.
56. Matsushita, K., Takaki, Y., Shinagawa, E., Ameyama, M., and Adachi, O. (1992).

Ethanol oxidase respiratory chain of acetic acid bacteria. Reactivity with ubiquinone of pyrroloquinoline quinone-dependent alcohol dehydrogenases purified from *Acetobacter aceti* and *Gluconobacter suboxydans*. *Biosci. Biotech. Biochem. 56*: 304.
57. Matsushita, K., Shinagawa, E., Adachi, O., and Ameyama, M. (1987). Purification, characterization and reconstitution of cytochrome *o*-type oxidase from *Gluconobacter suboxydans*. *Biochim. Biophys. Acta 894*: 304.
58. Matsushita, K., Shinagawa, E., Adachi, O., and Ameyama, M. (1990). Cytochrome a_1 of *Acetobacter aceti* is a cytochrome *ba* functioning as ubiquinol oxidase. *Proc. Natl. Acad. Sci. USA 87*: 9863.
59. Matsushita, K., Ebisuya, H., Ameyama, M., and Adachi, O. (1992). Change of the terminal oxidase from cytochrome a_1 in shaking cultures to cytochrome *o* in static cultures of *Acetobacter aceti*. *J. Bacteriol. 174*: 122.
60. Ferguson, S. J. (1988). Periplasmic electron transport reactions. In *Bacterial Energy Transduction* (Anthony, C., ed.), Academic Press, London, pp. 151–182.
61. Ameyama, M., Matsushita, K., Ohno, Y., Shinagawa, E., and Adachi, O. (1981). Existence of a novel prosthetic group, PQQ, in membrane-bound electron transport chain-linked, primary dehydrogenases of oxidative bacteria. *FEBS Lett. 130*: 179.
62. Ameyama, M., Matsushita, K., Shinagawa, E., and Adachi, O. (1987). Sugar-oxidizing respiratory chain of *Gluconobacter suboxydans*. Evidence for a branched respiratory chain and characterization of respiratory chain-linked cytochromes. *Agric. Biol. Chem. 51*: 2943.
63. Anthony, C. (1988). Quinoproteins and energy transduction. In *Bacterial Energy Transduction* (Anthony, C., ed.), Academic Press, London, pp. 293–316.

4

Quinoprotein Aldehyde Dehydrogenases in Microorganisms

Kazunobu Matsushita and Osao Adachi

Yamaguchi University, Yamaguchi, Japan

I. INTRODUCTION

Unlike NAD- and NADP-dependent aldehyde dehydrogenases, quinoprotein aldehyde dehydrogenases catalyze aldehyde oxidation in the presence of artificial dyes in vitro. Several quinoprotein aldehyde dehydrogenases have been reported with properties that vary according to the source of the enzyme. The subcellular localization of the enzyme also varies. Some aldehyde dehydrogenases are found tightly bound to the cytoplasmic membrane, and others occur in the soluble fraction of the host microorganism. It has not been confirmed whether the soluble enzymes are localized in the cytoplasm or in the periplasmic space. In this review, the physical and enzymatic properties of aldehyde dehydrogenases and some potential applications of the enzymes are discussed.

II. MEMBRANE-BOUND QUINOPROTEIN ALDEHYDE DEHYDROGENASE

The aldehyde dehydrogenases of acetic acid bacteria are typical of the membrane-bound dehydrogenases and act on a wide range of aliphatic aldehydes except for formaldehyde. Aldehydes that have a carbon chain length of 2–4 are oxidized most rapidly by the enzymes from genera of both *Acetobacter* and *Gluconobacter*. The enzymes are localized on the outer (periplasmic) surface of the cytoplasmic membrane of the organisms, and the oxidation of the aldehyde substrate is linked to its respiratory chain and coupled to energy transduction. Acetic acid bacteria characteristically exhibit a high content of this enzyme in the bacterial membrane. This enzyme plays a key role in vinegar production in acetic acid bacteria by functioning in sequence with a membrane-bound quinoprotein alcohol dehydrogenase. During alcohol oxidation, no aldehyde liberation is observed, indicating that alcohol dehydrogenase and aldehyde dehydrogenase function as a multienzyme complex to produce acetic acid.

Characterization of aldehyde dehydrogenase was first achieved in 1980 after solubilization and purification of the enzyme from *Gluconobacter suboxydans* by Adachi et al. [1]. Purification of the enzyme can be performed by a procedure involving solubilization of the enzyme from the membrane fraction with Triton X-100 and subsequent fractionation by chromatography over DEAE-cellulose and CM-cellulose columns. A *c*-type cytochrome is tightly bound to the dehydrogenase protein and exists as an enzyme-cytochrome complex. The cytochrome is always present in an almost oxidized form. The visible absorption spectra of this species exhibit absorption maxima at the wavelengths of 551, 523, and 418 nm for the reduced enzyme and an absorption maximum at 410 nm for the oxidized enzyme. The molecular weight of the enzyme is estimated to be about 140,000, and sodium dodecyl sulfate polyacrylamide gel electrophoresis (SDS-PAGE) shows the presence of two subunits having molecular weights of 86,000 and 55,000. The larger subunit is the dehydrogenase that possesses pyrroloquinoline quinone (PQQ) as the prosthetic group, but not the cytochrome moiety. One mole of noncovalently bound PQQ is present per mole of the dehydrogenase subunit. PQQ can be removed from the enzyme by treating the enzyme with 90% methanol, and the PQQ that is liberated can be quantitated by assaying its ability to reconstitute an apo-quinoprotein glucose dehydrogenase [2]. The smaller subunit is a *c*-type cytochrome. Aliphatic aldehydes except for formaldehyde are oxidized very rapidly with the enzyme in the presence of dyes, such as 2,6-dichlorophenolindophenol, phenazine methosulfate, or potassium ferricyanide; neither NAD^+, $NADP^+$, nor oxygen are able to function as an electron acceptor. The optimum pH for aldehyde oxidation is observed at 4.0. The enzyme exhibits stability over a broad range of pH and also exhibits thermal stability when present in the unsolubilized membrane fraction. However, once the enzyme has been solubilized from the membrane with detergents, the enzyme becomes labile like the other membrane-bound enzymes. The stable pH range becomes narrower—about pH 5–6. It has been observed, however, that the loss of enzyme activity during storage after solubilization can be overcome by the inclusion of sucrose and benzaldehyde with the enzyme.

The first successful isolation of a quinoprotein aldehyde dehydrogenase from the genus *Acetobacter* was performed with *Acetobacter aceti* IFO 3284, another strain famous for vinegar production [3]. Solubilization and further purification of the enzyme was achieved after treatment of the membrane fraction with 5% Triton

X-100 and 0.2% cetylpyridinium chloride (CPC) in the presence of 10 mM benzaldehyde as a stabilizing agent. The enzyme was also shown to be a quinoprotein dehydrogenase-cytochrome *c* complex. The visible spectrum of the oxidized enzyme shows an absorption maximum at 411 nm, and the reduced enzyme exhibits absorption maxima at 551, 523, and 417 nm. The enzyme is composed of three subunits with molecular weights of 78,000, 45,000, and 14,000 as determined from SDS-PAGE, and the molecular weight determination by gel filtration indicated a native molecular weight of 150,000. The largest subunit contains PQQ and possesses dehydrogenase activity, and the 45,000 molecular weight subunit is a *c*-type cytochrome. The physiological and enzymatic function of the smallest subunit has not been resolved. The catalytic and physicochemical properties of this enzyme showed a great similarity to those of the aldehyde dehydrogenase from *G. suboxydans*.

Recently, an interesting quinoprotein aldehyde dehydrogenase occurring in the membrane fraction of *Acetobacter polyoxogenes* NBI 1028 was solubilized by detergents and further purified to homogeneity by Fukaya et al. [4]. The catalytic properties of the enzyme were essentially the same as others obtained from acetic acid bacteria, except that the enzyme lacked a heme-containing component. The total molecular weight was 90,000, which was divided into two subunits of 75,000 and 19,000. The absorption spectrum of this enzyme lacks the absorption peaks that are characteristic of a cytochrome component and is similar to that of the alcohol dehydrogenase, which also lacks a cytochrome component and which occurs in the soluble fraction of alcohol-grown *Pseudomonas aeruginosa* [5,6]. The existence of PQQ in the enzyme is readily confirmed by measuring its fluorescence spectrum, which is almost identical to that of authentic PQQ [7].

Hommel and Kleber [8] have reported the purification of a quinoprotein aldehyde dehydrogenase from the membrane fraction of *Acetobacter rancens* CCM 1774. It is a dimeric enzyme of the total molecular weight of 145,000, which includes a cytochrome *c* component.

III. SOLUBLE QUINOPROTEIN ALDEHYDE DEHYDROGENASE

Quinoprotein aldehyde dehydrogenases, which are located in the soluble fraction of cells, can be divided into two species. One occurs as a quinohemoprotein, and the other is a quinoprotein that lacks a heme component. Patel et al. [9,10] reported a heme-containing aldehyde dehydrogenase from obligate methylotrophs. The enzyme from *Methylomonas methylovora* [9] catalyzes the oxidation of straight-chain aldehydes, aromatic aldehydes, glyoxylate, and glyceraldehyde. The molecular weight of the enzyme as estimated by gel filtration is 45,000, and the subunit molecular weight as determined by SDS-PAGE is 23,000. The purified enzyme is light brown in color and exhibits an absorption maximum at 410 nm. On reduction, absorption maxima at 552 and 523 nm are also observed. A stoichiometric determination of heme content, however, has not been performed. Another heme-containing aldehyde dehydrogenase from *Methylosinus trichosporium* [10] shows similar properties to those of the enzyme from *M. methylovora* in both catalytic and physicochemical characteristics. Given our present knowledge, it is believed that such enzymes must be quinoproteins, though no characterization of the prosthetic group of these enzymes has been performed to confirm whether they contain PQQ.

Alcohol dehydrogenases having the capability to oxidize various kinds of aldehydes were purified from *Comamonas testosteroni* [11] and *Pseudomonas putida* [12]. The enzyme from *C. testosteroni* was purified as an apoenzyme, whereas the enzyme from *P. putida* was purified as the holoenzyme in which PQQ was noncovalently associated. These enzymes are synthesized only when the host organisms are grown in a medium containing alcohols. These enzymes show a high reactivity to primary alcohols as well as aldehydes. These enzymes occur as monomers, and the molecular weight of these enzymes is estimated to be 67,000–70,000. The apoenzyme from *C. testosteroni* is a monomer with an absorption spectrum similar to that of oxidized cytochrome *c*. After reconstitution to the holoenzyme by the addition of PQQ, addition of substrate causes a change of absorption spectrum to that of reduced cytochrome *c*, indicating that the cytochrome is coupled to the dehydrogenase activity of the enzyme. The enzyme from *P. putida* catalyzes the oxidation of alcohols, as well as aliphatic and aromatic aldehydes. No liberation of aldehyde occurs during alcohol oxidation. The enzyme reaction yields the corresponding carboxylic acid to the reaction product, indicating that aldehyde oxidation is also mediated by the enzyme.

Judging from the physiological functions of these quinohemoprotein aldehyde dehydrogenases found in the soluble fraction, they must be localized on the outer surface of cytoplasmic membrane or in the periplasmic space. This seems reasonable because the heme component should have an interaction with the respiratory chain of the organisms on the periplasmic face of the cytoplasmic membrane in order to couple substrate oxidation to energy transduction.

With respect to nonheme quinoprotein aldehyde dehydrogenases, the enzymes from methylotrophic bacteria [13] and *Rhodopseudomonas* [14] have been purified and characterized. The enzyme from an obligate methylotroph, *Methylobacillus glycogenes*, shows no ability to oxidize formaldehyde when compared with the enzymes from *M. methylovora* and *M. trichosporium*. The enzyme has a molecular weight of 140,000 and is composed of two identical subunits of 70,000. The visible absorption spectrum of the enzyme shows no characteristic peaks due to the presence of a heme component. The enzyme has been concluded to contain covalently bound PQQ. No liberation of the prosthetic group is achieved by 90% methanol treatment. After acid hydrolysis and subsequent isolation of the chromophore, spectral data indicated the presence of covalently bound PQQ.

In the soluble fraction of *Rhodopseudomonas acidophila* M402, a dye-linked aldehyde dehydrogenase has been indicated by Yamanaka et al. [14], which is different from the dye-linked aromatic alcohol dehydrogenase of that organism [15]. Although the aromatic alcohol dehydrogenase catalyzes the oxidation of aromatic and aliphatic alcohols and aldehydes, the dye-linked aldehyde dehydrogenase is active specifically on straight-chain aldehydes (C_3–C_{10}) rather than on benzaldehyde and substituted benzaldehydes. Aromatic and aliphatic alcohols are not substrates for this enzyme. The enzyme has been purified and characterized to be an enzyme of molecular weight of 70,000, which is composed of two identical subunits of molecular weight of 35,000. There is no heme component in the enzyme. Phenazine methosulfate, 2,6-dichlorophenolindophenol, and nitroblue tetrazolium are active as electron acceptors for aldehyde oxidation, as was observed with the other aldehyde dehydrogenases described in this chapter. It is possible that this enzyme also contains covalently bound PQQ, although this cannot be conclusively proven judging from the data obtained thus far.

IV. APPLICATIONS OF QUINOPROTEIN ALDEHYDE DEHYDROGENASE

The potential applicability of quinoprotein aldehyde dehydrogenases in diagnostic tests has been indicated [1,16]. NAD-dependent aldehyde dehydrogenase has already been used for such purposes. Certain characteristics of these enzymes, however, should be noted. The preparation of pure NAD-dependent enzymes is tedious and rather expensive, and these enzymes require the expensive NAD^+ cofactor. The advantages of using quinoprotein-based enzyme systems can be seen below.

Middle-chain aliphatic aldehydes such as *n*-hexanal and *n*-pentanal occur in foodstuffs as the result of chemical degradation of peroxidates of fat oxidation, and these compounds give rise to an off-flavor. Since the threshold of such compounds is extremely low when compared to the corresponding carboxylates, only a trace amount may cause foods to smell bad. The elimination of such off-flavors by the use of the NAD-dependent aldehyde dehydrogenase from bovine liver has been tried, and from the results appeared to have been useful. However, in actual practice it is difficult to employ such an enzyme because of the requirement for the expensive NAD^+ cofactor. On the other hand, the applicability of the membrane-bound quinoprotein aldehyde dehydrogenase for these purposes has been reported [17]. When a small amount of bacterial cells, such as *A. aceti* or *G. suboxydans*, is incubated briefly with defatted soy milk or wheat dough, a remarkable and immediate decrease of off-flavors in these food materials can be observed in sensory analysis, as well as in gas chromatographic and gel filtration chromatographic profiles. *n*-Hexanal is eliminated by acetic acid bacteria over a wide range of pH and temperature conditions with no damage to the original materials.

Elimination of off-flavors of liquid foods, which are caused by aliphatic aldehydes, has been investigated with immobilized acetic acid bacteria entrapped in κ-carrageenan gel [18]. When the immobilized cells are in contact with defatted soy milk or other off-flavored food materials, reduction of the bad smell occurs readily. The optimum conditions for the immobilized cells to oxidize *n*-hexanal is 50°C and pH 5.0. The aldehyde oxidase activity of the immobilized cells remains after 65 days of storage at 5°C. The immobilized cells can be used repeatedly more than 60 times without a loss in efficiency, and the beads containing the bacteria remain intact.

When cells or membranes of acetic acid bacteria, such as *A. aceti* or *G. suboxydans* were treated at 60°C for 10 minutes, more than 50% of the original activity of the quinoprotein aldehyde dehydrogenase remained, while the quinoproteins alcohol dehydrogenase and glucose dehydrogenase were completely inactivated [19]. About 60% of the terminal oxidase activity of the organisms also remained after heating. The heat-treated preparations were found to be useful for the diagnostic microdetermination of aldehydes and the selective elimination of off-flavors due to aldehydes from food materials.

The use of such heat-treated membranes for enzymatic aldehyde determination instead of a purified quinoprotein aldehyde dehydrogenase is advantageous. Aldehyde determination by means of the rate assay as well as endpoint determination can be successfully performed. With the use of the heat-treated membrane, the instability of solubilized and purified quinoprotein aldehyde dehydrogenase can be overcome, and an ideal aldehyde determination has become possible.

Another interesting application of the quinoprotein aldehyde dehydrogenase of acetic acid bacteria relates to the enzymatic or microbial production of sorbic acid from sorbaldehyde [20]. Unlike a chemical condensation of ketene and crotonaldehyde, in which many troublesome problems are encountered, under optimum conditions about 60 g per liter of sorbic acid can be prepared with 100% conversion rate with resting cells of *A. rancens*.

REFERENCES

1. Adachi, O., Tayama, K., Shinagawa, E., Matsushita, K., and Ameyama, M. (1980). Purification and characterization of membrane-bound aldehyde dehydrogenase from *Gluconobacter suboxydans*. *Agric. Biol. Chem. 44*: 503.
2. Ameyama, M., Nonobe, M., Shinagawa, E., Matsushita, K., and Adachi, O. (1985). Methods of enzymatic determination of pyrroloquinoline quinone. *Anal. Biochem. 151*: 263.
3. Ameyama, M., Osada, K., Shinagawa, E., Matsushita, K., and Adachi, O. (1981). Purification and characterization of aldehyde dehydrogenase of *Acetobacter aceti*. *Agric. Biol. Chem. 45*: 1889.
4. Fukaya, M., Tayama, K., Okumura, H., Kawamura, Y., and Beppu, T. (1989).

Purification and characterization of membrane-bound aldehyde dehydrogenase from *Acetobacter polyoxogenes* sp. nov. *Appl. Microbiol. Biotechnol. 32*: 176.
5. Groen, B. W., Frank, J., and Duine, J. A. (1984). Quinoprotein alcohol dehydrogenase from ethanol-grown *Pseudomonas aeruginosa. Biochem. J. 223*: 921.
6. Rupp, M., and Gorisch, H. (1988). Purification, crystallization and characterization of quinoprotein ethanol dehydrogenase from *Pseudomonas aeruginosa. Biol. Chem. Hoppe-Seyler 369*: 431.
7. Ameyama, M., Hayashi, M., Matsushita, K., Shinagawa, E., and Adachi, O. (1984). Microbial production of pyrroloquinoline quinone. *Agric. Biol. Chem. 48*: 561.
8. Hommel, R., and Kleber, H. P. (1990). Properties of the quinoprotein aldehyde dehydrogenase from *Acetobacter rancens. J. Gen. Microbiol. 136*: 1705.
9. Patel, R. N., Hou, C. T., and Felix, A. (1979). Microbial oxidation of methane and methanol: Purification and properties of a heme-containing aldehyde dehydrogenase from *Methylomonas methylovora. Arch. Microbiol. 122*: 241.
10. Patel, R. N., Hou, C. T., Derelanko, P., and Felix, A. (1980). Purification and properties of a heme-containing aldehyde dehydrogenase from *Methylosinus trichosporium. Arch. Biochem. Biophys. 203*: 654.
11. Groen, B. W., van Kleef, M. A. G., and Duine, J. A. (1986). Quinoprotein alcohol dehydrogenase apoenzyme from *Pseudomonas testosteroni. Biochem. J. 234*: 611.
12. Adachi, O., Sanada, H., Hokazono, M., Shinagawa, E., Ameyama, M., and Matsushita, K. (1991). Middle chain alcohol dehydrogenase from *Pseudomonas putida*. In *Proceedings of Bioscience and Biotechnology '91*, Kyoto, p. 286.
13. Adachi, O., Matsushita, K., Shinagawa, E., and Ameyama, M. (1990). Purification and properties of methanol dehydrogenase and aldehyde dehydrogenase from *Methylobacillus glycogenes. Agric. Biol. Chem. 54*: 3123.
14. Yamanaka, K., Iino, H., and Oikawa, T. (1991). Purification and properties of dye-linked aldehyde dehydrogenase in *Rhodopseudomonas acidophila* M402. *Agric. Biol. Chem. 55*: 989.
15. Yamanaka, K., and Tsuyuki, Y. (1983). Occurrence of dehydrogenases for the metabolism of vanillyl alcohol in *Rhodopseudomonas acidophila* M402. *Agric. Biol. Chem. 47*: 1361.
16. Ameyama, M. (1982). Enzymatic microdetermination of D-glucose, D-fructose, D-gluconate, 2-keto-D-gluconate, aldehyde and alcohol with membrane-bound dehydrogenases. *Methods Enzymol. 89*: 20.
17. Nomura, Y., Sugisawa, K., Adachi, O., and Ameyama, M. (1987). Reduction of off-flavors in food materials with acetic acid bacteria. *Nippon Nogeikagaku Kaishi 61*: 1079.
18. Nomura, Y., Sugisawa, K., Adachi, O., and Ameyama, M. (1988). Reduction of off-flavors in liquid foods with immobilized acetic acid bacteria. *Nippon Nogeikagaku Kaishi 62*: 143.
19. Adachi, O., Shinagawa, E., Matsushita, K., and Ameyama, M. (1988). Preparation of cells and cytoplasmic membranes of acetic acid bacteria which exclusively contain quinoprotein aldehyde dehydrogenase. *Agric. Biol. Chem. 52*: 2083.
20. Nagasawa, T. (1991). Applications of quinoprotein aldehyde dehydrogenase. In *Proceedings of Isolation and Identification of Quinoproteins*, Yamaguchi, pp. 69–72.

5

Methylamine Dehydrogenase

Victor L. Davidson

The University of Mississippi Medical Center, Jackson, Mississippi

I. INTRODUCTION

Methylamine dehydrogenase (MADH) is a periplasmic enzyme that has been purified from several gram-negative methylotrophic and autotrophic bacteria [1–8]. It catalyzes the oxidation of methylamine to formaldehyde and ammonia and in the process transfers two electrons from the substrate to some electron acceptor:

$$CH_3NH_2 + H_2O + 2A \rightarrow CH_2O + NH_3 + 2A^- + 2H^+ \quad (1)$$

This reaction is the first step in the assimilation of methylamine, which serves as a carbon source for these bacteria. When MADH is assayed in vitro, small redox active species such as phenazine ethosulfate (PES) or phenazine methosulfate (PMS) are routinely used as the electron acceptor. In vivo, the natural electron acceptor, for MADH is a periplasmic type I or "blue" copper protein, amicyanin, which mediates electron transfer from MADH to *c*-type cytochromes [5,9–11].

All MADHs possess an $\alpha_2\beta_2$ subunit structure in which α and β refer respectively, to the heavier and lighter subunits. These are also referred to as the H and L subunits by some authors. The smaller β subunits each possess a covalently bound prosthetic group, which has recently been identified as tryptophan tryptophylquinone (TTQ) [12] (Fig. 1). This enzyme is unique among bacterial quinoprotein dehydrogenases in its possession of a covalently bound quinone, rather than dissociable pyrroloquinoline quinone (PQQ), as a redox cofactor. In

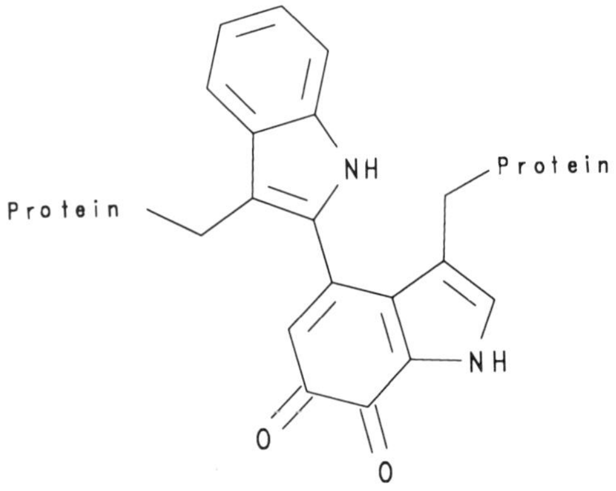

Figure 1 The tryptophan tryptophylquinone prosthetic group of MADH.

this respect, it is similar to the eukaryotic quinoprotein amine oxidases, which also contain covalently bound quinone prosthetic groups.

II. PHYSICAL PROPERTIES OF METHYLAMINE DEHYDROGENASE

A. TTQ, the Prosthetic Group of Methylamine Dehydrogenase

The nature of the covalently bound prosthetic group of MADH has been a controversial issue, which has only recently been resolved. In 1980, de Beer et al. [13] described electron spin resonance and electron-nuclear double resonance studies, which indicated that the prosthetic group of MADH from *Methylobacterium extorquens* AM1 (formerly *Pseudomonas* sp. AM1) was a quinone that also possessed two nitrogen atoms. This implicated PQQ as the prosthetic group. This possibility seemed particularly reasonable and attractive at that time in that each of the bacteria that were known to synthesize MADH also were known to synthesize the PQQ-bearing quinoprotein, methanol dehydrogenase. It seemed logical that both enzymes could possess the same cofactor. Direct evidence for the precise structure of the prosthetic group was hampered by several technical problems. Because the prosthetic group was covalently attached, extensive proteolytic treatment and subsequent chromatography under denaturing conditions were required to isolate relatively small prosthetic group–containing peptides, which were used in the subsequent analyses of structure. Given the highly reactive nature of the *o*-quinone function, it was necessary to derivatize the prosthetic group prior to these treatments to prevent side reactions with amino acids and other amino-containing compounds. Furthermore, the prosthetic group appeared to be attached at two different sites on the polypeptide chain, and the identities of those residues to which it was attached were unknown. Chemical sequencing of the β subunit of the enzyme from *M. extorquens* AM1 indicated the positions of the residues in the polypeptide chain that were thought to bind PQQ, but did not yield their identities [14].

In 1986, McIntire and Stults [15] reported that the prosthetic group of bacterium W3A1 MADH was not PQQ but a decarboxylated form of the cofactor, which was bound to the polypeptide via a serine oxygen and a cysteine sulfur. This claim was based upon fast atom bombardment mass spectroscopy of a semicarbazide-derivatized cofactor-containing peptide. This finding was disputed in subsequent reports by van der Meer et al., which stated that the cofactor of *Thiobacillus versutus* MADH was true PQQ, based upon HPLC analysis of phenylhydrazone [16] and dihexyl ketal [17] derivatives of the isolated cofactor. In 1989 the structure of *T. versutus* MADH was reported based on X-ray crystallographic studies [18,19]. It was not possible, however, to fit PQQ to the electron density of that structure. To resolve this inconsistency, the authors proposed yet another alternative model for the prosthetic group of MADH, which was termed pro-PQQ, in which an indole quinone was attached to a glutamic acid and an arginine residue.

A major discovery that finally allowed resolution of the debate concerning the precise identity of the cofactor of MADH was the sequencing of the gene that encoded the β subunit of the enzyme from *M. extorquens* AM1 [20]. Remarkably, the two amino acids that had been thought to covalently bind the quinone prosthetic group were revealed to both be tryptophan residues. Given this finding, all previous models of the cofactor had to be dismissed. None of the existing data could accommodate a structure consisting of an indole quinone species in addition to the two large tryptophans residues. The only reasonable conclusion that could be drawn was that the prosthetic group must in fact be derived from those gene-encoded tryptophans by some unknown posttranslational modification. In 1991, McIntire et al. [12] proposed the TTQ prosthetic group structure (see Fig. 1) based upon mass spectroscopy and nuclear magnetic resonance studies of derivatized prosthetic group-containing peptides from the bacterium W3A1 enzyme. It was also shown that TTQ provided a nearly perfect fit to the electron densities that were obtained from X-ray crystallographic studies of the enzymes from *T. versutus* and *Paracoccus denitrificans* [21]. Resonance Raman spectroscopy studies [22] further confirmed that the MADHs from *T. versutus*, *P. denitrificans*, and bacterium W3A1 possessed identical prosthetic groups. Thus, it is now universally agreed that TTQ is the prosthetic group of all MADHs.

A question that was raised by the discovery of TTQ, which is as yet unanswered, is how it is formed. Are the extensive posttranslational modifications that are necessary to transform two tryptophan residues into TTQ autocatalytic events, or are other enzymes involved? Do these modifications occur prior to the secretion of the polypeptide into the periplasm and assembly into the holoenzyme or during or after these processes? The elucidation of the mechanisms of post-translational modification that transform the two gene-encoded tryptophans into TTQ will be an important advance in our understanding of protein biosynthesis.

B. Features of the Structural Protein

The physical properties of the MADHs that have been characterized thus far are summarized in Table 1. It is clear from these data that MADHs are a reasonably conserved class of enzymes. Each MADH is a tetramer of two identical larger (α) subunits of molecular weight of 40,000–50,000 and two identical smaller (β) subunits of molecular weight of approximately 15,000.

1. Amino Acid Composition and Sequence

The amino acid compositions of five enzymes have been reported [6,18,20,23,24] and exhibit similar features, suggesting a high degree of conservation among this family of proteins. The amino acid compositions of the TTQ-bearing β subunits are very similar and rather unusual in their very high cysteine and proline contents. A complete amino acid sequence for both subunits of a single MADH has not yet been reported. The complete sequences of the TTQ-bearing β subunits of three enzymes are known [14,18,20,25,26] and, as shown in Table 2, further indicate a high level of homology.

2. Three-Dimensional Structure

The structure of the MADHs from *T. versutus* [18] and *P. denitrificans* [27] have recently been solved by X-ray crystallographic studies. MADH was shown to be a symmetrical molecule with extensive interactions between α and β subunits and relatively little interaction between like subunits. The active site of the enzyme is

Table 1 Properties of Methylamine Dehydrogenases from Various Sources

Source of enzyme	Type of methylotroph	Mol. wt. α subunit	Mol. wt. β subunit	pI	Ref.
Bacterium W3A1	Obligate	45	15.5	>8	4,28
Methylomonas sp. J	Obligate	40	13	9.0	23
M. methylotrophus	Obligate	42.7	15.9	ND	3
P. extorquens AM1	Facultative	40	13	5.2	29
P. denitrificans	Facultative	46.7	15.5	4.3	6
T. versutus	Facultative	47.5	12.9	3.9	7

ND, not determined.

Table 2 Amino Acid Sequences of the β Subunit of Methylamine Dehydrogenase[a]

	10 20 30 40
M. extorquens AM1:	A E S A G D P R G K W K P Q D N D V Q S C D Y W R H C S I D G N I C D C S G G S
T. versutus:	V D P R A K W Q P Q D N D I Q A C D Y W R H C S I A G N I C D C S A G S
P. denitrificans:	A D A P A G T D P R A K W V P Q D N D I Q A C D Y W R H C S I D G N I C D C S G G S
	50 60 70 80
M. extorquens AM1:	L T S C P P G T K L A S S S W V A S C Y N P T D K Q S Y L I S Y R D C C G A N V
T. versutus:	L T S C P P G T L V A S G S W V G S C Y N P P D P N K Y I T A Y R D C C G Y N V
P. denitrificans:	L T N C P P G T K L A T A S W V A S C Y N P T D G Q S Y L I A Y R D C C G Y N V
	90 100 110 120
M. extorquens AM1:	S G R C A C L N T E G E L P V Y R P E F G N D I I W C F G A E D D A M T Y H C T
T. versutus:	S G R C A C L N T E G E L P V Y N K D A – N D I I W C F G G E D – G M T Y H C S
P. denitrificans:	S G R C P C L N T E G E L P V Y R P E F A N D I I W C F G A E D D A M T Y H C T
M. extorquens AM1:	I S P I V G K A S
T. versutus:	I S P – V S G A –
P. denitrificans:	I S P I V G K A S

[a]The amino acid sequences MADH from *M. extorquens* AM1 [20] and *P. denitrificans* [26] were determined from the nucleotide sequences of their respective genes. The sequence of *T. versutus* [18] is an "X-ray" sequence determined from analysis of the crystal structure of the enzyme. The two gene-encoded tryptophan residues from which the TTQ prosthetic group are derived are underlined in each sequence.

located in a hydrophobic channel between the α and β subunits. The two indole rings that comprise the TTQ structure are not coplanar but at a dihedral angle of approximately 40–45°. The *o*-quinone of TTQ is present in the active site, whereas the edge of the second indole ring, which does not contain the quinone, is exposed on the surface of MADH. The details of these X-ray crystallographic investigations of the structure of MADH are discussed elsewhere [18,21,27].

3. Electrostatic Properties of and Immunological Cross-Reactivity Between Methylamine Dehydrogenases and Their Isolated Subunits

One physical property that does vary among MADHs is the isoelectric point (pI) of the enzyme (see Table 1). MADHs exhibit a wide range of pI values from 3.9 to 9.0. The enzymes that have been isolated from facultative autotrophic and methylotrophic bacteria exhibit acidic pI values, while the enzymes that have been isolated from obligate and restricted methyotrophs exhibit basic pI values. The basis for this diversity with respect to pI value has been examined by physical and immunochemical studies that were performed on the resolved subunits of MADHs [28]. Antibodies, which were raised to the isolated subunits of the MADHs of *P. denitrificans* and bacterium W3A1, were used to demonstrate immunological cross-reactivity between the β subunits of these enzymes. The β subunits of these enzymes also exhibited nearly identical pI values of 4.5 and 4.4. Conversely, the α subunits of the *P. denitrificans* and bacterium W3A1 enzymes did not cross-react immunologically and exhibited, respectively, very acidic and very basic pI values, which reflected those of their respective holoenzymes. Antibodies specific for the α subunit of the *P. denitrificans* enzyme, but not the bacterium W3A1 enzyme, cross-reacted with the analogous proteins in cell extracts of *T. versutus* and *M. extorquens* AM1. These data suggested that the relatively conserved family of MADHs could be further divided into two subclasses based upon pI value and immunological cross-reactivity of the α subunits. The particular type of MADH present in a particular species of bacteria appears to correspond to the range of substrates on which that bacterium is capable of growth (see Table 1). Such a distinction of subclasses of MADH is further supported by data from resonance Raman spectroscopy studies, which suggest that the environment of the protein-bound TTQ is different in the bacterium W3A1 enzyme than it is in the enzymes from *P. denitrificans* and *T. versutus* [22].

4. Stability

Most MADHs exhibit remarkable stability against denaturation by extremes of temperature and pH. For example, *P. denitrificans* MADH [6] retained 100% of its activity after a 30-minute incubation at 70°C, and approximately 65% of its activity after incubation at 80°C. At temperatures above 80°C, denaturation occurred more readily. When incubated for 48 hours at 30°C in buffers with pH values ranging from 3.0 to 10.5, no decrease in activity was observed at any of the pH values. Retention of significant activity after exposure to temperatures up to 70°C has also been observed for the enzymes that were isolated from *M. extorquens* AM1 [1,29], *Methylophilus methylotrophus* [3], and *T. versutus* [3]. Stability against exposure to extreme values of pH has also been reported for the enzymes of *M. extorquens* AM1 [1,29] and *Methylomonas* sp. J. [2].

C. Spectral Properties

All MADHs have characteristic visible absorption spectra that are similar to those of the *P. denitrificans* enzyme shown in Figure 2. These absorption spectra are quite different from those of methanol and glucose dehydrogenases, which contain noncovalently associated PQQ, and from those of eukaryotic quinoproteins with covalently bound prosthetic groups. For *P. denitrificans* MADH, the absorption spectrum of TTQ is perturbed at alkaline pH values [30,31]. A pH titration of the spectral changes observed for the oxidized enzyme yielded a pK value of 8.2 [31]. Resonance Raman spectroscopy studies have suggested that this spectral perturbation could be due to the formation of a hydroxy adduct of the quinone at high pH [22]. When MADHs are resolved into their individual subunits, only the β subunits exhibit a visible absorption spectrum. That spectrum was somewhat different from that of the holoenzyme [6,32], suggesting that the protein environment surrounding TTQ is influenced either directly by the α subunit or by interactions between the α and β subunits.

The fluorescence spectra of MADHs from *M. extorquens* AM1 [33] and *Methylomonas* sp. J. [2] have been reported. In each case excitation of the oxidized

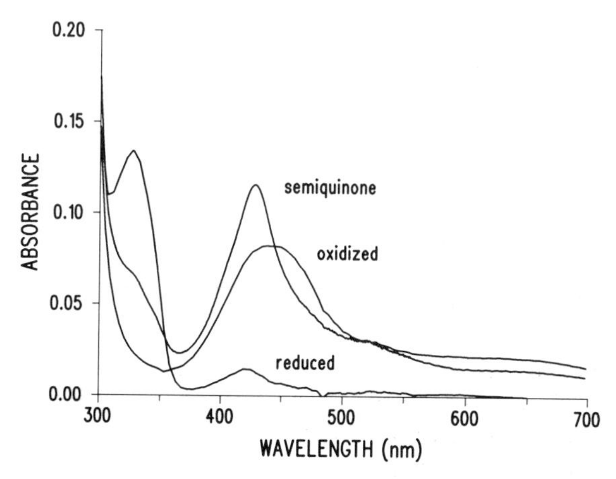

Figure 2 The visible absorption spectra of *P. denitrificans* MADH.

enzyme at 280–290 nm yielded a fluorescence emission maximum at 330–350 nm. After addition of methylamine to the enzyme, excitation at 330 nm yielded a fluorescence emission maximum at 380 nm. The fluorescence spectra of the resolved TTQ-bearing subunits of MADHs from *M. extorquens* AM1 [29] and *Methylomonas* sp. J. [23] have also been reported. In each case excitation of the subunit at 318 nm yielded a fluorescence emission maximum at approximately 382 nm. The fluorescence spectrum of a TTQ-bearing peptide, which was prepared after proteolytic digestion of the enzyme from bacterium W3A1 [4], has been reported as well and exhibited excitation and emission maxima at 320 nm and 410 nm, respectively.

In addition to absorption and fluorescence spectroscopy, other biophysical techniques have been used to probe the nature of the prosthetic group of MADH. Electron spin resonance spectra of the radical, semiquinone, forms of enzymes from *M. extorquens* AM1 [18] and bacterium W3A1 [4] have been reported, as has the electron-nuclear double resonance spectrum of the *M. extorquens* AM1 enzyme [13]. Detailed information has also been obtained from resonance Raman spectroscopy of MADHs from *P. denitrificans* [22] and bacterium W3A1 [32]. In the latter studies it was possible to assign several of the features of the resonance Raman spectrum to specific structural features of the protein-bound TTQ. For the *P. denitrificans* enzyme it was also possible to obtain resonance Raman spectra of the semiquinone form of the enzyme and of hydroxy and carbinolamine adducts.

D. Redox Properties

The three redox states of MADH [4,34] have been clearly characterized with regard to their spectral properties. The absorption spectra of the three redox states of *P. denitrificans* MADH [34] are shown in Figure 2. Similar spectra with a bit different absorption maxima have been reported for the enzyme from bacterium W3A1 [4]. Anaerobic reductive titration of the *P. denitrificans* enzyme with sodium dithionite [34] proceeded through the semiquinone intermediate, and reduction of the semiquinone during the second half of that titration was quite slow. These data indicated that the semiquinone was stable once formed and suggested a kinetic barrier to further reduction. At neutral pH, the semiquinone was slowly reoxidized under aerobic conditions, and the fully reduced enzyme was stable in the presence of oxygen and only slowly reoxidized by ferricyanide. Reductive titration of *P. denitrificans* MADH with methylamine at neutral pH proceeded directly to the fully reduced form of the enzyme without detectable formation of the semiquinone. That the spectrum shown in Figure 2 is truly that of the semiquinone species has been confirmed by pulse radiolysis studies [35]. Electrochemical titrations of this enzyme yielded an overall oxidation-reduction midpoint potential (E_m) value for the two-electron couple (fully oxidized/fully reduced) of +100 mV [34].

The factors that influence the formation and stability of the semiquinone form of TTQ in *P. denitrificans* MADH have been studied in detail [31]. The quinone prosthetic groups exhibited a pH-dependent redistribution of electrons from the 50% reduced plus 50% oxidized to the 100% semiquinone redox form. This phenomenon was only observed at pH values greater than 7.5. Once formed at pH 9.0, no change in redox state from 100% semiquinone was observed when the pH was shifted to 7.5, suggesting that the requirement of high pH was for formation and not stability of the semiquinone. The rate of semiquinone formation exhibited a first-order dependence on the concentration of MADH, indicating that this phenomenon was a bimolecular process involving intermolecular electron transfer between reduced and oxidized prosthetic groups. The rate of semiquinone formation also decreased with decreasing ionic strength, suggesting a role for hydrophobic interactions in facilitating electron transfer between MADH molecules. The suggestion that electron transfer between prosthetic groups was an intermolecular rather than an intramolecular process is reasonable in light of X-ray crystallographic studies of MADHs [18,27]. These studies indicated that the intramolecular distance between the prosthetic groups of this symmetrical molecule is approximately 45 Å, whereas the edge of one of the indole rings of TTQ is apparently exposed at the surface of the enzyme.

III. REACTION MECHANISM OF METHYLAMINE DEHYDROGENASE

A. Kinetic Studies

1. Steady-State Studies with Artificial Electron Acceptors

Steady-state kinetic studies of many MADHs have been performed using either phenazine methosulfate (PMS) or phenazine ethosulfate (PES) as electron acceptors [2,30,33,36,37]. The enzymes exhibited K_m values for methylamine in the range of 5–20 μM and specific activities in the range of 4–16 μmol/mg/min. The enzymes generally reacted with a variety of alkylamines but not with secondary, tertiary, or aromatic amines or amino acids. For *P. denitrificans* MADH, trimethylamine [38] and benzylamines [39] have been shown to be inhibitors of the enzyme when assayed with PES. The range of substrate specificity varies somewhat with enzymes from different sources, and substrate specificity was dictated primarily by the K_m value exhibited by the particular amine. Most enzymes exhibited a narrow pH optimum at approximately pH 7.5

A ping pong mechanism for the reaction of amine substrates and PES or PMS with MADH has been demonstrated with the enzymes from *M. extorquens* AM1

[33], *Methylomonas* sp. J [2], *P. denitrificans* [30], and bacterium W3A1 [37]. In each case it has been proposed that the aldehyde product is released prior to the reaction of the enzyme with the artificial electron acceptor, and that release of the second product, ammonia, occurs concomitant with the reoxidation of the enzyme.

For *P. denitrificans* MADH, a deuterium kinetic isotope effect of $^{D}V_{max}$ of 3.0 was observed for methylamine oxidation in experiments using PES as an electron acceptor [30]. These data are consistent with a mechanism in which removal of a methyl proton is the rate-determining step in methylamine oxidation.

2. Rapid Kinetic Studies

Stopped-flow rapid kinetic experiments have been performed on the reductive half-reaction of bacterium W3A1 MADH [40]. Two kinetically significant intermediates were observed. The first, relatively fast, transition was attributed to the reduction of the enzyme-bound quinone to the quinol. The second, relatively slow, transition was attributed to the release of the aldehyde product. No evidence for semiquinone formation was observed in this study of the reductive half reaction. Using perdeuteromethylamine as a substrate, a kinetic isotope effect of approximately 20 was reported for the rate constant attributed to the reductive step. A similar very large deuterium kinetic isotope effect has been reported for the quinoprotein plasma amine oxidase [41]. No detailed rapid kinetic study of the oxidative half-reaction of MADH has been reported.

B. Interactions with Ammonia

The TTQ prosthetic group of bacterium W3A1 MADH reacts reversibly with ammonia, acting as an activator (K_A = 2 mM) of the enzyme at low concentrations and as a competitive inhibitor (K_i = 16 mM) at high concentrations [37]. The addition of ammonia to this MADH also caused changes in its visible absorption spectrum, shifting the absorption maximum from 429 to 491 nm [4]. This was attributed to the formation of an iminoquinone adduct of that enzyme. Ammonia has also been shown to be a reversible competitive inhibitor of *P. denitrificans* MADH (K_i = 20 mM), but no activation of that enzyme was observed at lower concentrations [42]. Addition of ammonia to the MADHs from *P. denitrificans* and *T. versutus* caused changes in the visible absorption spectra of these enzymes that were similar to each other but quite different from that observed with the bacterium W3A1 MADH. For these enzymes, the absorption maximum was shifted to a lower wavelength, from 440 to 425 nm, with a small shoulder forming at 470 nm [22]. The resonance Raman spectrum of the ammonia adduct of the *P. denitrificans* enzyme indicated that these spectral perturbations probably reflected the formation of a carbinolamine adduct formed between ammonia and the reactive carbonyl of the TTQ prosthetic group. This is in contrast to the iminoquinone adduct, which was proposed with the bacterium W3A1 enzyme. As each of these enzymes possesses the identical prosthetic group, it has been proposed that the differences in the reactivity of these enzymes towards ammonia is probably due to differences in the active site environment surrounding TTQ in the respective proteins [22].

C. Inhibition by Nucleophilic Amines

Certain nucleophilic amines have been shown to alter the visible absorption spectrum of MADHs and to irreversibly inhibit the enzyme [4,33,42]. Systematic studies of the mechanisms of interaction of these amines with the TTQ prosthetic group were performed with the enzymes from *P. denitrificans* [42], and bacterium W3A1 [4]. Phenylhydrazine, semicarbazide, aminoguanidine, hydrazine, and hydroxylamine each irreversibly inactivated *P. denitrificans* MADH and caused changes in the absorbance spectrum of TTQ. Different spectral perturbations were observed on reaction with each of these inactivators, and in each case a stoichiometry of 2 moles per mole of enzyme (1:1 per TTQ) was required to observe complete modification of the absorbance spectrum. Hydrazine, hydroxylamine, and semicarbazide also caused changes in the absorption spectrum of bacterium W3A1 MADH [4]. The spectral changes reported in that study were similar but not identical to those which were observed with the *P. denitrificans* enzyme. Furthermore, the spectral perturbations reported with the bacterium W3A1 enzyme were maximal at a 1:1 ratio of reagent to enzyme (0.5:1 per TTQ). These findings suggest that the TTQ prosthetic groups of the bacterium W3A1 enzyme may exhibit differential half-site reactivities, whereas the prosthetic groups of the *P. denitrificans* enzyme are apparently equivalent.

Kinetic analysis of the inactivation of *P. denitrificans* MADH indicated that the reactions of hydrazine and hydroxylamine were very rapid, with stoichiometric inactivation occurring in less than 30 seconds. Phenylhydrazine and semicarbazide inhibited in a time-dependent manner and exhibited apparent bimolecular kinetics and second-order rate constants for inactivation, respectively, of 25 min^{-1} mM^{-1} and 39 $min^{-1}mM^{-1}$. In contrast, inactivation by aminoguanidine exhibited saturation behavior and kinetic parameters of $K_I = 2.5$ mM and $k_{inact} = 0.5$ min^{-1} were obtained.

The probable mechanisms of interaction of each of these inactivators is shown in Figure 3. In each case an inactive covalent adduct is formed between TTQ and the amine. It is proposed that for hydroxylamine an oxime adduct is formed, and in the other cases the corresponding hydrazone is generated. Only in the reaction with aminoguanidine was kinetic evidence for a reversible enzyme-inactivator complex obtained.

D. Inhibition by Cyclopropylamine

Cylclopropylamine acted as a mechanism-based inhibitor of the quinoprotein MADH from *P. denitrificans* [43]. Although TTQ was rapidly reduced by addition

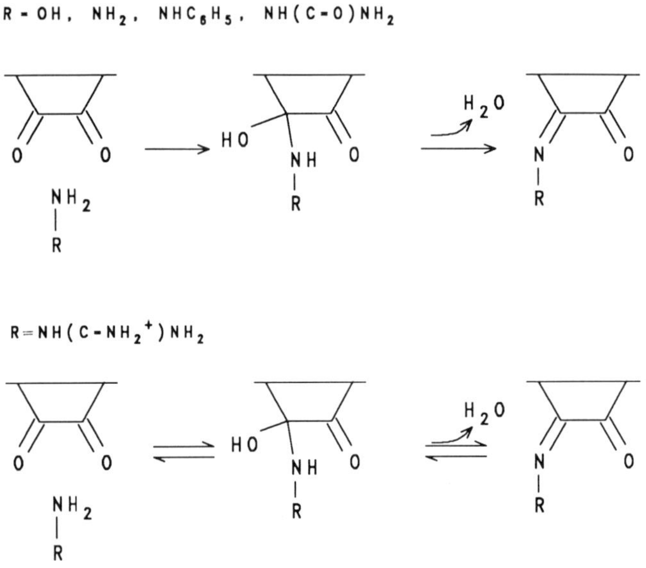

R = OH, NH_2, NHC_6H_5, $NH(C{=}O)NH_2$

R = $NH(C{-}NH_2^+)NH_2$

Figure 3 The proposed mechanisms for the inactivation of MADH by nucleophilic amines.

of a stoichiometric amount of cyclopropylamine, this compound did not serve as a substrate for the enzyme in the steady-state kinetic assay. In contrast to the above-mentioned inactivators, inactivation of the enzyme by cyclopropylamine was observed only in the presence of a reoxidant. Saturation behavior was observed and values for K_I of 3.9 μM and K_{inact} of 1.7 min^{-1} were determined. Enzyme inactivation was irreversible, as no restoration of activity was evident after gel filtration of MADH that had been incubated with cyclopropylamine in the presence of a reoxidant. The inactivated enzyme exhibited an altered absorption spectrum indicating that TTQ had been covalently modified. Electrophoretic analysis of the inactivated enzyme further indicated that covalent cross-linking of the α and β subunits had occurred. The proposed mechanism for this inhibition is shown in Figure 4.

The data and mechanism presented to explain the inhibition of MADH by cyclopropylamine required that some portion of the non–TTQ-bearing α subunit of MADH contribute to the environment that surrounds the active site. Such a suggestion is consistent with X-ray crystallographic data [18,27], which indicated that TTQ is located in a narrow channel at an interface between subunits. An intriguing question in the study of redox enzymes is the nature of the precise role of *noncatalytic* subunits—those that do not possess a redox center and have no

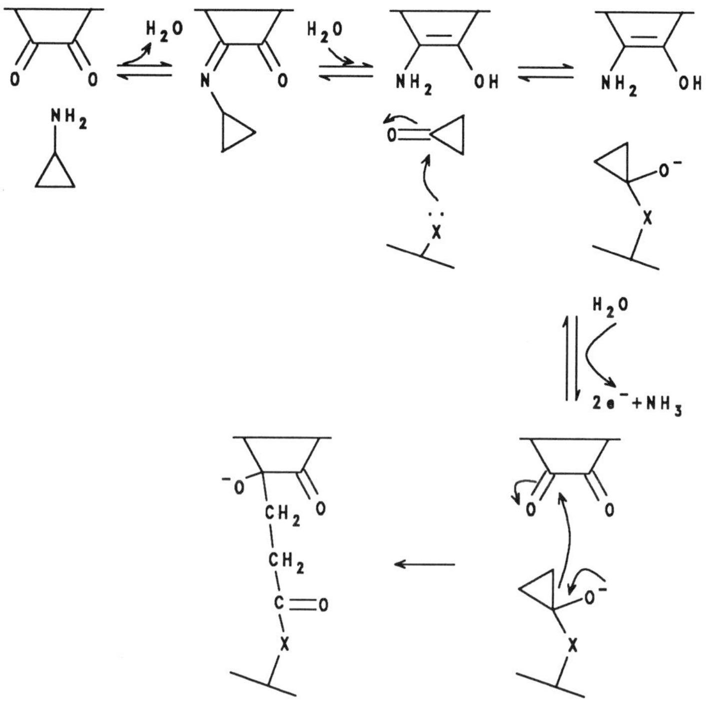

Figure 4 The proposed mechanism for the reaction of cyclopropylamine with MADH.

obvious role in catalysis. These data suggest that for MADH, the α subunit may play a critical role in establishing the functional environment at the active site of this enzyme.

E. Reactions with Benzylamines

It has been generally accepted that aromatic amines are not substrates for MADH [2,30,33,37]. It has, however, been recently reported [39] for *P. denitrificans* MADH that while benzylamine-dependent activity was not observed in the steady-state assay of this enzyme with PES, benzylamines did stoichiometrically reduce the enzyme-bound TTQ prosthetic group. Furthermore, benzylamines acted as reversible competitive inhibitors of methylamine oxidation. This was observed for a variety of *p*-substituted benzylamines. The efficiency of these substituted benzylamines to act as reductants of TTQ and inhibitors of methylamine oxidation increased with increasing electronegativity of the *p*-substituent. A Hammett plot of the rate constant for the reduction of the enzyme-benzylamine complex against substituent constants exhibited a positive slope, which suggested that the oxidation of these amines by methylamine dehydrogenase proceeds through a carbanionic reaction intermediate.

The reduction and inhibition by benzylamines of methylamine dehydrogenase is particularly interesting given similar results that have been observed with eukaryotic quinoprotein amine oxidases. Benzylamine derivatives with strongly electronegative *p*-substituents were shown to be effective ground state inhibitors of lysyl oxidase [44]. Furthermore, reactions of *p*-substituted benzylamines with plasma amine oxidase [45] exhibited similar correlations between rate constants for enzyme reduction and *p*-substituent constants. These data suggest that the mechanisms by which the reductive half-reaction of amine oxidation is catalyzed by methylamine dehydrogenase and the eukaryotic quinoprotein amine oxidases are very similar.

F. Summary of Reaction Mechanisms

At this point in time it is not possible to assign with certainty a complete reaction mechanism for MADH. For the reductive half-reaction of the enzyme with methylamine, the reaction mechanism presented in Figure 5 seems quite likely. The reaction is initiated by a nucleophilic attack by the amine nitrogen on one of the quinone carbonyls. This results in formation of a carbinolamine intermediate,

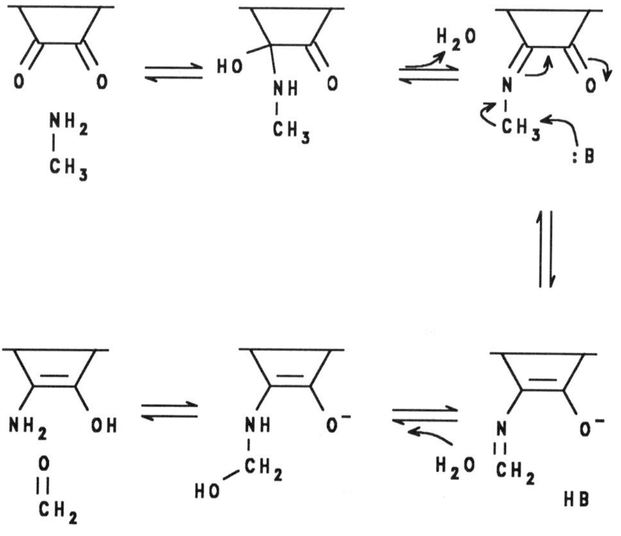

Figure 5 The proposed mechanism for the reductive half-reaction of MADH with methylamine.

which loses water to form an imine intermediate. Next, an active site nucleophile abstracts a proton from the methyl carbon, thus forming a carbanionic intermediate concomitant with the reduction of the prosthetic group. The resulting imine intermediate, which now possesses a double bond between the methyl carbon and amine nitrogen, is hydrolyzed to yield the aldehyde product and the reduced aminoquinol form of TTQ. Whereas the mechanism that has been proposed for the reductive half-reaction of MADH is relatively straightforward and consistent with the existing data, any proposed mechanism for the oxidative half-reaction with its natural electron acceptor is purely speculative. This is discussed in detail below.

IV. INTERACTIONS BETWEEN METHYLAMINE DEHYDROGENASE AND ITS PHYSIOLOGICAL ELECTRON ACCEPTORS

A. Amicyanin

Many studies have been performed on the interactions of methylamine dehydrogenases with other isolated periplasmic redox proteins from their host bacteria. From these data [5,9–11,46] it seems certain the primary electron acceptor for this enzyme is a periplasmic type I blue copper protein, amicyanin (Table 3). Most of the physical, redox, and spectroscopic properties of amicyanin are very similar to those of other small blue copper proteins such as azurin and plastocyanin. Available amino acid sequence data indicate that amicyanins are a separate and unique class of copper proteins [47,48]. Resonance Raman spectroscopy studies [49] and two-dimensional NMR studies [50] support this claim.

The specificity of the interaction between amicyanin and methylamine dehydrogenase is best demonstrated by studies of the isolated proteins from *P. denitrificans*. Each of these proteins is induced in this bacterium only during growth on methylamine as a carbon source [6,10]. Furthermore, the amicyanin gene is located immediately downstream of that for the β subunit of MADH, and inactivation of the former by means of gene replacement resulted in complete loss of the ability to grow on methylamine [25].

Table 3 Properties of Amicyanins

Organism	Mol. wt.	pI	E_m (mV)	Ref.
M. extorquens AM1	10,800	9.3	280	9
Organism 4025	11,500	5.3	294	5
P. denitrificans	12,500	4.8	294	10,46
T. versutus	13,800	4.7	260	11

B. Kinetic Studies of Methylamine Dehydrogenase with Physiological Electron Acceptors

The physiological role of amicyanin is not only to reoxidize MADH but also to mediate the transfer of electrons from that enzyme to soluble *c*-type cytochromes, which ultimately donate electrons to a cytochrome oxidase [51]. For *P. denitrificans*, in vitro studies have shown that cytochrome *c*-551i is the most efficient electron acceptor in this amicyanin-mediated process [52,53]. The mechanisms of catalysis and electron transfer by these proteins have been studied in solution using a steady-state kinetic approach [54]. Methylamine, oxidized amicyanin, and oxidized cytochrome *c*-551i were treated as reactants for MADH, and the initial rates of reaction were monitored by following the initial rate of reduction of cytochrome *c*-551i. A k_{cat} of 1100 min^{-1} for the methylamine-dependent reduction of cytochrome, and apparent K_m values of 2.5 μM for methylamine, 0.22 μM for amicyanin, and 1.3 μM for cytochrome *c*-551i were reported. The actual meaning of these values requires some discussion. It has been shown that cytochrome *c*-551i will only accept electrons from amicyanin in the presence of MADH [52–55]. It is not known whether cytochrome *c*-551i binds to one or both of these proteins in that process. As binding of cytochrome *c*-551i to either MADH alone or amicyanin alone would not result in electron transfer, it is reasonable to assume that the K_m value for cytochrome *c*-551i reflects its affinity for the MADH-amicyanin complex. The K_m value for amicyanin most likely reflects its affinity for MADH.

No detailed rapid kinetic studies of the oxidation of MADH by amicyanin have yet appeared in the literature. Without such data it is not possible to assign a reaction mechanism for the oxidative half-reaction of this enzyme with its physiological electron acceptor. A critical point to be addressed is that each reduced TTQ of MADH possesses two electrons that must be donated to a one electron carrier, amicyanin. X-ray crystallographic studies (discussed below) indicate that the stoichiometry of the electron transfer complex that is formed between these proteins is two amicyanin to one MADH, or one amicyanin to one TTQ. This suggests that the reoxidation of MADH must proceed through the semiquinone form of the enzyme (Fig. 6). As yet, however, there is no experimental data to support this, and the possible mechanism shown in Figure 6 should only be viewed at this point as a working hypothesis.

C. Complex Formation Between Methylamine Dehydrogenase and Amicyanin

Although MADH, amicyanin, and cytochrome *c*-551i from *P. denitrificans* are isolated as individual soluble proteins, they must form a ternary complex in order to perform their physiologically relevant function [55]. Direct electron transfer from reduced amicyanin to oxidized cytochrome *c*-551i does not occur in the absence of MADH. This has been explained in thermodynamic terms. The E_m

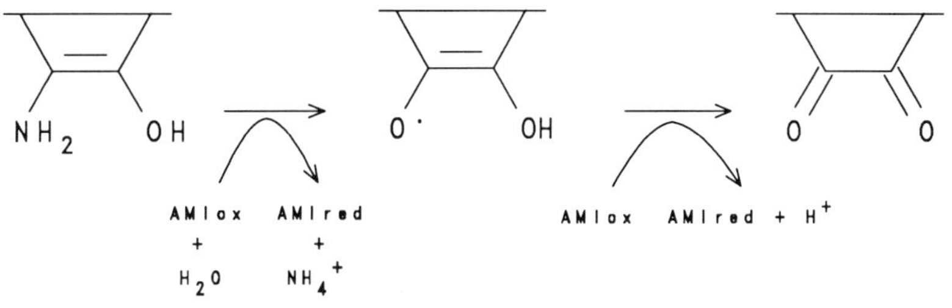

Figure 6 A possible mechanism for the oxidative half-reaction of MADH.

values for cytochrome *c*-551i and amicyanin are 190 mV and 294 mV, respectively [46]. One would not, therefore, expect amicyanin to reduce the cytochrome. It was demonstrated, however, that amicyanin and MADH formed a weakly associated complex which caused a perturbation of the absorption spectrum of TTQ and a 73 mV shift in the redox potential of amicyanin to 221 mV [55]. This complex-dependent shift in potential is critical in that it facilitates what would be an otherwise thermodynamically unfavorable electron transfer between amicyanin and the cytochrome to which it subsequently donates electrons.

The MADH-amicyanin complex has been further characterized by chemical cross-linking [56], resonance Raman spectroscopy [22], and X-ray crystallographic studies [57,58]. The data suggest that the stoichiometry of the complex is two amicyanin per MADH or one amicyanin per TTQ. The amicyanin is in contact with both the α and β subunits of MADH. The above-mentioned complex-dependent perturbations in the absorption spectrum of TTQ and the redox potential of amicyanin were only observed at low ionic strength [55]. Furthermore, in the steady-state kinetic assay described above activity was optimal at low ionic strength. When the assay was performed in the presence and absence of 0.2 M NaCl, the apparent K_m value for amicyanin increased approximately 10-fold at the higher ionic strength [54]. These data are consistent with the suggestion that electrostatic interactions between amicyanin and MADH are involved in the proper orientation of the proteins that facilitate intermolecular electron transfer. The chemical cross-linking and X-ray crystallographic studies, however, indicated that the association between proteins was stabilized primarily by hydrophobic interactions. A comparison of resonance Raman spectra of MADH and amicyanin free and in complex [22] revealed no significant changes, suggesting that neither the TTQ nor copper redox center is structurally altered during complex formation, and that the observed complex-dependent changes in their physical properties must be due to electrostatic interactions that accompany complex formation. Thus, efficient complex formation must involve some combination of electrostatic and hydrophobic interactions. The precise mechanism of the specific protein-protein association and the exact path of electron transfer between redox centers have not yet been completely determined. The resolution of these questions will impact greatly on our understanding of the process of intermolecular electron transfer between proteins.

V. RELATIONSHIP TO EUKARYOTIC AMINE OXIDASES

It has been assumed for some time that MADH was closely related to eukaryotic quinoprotein amine oxidases such as plasm amine oxidase and lysyl oxidase. This presumption was based for the most part on the erroneous claims that each of these enzymes possessed covalently bound PQQ as a prosthetic group [16,59–62]. As discussed above, it is now known that MADH possesses TTQ as a prosthetic group. Plasma amine oxidase has been shown to possess topa quinone, a post-translationally modified tyrosine residue, at its active site [63]. The precise nature of the lysyl oxidase prosthetic group is as yet uncertain. Now knowing that these enzymes do not possess the same prosthetic group, the question arises as to whether there is any legitimate basis for discussing them in the same context.

These enzymes may still be considered very similar with regard to their kinetic and reaction mechanisms. Steady-state kinetic studies have shown that, like MADH, lysyl [64] and plasma amine [65] oxidases exhibit a ping-pong kinetic mechanism in which aldehyde release precedes reoxidation. Mechanistic studies further support the involvement of a carbanionic intermediate in the reaction mechanisms of lysyl oxidase [44,66], plasma amine oxidase [45], and MADH [39]. Plasma amine oxidase and MADH are also similar in that each exhibits an unusually large primary deuterium kinetic isotope effect [40,41]. Thus, while it now appears that each of these enzymes possesses a different quinone prosthetic group, there is evidence to suggest that the mechanisms by which the reductive half-reaction of amine oxidation is catalyzed by MADH and the eukaryotic quinoprotein amine oxidases are very similar.

Another common factor in the structure and function of MADH and the eukaryotic quinoprotein amine oxidases is the role of copper. These amine oxidases possess bound copper in addition to the quinone, and both are contained on a single polypeptide chain. There is evidence to suggest that copper and the quinone interact, and that such an interaction is relevant to the oxidative half-reaction of amine oxidases with oxygen [67,68]. MADH does not possess bound copper. However, as described above, the electron acceptor for the quinone is the copper center of amicyanin. These two proteins form a physiologically relevant complex, and when in complex, interactions between the quinone and copper appear to occur and influence the physical properties of the two redox centers [55]. X-ray crystallographic data indicate that the copper of amicyanin and the closest portion of the TTQ prosthetic group are within 10 Å of each other in the complex formed by these proteins [58]. There is no evidence thus far from primary sequence analysis to suggest an evolutionary link between MADH and amicyanin and any of the amine oxidases. It is tempting, however, to speculate that the similarities between the two-protein bacterial amine oxidizing complex and the single protein eukaryotic amine oxidases, each of which employs a posttranslationally modified amino acid that functions as a quinone redox center and each of which uses copper, is more than pure coincidence.

ACKNOWLEDGMENT

Work from the author's laboratory reviewed here has been supported by National Institutes of Health Grant GM-41574.

REFERENCES

1. Eady, R. R., and Large, P. J. (1968). Purification and properties of an amine dehydrogenase from *Pseudomonas* AM1 and its role in growth on methylamine. *Biochem. J. 106*: 245.
2. Matsumoto, T. (1978). Methylamine dehydrogenase of *Pseudomonas* sp. J. Purification and properties. *Biochim. Biophys. Acta 522*: 291.
3. Haywood, G. W., Janschke, N. S., Large, P. J., and Wallis, J. M. (1982). Properties and subunit structure of methylamine dehydrogenase from *Thiobacillus* A2 and *Methylophilus methylotrophus. FEMS Microbiol. Lett. 15*: 79.
4. Kenny, W. C., and McIntire, W. (1983). Characterization of methylamine dehydrogenase from bacterium W3A1. Interaction with reductants and amino-containing compounds. *Biochemistry 22*: 3858.
5. Lawton, S. A., and Anthony, C. (1985). The role of blue copper proteins in the oxidation of methylamine by an obligate methylotroph. *Biochem. J. 228*: 719.
6. Husain, M., and Davidson, V. L. (1987). Purification and properties of methylamine dehydrogenase from *Paracoccus denitrificans. J. Bacteriol. 169*: 1712.
7. Vellieux, F. M. D., Frank, J. Jzn., Swarte, M. B. A., Groendijk, H., Duine, J. A., Drenth, J., and Hol, W. G. J. (1986). Purification, crystallization and preliminary X-ray investigation of quinoprotein methylamine dehydrogenase from *Thiobacillus versutus. Eur. J. Biochem. 154*: 383.
8. Davidson, V. L. (1990). Methylamine dehydrogenases from methylotrophic bacteria. *Methods Enzymol. 188*: 241.
9. Tobari, J., and Harada, Y. (1981). Amicyanin: An electron acceptor of methylamine dehydrogenase. *Biochim. Biophys. Res. Commun. 101*: 502.
10. Husain, M., and Davidson, V. L. (1985). An inducible periplasmic blue copper protein from *Paracoccus denitrificans*: Purification, properties, and physiological Role. *J. Biol. Chem. 260*: 14626.
11. van Houwelingen, T., Canters, G. W., Stobbelaar, G., Duine, J. A., Frank, J. Jzn., and Tsugita, A. (1985). Isolation and characterization of a blue copper protein from *Thiobacillus versutus. Eur. J. Biochem. 153*: 75.
12. McIntire, W. S., Wemmer, D. E., Christoserdov, A. Y., and Lidstrom, M. E. (1991). A

new cofactor in a prokaryotic enzyme: Tryptophan tryptophylquinone as the redox prosthetic group in methylamine dehydrogenase. *Science 252*: 817.

13. De Beer, R., Duine, J. A., Frank, J., and Large, P. J. (1980). The prosthetic group of methylamine dehydrogenase from *Pseudomonas* AM1. Evidence for quinone structure. *Biochim. Biophys. Acta 622*: 370.
14. Ishii, Y., Hase, T., Fukumori, Y., Matsubara, H., and Tobari, J. (1983). Amino acid sequence studies of the light subunit of methylamine dehydrogenase from *Pseudomonas* AM1: Existence of two residues binding the prosthetic group. *J. Biochem. 93*: 107.
15. McIntire, W. S., and Stults, J. T. (1986). On the structure and linkage of the covalent cofactor of methylamine dehydrogenase from the methyltrophic bacterium W3A1. *Biochem. Biophys. Res. Comm. 141*: 562.
16. van der Meer, R. A., Jongejan, J. A., and Duine, J. A. (1987). Phenylhydrazine as probe for cofactor identification in amine oxidoreductases. Evidence for PQQ as the cofactor in methylamine dehydrogenase. *FEBS Lett. 221*: 299.
17. van der Meer, R. A., Mulder, A. C., Jongejan, J. A., and Duine, J. A. (1989). Determination of PQQ in quinoproteins with covalently-bound cofactors and in PQQ-derivatives. *FEBS Lett. 254*: 99.
18. Vellieux, F. M. D., Huitema, F., Groendijk, H., Kalk, K. H., Frank, J. Jzn., Jongejan, J. A., Duine, J. A., Petratos, K., Drenth, J., and Hol, W. G. J. (1989). Structure of quinoprotein methylamine dehydrogenase at 2.25 Å resolution. *EMBO. J. 8*: 2171.
19. Vellieux, F. M. D., and Hol, W. G. J. (1989). A new model for the pro-PQQ cofactor of quinoprotein methylamine dehydrogenase. *FEBS Lett. 255*: 460.
20. Christoserdov, A. Y., Tsygankov, Y. D., and Lidstrom, M. E. (1990). Cloning and sequencing of the structural gene for the small subunit of methylamine dehydrogenase from *Methylobacterium extorquens* AM1: Evidence for two tryptophan residues involved in the active center. *Biochem. Biophys. Res. Comm. 172*: 211.
21. Chen, L., Mathews, F. S., Davidson, V. L., Huizinga, E., Vellieux, F. M. D., Duine, J. A., and Hol, W. G. J. (1991). Crystallographic investigations of the tryptophan-derived cofactor in quinoprotein methylamine dehydrogenase. *FEBS Lett. 287*: 163.
22. Backes, G., Davidson, V. L., Huitema, F., Duine, J. A., and Sanders-Loehr, J. (1991). Characterization of the tryptophan-derived quinone cofactor of methylamine dehydrogenase by resonance Raman spectroscopy. *Biochemistry 30*: 9201.
23. Matsumoto, T., Hiroka, B. Y., and Tobari, J. (1978). Methylamine dehydrogenase of *Pseudomonas* sp. J. Isolation and properties of the subunits. *Biochim. Biophys. Acta 522*: 303.
24. Kenny, W. C., and McIntire, W. (1984). Properties of methylamine dehydrogenase from bacterium W3A1. In *Microbial Growth on C_1 Compounds* (R. L. Crawford and R. S. Hanson, eds.), American Society for Microbiology, Washington, DC, p. 165.
25. van Spanning, R. J. M., Wansell, C. W., Reijnders, W. N. M., Oltmann, L. F., and Stouthamer, A. H. (1990). Mutagenesis of the gene encoding amicyanin of *Paracoccus denitrificans* and the resultant effect on methylamine oxidation. *FEBS Lett. 275*: 217.
26. Christoserdov, A. Y., Lidstrom, M. E., and Mathews, F. S. (1991). Personal communication.

27. Chen, L., Mathews, F. S., Davidson, V. L., Huizinga, E., Vellieux, F. M. D., and Hol, W. G. J. (1991). Three-dimensional structure of quinoprotein methylamine dehydrogenase from *Paracoccus denitrificans* determined by molecular replacement at 2.8 Å. *Proteins* (in press).
28. Davidson, V. L., and Neher, J. W. (1987). Evidence for two subclasses of methylamine dehydrogenases with distinct large subunits and conserved PQQ-bearing small subunits. *FEMS Microbiol. Lett. 44*: 121.
29. Shirai, S., Matsumoto, T., and Tobari, J. (1978). Methylamine dehydrogenase of *Pseudomonas* AM1. A subunit enzyme. *J. Biochem. 83*: 1599.
30. Davidson, V. L. (1989). Steady-state kinetic analysis of the quinoprotein methylamine dehydrogenase from *Paracoccus denitrificans*. *Biochem. J. 261*: 107.
31. Davidson, V. L., Jones, L. H., and Kumar, M. A. (1990). pH-dependent semiquinone formation by methylamine dehydrogenase from *Paracoccus denitrificans*. Evidence for intermolecular electron transfer between quinone cofactors. *Biochemistry 29*: 5299.
32. McIntire, W. S., Bates, J. L., Brown, D. E., and Dooley, D. M. (1991). Resonance Raman spectroscopy of methylamine dehydrogenase from bacterium W3A1. *Biochemistry 30*: 125.
33. Eady, R. R., and Large, P. J. (1971). Microbial oxidation of amines. Spectral and kinetic properties of the primary amine dehydrogenase of *Pseudomonas* AM1. *Biochem. J. 123*: 757.
34. Husain, M., Davidson, V. L., Gray, K. A., and Knaff, D. B. (1987). Redox properties of the quinoprotein methylamine dehydrogenase from *Paracoccus denitrificans*. *Biochemistry 26*: 4139.
35. McWhirter, R. B., and Klapper, M. H. (1990). Semiquinone radicals of methylamine dehydrogenase, methoxatin, and related *o*-quinones: A pulse radiolysis study. *Biochemistry 29*: 6919.
36. Chandrasekar, R., and Klapper, M. H. (1986). Methylamine dehydrogenase and cytochrome c-552 from the bacterium W3A1. *J. Biol. Chem. 261*: 3616.
37. McIntire, W. S. (1987). Steady-state kinetic analysis for the reaction of ammonium and alkylammonium ions with methylamine dehydrogenase from bacterium W3A1. *J. Biol. Chem. 262*: 11012.
38. Davidson, V. L., and Kumar, M. A. (1990). Inhibition by trimethylamine of methylamine oxidation by *Paracoccus denitrificans* and bacterium W3A1. *Biochim. Biophys. Acta 1016*: 339.
39. Davidson, V. L., Jones, L. H., and Graichen, M. E. (1992). Reactions of benzylamines with methylamine dehydrogenase. Evidence for a carbanionic reaction intermediate and reaction mechanism similar to eukaryotic quinoproteins. *Biochemistry 31*: 3385.
40. McWhirter, R. B., and Klapper, M. H. (1989). Mechanism of the methylamine dehydrogenase reductive half-reaction. In *PQQ and Quinoproteins* (Jongejan, J. A. and Duine, J. A., eds.), Kluwer Academic Publishers, Dordrecht, The Netherlands, p. 259.
41. Palcic, M. M., and Klinman, J. P. (1983). Isotopic probes yield microscopic constants: Separation of binding energy from catalytic efficiency in the bovine amine oxidase reaction. *Biochemistry 22*: 5957.

42. Davidson, V. L., and Jones, L. H. (1991). Cofactor-directed inactivation by nucleophilic amines of the quinoprotein methylamine dehydrogenase from *Paracoccus denitrificans*. *Biochim. Biophys. Acta* (in press).
43. Davidson, V. L., and Jones, L. H. (1991). Inhibition by cyclopropylamine of the quinoprotein methylamine dehydrogenase is mechanism-based and causes covalent cross-linking of α and β subunits. *Biochemistry 30*: 1924.
44. Williamson, P. R., and Kagan, H. M. (1987). Electronegativity of aromatic amines as a basis for the development of ground state inhibitors of lysyl oxidase. *J. Biol. Chem. 262*: 14520.
45. Hartmann, C., and Klinman, J. P. (1991). Structure-function studies of substrate oxidation by bovine serum amine oxidase: Relationship to cofactor structure and mechanism. *Biochemistry*, 30: 4605.
46. Gray, K. A., Knaff, D. B., Husain, M., and Davidson, V. L. (1986). Measurement of the oxidation-reduction potentials of amicyanin and c-type cytochromes from *Paracoccus denitrificans*. *FEBS Lett. 207*: 239.
47. Ambler, R. P., and Tobari, J. (1985). The primary structures of *Pseudomonas* AM1 amicyanin and pseudoazurin. *Biochem. J. 232*: 451.
48. van Beeumen, J., van Bun, S., Canters, G. W., Lommen, A., and Chothis, C. (1991). The structural homology of amicyanin from *Thiobacillus versutus* to plant plastocyanins. *J. Biol. Chem. 266*: 4869.
49. Sharma, K. D., Loehr, T. M., Sanders-Loehr, J., Husain, M., and Davidson, V. L. (1988). Resonance Raman spectroscopy of amicyanin, a blue copper protein from *Paracoccus denitrificans*. *J. Biol. Chem. 263*: 3303.
50. Lommen, A., Canters, G. W., and van Beeumen, J. (1988). A ^{1}H-NMR study on the blue copper protein amicyanin from *Thiobacillus versutus*: Resonance identifications, structural arrangements and determination of the electron self exchange rate constant. *Eur. J. Biochem. 176*: 213.
51. Anthony, C. (1993). The role of quinoproteins in bacterial energy metabolism. In *Principles and Applications of Quinoproteins* (V. L. Davidson, ed.), Marcel Dekker, New York, p. 223.
52. Husain, M., and Davidson, V. L. (1986). Characterization of two inducible periplasmic *c*-type cytochromes from *Paracoccus denitrificans*. *J. Biol. Chem. 261*: 8577.
53. Davidson, V. L., and Kumar, M. A. (1989). Cytochrome *c*-550 mediates electron transfer from inducible periplasmic *c*-type cytochromes to the cytoplasmic membrane of *Paracoccus denitrificans*. *FEBS Lett. 245*: 271.
54. Davidson, V. L., and Jones, L. H. (1991). Intermolecular electron transfer from quinoproteins and its relevance to biosensor technology. *Anal. Chim. Acta 249*: 235.
55. Gray, K. A., Davidson, V. L., and Knaff, D. B. (1988). Complex formation between methylamine dehydrogenase and amicyanin from *Paracoccus denitrificans*. *J. Biol. Chem. 263*: 13987.
56. Kumar, M. A., and Davidson, V. L. (1990). Chemical cross-linking study of complex formation between methylamine dehydrogenase and amicyanin from *Paracoccus denitrificans*. *Biochemistry 29*: 5299.
57. Chen, L., Lim, L. W., Mathews, F. S., Davidson, V. L., and Husain, M. (1988). Preliminary X-ray crystallographic studies of methylamine dehydrogenase and a

methylamine dehydrogenase-amicyanin complex from *Paracoccus denitrificans*. *J. Mol. Biol. 203*: 1137.

58. Chen, L., Durley, R., Poloks, B. J., Hamada, K., Chen, Z., Mathews, F. S., Davidson, V. L., Satow, Y., Huizinga, E., Vellieux, F. M. D., and Hol, W. G. J. (1992). Crystal structure of an electron-transfer complex between methylamine dehydrogenase and amicyanin. *Biochemistry* (in press).
59. Lobenstein-Verbeek, C. L., Jongejan, J. A., Frank, J. J., and Duine, J. A. (1984). Bovine serum amine oxidase: a mammalian enzyme having covalently-bound PQQ as a prosthetic group. *FEBS Lett. 170*: 305.
60. Moog, R. S., McGuirl, M. A., Cote, C. E., and Dooley, D. M., (1986). Evidence for methoxatin (pyrroloquinoline quinone) as the cofactor in bovine plasma amine oxidase from resonance Raman spectroscopy. *Proc. Natl. Sci. USA 83*: 8435.
61. van der Meer, R. A., and Duine, J. A. (1986). Covalently bound pyrroloquinoline quinone is the organic prosthetic group in human placental lysyl oxidase. *Biochem. J. 239*: 789.
62. Williamson, P. R., Moog, R. S., Dooley, D. M., and Kagan, H. M. (1986). Evidence for pyrroloquinoline quinone as the carbonyl cofactor in lysyl oxidase by absorption and resonance Raman spectroscopy. *J. Biol. Chem. 261*: 16302.
63. James, S. M., Mu, D., Wemmer, D., Smith, A. J., Kaur, S., Burlingame, A. L., and Klinman, J. P. (1990). A new redox cofactor in eukaryotic enzymes: 6-Hydroxydopa at the active site of bovine serum amine oxidase. *Science 248*: 981.
64. Williamson, P. R., and Kagan, H. M. (1986). Reaction pathway of bovine aortic lysyl oxidase. *J. Biol. Chem. 261*: 9477.
65. Ruis, F. X., Knowles, P. F., and Petterson, G. (1984). The kinetics of ammonia release during the catalytic cycle of pig plasma amine oxidase. *Biochem. J. 220*: 767.
66. Williamson, P. R., and Kagan, H. M. (1987). α-Proton abstraction and carbanion formation in the mechanism of lysyl oxidase. *J. Biol. Chem. 262*: 8196.
67. Dooley, D. M., McIntire, W. S., McGuirl, M. A., Cote, C. E., and Bates, J. L. (1990). Characterization of the active site of *Arthobacter* P1 methylamine oxidase: Evidence for copper-quinone interactions. *J. Amer. Chem. Soc. 112*: 2782.
68. Dooley, D. M., McGuirl, M. A., Brown, D. E., Turowski, P. N., McIntire, W. S., and Knowles, P. F. (1991). A Cu(I)-semiquinone state in substrate-reduced amine oxidases. *Nature 349*: 262.

6

Copper-Containing Amine Oxidases

William S. McIntire

Department of Veterans Affairs Medical Center and University of California, San Francisco, California

Christa Hartmann*

University of California, San Francisco, California

I. INTRODUCTION

Prior to 1990, most biochemists did not question the notion of a quinoprotein as an enzyme containing 2,7,9-tricarboxypyrroloquinoline quinone (PQQ) or a derivative thereof. Because rigorous proof of the cofactor structures had not been provided for enzymes reputed to contain covalently bound PQQ, a few investigators remained skeptical. Out of this skepticism, not to mention hard work, three classes of supposed quinoprotein were shown to contain previously unknown cofactors, two other classes contain no organic cofactor at all (dopamine β-hydroxylase, soy bean lipoxygenases [1,2]), and glutamate decarboxylase only contains pyridoxal phosphate [3]. The three new cofactors are 2,4′-bitryptophan-6′,7′-dione (tryptophan tryptophylquinone) in methylamine dehydrogenase [4], 3′-*S*-cysteinyltyrosine in galactose oxidase [5] and 6-hydroxydopa quinone, or topa quinone (TQ) in copper-containing amine oxidases [6] (Fig. 1). All are redox cofactors derived by modification of amino acyl side groups in the polypeptide chain of the respective enzymes. The last of these cofactors is of immediate concern in this article. Obviously, reinterpretation of the physical, chemical, structural, and mechanistic data for the amine oxidases is required. Since both are *ortho*-quinones (see tautomeric structures in Fig. 1), some of what has been

---

**Current affiliation*: Temple University, Philadelphia, Pennsylvania.

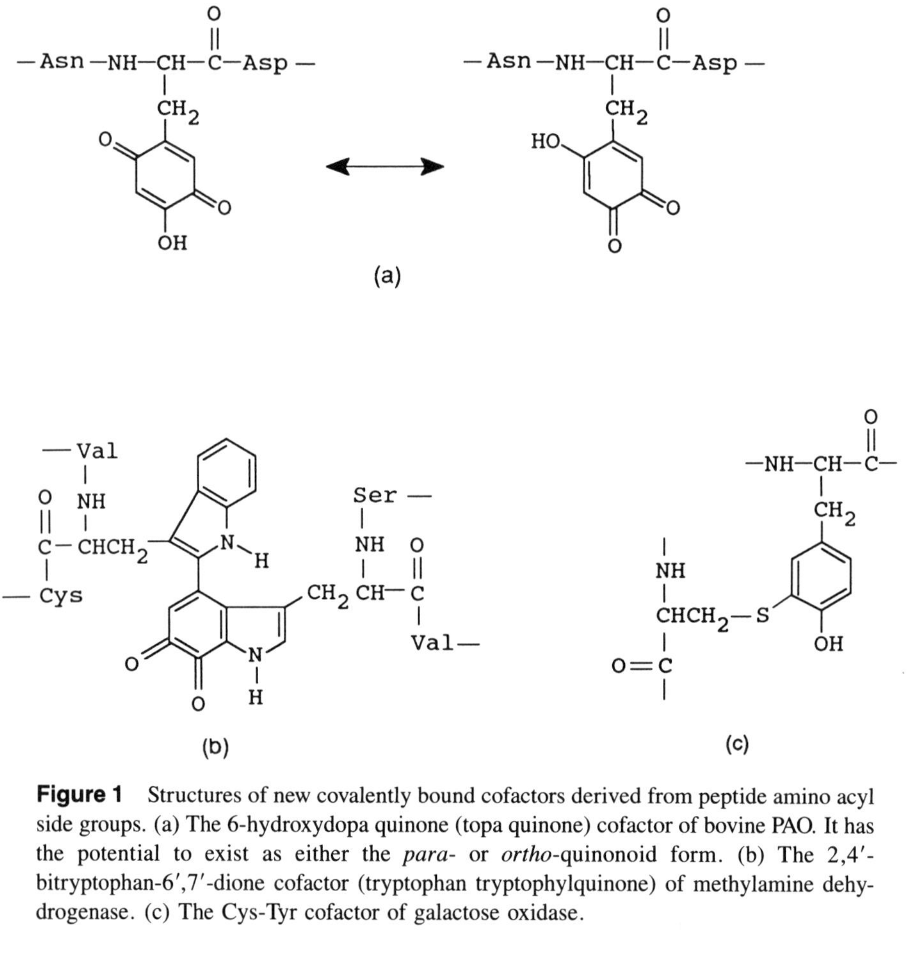

(a)

(b)

(c)

Figure 1 Structures of new covalently bound cofactors derived from peptide amino acyl side groups. (a) The 6-hydroxydopa quinone (topa quinone) cofactor of bovine PAO. It has the potential to exist as either the *para-* or *ortho-*quinonoid form. (b) The 2,4′-bitryptophan-6′,7′-dione cofactor (tryptophan tryptophylquinone) of methylamine dehydrogenase. (c) The Cys-Tyr cofactor of galactose oxidase.

learned about the properties of PQQ will be applicable to TQ, although it will certainly have its own unique properties.

As a class, the copper-containing amine oxidases (EC 1.4.3.6) include representatives from a diverse group of organisms (Table 1). To cover in detail all aspects of each of these enzymes is a daunting task beyond the scope of this treatise. Our purpose is to describe the physical, chemical, and biochemical properties that we believe are applicable to all.

Unfortunately, confusion exists when the mammalian copper amine oxidases are considered. These have been given various names: benzylamine oxidase (BAO), clorgyline-resistant amine oxidase, semicarbazide-sensitive amine oxidase, serum amine oxidase, plasma amine oxidase (PAO), monoamine oxidase, plasma monoamine oxidase, diamine oxidase (DAO), and histaminase. The first seven names are reserved for the monoamine oxidizers, although PAO can also oxidize putrescine and polyamines. On the other hand, various diamine oxidases have modest or no monoamine oxidase activity, and mammalian DAO and histaminase are synonymous. While BAO and semicarbazide-sensitive amine oxidase from blood and various solid tissues in mammals have been identified and studied, it is not always certain if these are tissue-specific forms [120] or the same enzyme [121]. Usually, distinctions are made between apparently similar oxidases on the basis of substrate specificity, although an agreed-upon set of criteria has not been established. Properties (e.g., steady-state parameters) of a particular oxidase often vary from preparation to preparation. This makes it difficult to come to a definitive conclusion regarding its identity.

Copper-containing amine oxidases are distinct from other mammalian amine oxidases. Outer mitochondrial membrane-bound monoamine oxidases, MAO-A and MAO-B, contain covalently bound flavin adenine dinucleotide (FAD) as the only cofactor. It prefers primary aromatic amine substrates, but will act on primary aliphatic and secondary and tertiary amines. The tissue localization, physiological function, structure, and mechanism of MAO-A and MAO-B are discussed in a recent review [122]. Polyamine oxidases have been identified in bacteria, fungi, plants, insects, and mammals [123–125]. The mammalian enzyme contains noncovalently bound FAD and possibly Fe^{2+}. It is required for the conversion of N^1-acetylspermine to spermidine (Spd) and N^1-acetylspermidine to putrescine (Put) with the production of *N*-acetyl-3-aminoproprionaldehyde; thus, this enzyme oxidizes at a secondary amine site [123,126–128]. The plant enzymes, isolated from monocots, barley, maize, millet, and water hyacinth as well as the enzyme from *Pseudomonas putida*, are primary monoamine-oxidizing and FAD-containing enzymes which convert spermine (Spm) and spermidine to 1-(3-aminopropyl)-4-aminobutyraldehyde and 4-aminobutyraldehyde, respectively. The enzyme from oat does not appear to contain FAD [123,129–131]. Fungal amine oxidases from the mycelia of *Aspergillus terreus* and two *Penicillium* species also contain FAD and oxidize spermidine or spermine to putrescine and 3-aminoproprionaldehyde. A polyamine oxidase has been partial purified from *Candida boidinii* peroxisomes, but it was not reported whether this was a flavoprotein [124,132]. Putrescine oxidase from *Micrococcus rudens* [133] and tyramine oxidase from *Sarcina lutea* [134] have been isolated, and both contain noncovalently bound FAD. Amine oxidases from the cockroaches can oxidize monoamines and diamines. While the prosthetic groups for these oxidases were not identified, they were not inhibited by semicarbazide (i.e., they are not TQ-containing amine oxidases) [125]. *Escherichia coli* may synthesize a flavoprotein 2-phenylethylamine oxidase [135]. The interesting fructosyl amine oxidases from *Aspergillus* sp. 1005, *Corynebacterium* sp. 2-4-1, and a *Penicillium* species have also been characterized. These enzymes contain noncovalently bound FAD and convert fructosylamines and fructosylamino acids to the corresponding keto-aldehyde and free amine, therefore, these are classified as secondary amine oxidases [136].

Table 1 Properties of Copper-Containing Amine Oxidases[a]

Source	Enzyme	M_r (kDa)	Subunit M_r (kDa)	Cu(II) (g-atom/mol)	TQ (mol/mol)	Carbohyd.
Bacteria						
Arthrobacter P1	MAOx	168	82	2	2	None
Arthrobacter globoformis	PEAO	nd	nd	+	+	nd
Fungi						
Aspergillus niger	DAO	255	137	2	2	0.47%
		252	85	nd	nd	nd
Candida boidinii	MAOx	150[d]	81	+	+	nd
	BAO	136[e]	79	+	+	nd
Candida utilis	MAOx	510	85	+	+	nd
	BAO	190	76	+	+	nd
Hansenula polymorpha	MAOx	nd	77.435	nd	nd	nd
Kluyveromyces fragalis	BAO	nd	nd	nd	nd	nd
Lyophyllum aggregatum	PEAO	nd	nd	+	+	nd
Penicillium chrysogenum	agmatine oxidase	160	80	2	+	nd
Pishia pastoris	MAOx	168[e]	80	+	+	nd
	BAO	265	116	+	+	nd
	BAO	nd	106	2	2	nd
Trichosporon cutaneum X_4	BAO	153	26 & 19[g]	+	+	nd
Miscellaneous	AO	nd	nd	nd	nd	nd
yeast	AO	nd	nd	nd	nd	nd
Plants						
Camellia sinensis (tea)	BAO	162	81	+	+	nd
Canavalia ensiformis (jack bean)	DAO	nd	nd	+	+	nd
Arachis hypogea (groundnut)	DAO	nd	nd	nd	nd	nd
Cicer arietinum	DAO	nd	nd	+	+	nd
Cucumis sativus[l]	DAO	87	nd	+	+	nd
(cucumber)	DAO	100	nd	+	+	nd
Euphorbia characias	DAO	144	69.5	2	2	12% neutral sugar

AAA	pI[b]	Pure	Location	Subst. specif.	λ_{max}(nm) [ϵ; $mM^{-1}cm^{-1}$][c]	Ref.
nd	4.6	Homog.	Cytoplasm	yes	486 [5.24]	7,8
nd	<7.0	Partial	Cytoplasm	yes	nd	9
nd	<7.0	Homog.	Mycelia	yes	480 [4.51]	10–12
					480 [6.4]	13,14
nd	<7.0	Homog.	Mycelia	nd	490 [5.5]	15
nd	<7.0	~Homog.	nd	yes	nd	16,17
nd	<7.0	~Homog.	nd	yes	nd	
nd	<7.0	~Homog.	nd	yes	nd	18
nd	<7.0	~Homog.	nd	yes	nd	
yes[f]	nd	nd	Peroxisomes	nd	nd	19
nd	nd	nd	nd	nd	nd	20
nd	nd	nd	Cytoplasm	yes	nd	21
nd	5.7	Homog.	Mycelia	yes	nd	22,23
nd	<7.0	~Homog.	nd	yes	nd	17,18
nd	<7.0	~Homog.	nd	yes	nd	
nd	<7.0	Homog.	nd	yes	480 [2.85]	24
nd	<7.0	Homog.	nd	yes	nd	25
nd	nd	nd	Mycelia	nd	nd	10
nd	nd	nd	nd	nd	nd	18,20,26
nd	5.0	Homog.	Young leaves	yes	490 [5.7]	27
nd	nd	nd	[h,i]	nd	nd	28
nd	nd	Partial	4-day-old seedlings	yes	nd	29
nd	nd	~Homog.	[j,k]	yes	nd	30
nd	<7.0	~Homog.	7-day-old	yes	nd	31
nd	>7.0	~Homog.	seedlings[m]	yes	nd	
yes	5.2	Homog.	Latex	yes	490 [2.0]	32

Table 1 Continued

Source	Enzyme	M_r (kDa)	Subunit M_r (kDa)	Cu(II) (g-atom/mol)	TQ (mol/mol)	Carbohyd.
Glycine max (soybean)	DAO	?	25	+	+	nd
	DAO	nd	nd	+	+	nd
Hordeum vulgare (barley)	DAO	150	75	2	+	14% neutral sugar
Hyosyamus niger	DAO	135	66	+	+	nd
Lathyrus cicera	DAO	150	75	2	+	14% neutral sugar
Lathyrus sativus (chick pea)	DAO	148	75	2.7	+	None
Lens culinaris (lentil)	DAO	154	78	2	1	14% neutral sugar
Lupinus luteus (lupine)	DAO	nd	nd	+	+	nd
Nicotiana tabacum (tobacco)	DAO	nd	70	+	+	nd
Onobrychis viciifolia	DAO	?	22–55	+	+	nd
Phaseolus vulgaris (kidney bean)	DAO	150	75	2	+	14% neutral sugar
Pisum sativum (pea)	DAO	nd	77	2	nd	nd
		185	nd	+	+	nd
		174	nd	nd	nd	nd
		nd	nd	nd	nd	nd
		120	60	+	+	nd
		178	94	2	nd	nd
		180	85	nd	1	13% neutral sugar
Trifolium subterraneum (clover)	DAO	150p	80	1.5	+	nd
Vicia faba (fava bean)	DAO	126	74	2	1	nd

AAA	pI[b]	Pure	Location	Subst. specif.	λ_{max}(nm) [ϵ; $mM^{-1}cm^{-1}$][c]	Ref.
nd	nd	~Homog.	6-day-old seedlings[k]	yes	nd	33
nd	nd	Partial	3 day-old seedlings[k]	yes	nd	34
nd	6.8	Homog.	8-day-old seedlings[k,m]	yes	498 [2.4]	35
nd	<7.5	~Homog.	Cultured roots	yes	nd	36
nd	<7.0	Homog.	8-day-old seedlings[k]	yes	498 [2.4]	37
yes	<7.0	Homog.	5-day-old seedlings[n]	yes	nd	38
yes	<7.0	Homog.	8-day-old seedlings[k]	yes	498 [3.3] ~490 [3.5]	39,40 41
no	nd	Partial	13-day-old seedlings	yes	nd	42
nd	<7.0	Partial	4-week-old roots	yes	nd	43,44
nd	8–9	Partial	7-day-old roots and shoots[t]	yes	nd	45
nd	<7.0	Homog.	8-day-old seedlings[k]	yes	498 [2.4]	37
nd	<7.0	Homog.	Seedlings	yes	nd	46
nd	<7.0	Homog.	5-day-old seedlings[k,o]	yes	500 [2.7]	47
nd	<7.0	~Homog.	7-day-old seedlings[k]	nd	495 [3.14]	48
yes	nd	~Homog.	Seedlings	nd	nd	49
nd	nd	nd	nd	nd	nd	50
nd	6.5	Homog.	12-day-old seedlings[k]	nd	nd	51
yes	7.4	Homog.	8-day-old seedlings	nd	500	52
yes	nd	Homog.	Young folded leaves	yes	490 [2.0]	53
yes	7.2	Homog.	14-day-old leaves	yes	nd	54,55

Table 1 Continued

Source	Enzyme	M_r (kDa)	Subunit M_r (kDa)	Cu(II) (g-atom/mol)	TQ (mol/mol)	Carbohyd.
Mammalian						
Human	PAO	nd	nd	+	+	nd
	DAO	nd	nd	+	+	nd
	DAO	nd	nd	+	+	nd
	DAO	125[u]	nd	+	+	nd
	DAO	nd	90	+	+	nd
	DAO	235[q]	70	2.0[r]	nd	nd
	DAO	280,300	170	nd	nd	nd
	DAO	nd	90	nd	nd	yes
	DAO	182	nd	nd	+	nd
	DAO	nd	105	+	+	nd
	DAO	nd	nd	nd	nd	nd
	DAO	Spleen, liver, eosinophil and neutrophil granulocytes, lymph, amniotic fluid, urine				
	BAO	Neonatal phararyngeal aspirate, and amniotic fluid				
Bovine	PAO	nd	nd	nd	+	nd
	PAO	170	87	1–2	+	yes
	PAO	180	90	2	1	nd
	PAO	190	nd	2	nd	nd
	PAO	nd	nd	nd	1.64[s]	nd
	BAO	170	nd	+	?	nd
	BAO	?	80	2	?	nd
	BAO	nd	nd	?	+	nd
	BAO	nd	nd	?	+	nd
Porcine	DAO	120–135	87	2	2	nd
	DAO	185	nd	2.17	nd	nd
	DAO	nd	85–90	nd	nd	yes
	DAO	172	91	2	nd	nd
	DAO	170,220	130	nd	nd	nd
	PAO	195	nd	~4	nd	nd
	PAO	nd	nd	3	3	nd
	PAO	nd	nd	2–3	3–4	nd
	PAO	nd	nd	2	1	nd
	PAO	186	97	2	nd	nd
	PAO	196	95	2	1	yes
	BAO	nd	130	+	+	nd
	BAO[v]	nr	nr	?	+	nr

AAA	pI[b]	Pure	Location	Subst. specif.	λ_{max}(nm) [ϵ; $mM^{-1}cm^{-1}$][c]	Ref.
nd	nd	90%	Adult plasma	yes	nd	56
nd	nd	nd	Pregnancy plasma	yes	nd	57
nd	<8.6	Partial	Placentae	yes	nd	58,59
nd	6.0	Homog.	Placentae	yes	nd	60
nd	7.1	Homog.	Placentae	yes	nd	57
nd	6.5	Homog.	Placentae	nd	nd	61
nd	nd	~Homog.	Placentae	nd	nd	62
yes	5.3–6.6[p]	Homog.	Placentae	nd	nd	63
nd	<7.4	Homog.	Semen	minimal	nd	64
nd	6.0	Homog.	Kidney	yes	nd	65
nd	<7.6	Partial	Intestinal mucosa cytoplasm	yes	nd	66
						67
						68
nd	nd	Partial	Plasma	yes	nd	69
yes	4.5	Homog.	Plasma	yes	480	70–75
nd	nd	~Homog.	Plasma	yes	480	76–78
nd	nd	Homog.	Plasma	nd	476 [3.8]	79
nd	nd	Homog.	Plasma	nd	nd	80
nd	<7.4	partial	Dental pulp	yes	nd	81
nd	nd		Aorta	?	?	82
nd	nd	nd	Retina, optic nerve	nd	nd	83
nd	nd	nd	Lung	yes	nd	84
nd	<7.2	~Homog.	Kidney	yes	408 and 480	85–86
nd	<7.0	Homog.	Kidney	nd	470 [~3.7]	87
yes	nd	~Homog.	Kidney	nd	nd	88
nd	<7.2	~Homog.	Kidney	nd	nd	89
nd	nd	Homog.	Kidney	nd	nd	62
nd	<7.0	~Homog.	Plasma	yes	470 [~3.4]	90
nd	nd	nd	Plasma	nd	nd	91
nd	nd	~Homog.	Plasma	nd	~480 [~4.9]	92
nd	nd	nd	Plasma	nd	nd	93
nd	nd	Homog.	Plasma	nd	nd	94
yes	4.5–5.0[p]	Homog.	Plasma	nd	nd	95,96
nd	<7.0	~Homog.	Aorta smooth muscle	nd	nd	97
nr	nr	nr	Dental pulp	yes	nr	98

Table 1 Continued

Source	Enzyme	M_r (kDa)	Subunit M_r (kDa)	Cu(II) (g-atom/mol)	TQ (mol/mol)	Carbohyd.
Equine	DAO	nd	nd	+	+	nd
Sheep	PAO	nd	nd	+	+	nd
Rabbit	PAO	nd	nd	nd	nd	nd
	PAO	nd	nd	nd	+	nd
	DAO	nd	nd	nd	+	nd
	DAO	nd	nd	nd	nd	nd
Rat	BAO	183	nd	?	+	nd
	BAO	nd	nd	?	+	nd
	BAO	nd	nd	nd	+	nd
	BAO	nd	nd	nd	nd	nd
	DAO	nd	nd	nd	nd	nd
Other mammals: cat, dog, monkey, mouse, horse, goat, guinea pig, dogs, marmoset	DAO	nd	nd	nd	nd	nd
Bird						
Coturnix coturnix japonica (quail)	DAO and BAO	nd	nd	nd	+	nd
Fish						
Seriola quinqueradiata (yellowtail)	DAO	140	80 & 60	+	+	nd
Scomber japonicus (common mackerel)	DAO	nd	nd	nd	+	nd
Various teleostean fish	DAO	nd	nd	+	+	nd
Parasiluris astus (catfish)	BAO	nd	nd	nd	+	nd

AAA	pI[b]	Pure	Location	Subst. specif.	λ_{max}(nm) [ϵ; $mM^{-1}cm^{-1}$][c]	Ref.
nd	nd	partial	Kidney	nd	nd	99
nd	nd	partial	Plasma	yes	nd	100
nd	nd	~Homog.	Plasma	yes	nd	101
nd	nd	nd	Lung and heart	yes	nd	102
nd	nd	nd	Liver	nd	nd	103
nd	nd	nd	Kidney	nd	nd	104
yes	nd	Partial	Brown adipose tisue	nd	nd	105
nd	nd	nd	White adipose tissue	yes	nd	106
nd	nd	nd	Liver microsomes, vascular smooth muscle	yes	nd	107,108
nd	nd	nd	Aorta, skull, heart, lung, testis	for some	nd	109,110
nd	nd	nd	Lymph, blood, intestinal mucosa, uterus	nd	nd	111–113
nd	nd	nd	Various tissues	nd	nd	111–114
nd	nd	nd	Liver microsomes	yes	nd	107
nd	<7.0	Homog.	Pyloric ceca	yes	nd	115
nd	nd	nd	Pyloric ceca and intestine	yes	nd	116
nd	nd	nd	Intestine	yes	nd	117
nd	nd	nd	Intestine, liver, etc.	yes	nd	118

Table 1 Continued

Source	Enzyme	M_r (kDa)	Subunit M_r (kDa)	Cu(II) (g-atom/mol)	TQ (mol/mol)	Carbohyd.
Mugil cephalus (mullet)	PAO	190	112–90	2	+	20% neutral sugar
Salmo gairdneri	BAO	nd	nd	nd	+	nd
Other						
Mytilus gallo-provincialis (mussel)	DAO and BAO	nd	nd	nd	+	nd

[a]nd = Not determined or not done. nr = Not reported. The + symbol indicates that the oxidase tested positive using various reagent specific for Cu(II) (e.g., *N,N*-diethyldithiocarbamate) or the quinone cofactor (e.g., phenylhydrazine or semicarbazide). The numerical values were determined by atomic absorption or EPR measurements for Cu(II), and titration with a carbonyl reagent (usually phenylhydrazine) for he quinone. In order to be complete and unbiased, some enzymes from a single source have multiple listings and references. The listings are in chronological order. PAO = plasma amine oxidase; MAOx = methylamine oxidase; DAO = diamine oxidase; PEAO = 2-phenylethylamine oxidase; BAO = benzylamine oxidase.

[b]In most cases, the pI was determined to be greater or less than the pH value given based on the ability to bind to a cation or anion exchange resin.

[c]Some values were estimated from spectra reproduced in the references.

[d]Aggregates to tetramer and octomer.

[e]Aggregates to tetramer.

[f]The gene has been cloned and sequenced and an amino acid sequence deduced.

[g]A_3B_3 structure.

[h]Highest activity in internodes and petioles of 13-day-old plants, and lower amounts in older regions of plant.

[i]Grown under continuous light.

II. RAISON D'ETRE AND MEDICAL AND PHARMACOLOGICAL ASPECTS OF THE COPPER-CONTAINING AMINE OXIDASES

The function of amine oxidases in prokaryotes is obvious. These enzymes allow the organism to use the appropriate alkyl- or arylamine as a carbon source for growth. In higher organisms, the roles these oxidases play are not always well defined. One purpose of PAO and DAO in plants and animals is the catabolism of the diamine Put and the polyamines Spd and Spm. These amines are important agents in various fundamental cellular processes, e.g., tissue differentiation, cellular proliferation, growth of tumors and transformation of cultured cells, wound healing, and possibly programmed cell death [123,126–129,131,137–143]. During these processes, there are dramatic changes in the concentrations of the

AAA	pI[b]	Pure	Location	Subst. specif.	λ_{max}(nm) [ϵ; $mM^{-1}cm^{-1}$][c]	Ref.
nd	nd	Homog.	Serum	no	480 [2.9]	119
nd	nd	nd	Liver microsome	yes	nd	107
nd	nd	nd	Liver microsome	yes	nd	107

[j]10-day-old shoots and cotyledons.
[k]Grown in the dark.
[l]Contains two copper-amine oxidases.
[m]Grown under light/dark cycle.
[n]Highest activity in the cytosol of the embryo axis, lower in the cotyledons.
[o]Isolated from epicotyls.
[p]Isozymes are present.
[q]Determined by sedimentation-equilibrium ultracentrifugation. Value of M_r may be due to rapid equilibrium between dimer and tetramer.
[r]Claimed to contain 2.4 g-atoms of manganese/mol in addition to copper.
[s]Linear relationship between mol of phenylhydrazine/mol subunit and activity was found. Extrapolation to maximal activity of 0.44 μmol benzylamine/(min × mg) gives ~2 mol of cofactor/mol of PAO.
[t]Grown in light and dark.
[u]Aggregates to dimer, trimer, and tetramer.
[v]Porcine dental pulp contains two soluble BAOs, one similar to PAO, a tissue-bound oxidase active towards benzylamine, tryptamine, tyramine, and phenethylamine, and a tissue-bound oxidase active towards serotonin.

enzymes responsible for biosynthesis and catabolism of these amines, resulting in alterations of cellular levels of Spd and Spm.

Polyamines are important or essential for transcription and translation via direct interaction with DNA, mRNA, tRNA, and rRNA [123,126,142]. Spd is attached to a lysyl residue and converted to the novel amino acid, hypusine, in initiation factor eIF-4D in all eukaryotic cells [123,126,143,144]. Cell growth and proliferation are regulated by several posttranslational modification processes directly influenced by polyamines. These include protein phosphorylation by cyclic-AMP–independent protein kinases [123,145,146], hypusine formation, and transglutaminase formation of mono- and bis-γ-glutamyl-polyamine derivatives of various proteins [123,126,147]. Polyamines also participate in regulation of hormone-dependent metabolism [148] and mediate other hormonal action [149]. Additional functions include stimulation of transport of a precursor protein into mitochondria [150], involvement in membrane structure and function [142], and conversion of Put and Spd to γ-aminobutyric acid [151].

Although Spd and Spm are directly involved in cell growth, proliferation, and differentiation, Put is considered the crucial regulator of these processes [152,153]. Most studies of these processes focus on biosynthesis of Put, Spd, and Spm and usually involve inhibition of ornithine decarboxylase by the specific, nontoxic drug difluoromethylornithine (DFMO) [123,126,138,140,154]. These amines are catabolized by two kinds of oxidases. The first is the flavoprotein polyamine oxidase, which oxidizes N^1-acetyl-Spm to Spd and N^1-acetyl-Spd to Put [126,148,153]. (The *N*-acetylated polyamines constitute a significant portion of the cellular polyamine pool and are the excreted forms). The second type include BAO, DAO, and PAO [127]. Put is the preferred substrate for DAO, whereas Spd, Spm, N^1-acetyl-Spm, and N^1-acetyl-Spd are better substrates for PAO [153,155]. In normal rats, 40% of the amines are excreted from the cellular pool, whereas 60% are oxidized by DAO and BAO [156], making these enzymes important in controlling cellular polyamine levels.

In tumors and growing cultured cancer cells, the levels of polyamines and DAO increase dramatically compared to normal or nongrowing cells [152,155–157]. When levels of Put, Spd, and Spm are diminished by treatment with DFMO, cell proliferation stops, but resumes if endogenous amines are added [127,138,158]. These observations indicate that the high levels of these amines are causative agents and not a result of tumor growth. Similar studies with normally growing or differentiating cells provide the same conclusion [127,138,139,159].

From the foregoing, the pharmacological importance of BAO, DAO, and PAO is apparent [127]. Drugs that interact specifically with these enzymes could have clinical applications. No drugs are absolutely specific for any of these oxidases. 3-Hydroxybenzyloxyamine (a potent BAO and PAO inhibitor) and aminoguanidine (a potent BAO, DAO, and PAO inhibitor) come closest to fulfilling this requirement [127]. Azomethane-induced large bowel tumors were promoted when aminoguanidine was administered to rats. When it was withdrawn, tumor proliferation ceased, DAO activity was diminished in the proliferating cells [138].

Use of inhibitors of polyamine catabolism in concert with antineoplastic drugs has met with success [158,160]. Combining methylglyoxal bis(guanylhydrazone) (MGBG) with DFMO for treatment of tumor-bearing mice resulted in greater tumor growth inhibition than with either drug alone [161]. Both drugs increase cellular uptake of the amines. Originally, MGBG was considered an inhibitor of only *S*-adenosylmethionine decarboxylase, but it is a more potent DAO inhibitor. Thus, polyamines were not oxidized and the levels of these in the tumor cells "normalized."

Since PAO is a plasma enzyme, it would seem to have little to do with cellular Spd and Spm levels, oxidizing these amines after they leave the cell. However, these amines can be reabsorbed by cells, and their levels in blood could be important. During pregnancy, the fetus releases large amounts of Put, Spd, and Spm. It is known that human maternal DAO activity in placenta and maternal blood increases 500- to 1000-fold compared to normal by 16 weeks of gestation and is maintained at these levels to term [156,162]. Since Put is a good substrate for DAO, and PAO prefers Spd and Spm, PAO may also be important in maintaining proper blood levels of these amines.

Direct pharmacological uses of the oxidases have been reported. When either DAO or PAO encased in reconstituted Sendai virus envelopes were microinjected into transformed fibroblasts, the cells were severely damaged [163]. DAO bound to concanavalin A Sepharose inhibited tumor growth when injected intraperitoneally into mice with Ehrlich ascites tumors [155]. Since these cells contain elevated levels of di- and polyamines, injected PAO and DAO produced cytotoxic aminoaldehydes, dialdehydes, and H_2O_2.

There are other roles for DAO in mammals. In small intestines, it is involved in catabolism of potentially toxic amines produced by enteric bacteria [137]. DAO (alias histaminase) is an important enzyme in the catabolism of histamine [120]. Although the role of histamine in cell growth, proliferation, and differentiation is uncertain, its role in inflammatory, allergic, and ischemic phenomena is well documented [164].

As concerns the pharmacology of DAO and PAO, an important caveat exists: The best inhibitors of these enzymes are not specific. For example, aminoguanidine is also an inhibitor for lysyl oxidase (EC 1.4.3.13), an enzyme essential for the formation of elastin and collagen [165]. (Like BAO, PAO, and DAO, lysyl oxidase is a copper- and quinone-containing amine oxidase. (see Chapter 7 for a review of this enzyme.) Conversely, many inhibitors of other enzymes also inhibit PAO and DAO. The lathyrogens β-aminopropionitrile, aminoacetonitrile, phenylhydrazine, and semicarbazide, which inhibit lysyl oxidase, also inhibit PAO and DAO, as do many of the inhibitors of the mitochondrial flavoproteins MAO-A and -B [166,167]. The clinical importance of MAO inhibitors in treatment of neurological diseases and disorders is well established [167]. Because of a lack of absolute specificity, using inhibitor-based drugs for these various enzymes could lead to unintentional side effects. Additionally, being a blood enzyme, PAO could oxidize a drug designed to act on another enzyme (e.g., MAO), thus reducing the drug's efficacy.

The neutral compound 1-methyl-4-phenyl-1,2,3,6-tetrahydropyridine (MPTP) passes the blood-brain barrier, where MAO oxidizes it to the cationic neurotoxin 1-methyl-4-phenylpyridinium (MPP^+). MPP^+ causes parkinsonism via destruction of the substantia nigra in human brain [168]. A recent report describes the oxidation of MPTP by human PAO [169]. This may explain the resistance of some species to the neurotoxic effect of MPTP [168]. If they have high levels or very active PAO in the blood, conversion of MPTP to MPP^+ would not allow the latter to penetrate the blood-brain barrier. Another possible role for PAO, therefore, may be the detoxification of harmful amines in blood. In another recent report, analogs of MPTP have been shown to be weak inhibitors of human PAO [170]. Conversely, some amines may also be converted to toxic compounds (e.g., aldehydes) by the action of PAO. Thus, xenobiotic amines could be oxidized by plasma amine oxidase to produce toxic compounds. An example of this is the oxidation of allylamine to the cardiovascular toxin acrolein [171,172].

As mentioned earlier, the function of DAO in plants may be for the catabolism and control of the levels of Put [129]. The highest amounts of DAO are found in seedling or young leaves, and in these, the oxidases are found in the youngest tissue (see Table 1) [28,129,173]. For both lentil and pea seedlings, much more DAO is found in etiolated (dark-grown) tissue than in tissue grown in light [174]. In lentil shoots, DAO levels peak at ~7 days and decline slowly in dark-grown but more rapidly in light-grown tissue [175,176]. Evidence exists that levels (activities) of DAO in seedlings are regulated by the plant hormones, auxins [185]. Naturally occurring inhibitors of DAO have been identified in groundnut (*Arachis hypogea*), *Lupinus luteus*, and *Sorghum vulgare* seedlings and may explain why DAO has not been detected in some seedlings or in older plant tissues [42,178]. The drop in apparent DAO levels in lentil seedlings after 7 days may be due to a progressive accumulation of an inhibitor in seedlings.

It has been proposed that DAO-produced aminoaldehydes would be converted to succinate to be used as an energy source in etiolated Leguminosae seedlings. Thus, seedlings would be "sustained" during nonphotosynthetic growth. However, exposure of etiolated lentil seedlings to red, far red, or blue light indicate that DAO levels are modulated by a phytochrome-mediated process. It was concluded that the levels of DAO were a result of photomorphogenesis rather than the seedlings capacity for photosynthesis [179].

There is a direct relationship between DAO levels and the levels of Put or cadaverine in various regions from the base of the root to the apex of etiolated or light-grown pea seedlings. However, there is no correlation between DAO levels with either Spm or Spd [175,176]. Although DAO is very soluble, in various regions of pea and lentil seedlings the enzyme is bound to membranes and cell walls [180,181]. The greatest concentration of DAO in the "cell wall is in the lamellar region and in cells with highly lignified walls" for pea and maize seedlings [198]. It has been proposed that one of the physiological functions of DAO is to provide H_2O_2 in the cell wall for a peroxidase-dependent "lignification and cell wall stiffening" [181].

In some plants (e.g., *Hyoscyamus niger*, *Lupinus luteus*, and *Nicotiana tabacum*), DAO has been shown to be involved in alkaloid biosynthesis [36,42–44]. In soybean radicles, cadaverine levels parallel those of "uricase" activity. A partially purified preparation indicates that this activity is derived from a DAO/peroxidase system. Oxidation of cadaverine supplies H_2O_2 for the conversion of urate to allantoin [183]. Allantoin is an important nitrogen storage compound in plants [184]. Whether the DAO portion of this system is a copper- and TQ-containing amine oxidase is unknown. Numerous other roles for DAO in plants have been proposed [129,185].

In most mammals, the highest levels of DAO activity are found in intestine [162,186]. In general, it also seems that herbivores have the highest levels of DAO in their blood. There is, however, no correlation between the amount of blood histamine levels and the DAO activity in various animals [112]. Perhaps the high level of DAO in these fluids and tissues are, at least in part, a response to ingested and absorbed plant DAO inhibitors.

III. LOCALIZATION AND BIOSYNTHESIS OF THE OXIDASES

Antibodies have been extremely useful for immunohistochemical localization of amine oxidases in various tissues of an organism and intracellularly. More traditional biochemical and histochemical methods have also been used for this purpose [186].

A. Mammalian Amine Oxidases

The reader is directed to two reviews of the literature published prior to 1984 which deal with the localization of copper-containing amine oxidases in mammals [120,186]. We will only reiterate the most salient points and add new findings from reports published after 1984 concerning these oxidases. A detailed discussion of the change of activity levels of human PAO and DAO in physiological and pathological conditions can be found in Ref. 67.

1. Small Intestines

The highest levels of DAO in the gastrointestinal tract occur in small intestines, and only canine large intestine has high DAO levels. Carnivores and omnivores have higher intestinal levels than herbivores, suggesting a protection by DAO against biogenic amines. In the intestinal tract of rat, the highest levels of DAO are found in the ileum, with progressively lower amounts in the jejunum, caecum, and colon. There is no correlation between the levels of DAO and histamine in these tissues, but the levels of histamine and histidine decarboxylase activity are highly correlated, indicating the latter enzyme controls intestinal histamine levels [187]. In small intestine epithelium, enterocytes of the apex of the villi have the most DAO, and crypt cells the least. The oxidase is transported from the enterocytes to a binding site of the intestinal vasculature [188]. DAO is released into blood from endothelial cells of the microvasculature of numerous mammals. For guinea pig, the source of plasma DAO is liver. Treatment with heparin causes release of this oxidase into blood from the intestine [188].

2. Kidney

DAO is found in the epithelium of convoluted tubules and possibly the epithelium of Bowman's capsule of renal cortex of porcine kidney. Intracellular locale for DAO is the membrane of endoplasmic reticulum of rabbit kidney cortex.

3. Liver

DAO seems to be present in purified hepatocytes but not Kupffer's cells of rabbit liver. Intracellularly, DAO is associate with the cytosolic surface of microsomal membranes [189]. BAO is found in liver microsomes of rat, trout, quail, and mussel. However, DAO is found only in microsomes of the latter two species [107].

4. Placenta

This tissue seems to contain more DAO than all other tissues. It also is the source of serum DAO in pregnant women. DAO is localized in the cytoplasm of cells of the decidua basalia, which is the maternal portion of the placenta and is absent from all fetal placenta tissue.

5. Serum

As mentioned above, both intestine and placenta release DAO into blood. PAO does not seem to be associated with the walls of blood vessels or blood cells. DAO is also found in lymph serum in rats, guinea pigs, and rabbits [113].

6. Aorta, Arteries, Vascular Tissue, and Heart

A BAO is located in plasma membrane of rat aorta. Aortic smooth muscle cells of the tunica media were found to contain the highest levels of the oxidase in rats and humans [110,190,191]. Another BAO has also been detected in rat superior mesenteric and femoral arteries. The mesenteric enzyme is similar to PAO [108]. Rabbit aorta also contains a BAO with properties similar to the PAO from this animal. Microvessels of bovine brain contain considerably more BAO than bovine aorta. Rabbit heart contains BAO [102]. BAO from cultured rat smooth muscle cells has properties nearly identical to rat BAO [192].

All human vascular tissue contains high levels of BAO, although blood vessels contained the highest. Nonvascular smooth muscle cell contained lesser amounts, but higher than connective tissue or striate muscle tissue. Again, vascular smooth muscle BAO is similar to PAO [191]. The tunica media of human umbilical arteries is also a source of BAO [193]. BAO has been purified from aortal tissue [194] and cultured aorta smooth muscle cells of pig [97]. It, too, is located in medial smooth muscle. Its antibodies cross-react with porcine PAO, indicating that these enzymes are similar.

Because of the similarity of vascular smooth muscle BAO and PAO, it has been suggested that this tissue is the source of the plasma enzyme [191]. If this is true, then it would be the first plasma protein that is excreted from the arterial or vascular tissue. All plasma proteins are synthesized in the liver, with the exception of immunoglobulins, which are produced in plasma cells and B lymphocytes [195].

7. Lung

Published studies report DAO in this tissue for rabbit [102,196] and cow [84]. For rabbit, the enzyme is situated in the vascular endothelium, while that for cow is associated with a microsomal fraction. Another study indicates that neither rabbit nor rat lung contains DAO [197].

8. Adipose Tissue

The cell membrane of rat intercapsular brown adipose tissue harbors BAO [105]. Rat white adipose tissue also contains BAO [106], as does cultured preadipocytes [182].

9. Thymus

DAO in this tissue of rats is not located in lymphocytes. In inflammatory diseases, the activity of the oxidase increases in eosinophils.

10. Fibroblasts

DAO is found in these cells. Levels are highest during rapid proliferation of cultured human fibroblasts.

11. Salivary Glands

DAO activity has been detected in salivary glands and gastric mucosa of several mammals.

12. Dental Pulp

Bovine and porcine dental pulp contain BAO. For the porcine source, there are two soluble BAOs, one of which is similar to PAO, and two membrane-bound oxidases, one of which is a serotonin oxidase [98].

13. Seminal Fluid

DAO was purified from human source. The enzyme is very similar to placental DAO. Spermatozoa have a DAO distinct from the seminal fluid enzyme.

14. Other

Human pregnancy fluid and newborn pharyngeal aspirate contains BAO [68]. Rat skull contains a BAO [109]. Rat testes and uterus have DAO activity, as do various tissues of cat, dog, goat, and marmosets (Table 1).

B. Plant DAO

1. Piseum sativum *(Pea)*

As stated earlier, levels of DAO from the root to the apex of young etiolated pea and lentil seedling epicotyls are related to levels of endogenous Put and cadaverine [175,176]. The level of DAO peaks 1–2 weeks after germination [52,199]. The highest levels are within the cell wall, although the oxidase is solubilized during isolation [180,181,198]. Much higher levels of DAO are present in dark grown vs. light grown seedlings [174], and higher levels of DAO are found in the remaining tissues when one or both cotyledons are removed [199]. The cotyledon and embryo of 6-day-old seedlings contain different forms of DAO (DAO_c and DAO_{em}) [200]. The enzyme has been purified from epicotyl (DAO_{ep}) [47,50,52], where it is located in the apoplast [181]. Seedling cotyledons have higher DAO activities than epicotyls, and both have higher levels than roots [199].

Numerous factors affect DAO activity (and presumably the amount of enzyme biosynthesized) in the seedlings. For example, Put, Spd, and ornithine induce DAO_c, whereas auxins depress its levels only when the embryo is present. 5-Fluorouracil depresses synthesis of DAO from both tissues, but cycloheximide effects only DAO_c [201]. Phytic acid induces DAO_c but has no effect on DAO_{em} [202]. Phenolic compounds, which regulate indolacetic acid levels, also decrease DAO_c and DAO_{em} activities [203]. It has been demonstrated that O_2 supplied through the embryonic axis is essential for induction of DAO_c biosynthesis [204]. Both kinetin and gibberellic acid retard the fall off of DAO activity in pea leaves, but benzimidazole increases the rate. These substances accelerate the decrease in Put levels [205]. Although K^+ deficiency decreases the level of Put in pea plants, the activity of DAO remains unaffected [206].

2. Lens culinaris *(Lentil)*

As with pea seedlings, etiolated lentil seedlings produce more DAO than when grown in light, and the level peaks after 1–2 weeks. Antibodies against lentil DAO indicate that it is located in all seedling tissues with the highest amounts in the cotyledon. Although the oxidase is very soluble, it seems to be associated with membranes and cell wall [180]. DAO activity and immunoreactivity parallel each other in dark- or light-grown cells. This indicates that an endogenous inhibitor is not responsible for changes in activity levels [176].

3. *Other Plants*

DAO has been obtained from the embryos of *Arachis hypogea* (groundnut) seedlings. Removal of both cotyledons causes a decrease in the activity of DAO, however, removal of one has no effect. Several plant hormones reduce the activity of DAO initially (2-day-old seedlings), but the inhibition diminishes in older seedlings. A similar effect was observed with Put, Spd, and Spm [29].

DAO activity was detected in all parts of light-grown *Canavalia ensiformis* (jackbean) seedlings, with the exception of roots and embryo. The highest levels were detected in young internodes and petioles. In these regions, DAO activity was in regions where lignin would later form, however, the authors of this work point out that roots also contain lignin. In the cotyledons, the activity was "limited and transient." The lowest DAO activity is found in the oldest tissue [28].

It has been suggested that DAO in roots of *Hyoscyamus niger* is involved in biosynthesis of tropane alkaloids (e.g., hysocyamine and scopolamine) via oxidation of *N*-methylputrescine. This diamine is the best substrate for this DAO. DAO activity was also detected in shoots of this plant [36]. In ethiolated *Lathyrus sativus* seedlings, DAO activity increased to a maximum after ~4 days and then gradually declines to a constant level after ~10 days. More of the oxidase activity was found in the embryo axis than the cotyledons. Within the cells, DAO was in the cytosol [38].

Lupinus luteus seedlings have more DAO activity in the roots, with slightly lower activity in the shoots, much lower in young leaves, and none in the cotyledons. The oxidase was implicated in the biosynthesis of lupine alkaloids, and it is inhibited by the alkaloids lupinine and spartein and other small molecules. This inhibition explains the dramatic drop in DAO activity 14 days after germination [42].

N-Methylputrescine oxidase is required for the synthesis of nicotine in *Nicotiana tabacum*. Activity is detected in the roots of intact or decapitated plants. No activity was found in extracts of leaves or cultured callus tissue [44].

DAO activity in the legume *Onobrychis viciifolia* is about the same in light- or dark-grown seedlings. In roots, activity rises rapidly to a maximum at ~7 days, then decreases rapidly. Activity in shoots, albeit lower than in roots, does not change during growth [45].

Using anti-DAO antibodies, it was shown that the enzyme is present only in the leaf primordia and the youngest leaves of *Trifolium subterraneum* (clover). No DAO was synthesized in copper-deficient plants. The synthesis of DAO paralleled the copper content in leaves up to ~10 μg/g of dry weight. Addition of copper to copper-deficient plants does not restore the enzyme [207].

DAO of *Vicia faba* (fava bean) is found in the apical parts and leaves of 18-day-old seedlings. No activity was detected in the cotelydons or roots [55].

Unlike the copper-containing amine oxidases from other plants, the enzyme from mature tea leaves prefers primary monoamines as substrates. Ethylamine is found in large amounts in the leaves and is required for the biosynthesis of the flavor component theanine and incorporated into catechins in tea seedlings. As the plant ages, the level of ethylamine decreases, probably via the action of the oxidase [27].

Benzylamine and 2-phenylethylamine activities have been detected in other plants: *Brassica campestris* (turnip), *Brassica oleracea* (Brussels sprouts), *Perilla frutescens* (beefsteak plant), *Chrysanthemum coronarium* [21]. Whether these are copper- and TQ-containing oxidases has not been established.

C. Yeast and Fungi Amine Oxidases

Copper-containing amine oxidases are found in the mycelia when yeast or fungi are grown on monoamines and diamines: *Aspergillus niger*, *Fusarium bulbigenum*, *Monascus anka*, *Penicillium chrysogenum* [23,26]. Amine oxidase activity has been measured in the mycelia of other fungi and yeast, although the nature of the oxidases has not been defined [10]. Copper- and TQ-containing amine oxidases are peroxisomal enzymes in other species: *Aspergillus niger*, *Candida boidinii*, *Candida utilis*, *Hansenula polymorpha* [16,18]. Numerous other yeasts, molds, mushrooms, and fungi have amine oxidase activity, however, the identity and intracellular location of the oxidases have not been determined [18,20,21,25].

D. Fish Amine Oxidases

In catfish (*Parasilurus asotus*) the highest BAO activity was found in intestine, with lesser activity in liver, low activity in skin and heart tissue, and none in kidney [118]. In mullet (*Mugil cephalus*) BAO was isolated from plasma [119]. Other studies identified DAO activity in various digestive organs of numerous fishes [115–117].

IV. GENERAL PROPERTIES

The first studies of a copper-containing amine oxidase appeared in the 1930s and dealt with histaminase from porcine kidney. In the 1960s it was established that it and DAO are one and the same. (For work published before circa 1975, readers are referred to Refs. 137,208–210.) Although a considerable amount of work on these enzymes appeared from 1930 through the 1950s, it wasn't until the late 1950s and 1960s that the first reports of highly purified preparations of several of the oxidases appeared. Only then could progress be made in understanding the basic biochemistry of these enzymes.

Table 1 presents information regarding fundamental biochemical properties of numerous amine oxidases. This table is comprehensive in that it includes all examples where some evidence exists that each is a copper- and TQ-containing amine oxidase. In many cases, more work is required to substantiate these suppositions. In general, these oxidases have α_2 structures with molecular weights in the range of 150–200 kDa, with some obvious exceptions. Oxidases from the yeast *Trichosporon cutaneum* X_4 and the fish *Seriola quinqueradiata* (yellowtail) are reported to have $\alpha\beta$ structures. DAO from *Aspergillus niger* and BAO from *Pishia pastoris* have unusually high molecular weights, whereas the molecular weights for oxidases from cucumber are quite low.

Bovine PAO does not dissociate into subunits when treated with acid, base, 8 M urea, or 6 M guanidine hydrochloride. If 2-mercaptoethanol is present with 6 M guanidine hydrochloride, the subunits easily dissociate. It was concluded that the subunits are cross-linked by disulfide bond(s) [71,76]. For porcine kidney DAO [88], porcine PAO [94], and *Euophorbia characias* latex DAO [218], the subunits dissociate on sodium dodecylsulfate polyacrylamine electrophoresis gels without pretreatment with 2-mercaptoethanol. For many other amine oxidases where the subunit molecular weights were reported (Table 1), either 2-mercaptoethanol was not present or its presence or absence was not reported; thus, for most of these enzymes it has not been established whether the subunits are cross-linked.

A large number of the copper-containing amine oxidases have been purified to homogeneity (Table 1). Several have been crystallized: bovine PAO [70,75,211], porcine PAO [90], porcine kidney DAO [87], *Aspergillus niger* DAO [10,14], *Penicillium chrysogenum* agmatine oxidase [22,23]. To date no one has reported the preparation of crystals that diffract X-rays.

Most, if not all, of the eukaryotic copper-containing amine oxidases are glycoproteins, although this has not been established for any of the purified oxidases from fungi. MAOx from the gram-positive bacterium *Arthrobacter* P1 is not a glycoprotein [7]. In general they are acidic proteins (i.e., $pI < 7.0$), however, some have pI values ≈ 7.0. Although many have been purified to homogeneity, the purification procedure used may modify the properties of the oxidases, particularly as pertains to reactivity of the quinone cofactor [80] (see Section V). Substrate specificity for many of these oxidases has been determined. Space does not permit discussion of this aspect for each enzyme, so the interested reader is directed to the references listed in the Table 1 and noted elsewhere in this treatise. Much work has been devoted to designing specific inhibitors of a number of the mammalian enzymes, with the hope of finding compounds of pharmacological importance. Again, this aspect is beyond the scope of this review and the reader can find further discussion in several recent papers and reviews [67,121,166, 212–214].

Amino acid compositions of many of these oxidases are known (see Table 1), however, amino acid sequence data has been published for only three of these: (1) the entire sequence of the amine oxidase from *Hansenula polymorpha* as deduced form the DNA sequence of the cloned structural gene [19], (2) the sequences of the two regions of MAOx from *Arthrobacter* P1 [215], and (3) the sequence around the cofactor site of bovine PAO [6]. Figure 2 shows a comparison of the sequences of the *Hansenula polymorpha* amine oxidase and the N-terminus and a cyanogen bromide cofactor–containing peptide of MAOx of *Arthrobacter* P1. Notice that there is fairly high homology in several regions of these sequences. The sequence of the cofactor peptide of bovine PAO is -Leu-Asn-TOPA-Asp-Tyr- [6]. There is a corresponding sequence for the *Hansenula* oxidase starting at residue 403: -Ala-Asn-*Tyr*-Glu-Tyr- [19] (Fig. 2). Since it is assumed that TQ is derived from tyrosine, these sequences are indeed similar. Attempts to sequence the N-terminus of intact bovine PAO by chemical methods failed, which indicated that the N-terminal amino acid is blocked, but a carboxypeptidase A digest released a tyrosine residue from its C-terminus [209]. N-terminal analysis of porcine kidney DAO yielded a threonine residue [88].

Bovine PAO contains two free buried sulfhydryl groups (23 cysteinyl residues/ 180,000 [209]), which become reactive at pH 4.6 or in the presence of 8 M urea or

```
        1                                           10
YAO     Met-Glu-Arg-Leu-Arg-Gln-Ile-Ala-Ser-Gln-Ala-Thr-Ala-Ala-Ser-Ala-Ala-


                 20                                          30
YAO     Pro-Ala-Arg-Pro-Ala-His-Pro-Leu-Asp-Pro-Leu-Ser-Thr-Ala-Glu-Ile-Lys-Ala-Ala-
                            o   •   •   •   •   •   •   •   o       •   •       •
MAOx-1  Leu-Val-Gly-Val-Ser-His-Pro-Leu-Asp-Pro-Leu-Ser(Ser)Val-Glu-Ile-Ala-???-Ala-
        1                                   10


                     40                                      50
YAO     Thr-Asn-Thr-Val-Lys-Ser-Tyr-Phe-Ala-Gly-Lys-Lys-Ile-Ser-Phe-Asn-Thr-Val-Thr
                    o   •               •                   •   •           o   o
MAOx-1  Val-Ala-Ile-Leu-Lys-Glu-Gly-Pro-Ala-Ala-Ala-???-----Ser-Phe(Phe)Phe-Ile-Ser-
        20                                      30


                         60                                                70
YAO     Leu-Arg-Glu-Pro-Ala-Arg-----Lys-Ala-Tyr-Ile-Gln-Trp-Lys-Glu-Gln-→→→
        o                           •
MAOx-1  Val(Val)Leu-???-???-Pro-???-Lys-???-???-
                40                  45


                                             380                                 390
YAO     Asp-Phe-Lys-His-Ser-Asp-Phe-Arg-Asp-Asn-Phe-Ala-Thr-Ser-Leu-Val-Thr-Arg-Ala-
                    •           •   •   •   o           •                   •   •
MAOx-2  ???-???-Met-His-Phe-Asp-Phe-Arg-Glu-Gly-Thr-Ala-Glu------------Thr-Arg-Arg-


                                                 400
YAO     Thr-----Lys-Leu-Val-Val-Ser-Gln-Ile-Phe-Thr-Ala-Ala-Asn-Tyr-Glu-Tyr-→→→
        o       •   •   •   o           •       •
MAOx-2  Ser(Arg)Lys-Leu-Val-Ile(Cys)Phe-Ile-Ala-Thr-???-???-???-???-???-???-
```

Figure 2 YAO denotes the sequence from yeast (*Hansenula polymorpha*) amine oxidase (deduced from cDNA sequence [19]), MAOx-1 denotes the sequence from the N-terminus of the intact MAOx from *Arthrobacter* P1 [215], and MAOx-2 denotes the sequence from a cofactor-containing cyanogen bromide peptide from MAOx from this organism. *Met-His* is the point of cyanogen bromide cleavage. The sequence of MAOx-2 courtesy of Dr. David M. Dooley. The underlined amino acids indicate the part of the *Hansenula* amine oxidase sequence that corresponds to the TQ-containing peptide isolated from bovine PAO [6]. The symbols •, ○, ---, ???, and →→→ denote identical residues, conservative substitutions, inserted spaces to maintain optimal overlap, unidentified residues, and continuation of the YAO sequence to the C-terminus, respectively.

5 M guanidine hydrochloride. Carboxymethylated or *p*-chloromercaptobenzoate-derivatized PAO has no less than ~75% of the normal activity [216]. Two sulfhydryl groups/dimer become reactive towards *N*-ethylmaleimide on treatment of bovine PAO with substrate. Two SH groups/dimer are labeled (along with other groups) when the suicide substrate 2-bromoethylamine is presented to PAO. It was proposed in 1978 that the SH group is sequestered in the oxidized enzyme and becomes reduced on exposure to substrate, i.e., it is bound to an unidentified cofactor in the enzyme [217]:

$$\text{-S-X} \xrightarrow[2H^+]{2e^-} \text{-SH} + \text{XH}$$

Since we now know the nature of the organic cofactor (i.e., TQ), this proposal is untenable. By analogy with BAO from *Euphorbia characias*, latex the SH groups released on substrate reduction are likely those that are exposed when the oxidase is denatured.

DAO from *Euphorbia characias* latex has two free sulfhydryl groups/dimer (14 cysteinyl residues/dimer), which are reactive only after treatment with 8 M urea or anaerobic reduction with substrate. The activity was reduced to ~2% with 4,4′-dithiodipyridine treatment, but retained 45% of maximal activity when *N*-ethylmaleimide was used [218]. It can be concluded that these free sulfhydryl groups are not ligands for the Cu(II) nor essential active site residues.

Photoinactivation of bovine PAO using Rose Bengal, in the presence and absence of substrate, provided evidence for an essential active site histidine. It was not established if this histidine is one of those liganded to Cu(II) or truly an active site residue [219].

Reaction of porcine kidney DAO with phenylglyoxal or 2,3-butadione totally inactivates the enzyme. These reagents are specific for arginine residues. It was not determined how many arginines are derivatized upon inactivation of the oxidase or whether they are active site residues [220].

Several oxidases are reported to exist as isozymes as judged by chromatographic and electrophoretic mobility of native or denatured enzymes: human PAO and BAO in other human tissue [221], human placental DAO [63], bovine PAO [70,75,76,209], porcine PAO [90,95,96], lentil seedling DAO [40], pea seedling [48], clover leaf DAO [53], MAOx from *Candida boidinii* [16], MAOx from *Arthrobacter* P1 [8]. Many of the oxidases aggregate, and this may in part explain the heterogeneity seen for undenatured forms. Those that aggregate are human placental DAO [60,61,120], pig kidney DAO [88,89,222], bovine PAO [71,223], porcine PAO [223], and the yeasts *Aspergillus niger* [15], *Candida bodinii* [16], and *Trichosporon cutaneum* [25]. A more likely explanation of the heterogeneity is protein modification. It has been suggested that chromatographic properties of bovine PAO are altered by different carbohydrate content of subspecies [209]. In a very careful study with porcine PAO, there was a definite difference in the amount of fructose, glucosamine, mannose, and galactose in the carbohydrate component of five isozymes, although the amino acid content for all were essentially the same [96]. Different carbohydrate content cannot explain the heterogeneity for MAOx from *Arthrobacter* P1 since the oxidase is not a glycoprotein [7].

Other possible causes for multiple forms of the enumerated oxidases could arise from proteolytic cleavage or direct modification of amino acyl side groups. Deamidation of glutamyl and/or asparagyl residues could be one type of modification. Some of the oxidases have lysyl oxidase activity, i.e., they can oxidize lysine, lysine methylester, lysine-containing peptides, and lysyl residues of proteins. These are bovine PAO [81,211], bovine dental pulp BAO [81], sheep BAO [100], seedling DAO of pea [46], of *Vicia faba* (fava beans) [55], of *Lathyrus sativus* (chick pea) [38], and of *Glycine max* (soybean) [33], and BAO from the yeast *Pichia pastoris* [17,24]. During purification and storage, highly concentrated solutions would cause molecules of the oxidases to come into close contact with one another. Even if the rate is slow or not measurable in normal assays, oxidation of surface lysyl residues of the oxidases could occur to a substantial degree over long periods of time. This phenomenon would definitely change the mobility of a particular oxidase during ion exchange chromatography. If this does occur for some of the oxidases, then one might expect unusually high apparent cofactor stoichiometries when the oxidases are titrated with phenylhydrazine, because of the aldehyde (allysine) formed on oxidation of the lysyl residues. As is apparent from Table 1, the values are between 1 and 2 mol of quinone cofactor/mol of oxidase, which are not unexpected. Since substoichiometric amounts of the titrant are employed for titrations, perhaps the reaction of allysine with phenylhydrazine with cofactor is so much slower that it does not affect the apparent stoichiometry. (Existence of allysyl residue on the surface of the oxidase might contribute the aggregation phenomenon; allysyl residues on one molecule of oxidase could form a Schiff's base with lysyl residues of another.) Reaction of an oxidase with excess, possibly radioactive phenylhydrazine, followed by proteolytic digestion and isolation and identification of allysylphenylhydrazone would offer proof that self oxidation does take place.

Antibodies for several oxidases have been obtained: human placental DAO, both polyclonal [62] and monoclonal [224]; bovine PAO [225]; bovine aorta BAO [97]; porcine PAO [225]; porcine kidney DAO [62,225–228]; the yeast *Candida boidinii* MAOx [18]; lentil seedling (*Lens culinaris*) DAO [176,180]; chick pea seedling (*Lathyrus sativus*) DAO [38]; DAO from young clover leaves (*Trifolium subterraneum*) [207]. In a few cases, these antibodies have been used to determine the cross-reactivity with related antigens. Antibodies for BAO from cultured porcine aorta cells [97] and rat white adipose tissue [106] cross-reacts with porcine PAO [97]. Human placental DAO antibodies cross-react with porcine kidney DAO. Conversely, antiporcine kidney DAO antibodies do not cross-react with human placental DAO [62]. There is no cross-reactivity of antibodies for bovine PAO, porcine PAO, or porcine kidney DAO [225]. Antibodies against *C. boidinii* MAOx cross-react with crude extract of a number of yeasts grown on methylamine, however, similar extracts from other yeasts did not cross-react. These antibodies also cross-reacted with BAO from *Candida utilus* but not BAO from *C. boidinii* or *Pishia pastoris* [18]. Lentil seedling DAO antibodies react with DAO from *Cicer arietinum* and pea seedlings [30]. Antibodies for DAO from chick pea seedlings recognize pea seedling DAO [38], even though the latter is a glycoprotein but the former is not (see Table 1).

V. THE COFACTORS

For quite a long time, there has been universal agreement that Cu(II) is essential for catalysis and that the typical α_2 oxidases contain 1 g-atom of copper/subunit. Today, the structure of the Cu(II) is fairly well defined. Tetragonal Cu(II) is likely bound via three equatorial imidazole (histidyl) ligands and one equatorially and one axially bound H_2O/OH^-. Water or TQ as the sixth (axial) ligand of Cu(II) is still speculative. Additionally, evidence exists for an axial sulfhydryl ligand from cysteine. (See Section VI for a more complete discussion.)

In contrast, the identification and stoichiometry of a covalently bound organic cofactor has been subject to much disagreement and debate within the field. In 1940, Zeller suggested that this cofactor was a riboflavin (vitamin B_2) derivative in mammalian DAO. More recently, Hamilton speculated that an open ring form of riboflavin could serve as the cofactor [229]. In a paper published in 1949, Werle and van Pechmann proposed that it is a pyridoxal (vitamin B_6) derivative. This notion was also advanced by several other laboratories in the 1950s [209,230,231 and references therein]. Until 1979, this idea persisted in the literature even though the evidence was circumstantial. The identification relied on chemical and physical properties of the cofactor, not on direct structural analysis. As outlined in reviews by Yasunobu et al. in 1976 [209] and Knowles and Yadav in 1984 [231], not all chemical and physical data were consistent with pyridoxal phosphate. Evidence in support of a B_6 derivative was UV-visible absorbance of the oxidases, which was reminiscent of that of B_6, reaction and conversion of amines to the corresponding aldehydes by the enzymes, reaction of the cofactor with carbonyl reagents, reactivation of some of the oxidases by pyridoxal phosphate, reactivation of apo-B_6 enzymes with an acid hydrolysate of porcine PAO and DAO, detection of 2 and 4 mol phosphate/mol of enzyme for porcine and bovine PAO, respectively, and microbiological assays [231].

Although not normally recognized as a redox cofactor, a mechanism has been proposed for enzymic amine oxidation involving pyridoxal phosphate [232]. A pyridoxal-Cu(II) complex can catalyze the nonenzymic oxidation of alkylamines and amino acids [232a and references therein]. When O_2 is present in the model system, the Cu(II)-Schiff's base that forms converts to a coenzyme oxime-Cu(II) complex, after release of the aldehyde product. Hydroxylamine is released on reaction of the complex with a second molecule of amine. This model reaction differs in several crucial aspects when compared with the enzymic oxidation. First, as mentioned in several places in this chapter, although the prosthetic groups "communicate," it is not certain that there is a direct bonding between the organic cofactor and Cu(II) in the oxidases. Next, with the oxidases, a second molecule of substrate is not required for release of the nitrogen-containing product: NH_4^+ is released on reoxidation of the enzymes. Last, the products of the enzyme reaction are aldehyde, NH_4^+, and H_2O_2, not aldehyde and hydroxylamine as in the model system. It is possible that the oxidases divert the reaction away from formation of an oxime intermediate. This discussion is academic, since the evidence for TQ as the cofactor in copper-containing amine oxidases is very persuasive. It should be mentioned that a sulfenic amino acyl group has also been proposed as the cofactor for these oxidases [217].

The discovery of 2,7,9-tricarboxy-PQQ in 1979 as the noncovalently bound cofactor of methanol dehydrogenase from a methylotrophic gram-negative bacterium [233] led to speculation that a PQQ derivative was also the covalently bound prosthetic group in the copper-containing amine oxidases. This notion is obviously more aesthetically pleasing then B_6. A quinone would not only show many of the same chemical properties as B_6 (i.e., reaction with amines and carbonyl reagents), but it is a legitimate redox cofactor. It was not long before papers appeared that reported the release of derivatized authentic 2,7,9-tricarboxy-PQQ from various copper-containing amine oxidases: bovine PAO [177,235]; porcine kidney DAO [235,237]; pea seedling DAO [238]; lentil seedling DAO [239]; *Arthrobacter* P1 MAOx [7]. In all these cases the methods relied on derivatizing the cofactor with a number of reagents and studying its chemical and physical properties in the enzyme or after proteolytic release; for example, comparison of spectral properties and HPLC retention times with the same derivatives of 2,7,9-tricarboxy-PQQ. During this same time period, other studies [234,236] provided evidence inconsistent with pyridoxal as the cofactor in bovine and porcine PAO, but suggested that PQQ might be the prosthetic group. While these are good procedures for screening enzymes for bound PQQ, they cannot substitute for direct structural analysis. Unfortunately, most of the interested scientific community fell into the same trap set earlier by the belief that pyridoxal phosphate was the covalently bound cofactor in this class of oxidases. As mentioned in the introduction of this chapter, numerous enzymes believed to contain covalently bound PQQ either contain no organic cofactor or another heretofore unknown cofactor derived from modification of amino acyl side groups in the polypeptide chains of the respective enzymes (see Fig. 1).

For bovine PAO (and presumably all the other copper-containing amine oxidases), a recent study has shown that the cofactor is 6-hydroxydopa quinone (Fig. 1), also known as topa quinone (TQ) [6]. In this study, the cofactor in the native enzyme was first derivatized with phenylhydrazine, after which a pure, stable thermolytic cofactor peptide was obtained. A detailed UV, mass, and ^1H NMR spectral analysis of this peptide and a synthesized TQ-containing peptide analog was done. Since this study involved direct structural analysis, in addition to comparison of chemical and physical properties, there can be no doubt that the conclusion is unambiguous. (It should be pointed out that the idea that porcine PAO and porcine kidney DAO contained pyridoxal phosphate was still being advanced as late as 1990. In two publications [240,241], mass spectral evidence is presented as evidence for this contention.)

It would be safe to speculate that TQ is derived from a tyrosyl residue in the polypeptide chain of the amine oxidases. Several interesting aspects concerning biosynthesis present themselves. For example, are the modifications catalyzed by external enzymes, or are they self-catalytic? Since oxidation(s) of the *p*-alkylphenol portion of a tyrosyl residue is analogous to reactions catalyzed by known enzymes, added factors might be required as shown here [242]:

First Step

Tyrosine 3-Monooxygenase

L-Tyrosine + tetrahydropteridine + $O_2 \rightarrow$ 3,4-dihydro-L-phenylalanine + dihydropteridine + H_2O

p*-Hydroxybenzoate Hydroxylase*

p-Hydroxybenzoate + NADH + $O_2 \rightarrow$ 3,4-dihydroxybenzoate + NAD^+ + H_2O

Tyrosinase

Tyrosine + DOPA + $O_2 \rightarrow$ DOPA + DOPA-quinone + H_2O

p*-Coumarate Hydroxylase*

p-Cinnamate + ascorbate + $O_2 \rightarrow$ 3,4-dihydroxycinnamate + dehydroascorbate + H_2O

Second Step

2-Monooxygenase

Orcinol + NADH + $O_2 \rightarrow$ 2,3,5-trihydroxytoluene + NAD^+ + H_2O

Similar reaction(s) might require other endogenous factors (e.g., flavoprotein, ferredoxin, rubredoxin) [242].

A possible self-catalytic mechanism would involve the Cu(II) bound to the apo-TQ enzyme. Cu(II) could oxidize the phenolic portion of the tyrosyl residue, thus making it susceptible to nucleophilic attack by water. The mechanism might involve a species similar to the essential Cu(II)-tyrosyl radical found in galactose oxidase [5,243]. Reduction of O_2 to H_2O_2 at the copper site (as occurs for the holoenzymes) would remove the electrons obtained from the tyrosyl group. This self-catalytic mechanism requires direct bonding of Cu(II) and the tyrosyl group during the conversion process, and implies a direct Cu(II)/TQ interaction in the fully formed oxidases. Another interesting aspect of biosynthesis concerns the timing of the modifications. These could be co- or posttranslational or a combination of the two.

The question of cofactor stoichiometry is also of interest. Inspection of Table 1 indicates that these oxidases contain 1 or 2 mol of TQ/mol enzyme. Anaerobic oxidation of benzylamine by bovine PAO and putrescine by lentil seedling DAO produces 2 mol of aldehyde/mol of enzyme [41,80], whereas with porcine PAO, 1 mol benzaldehyde/mol was measured [92,244,245]. Since these enzymes are homodimers, one might be inclined to assume that the stoichiometry is 2 mol/mol, and the 1 mol/mol results might be attributed to inaccurate protein determinations. However, in general, the same procedures were used for the measure of the Cu(II) stoichiometries as for the TQ content. For bovine PAO, the incorporation of radioactive phenylhydrazine is a linear function of the activity, and low activity is probably the result of lengthy purification procedures [80]. For a "fully" active enzyme, a phenylhydrazine titration yielded ~2 mol/mol enzyme stoichiometry. The investigators in this study used a published extinction coefficient at 280 nm (20.8 for a 1% solution) and a published M_r (170,000). There is general agreement on the M_r (Table 1), although there is disagreement on the extinction coefficient. Being aware of this, these investigators checked protein concentrations by a chemical assay. Still, one should be cautious since M_r and protein determinations are notorious for their inaccuracies. The Cu(II) content of the enzyme offers an independent, internal standard for stoichiometries. The investigators of this study found 1.8 g-atom Cu(II)/mol PAO for the most active preparation. It appears that low stoichiometries for this and possibly other amine oxidases can result from enzyme inactivated during purification. A possible inactivation mechanism has been proposed [80].

Low stoichiometries would also result from impure preparations. If, for example, a preparation contained 10 contaminating proteins each contributing ~2% of the total protein, then the total contamination would be 20%. Two percent of any protein might easily be lost in the background in gel electrophoresis or molecular sieving chromatography runs.

The 1 mol/mol stoichiometries have been explained as half-site reactivity [95]. The two Cu(II) sites of porcine PAO are nonequivalent in the absence of substrate or inhibitor as determined by EPR measurements [94,246], although for half-site reactivity we would expect the two sites in this oxidase to be equivalent in the absence of substrate or inhibitor. The nonequivalence would present itself only after an oxidase is exposed to these. A more tenable explanation would be the irreversible inactivation of one of the two TQ groups, which causes a conformational change, or alters one of the copper sites via direct communication with the cofactor. To our knowledge, there is scant other evidence (e.g., kinetic [231]) that would support half-site reactivity (i.e., flip-flop mechanisms).

VI. SPECTRAL PROPERTIES

A. Electronic Spectral Properties

The pink color of these enzymes is their most obvious trait. To produce this color, they absorb in a broad region centered between 460 to 500 nm [22,185,247,248] (Fig. 3), but the reported extinction coefficients vary by a considerable amount depending on the enzyme (from $\epsilon = 2.0$ to $5.7\ mM^{-1}cm^{-1}$; see Table 1). There is

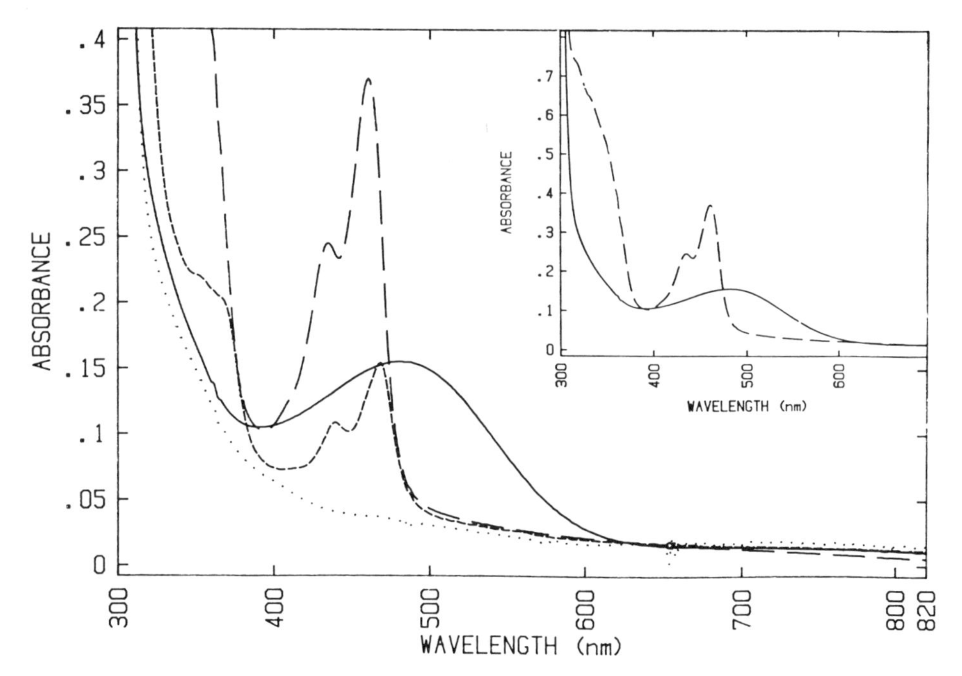

Figure 3 Spectra of MAOx (34 mM) in various redox states at pH 7.5. Spectrum of the oxidized enzymes (——); spectrum of the two-electron, substrate-reduced enzyme in the presence of Cl^- (···); spectrum of the two-electron, substrate-reduced enzyme in the absence of Cl^- (---); spectrum of substrate-reduced enzyme in the presence of CN^- (– – –). For MAOx reduced in the presence of Cl^-, TQ is fully reduced and Cu(II) is oxidized. For MAOx reduced in the presence of CN^-, TQ is one-electron reduced (topa semiquinone radical) and the copper exists as Cu(I). For enzyme reduced in the absence of Cl^-, a slow decrease in the absorbance of the two sharp peaks was probably caused by the small amount of Cl^- in the stoichiometric methylamine hydrochloride used to reduce MAOx.

even disagreement on the value of ϵ for a particular oxidase, as exemplified by pea seedling DAO. For bovine PAO, there is a linear relationship for moles of ^{14}C-phenylhydrazine incorporated into the enzyme (moles of reactive TQ) vs. activity [80]. This raises the possibility that the 480 nm absorbance may also be affected for PAO with activity of less than the maximum.

The 480 nm absorbance is attributed in large part to TQ, whereas the enzyme-bound Cu(II) provides broad, featureless absorbance spanning from ~600 nm to well above 800 nm (8,249–251). The Cu(II) electronic transitions are more evident in circular dichroism (CD) spectra [79,249,250,252]. A negative CD band at >760 nm and a positive one at approximately 560–760 nm are attributed to d-d transitions of tetragonal Cu(II). Negative CD bands in the region from approximately 390–560 nm are mostly due to TQ electronic transitions. Assignment of a weak ligand-to-metal charge transfer band at ~480 nm has resulted from studies of Cu(II)-depleted enzyme [79].

The spectra of both TQ and Cu(II) in these enzymes are altered by various reagents. It is known that N_3^-, SCN^-, OCN^-, HS^-, $HOC_2H_4S^-$, *N,N*-diethyldithio-carbamate, *t*-butylisocyanide, and CN^- are equatorial ligands for Cu(II) in the oxidases [249,250,253–255]. Dramatic changes in ligand field transitions occur when these anions are added to the oxidases and are most apparent on viewing CD and difference UV-visible spectra. These perturbations are attributed to ligand-to-metal charge transfer interactions. There is evidence that Cl^- also interacts with Cu(II) in MAOx [249], however, addition of this anion has no effect on the UV-visible spectrum of oxidized MAOx (McIntire, unpublished data) or PAO [250]. While N_3^- and SCN^- cause complex inhibition of bovine PAO and porcine DAO activity with respect to the amine substrate in steady-state kinetic experiments, Cl^- is an ineffective inhibitor of PAO [250,256]. On the other hand, preliminary research indicates Cl^- to be an inhibitor of MAOx and 2-phenylethylamine oxidase [9,249].

Although NH_4^+ is a product of amine oxidation by these enzymes and the simplest substrate analog, until 1990 no one, to our knowledge, reported the effect of this cation on the electronic spectrum of any of the oxidases. This has now been done for one of the oxidases, MAOx. At 140 mM NH_4^+/NH_3 the λ_{max} of MAOx shifts from 486 nm to 450 nm, with only a slight increase in absorbance. The electron paramagnetic resonance (EPR) spectrum of Cu(II) is unaffected, so the change is solely due to perturbation of the electronic transitions of TQ [249]. It is assumed that topa iminoquinone is formed in the reaction.

These oxidases are highly reactive towards carbonyl reagents. Among these are semicarbazide, phenylhydrazines, hydrazine, hydroxylamine [209], and hydrazides [257]. All form extremely stable complexes with TQ and cause dramatic changes of the electronic transitions of this quinone. Phenylhydrazines in particular have been useful in the study of the oxidases: they have been used to determine the content of reactive TQ [13,24,32,52,54,80,91–93,95]. Resonance Raman

(RR) spectral studies of phenylhydrazones of the several oxidases and their cofactor peptides have provided strong evidence against PQQ and pyridoxal phosphate as the catalytic prosthetic group. The resonance Raman studies also indicate that TQ is probably the cofactor in all copper-containing amine oxidases [258]. The RR spectrum of the phenylhydrazone of copper-depleted MAOx shows some changes compared to holo-MAOx-phenylhydrazone. It was suggested that this is a reflection of an interaction between TQ and Cu(II) [249]. However, removal of Cu(II) would cause the "hole" thus formed to collapse and force abnormal interaction of amino acyl groups. This might lead to conformational changes in the region of the quinone.

Fluorescent lifetimes were measured for 9-hydrazinoacridine-modified (TQ-substituted) and Co(II)-substituted (Cu(II) replaced) porcine PAO. Analysis of the data provided a minimal distance of 11.7 Å from the acridine to the metal. This distance could allow direct interaction of TQ and Cu(II) [259].

Changes in redox state of both TQ and Cu(II) are reflected in the spectra of these oxidases. Until 1990, it was thought that only plant oxidases formed a significant amount of topa semiquinone/Cu(I) on anaerobic substrate reduction [41,185,260,261] and that mammalian enzymes produced topa dihydroquinone/Cu(II) on substrate reduction [22,70,75,90,101,247,262–264] (amine substrates are obligatory two-electron reductants). Addition of substrate to mammalian oxidases bleaches the 480-nm absorption peak, while the same treatment of plant oxidases results in replacement of the broad 480-nm absorbance with two sharp peaks at 430 and 460 nm. If CN^- is added to any of the oxidases and then substrate-reduced, again the 480 nm is replaced by the 430- and 460-nm peaks. In fact, these new bands are more intense in the reduced plant oxidases with CN^--complexed Cu(I) [185]. (These spectral features were first reported by Mann, however, he did not realize they were a property of TQ-semiquinone radical [265].) Although, these spectral bands have been attributed to the topa semiquinone radical, some confusion existed since an organic radical signal was not detected by electron paramagnetic resonance (EPR) spectroscopy in frozen solution for any substrate-reduced oxidase when CN^- was excluded. (A low intensity organic radical ($g \sim 2$) signal, probably due to topa semiquinone, was observed by several investigators [82,266,267]. However, these investigators never attributed this signal to the enzyme-bound quinone cofactor.)

When MAOx is reduced anaerobically by methylamine in the presence of Cl^-, the 480-nm peak bleaches to ~10% of the original absorbance. However, when Cl^- is absent, substrate titration produces the 430- and 460-nm peaks, although they are not nearly as intense as those produced when MAOx is reduced in the presence of CN^- (Fig. 3) [249]. A plausible explanation for these observations would be that Cl^- lowers the redox potential of the Cu(II)/Cu(I) couple relative to the topa semiquinone/topa dihydroquinone couple, while CN^- has the opposite effect. Recent EPR experiments have confirmed that the mammalian enzymes form a small amount of topa semiquinone/Cu(I) on substrate reduction at ambient temperature and pH [252]. This was not discovered earlier because past EPR experiments were done with frozen solutions of substrate-reduced oxidases. Either large pH changes that occur on freezing phosphate and other buffers [268] or the temperature change causes large shifts of the copper and/or TQ redox potential. The latter seems most likely since lowering the temperature of liquid solutions of reduced MAOx significantly lowered the absorbance of the 430- and 460-nm peaks [252].

B. EPR and NMR Spectral Properties

The first EPR spectrum for a copper amine oxidase was reported by Yamada et al. in 1963 for bovine PAO [269]. The spectrum expected for a type 2 Cu(II) center (nonblue, weakly colored, EPR active [270]) was recorded and the g value of the maximum was 2.053 at pH 7.2 and 108 K. A more recent investigation provided data relevant to the structure of this site in porcine PAO [94,246]. It was apparent that two types of Cu(II) were present in equal amounts: one with nearly axial and the other with rhombic symmetry. (See Table 2 for a compilation of EPR parameters for various amine oxidases.) Interestingly, the single-copper form (made by adding one g-atom Cu(II)/mol of dimeric copper-depleted PAO) still contained a equimolar content of each type of Cu(II) site, and the enzyme had half the apparent maximal activity. Subtle changes were noted as a function of pH (7.0–8.9) and ionic strength (I = 0.9–0.16). There was no indication of spin-spin interaction between the Cu(II) sites, and NMR relaxation measurement of water protons at ambient temperature indicated rapid exchange of bound water molecules with bulk solvent. From the NMR and EPR data, it was concluded that each Cu(II) had two nitrogen-donating ligands and an axial and an equatorial oxygen-donating (both water-derived) ligand. The water ligands are deprotonated with $pK_a \approx 8$. Only equatorially bound water is displaced by either N_3^- or CN^-. These anions competitively inhibit reoxidation by O_2, so it was concluded that O_2 is reduced via coordination with Cu(II)/Cu(I). Imidazoles of histidyl residues were proposed as ligands, since they fulfill the requirement for sp^2 lone pair electrons of the nitrogen-donating ligands.

In contrast, the EPR spectra of bovine PAO, porcine kidney DAO, MAOx, and pea seedling DAO were consistent with a single type Cu(II) with axial symmetry (Fig. 4, Table 2). As expected, significant changes in the EPR spectra occurred when Cu(II) ligands were added to any of the enzymes, and N_3^- and CN^- each rendered the two porcine PAO Cu(II) sites identical with axial symmetry (Table 2). For all the oxidases, binding of these equatorial ligands does not alter the tetragonal structure of the Cu(II) sites.

Suzuki et al. [271] observed nitrogen superhyperfine splitting (= 1.4 mT) for bovine PAO only in the presence of >1 M $(NH_4)_2SO_4$. The splitting was not due to

Cu(II) coordination of NH_4^+, but the result of at least two imidazole copper ligands. Perhaps high NH_4^+ concentration induces structural changes around the Cu(II) site, possibly caused by NH_4^+ binding to TQ. Weak superhyperfine splitting was observed for porcine kidney DAO [86,255], which was also seen for the N_3^- complex, and strong superhyperfine splitting is observed for the CN^- and SCN^- complexes of this oxidase. The splitting is partly attributed to nitrogen-donating ligands (imidazole) equatorially bound.

The parameters in Table 2 would indicate that substrate reduction does not significantly affect the EPR spectra of Cu(II) of any oxidase. (Under the conditions used to obtain the spectra, TQ is fully, i.e., two-electron reduced, and the copper is oxidized.) This would suggest little or no direct interaction (bonding) between TQ and Cu(II). However, it was found that changes in peak resolution, shape, and width occur on substrate reduction [78,79,246], so some interaction between the redox centers may be taking place. A similar situation presents itself for amine oxidases modified with carbonyl reagents [79,257,266]. Treatment of bovine PAO with *N,N*-diethyldithiocarbamate (DDC) provides an EPR spectrum with significant differences when compared with the same enzyme treated with this reagent and either phenylhydrazine or 3-(1-pyrryl)phenylacetic hydrazide. DDC is thought to only react with Cu(II) [274], while the other two reagents only react with TQ [257,266]. This would argue for a direct interaction of the two prosthetic groups. Alternatively, structural changes around the Cu(II) could be induced when the substrate binding site is occupied or TQ is reduced.

Two sets of exchangeable, water-bound protons were detected by electron nuclear double-resonance (ENDOR) studies in D_2O for porcine PAO. Hyperfine coupling indicated that one water is bound equatorially and the other axially. Due to broad lines, only a single hyperfine coupling to ^{14}N was apparent, indicating similar coordination of all the nitrogen-donating ligands. Equatorial binding of N_3- eliminated the water couplings, but had little effect on the ^{14}N hyperfine interactions. Together with earlier EPR results, it was concluded that Cu(II) has three nitrogen-donating equatorial ligands, probably histidyl residues, in addition to one axial and one equatorial water. Evidence for or against a sixth (axial) ligand (water or TQ?) was not provided. Interestingly, the data from this study did not distinguish between the two types of Cu(II) centers in porcine PAO [275].

Pulsed electron-spin echo envelope modulation studies of the Cu(II) site of porcine kidney DAO and bovine PAO provided evidence for two different types of equatorial nitrogen-donating ligands. Histidyl side groups were confirmed as the ligand because coupling of the Cu(II) to the unliganded imidazole nitrogens could be seen. (Coupling of the directly bound nitrogens is too large to be detected by this procedure.) The method also confirmed the presence of one axial water (Cu(II)-O, bond length = 2.6 Å) and one equatorial water ligand (Cu(II)-O, bond length = 2.1 Å). Again, CN^- and N_3^- were found to displace the equatorial water, and the distinction between the two sets of imidazole ligands is lost on binding of

Table 2 EPR Parameters

Enzyme	Symmetry	g_{xx}	g_{yy}	g_{zz}	Hyperfine splitting (mT) A_{xx}	A_{yy}	A_{zz}	Temp. (°K)	pH[a]	Ref.
Bovine PAO										
Native	Axial	2.06		2.29			16.4	77	7.2	267
Native	Axial	2.06		2.29			16.1[b]	77	7.2	271
Native		2.06		2.30			16.4			
+ benzylamine	Axial	2.06		2.31			16.2	77	7.2	79
+ phenylhydrazine		2.08		2.31			15.6			
Native	Axial	2.04		2.25			15.5	[c]	8.0	253
+ sulfide	Axial	2.02		2.17			17.5[d]	[c]	8.0	253
Native		2.060		2.28			15.5			
+ azide	Axial	2.050		2.240			16.0	40	[c]	250
+ thiocyanate		2.055		2.255			17.0			
+ 3-(1-pyrryl)-phenylacetic hydrazide	Axial	Minor changes						77	7.2	257
+ *N,N*-diethyl-dithiocarbamate	Axial	Large changes + superhyperfine splitting								
Bovine aorta BAO										
Native	Axial	2.07		2.29			16.0	113	7.8	82
		2.08		2.30			12.0	113	3.0	82
Porcine PAO										
Native	Axial	2.06		2.266						
+ benzylamine	Axial	2.059		2.271			15.0			272
+ dithionite	Axial	2.058		2.268						
Native	Axial	2.078		2.286	1.0		15.5			
	Rhombic	2.039	2.065	2.294			14.0	150	7.0	94
+ azide	Axial	2.055		2.257			16.5			
+ cyanide	Axial	2.054		2.218			15.8			

Native	~Axial	2.059	2.030	2.2760	1.5		16.2			
	Rhombic	2.079	2.068	2.2862	2.0	2.0	14.8	[c]	[c]	246
+ benzylamine	~Axial	2.259	2.057	2.2760	1.5		16.2			
	Axial	2.072		2.2860	1.5	1.5	14.8			
Porcine kidney DAO										
Native	Axial	2.063[f]		2.294			14.9	98	7.4	86
+ ^{14}N-putrescine	Axial	Superhyperfine lines of main peak more pronounced relative to native oxidized DAO								86
+ ^{15}N-putrescine		Spectrum identical to that produced by ^{14}N-putrescine								86
Native		2.061		2.286			17.3			
+ azide	Axial	2.058[g]		2.224			15.6	—	—	255
+ thiocyanate		2.045[g]		2.210[g]			19.0			
+ cyanide		2.036[g]		2.195[g]			17.1			
Aspergillus niger DAO										
Native	Axial	2.07		2.31			16.2	93	7.0	273
MAOx										
Native	Axial	2.06		2.229			[c]	77	7.2	249
PSAO										
Native	Axial	2.06		2.29			16.0	77	7.2	[e]

[a]pH at room temperature in phosphate buffers. Differences in g and hyperfine splitting values may be due to pH changes that occur on freezing phosphate buffers [268].
[b]Superhyperfine splitting = 1.4 mT due to the nitrogen-donating ligands of Cu(II), which only becomes apparent in the presence of high $[NH_4^+]$.
[c]Not reported.
[d]Superhyperfine splitting = 1.55 mT.
[e]Dr. David M. Dooley, personal communication.
[f]Weak superhyperfine splitting seen.
[g]Weak superhyperfine splitting for the azide complex, and strong for the cyanide and thiocyanate complexes ($A_{\perp}$ = 1.45 mT and $A_{\parallel}$ = 1.1 mT for both).

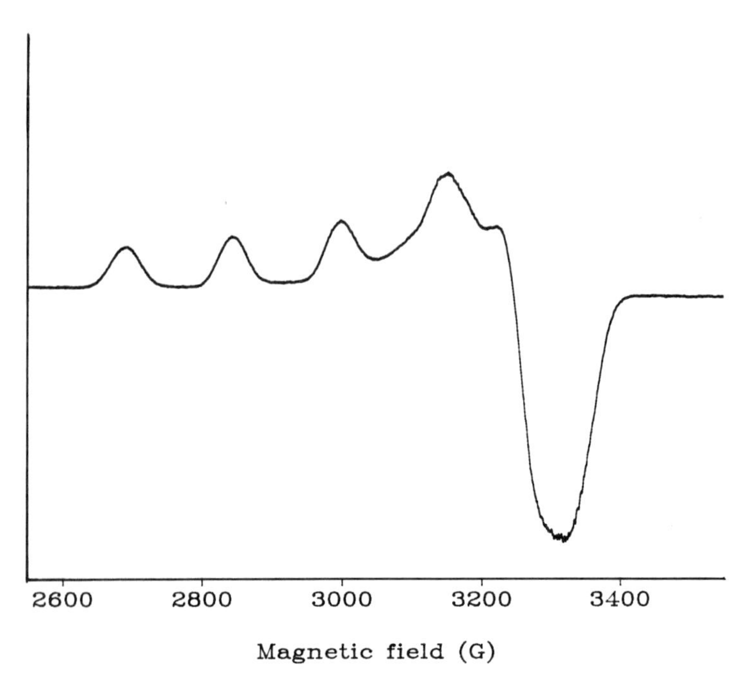

Figure 4 EPR spectrum of oxidized pea seedling DAO [0.7 mM Cu(II)] at 77 K, in 50 mM sodium phosphate, pH 7.0. Microwave power = 1 mW, frequency = 9.409 GHz, and modulation amplitude = 20 G. (Figure courtesy of Dr. David M. Dooley.)

either anion. This study also detected outer sphere ("ambient") water molecules [276]. The structure of Cu(II) in half-depleted (inactive) bovine PAO was found to be the same as in the native enzyme. Phenylhydrazine binding to, or substrate reduction of TQ has no effect on inner sphere Cu(II) coordination, but ambient water is displaced [223,267]. This indicates that phenylhydrazine does not bind to Cu(II), but suggests conformational changes around the Cu(II) when the substrate-binding site is occupied. It can be inferred that there is no direct interaction of the quinone cofactor and Cu(II).

The spin label 3-(maleimido)-2,2,5,5-tetramethyl-1-pyrrolinyloxyl reacts with two nonessential surface(?) cysteinyl residues/monomer of native, oxidized bovine PAO. Substrate reduction or phenylhydrazine derivatization causes the exposure of a single essential, active site sulfhydryl group, which can react with this spin label and derivatives thereof containing one to four CH_2 units between the maleimido and nitroxide group. EPR experiments found that this single cysteinyl residue is in a restricted environment, and that the label and Cu(II) do not interact [277]. Whether this cysteine is truly an active site residue is unproven.

Bovine and porcine PAO were exposed to 4-amino-2,2,6,6-tetramethylpiperidine-*N*-oxyl (4-amino-TEMPO) or 4-hydrazino-TEMPO and followed by $NaBH_4$ treatment, which trapped the spin labels at the reactive position of TQ. EPR measurements indicated that the very immobile TEMPO is in a sterically constricted environment and is not accessible to small anions [278]. It was concluded that the spin label is 13 Å from, and has no magnetic communication with, Cu(II). Furthermore, removal of Cu(II) had no effect on the EPR spectrum of the label.

Any interaction between the organic cofactor and Cu(II) might be reflected in changes in its EPR spectra on magnetic isotopic substitution of TQ. It is assumed that upon reduction, substrate-derived nitrogen is retained in topa aminoquinol (i.e., amino transferase mechanism is operating; see below); thus, porcine kidney DAO has been reduced with ^{14}N- and ^{15}N-putrescine. However, the EPR spectra of Cu(II) are identical in both cases [86], suggesting little communication between the prosthetic groups.

DDC plus a reducing agent removes a single copper ion from bovine PAO. The EPR spectrum of this enzyme is the same as for native PAO, except for an unexplained sharp singlet at g = 2.004. From very recent studies, this sharp peak is likely due to a small amount of TQ radical (see next paragraph). On reconstitution with Cu(II), the activity and EPR spectrum of untreated PAO was obtained [266]. In another study, PAO and PAO half and fully Cu(II)-depleted each bound only one phenylhydrazine/dimer. The first two of these had identical Cu(II) EPR spectra. It was stated that the K_d for phenylhydrazine decreased by a factor of 100 in Cu(II)-depleted PAO relative to native PAO.

As mentioned earlier, the EPR spectrum of TQ semiquinone radical (g ~ 2) of substrate-reduced Cu(I)-CN-complexed amine oxidases (Fig. 5) has been observed for some time, but only recently were EPR spectra recorded for the radical in substrate-reduced unmodified oxidases [252]. Unfortunately, all past EPR experiments involved frozen anaerobic solutions in phosphate buffers, whereas room temperature measurements are required (Fig. 6). Interestingly, the EPR spectra of the TQ radical in the oxidase are essentially the same whether CN^- is liganded to the copper or not. This offers evidence against direct interaction of TQ and copper.

The EPR spectra of ^{14}N- and ^{15}N-$CH_3NH_3^+$-reduced, CN^--complexed MAOx are quite different (Fig. 5). In order to explain this difference, substrate-derived nitrogen must remain bound to reduced TQ. Once formaldehyde is released on enzyme-mediated water hydrolysis, the aminoquinol form of TQ remains. Transfer of one electron to Cu(II) produces topa iminosemiquinone. The unpaired electron couples with the substrate-derived nitrogen (and hydrogens of TQ) to produce the spectra in Figure 5. These observations support the amino transferase mechanism for the oxidases. To date the complex hyperfine couplings seen in Figure 5 have not been deconvoluted.

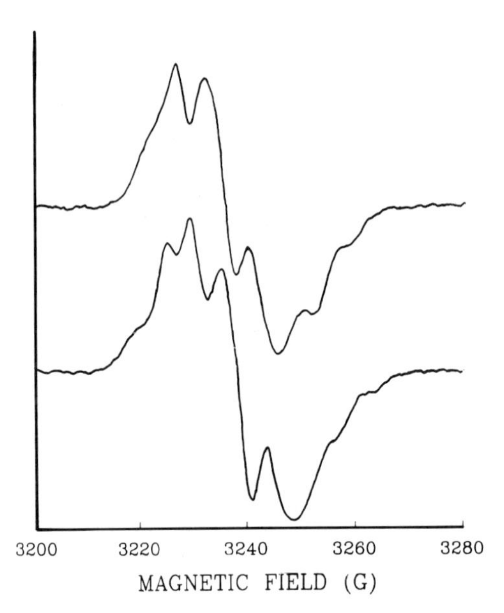

Figure 5 EPR spectra of substrate-reduced MAOx in the presence of CN^-, in 50 mM sodium bartibal, pH 8.0 at 173 K. Microwave power = 0.5 mW, frequency = 9.101 GHz, and modulation amplitude = 0.4 G. The upper spectrum is for MAOx reduced with ^{15}N-methylamine, and the lower is for enzyme reduced with ^{14}N-methylamine.

The longitudinal (T_1) and transverse (T_2) relaxation times of nitrogen and hydrogen of several substrate analogs were measured by NMR in the presence of porcine kidney DAO. Analysis of the data revealed that nonexchangeable protons of dimethylamine are ~8 Å from Cu(II). *N*(α),*N*(α),*N*(1)-Trimethylhistamine has a tight and a loose binding mode in the oxidase. When tightly bound, ring methyl protons are ~9 Å, and imidazole protons at positions 2 and 4 are ~8 Å from the copper, whereas when loosely bound these latter protons are ~4 Å away. This is close enough for imidazole nitrogens to ligand to Cu(II). Whether there are two different binding sites on each subunit or there is differential binding to each subunit is not known. Relaxation rates of residual protons in a 99.7% D_2O solution of DAO are affected by addition of *N*(α),*N*(α),*N*(1)-trimethylhistamine, but it was not certain if its loose binding displaces a water ligand of Cu(II). There was no indication of displacement of copper-bound water by O_2, nor was the linewidth of the nitrogen resonance of $^{15}NH_4^+$ altered in the presence of DAO. Because of the low magnetogyric ratio of ^{15}N, the nitrogen would need to be a few angstroms from Cu(II) for an effect to be seen [279].

T_1 and T_2 values were estimated for ^{19}F of 2-, 3-, and 4-trifluoromethylphenylhydrazine-derivatized TQ in porcine PAO and its Cu(II)-depleted form. These

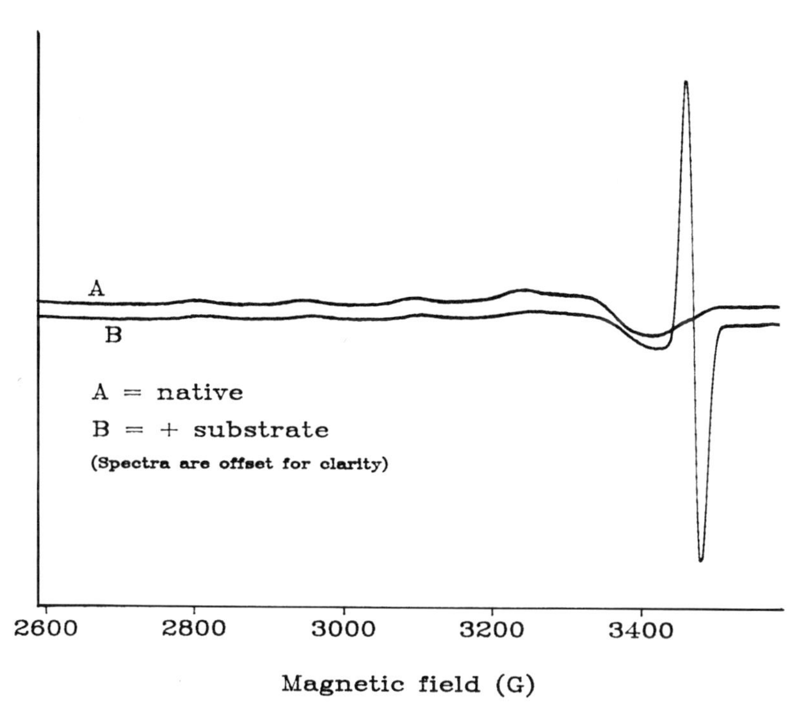

Figure 6 EPR spectrum of MAOx (0.8 mM) at room temperature before and after anaerobic reduction by benzylamine. The sample is in 50 mM barbitol, pH 8.0, in a 0.3 mm × 6 mm × 100 mm glass cell. The instrument conditions were microwave power = 20 mW, frequency = 9.8 GHz, and modulation amplitude = 20 G. (Figure courtesy of Dr. David M. Dooley.)

values provided distances of ^{19}F to Cu(II) of 10.9, 14.3, and 15.5. Å for the 2-, 3-, and 4-isomers, respectively, with Cu(II) in the plane of the aromatic ring. 3,5-Bis(trifluoromethyl)phenylhydrazine-derivatized enzyme gives two resolved ^{19}F peaks, each with identical values of T_1 [280]. Figure 7 presents a computer-generated model of the structure of the supposed TQ/Cu(II) site, based on these data.

T_1 values for water protons were determined as a function of magnetic field strength for various oxidases [223]. From the paramagnetic contribution to T_1, i.e., T_{1p}, bovine PAO, porcine PAO, porcine kidney DAO, and MAOx have one rapidly exchanging H_2O, whereas only porcine PAO and MAOx have a second exchangeable H_2O/OH^- (fast exchange at 25°C and slow exchange at 5°C), which is equatorially bound as determined by studies in the presence of N_3^-. Exchange of equatorial H_2O/OH^- in the other two oxidases is too slow to contribute to T_{1p}.

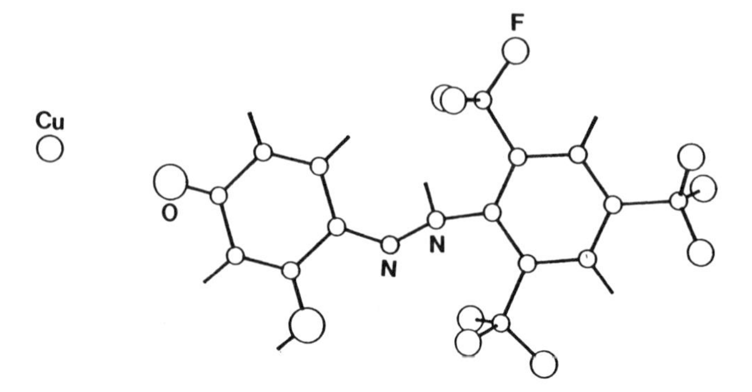

Figure 7 Computer-generated structure of the Cu(II)/TQ site for trifluorophenylhydrazine-derivatized porcine PAO. The constraints for the energy-minimized molecular mechanics model are the distances given in Ref. 280 (see text). The model depicts the location of *ortho* and *para* trifluoromethyl groups. Three trifluoromethyl groups were used to constrain the structure. The two *ortho* trifluoromethyl groups are equidistant from Cu(II). The model was generated using Quanta and Charmm software from Polygen Corp. (Figure courtesy of Dr. David M. Dooley.)

T_{1p} was unaffected by substrate for bovine PAO. Plots of either T_{1p}^{-1}/mol protein liter^{-1}), absorbance at 480 nm (removal of copper requires dithionite treatment which leaves TQ reduced in copper-depleted enzyme), or activity were linear with respect to copper content of bovine or porcine PAO. A linear plot of activity vs. Cu(II) content is in conflict with the notion of nonequivalent copper-binding sites. There was no evidence from this study that the two Cu(II) sites of porcine PAO are nonequivalent.

C. X-Ray Analysis

Although crystals of several of these oxidases have been obtained, to date no one has published X-ray crystallographic data for any of the oxidases. Several laboratories are attempting to obtain the X-ray structures of a number of these enzymes. A three-dimensional map of a copper-amine oxidase should be available in the near future. This invaluable information will help answer many questions concerning these enzymes, including: Are the copper and TQ site identical in both subunits? Is half-site reactivity a viable explanation for various observations? How do the subunits interact and communicate? Do these enzymes exhibit cooperative behavior? Are there one or two TQ groups/ dimer? What are the ligands for Cu(II)? Which carbonyl group of TQ is attacked by substrate? Does TQ interact directly (bond) with Cu(II), or is the communication between these groups longer range via conformational change? If there is a widespread separation between the two groups, what is the mechanism of electron transfer (e.g., electron tunneling), and what amino acyl groups are involved?

X-ray absorption spectroscopy has been useful in defining the structure around the Cu(II) site. This technique, called extended X-ray absorption fine structure (EXAFS), provides information about the kind and number of ligands surrounding metal ions. EXAFS of bovine PAO and $[Cu(imidazole)_4]^{2+}$ were quite similar and characteristic of imidazole-Cu(II) bonding. The data were interpreted in terms of three or four equatorial imidazole ligands of PAO, $\sim 1.99 \pm 0.02$ Å from tetragonal Cu(II). The data also argued against a sulfur-donating equatorial ligand [281].

Another EXAFS study provides a different picture of the Cu(II) site in porcine PAO. The evidence supported a sulfur-centered axial ligand at 2.38 Å, possibly from methionine, although Cl^- is a possibility. It was also concluded that the four equatorial ligands consist of two, indistinguishable imidazoles at 2.00 Å and two OH^- (or OH^- and phenolate, from tyrosine or TQ?) at 1.90 Å. Evidence for a sixth (axial water) ligand was not obtained [282]. It should be noted that CD and magnetic CD spectral studies do not support a sulfur-based ligand of the metal in Co(II)-substituted bovine PAO. These investigations indicated two nitrogen-donating and two oxygen-donating ligands of tetrahedral Co(II) [283]. Figure 8 provides a picture of the structure of the Cu(II) site in the amine oxidases.

VII. KINETIC AND CHEMICAL MECHANISM OF ACTION

Historically, mechanistic conclusions from studies on copper-containing amine oxidases had been based upon the current working hypothesis for the structure of the organic cofactor in this class of enzymes. Based upon the enzymes' high reactivity toward carbonyl reagents and the inability to isolate and characterize the

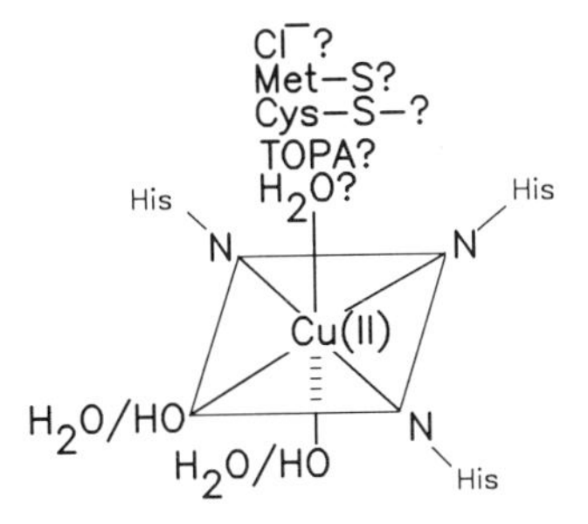

Figure 8 The deduced structure of the Cu(II) site in the amine oxidases. His-N represents imidazole nitrogen equatorial ligands of the copper. The axial ligands with a question mark are those that have been postulated by various groups as a possible sixth (axial) ligand.

cofactor, early studies assumed that the cofactor was pyridoxal phosphate. Despite failed efforts to incorporate tritium into the enzyme upon reduction with [^{3}H]$NaBH_4$ [217], a well-known property of pyridoxal derivatives, pyridoxal phosphate was still considered a potential cofactor candidate as late as 1990 [240]. Incorporation of [^{14}C]-labeled amine substrate into the enzyme was observed upon reduction with $NaBH_4$, but radiolabel was bound as an ε-*N*-alkyllysine derivative, the product of nonspecific trapping of product aldehyde [284]. In 1984, Lobenstein-Verbeek et al. and Ameyama et al. independently reported that bovine PAO contained PQQ [177,285]. However, it is now known that the cofactor in bovine PAO is TQ [6]. Prior to the identification of TQ, several other alternative proposals had been put forth in the literature, including the postulated role of a ring-modified flavin [229] or a sulfenic acid derivative [217]. The historical development of the mechanisms proposed for this class of enzymes will be presented.

A. Mechanism of Substrate Binding and Enzyme Reduction

Early work on bovine and porcine PAO [231,263,286,287] clearly established that these enzymes operate by a ping-pong bi ter mechanism and follow the minimal mechanism shown in Eqs. 1–3:

Enzyme Reduction:

$$E_{ox} + RCH_2NH_3^+ \underset{k_2}{\overset{k_1}{\rightleftharpoons}} E_{ox}\cdot RCH_2NH_3^+ \underset{k_4}{\overset{k_3}{\rightleftharpoons}} E_{red}\text{—}\overset{+}{N}H\text{=}CHR \tag{1}$$

Imine Hydrolysis:

$$E_{red}\text{—}\overset{+}{N}H\text{=}CHR + H_2O \xrightarrow{k_5} E_{red}\text{—}NH_2 + RCHO \tag{2}$$

Enzyme Reoxidation:

$$E_{red}\text{—}NH_2 + H_2O \xrightarrow[O_2]{k_7} E_{ox} + H_2O_2 + NH_4^+ \tag{3}$$

Steady-state kinetic studies with human placental DAO [57], pea seedling DAO [49], and porcine kidney DAO [288,289] also indicate that these enzymes obey a ping-pong type kinetic mechanism [231]. The mechanism is considered to occur in three steps: the first is described by substrate binding to TQ, followed by proton abstraction with concomitant enzyme reduction (Eq. 1); the second step, hydrolysis of product imine intermediate (Eq. 2), is followed by the final step, reoxidation of reduced cofactor by dioxygen to produce hydrogen peroxide (Eq. 3).

For some of the oxidases, the steady-state mechanism is somewhat more complex than a simple ping-pong. For example, several of these enzymes show amine substrate inhibition at high concentration: human placental DAO and PAO

[57,290], *Euphorbia* latex DAO [32]. In a study with porcine PAO, Eadie-Hofstee plots were curved for benzylamine and *p*-methylbenzylamine as the varied substrates ($[O_2]$ held constant) [291]. The equation describing this behavior has the form:

$$v = \frac{\alpha_1[S] + \alpha_2[S]^2}{\beta_0 + \beta_1[S] + \beta_2[S]^2}$$

This behavior can be explained by (1) a nonhomogeneous enzyme preparation, (2) reversible enzyme dissociation, (3) two nonequivalent active sites, and (4) the fact that an E·S·S species is involved in catalysis [291]. The third possibility is supported by EPR measurements, which indicate two nonidentical Cu(II) sites in porcine PAO and appears to be supported by the facts that anaerobic benzylamine oxidation releases 1 mol of benzaldehyde/mol of enzyme [93,244,245] (lentil seedling DAO and bovine PAO release 2 mol of aldehyde/mol [41,80]) and 1 mol of phenylhydrazine bound/mol of enzyme [93]. These results taken together seem to support half-site reactivity, at least for porcine PAO. However, the steady-state results indicated two sites with different activities functioning simultaneously, whereas half-site reactivity requires one site to be inactive when the other is active. It was concluded that the most probable explanation of the curved Eadie-Hofstee plots is offered by the fourth alternative [291].

For MAOx from *Candida boidinii* the steady-state mechanism appears to be of the ordered-binding type [16]. While 1/v vs. $1/[O_2]$ plots were parallel for different methylamine concentrations (as expected for a ping-pong type mechanism), 1/v vs. 1/[methylamine] plots intersected for different $[O_2]$.

As depicted by Eqs. 1–3, the substrate amino group is expected to participate in binding. If true, then complex formation between enzyme and amine should be dependent upon the protonation state of substrate. However, there are discrepancies in the literature concerning the exact nature of substrate binding to the enzyme [101,209,290,292]. From early pH/rate profiles, it was suggested that for porcine PAO, substrate binding requires the unprotonated form of the amine and the protonated form of a group in the enzyme ($pK_a = 8.8$) [245,293]. It was concluded that the substrate amino group contributes to binding through interaction with an electrophilic group at the amine-binding site. Conversely, bovine PAO has been reported to act on the protonated form of the substrate, to possess an amine-binding site in the enzyme with $pK_a = 7.1–7.3$, and to contain an attacking group on the enzyme with $pK_a = 8.5$ [209]. It was concluded that in order to efficiently attack the protonated amine, this active site base must also be protonated.

These latter results are similar to those reported in a more recent detailed study of pH-dependent isotope effects with the bovine enzyme [294]. Evidence indicates an essential basic residue with $pK_a = 8.0 \pm 0.1$ in the free enzyme, which had a $pK_a = 5.6 \pm 0.3$ in the E·S complex. Although the observed pK_a values for porcine PAO and bovine PAO are in reasonable agreement, the question of the protonation state of the amine substrate upon binding in porcine PAO has not been readdressed since the initial reports [245,293].

Arising from the belief that pyridoxal phosphate was the organic cofactor in amine oxidases, early mechanistic studies of amine oxidases focused on determining the intermediacy of Schiff base complexes. These studies involved the examination of nitrogen transfer from the substrate to the cofactor during the course of enzyme reduction. It has been reported [287,295] that NADH is oxidized under aerobic conditions in the presence of a NADH/oxoglutarate/glutamate dehydrogenase assay system with porcine PAO and bovine PAO. It was concluded that glutamate was formed from ammonia released in these assays. Later, Lindstrom ct al., reported that the apparent ammonia release neither required the presence of catalytically active porcine PAO, nor led to any consumption of amine substrate [296]. This discrepancy was ultimately resolved for porcine PAO in experiments by Ruis et al. in which the time course for ammonia release was shown to correlate with enzyme reoxidation and not with aldehyde release [297]. It was also discovered that anaerobically putrescine-reduced lentil seedling DAO releases ammonia only after exposure to O_2 [41]. A more recent study with bovine PAO has shown the absence of anaerobic ammonia release [80]. Product inhibition studies in steady-state experiments indicated that the order of product release from human placental DAO [57] and porcine DAO [289] is aldehyde, H_2O_2, and NH_4^+, which supports the notion that ammonia is not released from the reduced enzyme.

To date, no one has specifically addressed the possibility of ammonia release from the reoxidized enzyme at the end of the first cycle, before another molecule of substrate is bound, or that the substrate binds to the iminoquinone form of TQ after the first cycle in steady-state. Iminoquinones have higher redox potentials than the corresponding quinones, thus making the iminoquinone more reactive (i.e., more electrophilic). This aspect of the mechanism is particularly important when comparing results from steady-state and single-turnover (i.e., stopped-flow) experiments. If ammonia is released from a fraction of the oxidase at the end of the first cycle, some enzyme would contain TQ and some topa iminoquinone. These two forms of the enzyme could have different steady-state parameters, which might explain the curved Eadie-Hofstee plot obtained for the oxidation of benzylamine or *p*-methylbenzylamine by bovine PAO [291]. The data in Ref. 297 indicate that the apparent first-order rate constant for substrate binding/reduction is not that much larger than the rate constant for release of ammonia from porcine PAO during turnover. Thus, this scenario may be a required feature for the steady-state mechanism of these oxidases.

A direct approach for the detection of Schiff's base in bovine PAO was pursued by Hartmann and Klinman [298]. Primary/secondary α-deuterium kinetic isotope effects with bovine PAO were found to be quite large for the reduction of cofactor by benzylamines, phenethylamines, and the corresponding [α,α-^{2}H]-analogs (k_H/k_D = 13–15) [299], indicating the accumulation of the E·S complex prior to a rate-limiting bond cleavage step. It was anticipated that reduction of the enzyme with $NaCNBH_3$ (a specific reductant of Schiff's base complexes) in the presence of substrate would result in the trapping of the putative Schiff's base. Indeed, incubation with high concentrations of substrate led to irreversible and complete loss of enzyme activity. Concomitant with enzyme inactivation, [^{14}C]-substrate was found to be stoichiometrically incorporated into the enzyme. Analogous to earlier experiments with [^{3}H]-$NaBH_4$, reduction of the enzyme-substrate complex with [^{3}H]-$NaCNBH_3$ failed to give rise to labeling of the protein. This result, combined with incorporation of ^{14}C into the enzyme, ruled out the possibility of labeling of the enzyme by nonspecific interaction of product aldehyde with lysyl side chains.

This reductive trapping method was repeated for MAOx isolated from *Arthrobacter* P1, with similar results to those described above for the bovine enzyme [300]. These results led to the proposal of a transamination mechanism for substrate oxidation, facilitated by the quinone cofactor. The results of the double labeling experiments are fully consistent with the structure of TQ proposed by Janes et al. [6].

Stopped-flow methods have also been used to study the oxidation of substrates both aerobically and anaerobically. In 1973, Lindstrom et al. monitored the rate of oxidation of benzylamine by porcine PAO at 470 nm, a wavelength at which the enzyme exhibits a maximum in the oxidized form [263]. They reported a second-order rate constant of 1 sec^{-1} mM^{-1} for bleaching at 470 nm. The kinetic data were interpreted to indicate that substrate binding to porcine PAO was practically irreversible. Under aerobic conditions, addition of substrate to the enzyme led to biphasic changes in absorbance at 470 nm. The initial rapid decrease in absorbance was attributed to the formation of the $E_{red} \cdot P_{imine}$ complex, followed by a slower increase in absorbance, indicating a return of the enzyme to the oxidized state. Lindstrom and Pettersson later went on to monitor peroxide formation in the stopped-flow and found that peroxide dissociated in the oxygen-dependent step and was associated with the reappearance of the 470 nm absorption band [296].

It was reported that during the anaerobic reduction of porcine PAO with benzylamine, biphasic absorbance changes were observed at 360 nm concomitant with a single exponential decrease in absorbance at 470 nm [301]. However, when *p*-methoxybenzylamine was used as the substrate, biphasic absorbance changes at 470 nm were observed. Use of [α,α-^{2}H]-benzylamine significantly decreased the apparent accumulation at 360 nm. These authors concluded that the intermediate observed at 360 nm with benzylamine as substrate was the $E_{ox} \cdot S$ complex (Eq. 1) as was the intermediate at 470 nm observed with *p*-methoxybenzylamine.

In a recent rapid scanning stopped-flow study, spectral intermediates were characterized during the course of anaerobic reduction of bovine PAO with benzylamine, *p*-hydroxy-, and *p*-methoxybenzylamines [302]. Benzylamine caused transient formation of an intermediate at 340 nm and appearance of a spectral species absorbing at 310 nm formed at a rate corresponding to the rate of decay at 480 nm. *p*-Methoxybenzylamine produced the same spectral species at 310 nm, a transient intermediate at 340 nm, and an additional transient species centered at 430 nm. [α,α-^{2}H]-4-Hydroxybenzylamine yielded a spectrum similar to that observed with the *p*-methoxy substrate except that the species at 440 nm appeared to have a much larger half-life ($t_{1/2} \approx 250$ msec) and that the intermediate at 340 nm was obscured by the formation of product aldehyde a (λ_{max} = 330 nm; [217]). These results were interpreted as follows: (1) since the rate of formation of the species at 310 nm closely corresponded to the rate of enzyme reduction monitored at 480 nm, this species was most probably E_{red}-NH_2 (topa amino-quinol); (2) the species at 340 nm that formed and then decayed at a rate comparable to enzyme reduction was attributed to the formation and disappearance of the E_{ox}·S complex; and (3) the species that absorbs at 430 nm (p-methoxy) and 440 nm (p-hydroxy) was tentatively identified as the formation of a quinonoid intermediate formed upon proton abstraction from the substrate by the enzyme during catalysis (Fig. 9).

Some of the earliest studies of amine oxidases focused upon the mode of C-H bond activation. Abeles and coworkers examined mechanism-based inactivation of bovine PAO by 2-propynylamine and 2-chloroallylamine [303]. They demonstrated that these compounds were oxidized with concomitant inactivation of the enzyme, but allylamine behaved merely as a substrate. Protection from inactivation was observed in the presence of benzylamine, and covalent modification of enzyme was established using tritiated 2-propynylamine. These results were interpreted as being due to proton abstraction leading to allene formation followed by addition of nucleophile from the enzyme to form the adduct. Subsequently, these investigators also showed that glycine esters with good leaving groups could act as suicide inactivators, but that other glycine esters were merely substrates [304]. Proton abstraction leading to ketene formation was again invoked as the mechanism of inactivation. An important outcome of this work was the discovery that, unlike the aforementioned inhibitors, β-chlorophenethylamine was found to be a good substrate, leading exclusively to HCl and phenylacetaldehyde as products [305]. This result, in conjunction with those described above, suggested a carbanionic intermediate, which gives rise to an elimination mechanism in the presence of a good leaving group at the β-carbon.

2-Carboxy-PQQ reacts with glycinamide to produce a cyclic adduct (an oxazole) [306]. In a similar manner, the *ortho*-quinonoid form of TQ in PAO could react with some glycine esters to form the corresponding inert oxazoles (Fig. 10), rather than the reactive ketene intermediates as proposed originally [304].

Structure-reactivity correlations have also been used to aid in the determination of the mechanism of C-H bond activation with amine oxidases. Substituent effects for porcine PAO were determined by stopped-flow kinetic measurements. The effect of the ring substituent on the rate of benzylamine oxidation were quite small

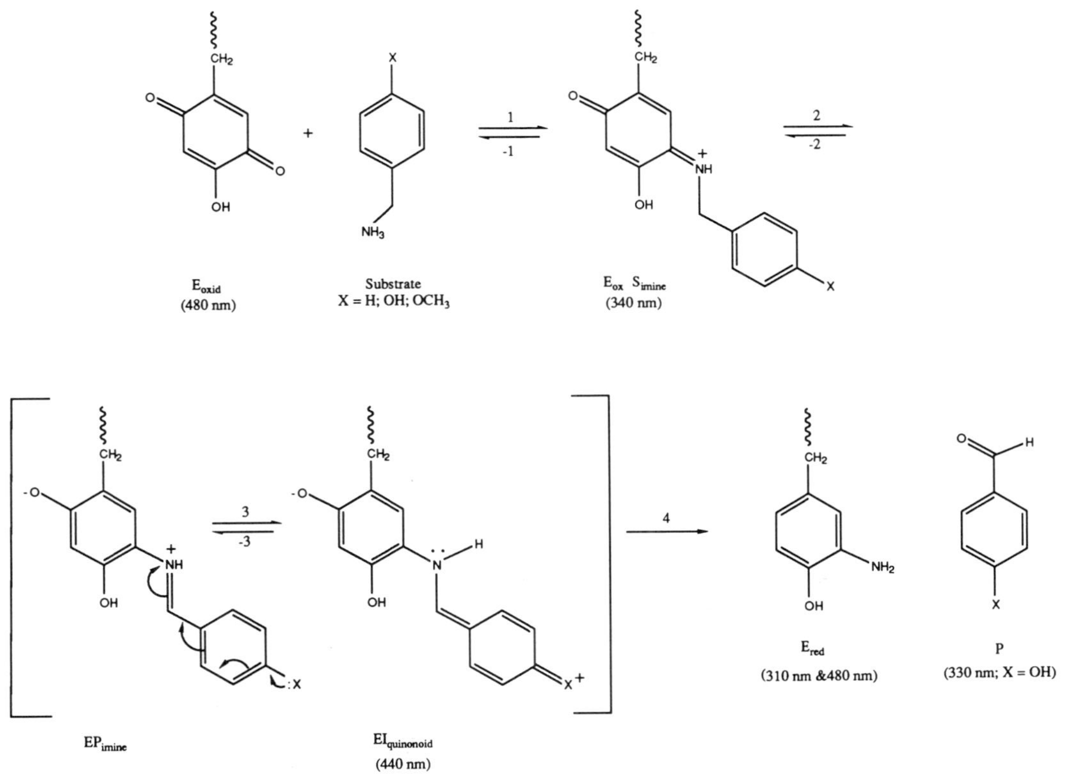

Figure 9 The postulated mechanism for the reaction of *p*-hydroxy- and *p*-methoxybenzylamine with copper-containing amine oxidases. The bracketed part shows the formation of a quinonoid species when X = OH or OCH_3.

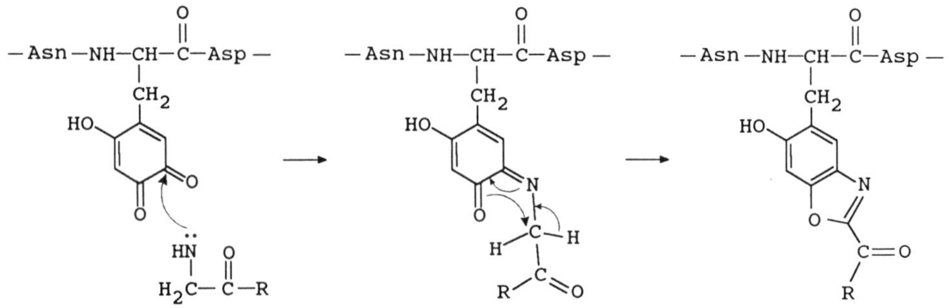

Figure 10 The postulated mechanism-based inhibition of copper-containing amine oxidases by glycine. The reaction of TQ with glycine produces an oxazole.

(ρ = 0.3), making distinction between homolytic and heterolytic mechanisms impossible [293]. This minor effect of structure on the observed rate may be a reflection of the limited range of substrates and/or only partial rate limitation by substrate oxidation since k_H/k_D = 2.8 with benzylamine and [α,α-^{2}H]-benzylamine [307].

A recent study with bovine PAO [308] combined structure-reactivity correlations with deuterium isotope effects, permitting direct calculation of microscopic rate constants for substrate binding and oxidation [299]. A wide range of *para*-substituted benzylamines was examined, allowing for a clear separation between electronic, hydrophobic, and steric effects. Results of double labeling experiments led to the proposal of a transamination mechanism for substrate oxidation, analogous to pyridoxal phosphate–dependent enzymes [298]. This transamination process was expected to be base catalyzed and lead to the formation of a resonance stabilized carbanionic intermediate. Multiple regression analysis of the data yielded an electronic effect of $\rho = 1.47 \pm 0.27$ for the bond cleavage step, providing the first direct evidence for a proton activation mechanism during the course of substrate oxidation, thus supporting the intermediacy of a carbanion species. An additional effect, determined from regression analysis, indicated inhibition of catalysis by hydrophobic substituents $\pi = 0.71 \pm 0.21$.

B. Stereochemistry of Substrate Oxidation

Lovenberg and Beaven [309] approached the question of C-H bond cleavage in another fashion while providing information concerning the stereochemical course of substrate oxidation. These authors reported that plasma amine oxidases, unlike the flavin-containing monoamine oxidases, catalyze the exchange of tritium from the β-carbon of phenylethylamines during substrate oxidation. The release of label from the β-carbon into solvent was ascribed to the reversible formation of an enamine off the main enzyme reaction pathway and provided a mechanistic distinction between the copper- and flavin-containing amine oxidases. This exchange process, rapid relative to other steps in the mechanism led to the conclusion that this stereospecific loss of hydrogen occurred from the imine intermediate (Eq. 2), prior to its hydrolysis, rather than from spontaneous exchange from the product [310,311]. The rapid exchange reaction contrasts with the fate of the hydrogen at the α-carbon, which has been proposed to be sequestered during the course of cofactor reduction [299]. These results, combined with the results of other observations, led to the proposal that bovine PAO functioned via a two-base mechanism which would allow for the active site residue catalyzing phenethylamine oxidation to remain protonated, while a second group catalyzed exchange from the β position [299].

Initially, it appeared that this proposal was a satisfactory explanation of the bovine PAO oxidation of amine substrates. However, subsequent pH-dependent isotope effect studies revealed identical pK_a values for exchange and substrate oxidation ($pK_a = 5.6 \pm 0.3$ and 5.5 ± 0.01, respectively), implying one residue performs both functions [294]. Support for a single catalytic residue catalyzing both oxidation and exchange comes from recent stereochemical probes which demonstrate a *syn*-relationship between the bonds cleaved at C(α) and C(β) of dopamine during the course of substrate oxidation and exchange [311]. The scheme in Fig. 9 integrates the most current available data with previous findings in the bovine PAO reaction. As shown, covalent attachment of amine substrate to the C(3)-carbonyl of active site TQ [6] leads to a Schiff's base intermediate, which can be trapped by reduction with $NaCNBH_3$ [298]. Base-catalyzed hydrogen abstraction leads to a carbanionic intermediate, with the resulting negative charge undergoing partial delocalization into the benzene ring of substrate and cofactor [308]. This delocalization is facilitated by electron-withdrawing substituents at the *para*-position of benzylamine substrates, leading to the formation of transiently formed quinonoid intermediates. Hydrolysis of the Schiff's base complex has been proposed to be facilitated by proton transfer from the active site base to the oxyanion at C(4) of the cofactor [308]. This proton transfer would allow the active site base to catalyze the observed tritium exchange reaction at the β-carbon of phenylethylamines [309] and to destabilize the product imine complex allowing for rapid hydrolysis to product [294]. Hydrolysis of the $E_{red} \cdot P_{imine}$ complex to product aldehyde results in the transfer of substrate nitrogen to the cofactor [80]. Although the reoxidation of the reduced cofactor has not been thoroughly characterized, it has been proposed to be facilitated by the active site copper. Consistent with distance-mapping experiments by Williams and Falk with porcine PAO [280], the active site copper is thought to be adjacent to the oxygen at C(6) of the cofactor.

In an elegant set of experiments, Palcic and coworkers have determined the stereochemical trends in copper amine oxidase reactions [312,313]. In general,

Table 3 Stereochemistry of Proton Abstraction and Solvent Exchange with Tyramine, Dopamine, or 2-Phenylethylamine Amine Oxidases

Enzyme	C(α) proton abstraction	C(β) solvent exchange	Ref.
PAO			
Porcine	pro-*R*	yes	312
Bovine	pro-*R*/pro-*S*	yes	310–312
Sheep	pro-*R*/pro-*S*	yes	313
Rabbit	pro-*R*/pro-*S*	yes	313
BAO			
Bovine aorta	pro-*S*	yes	[a]
Porcine aorta	pro-*S*	yes	[a]
DAO			
Porcine kidney	pro-*S*	no	313,315
Soybean seedling	pro-*S*	no	313
Chick pea seedling	pro-*S*	no	313
Pea seedling	pro-*S*	no	312,316–321

[a]Scaman, C. H., and Palcic, M. M., personal communication.

evolutionary pressures dictate the conservation of substrate geometry and the chemistry of transformation from substrate to product [314]. However, stereochemical uniformity in substrate oxidation is lacking in copper amine oxidases. Table 3 summarizes the current stereochemical trends of copper-containing amine oxidases.

Palcic et al. also investigated the solvent exchange properties of amine oxidases [312,313]. With dopamine and tyramine as substrates, they observed a strong correlation between the stereochemical course of proton abstraction at C(α) and the solvent exchange pathway into C(β) of products for all of the enzymes in Table 1. Since it was proposed that proton abstraction at C(α) and wash-in at C(β) appear to be catalyzed by a single active site base in bovine PAO [311], it will be interesting to see if the stereochemical relationship between these processes can be determined. The results shown in Table 3 indicate that the plant enzymes exhibit uniform stereochemistry but amine oxidases from mammalian sources do not. The source of the stereochemical trends is currently under more detailed investigation by Palcic and coworkers.

C. Proton Tunneling in the Bovine Serum Amine Oxidase Reaction

Early observations of large primary isotope effects exceeding the semiclassical limit [299] raised the possibility of hydrogen tunneling during the course of bovine PAO oxidation of benzylamine. In 1985, Saunders proposed that comparison of D/T and H/T isotope effects might provide a sensitive way to detect quantum mechanical tunneling in hydrogen transfer reactions [322] because the interdependence of mass (Swain-Schaad relation [323]) would break down under conditions where hydrogen tunneling might occur. Using benzylamine as the substrate, a systematic investigation of the steady-state primary and secondary D/T and H/T isotope effects on cofactor reduction was conducted, and the results are consistent with a single rate-limiting C-H bond cleavage step for V_{max}/K_m [299,324], and the secondary H/T and D/T kinetic effects correspond to those predicted by equilibrium isotope effects. However, the primary H/T and D/T isotope effects at 25°C, 35.2 ± 0.8 and 3.07 ± 0.07, respectively, were larger than semiclassical values and gave unusually low preexponential factor ratios, $A_H/A_T = 0.12 \pm 0.04$ and $A_D/A_T = 0.51 \pm 0.10$. While this type of Arrhenius behavior could arise in kinetically complex reactions, the combination of unusually large primary isotope effects and inverse isotope effects on Arrhenius preexponential factors could not be explained by semiclassical behavior. From these observations it was concluded that both protium and deuterium undergo significant tunneling during the course of substrate oxidation. This is the first example of quantum mechanical tunneling in an enzyme-catalyzed proton abstraction reaction.

D. Mechanism of Enzyme Reoxidation

It is generally agreed that copper is an essential cofactor of amine oxidases: there is evidence that copper plays a functional, rather than structural, role. The best evidence for this comes from addition of anions such as azide [325] or cyanide [251], which inhibit enzyme activity as well as modify the EPR spectrum. Chemical intuition would suggest that copper plays a redox role, therefore, the amine substrate might be expected to reduce Cu(II) to Cu(I), and O_2 would then restore the copper to its Cu(II) state. Numerous EPR experiments [86,261,269, 273,326], including rapid freeze-quench [327], failed to detect changes in the oxidation state of the copper in the presence of amine substrate. There has been one report that copper reduction might occur [328], but this has never been confirmed. Implication of copper in the reoxidation of substrate-reduced enzyme [46,249,329] was not supported by experimental evidence, which indicated no change in the oxidation state of copper. These results led to proposals that Cu(II) acts as a Lewis acid [330], that it has an indirect role in catalysis [267], or that copper serves a structural role [326].

As mentioned earlier, Dooley and coworkers have recently presented evidence for the generation of a Cu(I)-semiquinone state by substrate reduction of several amine oxidases under anaerobic conditions at room temperature [252]. EPR spectral changes accompanying the addition of amine to several amine oxidases under anaerobic conditions resulted in a relatively sharp signal at $g \sim 2$ signal (see Fig. 6) and show that the same radical was formed in different amine oxidases. It

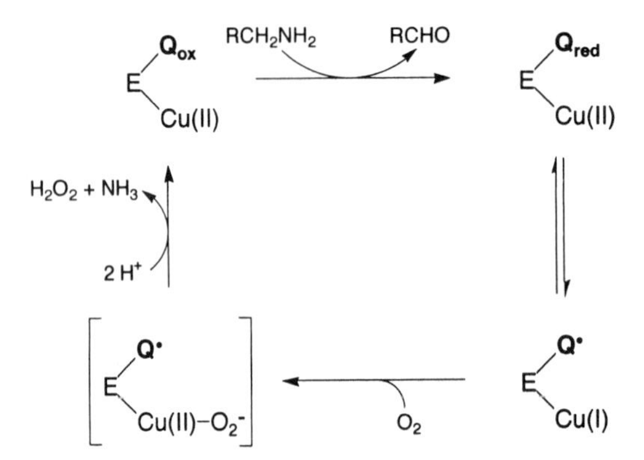

Figure 11 Possible catalytic cycle of copper-containing amine oxidases. The species in the brackets is hypothetical. Q_{ox} = TQ, $Q^{\cdot}$ = topa semiquinone, Q_{red} = topa aminoquinol. (Figure courtesy of Dr. David M. Dooley.)

was concluded that because the radical EPR spectrum was independent of the enzyme and substrate used, the radical must be associated with a moiety that is conserved among the amine oxidases included in the study. The assignment of the $g \sim 2$ signal to a semiquinone radical is supported from previous work by these same investigators [249,251]. The yield of the semiquinone in the absence of cyanide ranged from $<1\%$ of the theoretical maximum for porcine plasma amine oxidase to $\sim 20\%$ for pea seedling diamine oxidase and *Arthrobacter* P1 methylamine oxidase. These results led to the mechanism shown in Figure 11. Finally, evidence for a bound superoxide intermediate has been presented in several independent studies [256,331,332].

REFERENCES

1. Robertson, J. G., Kumar, A., Mancewicz, J. A., and Villafranca, J. J. (1989). Spectral studies of bovine dopamine β-hydroxylase. *J. Biol. Chem. 264*: 19916.

2a. Michaud-Soret, I., Daniel R., Chopard, C., Mansuy, D., Cucurou, C., Ullrich, V., and Chottard, J. C. (1990). Soybean lipoxygenase-1, -2a, -2b, and 2c do not contain PQQ. *Biochem. Biophys. Res. Commun. 172*: 1122.

2b. Veldink, G. A., Boelens, Maccarrone, M., van der Lecq, F., Vliegenthart, J. F. G., Paz, M. A., Flückiger, R., and Gallop, P. M. (1990). Soybean lipoxygenase-1 is not a quinoprotein. *FEBS Lett. 270*: 135.

3. De Biase, D., Maras, B., and John, R. A. (1991). A chromophore in glutamate dehydrogenase has been wrongly identified as PQQ. *FEBS Lett. 278*: 120.

4. McIntire, W. S., Wemmer, D. E., Chistoserdov, A., and Lidstrom, M. E. (1991). A new cofactor in a prokaryotic enzyme: tryptophan tryptophylquinone as the redox prosthetic group in methylamine dehydrogenase. *Science 252*: 817.

5. Ito, N., Phillips, S. E., Stevens, C., Ogel, Z. B., McPherson, M. J., Keen, J. N., Yadav, K. D. S., and Knowles, P. F. (1991). Novel thioether bond revealed by a 1.7 Å crystal structure of galactose oxidase. *Nature 353*: 87.
6. Janes, S. M., Mu, D., Wemmer, D., Smith, A. J., Kaur, S., Maltby, D., Burlingame, A. L., and Klinman, J. P. (1990). A new redox cofactor in eukaryotic enzymes: 6-Hydroxydopa at the active site of bovine serum amine oxidase. *Science 248*: 981.
7. van Iersel, J., van der Meer, R. A. and Duine, J. A. (1986). Methylamine oxidase from *Arthrobacter* P1. A bacterial copper-quinoprotein amine oxidase. *Eur. J. Biochem. 161*: 415.
8. McIntire, W. S. (1990). Methylamine oxidase from *Arthrobacter* P1. *Methods Enzymol. 188*: 227.
9. Shimizu, E., Ichise, H., and Yorifuju, T. (1990). Phenylethylamine oxidase, a novel enzyme of *Arthrobacter globiformis*, may be a quinoprotein. *Agric. Biol. Chem. 54*: 851.
10. Yamada, H., and Adachi, O. (1971). Amine oxidase (*Aspergillus niger*). *Methods Enzymol. 17B*: 705.
11. Kumagai, H., Kishimoto, N., and Yamada, H. (1978). Subunit structure of amine oxidase from *Aspergillus niger*. *Agric. Biol. Chem. 42*: 893.
12. Suzuki, H., Ogura, Y., Yamada, H., and Arima, K. (1975). Studies on sulfhydryl groups of *Aspergillus niger* amine oxidase. *Biochim. Biophys. Acta 430*: 23.
13. Suzuki, H., Ogura, Y., and Yamada, H. (1971). Stoichiometry of the reaction by amine oxidase from *Aspergillus niger*. *J. Biochem. 69*: 1065.
14. Yamada, H., Adachi, O., and Ogata, K. (1965). Amine oxidases of microorganisms. Part II. Purification and crystallization of amine oxidase of *Aspergillus niger*. *Agric. Biol. Chem. 29*: 649.
15. Yamada, H., Adachi, O., and Ogata, K. (1965). Amine oxidases of microorganisms. Part VII. An improved purification procedure and further properties of amine oxidase of *Aspergillus niger*. *Agric. Biol. Chem. 33*: 1707.
16. Haywood, G. W., and Large, P. J. (1981). Microbial oxidation of amines. Distribution, purification and properties of two primary-amine oxidases from the yeast *Candida boidinii* grown on amines as sole nitrogen source. *Biochem. J. 199*: 187.
17. Large, P. J., and Haywood, G. W. (1990). Amine oxidases from methylotrophic yeast. *Methods Enzymol. 188*: 427.
18a. Green, J., Haywood, G. W., and Large, P. J. (1983). Serological differences between the multiple amine oxidases of yeast and comparison of the specificities of the purified enzymes from *Candida utilis* and *Pichia pastoris*. *Biochem. J. 211*: 481.
18b. Zwart, K. B., and Harder, W. (1983). Regulation of metabolism of some alkylated amines in the yeast *Candida utilis* and *Hansanula polymorpha*. *J. Gen. Microbiol. 129*: 3157.
19. Bruinenberg, P. G., Evers, M., Waterham, H. R., Kuipers, J., Arnberg, A. C., and Greet, A. B. (1989). Cloning and sequencing of the peroxisomal amine oxidase gene from *Hansanula polymorpha*. *Biochim. Biophys. Acta 1008*: 157.
20. Sherlock, L. A., and Large, P. J. (1985). Control of benzylamine oxidase activity in *Kluyvermyces fragalis* grown on primary amines as nitrogen source. *FEMS Microbiol. Lett. 29*: 99.

21. Ueda, T., Ueda, T., Oguchi, K., Nishiwaki, A., and Yasuhara, H. (1985). Enzyme characterization of amine oxidase in *Lyophyllum aggregatum* KUHNEN ("Hon-shimeji"). *Chem. Pharm. Bull. 33*: 4957.
22. Kumagai, H., and Yamada, H. (1985). Bacterial and fungal amine oxidases. In *Structure and Function of Amine Oxidases* (Mondovi, B., ed.), CRC Press, Boca Raton, FL, p. 37.
23. Isobe, K., Tani, Y., and Yamada, H. (1982). Crystallization and characterization of agmatine oxidase form *Penicillum chrysogenum. Agric. Biol. Chem. 46*: 1353.
24. Tur, S. S., and Lerch, K. (1988). Unprecedented lysyl oxidase activity of *Pichia pastoria* benzylamine oxidase. *FEBS Lett. 238*: 74.
25. Large, P. J., and Sherlock, L. A. (1987). Characterization of the amine oxidase involved in the growth of *Trichosporon cutaneum* X_4 on ethylamine as a source of carbon, nitrogen and energy. *Arch. Microbiol. 147*: 64.
26. Yamada, H., Adachi, O., and Ogata, K. (1965). Amine oxidases of microorganisms. Part I. Formation of amine oxidase by fungi. *Agric. Biol. Chem. 29*: 117.
27. Tsushida, T., and Takeo, T. (1985). Purification and some properties of tea leaf amine oxidase. *Agric. Biol. Chem. 49*: 319.
28. Scorer, K. N., Caamal, R. M., Oropeza, C., and Loyola-Vargas, V. M. (1989). Amine oxidase activity in *Canavalia ensiformis* during early growth: A histochemical study. *J. Plant Physiol. 135*: 246.
29. Sindhu, R. K., and Desai, H. V. (1980). Partial purification and characterization of diamine oxidase from groundnut embryo. *Ind. J. Biochem. Biophys. 17*: 194.
30. Angelini, R., Di Lisi, F., and Federico, R. (1985). Immunoaffinity purification of diamine oxidase from Cicer. *Phytochemistry 24*: 2511.
31. Percival, F. W., and Purves, W. K. (1974). Multiple amine oxidases in cucumber seedlings. *Plant Physiol. 54*: 601.
32. Rinaldi, A., Floris, G., and Finazzi-Argo, A. (1982). Purification and properties of diamine oxidase from *Euphorbia* latex. *Eur. J. Biochem. 127*: 417.
33. Nikolov, I., Pavlov, V., Minkov, I., and Damjanov, D. (1990). Purification and some properties of an amine oxidase from soybean seedlings. *Experientia 46*: 765.
34. Suzuki, Y. (1973). Some properties of the amine oxidase of soybean seedlings. *Plant and Cell Physiol. 14*: 413.
35. Cogoni, A., Piras, C., Farci, R., Antonelli, M., and Floris, G. (1990). *Hordeum vulgare* seedling amine oxidase. *Plant Physiol. 93*: 818.
36. Hashimoto, T., Mitani, A., and Yamada, Y. (1990). Diamine oxidase from cultured roots of *Hyoscyamus niger*. Its function in tropane alkaloid biosynthesis. *Plant Physiol. 93*: 216.
37. Cogoni, A., Piras, C., Medda, R., Rinaldi, A., and Floris, G. (1989). Amine oxidase from *Lathyrus cicera* and *Phaseolus vulgaris*: Purification and properties. *Prep. Biochem. 19*: 95.
38a. Suresh, M. R., Ramakrishna, S., and Adiga, P. R. (1976). Diamine oxidase of *Lathyrus sativus* seedlings. *Phytochem. 15*: 483.
38b. Suresh, M. R., and Adiga, P. R. (1979). Diamine oxidase of *Lathyrus sativus* seedlings. Purification and properties. *J. Biosci. 1*: 109.

39. Floris, G., Giartosio, A., and Rinaldi, A. (1983). Diamine oxidase from *Lens esculenta* seedlings: Purification and properties. *Phytochemistry 22*: 1971.
40. Oratore, A., D'Andrea, G., Floris, G., Rinaldi, A., and Finazzi-Argo, A. (1987). Microheterogeneity of lentil seedling (*Lens culinaris*) amine oxidase. *Ital. J. Biochem. 36*: 92.
41. Rinaldi, A., Giartosio, A., Floris, G., Medda, R., and Finazzi-Argo, A. (1984). Lentil seedling amine oxidase: Preparation and properties of the copper-free enzyme. *Biochem. Biophys. Res. Commun. 120*: 242.
42. Schütte, H. R., Knöfel, D., and Heyer, O. (1966). Biosynthesis of lupine alkaloids. Enrichment of diamine oxidase from *Lupinus luteus* seedlings. *Z. Pflanzenphysiol. 55*: 110.
43. Davies, H. M., Hawkins, D. J., and Smith, L. A. (1989). Quinoprotein characterization of *N*-putrescine oxidase from tobacco roots. *Phytochemistry 28*: 1573.
44. Mizusaki, S., Tanabe, Y., Noguchi, M., and Tamaki, E. (1972). N-methylputrascine oxidase from tobacco roots. *Phytochemistry 11*: 2757.
45. Pec, P., Zajoncová, L., and Jilek, M. (1989). Diamine oxidase from sainfoin (*Onobrychis viciifolia*). Purification and some kinetic properties. *Biologia (Bratsava) 44*: 1177.
46. Hill, J. M., and Mann, P. J. G. (1964). Further properties of the diamine oxidase of pea seedling. *J. Biochem. 91*: 171.
47. McGowen, R. E., and Muir, R. M. (1971). Purification and properties of amine oxidase from epicotyls of *Pisum sativum*. *Plant Physiol. 47*: 644.
48. Macholán, L., and Haubrová, J. (1976). Isolation and some characteristics of diamine oxidase from etiolated pea seedlings. *Coll. Czech. Chem. Commun. 41*: 2987.
49. Nylen, U., and Szybek, P. (1974). Kinetics and other characteristics of diamine oxidase from pea seedlings. *Acta Chem. Scand. B 28*: 1153.
50. Akira, H., Tsuchiya, M., and Akatsuka, T. (1983). Purification and properties of diamine oxidase from epicotyls of *Pisum sativum*. *Sci. Rep. Fac. Agricul. Ibaraki Univ.* (1983): 47. Biol. Abstr., 77: Abstr. #54995.
51. Kluetz, M. D., Adamson, K., and Flynn, J. E. (1980). Optimized preparation and determination of pea seedling diamine oxidase. *Prep. Biochem. 10*: 615.
52. Yanagisawa, H., Hirasawa, E., and Suzuki, Y. (1981). Purification and properties of diamine oxidase from pea epicotyls. *Phytochem. 20*: 2105.
53. Delhaize, E., and Webb, J. (1987). Purification and characterization of diamine oxidase from clover leaves. *Phytochem. 26*: 641.
54. Matsuda, H., and Suzuki, Y. (1981). Purification and properties of the diamine oxidase from *Vicia faba* leaves. *Plant Cell Physiol. 22*: 737.
55. Matsuda, H., and Suzuki, Y. (1977). Some properties of the amine oxidase in *Vica fava* seedlings. *Plant Cell Physiol. 18*: 1131.
56a. McEwen, C. M. (1965). Human plasma amine oxidase. I. Purification and identification, *J. Biol. Chem. 240*: 2003.
56b. McEwen, C. M. (1971). Monoamine oxidase (human serum or plasma). *Methods Enzymol. 17B*: 692.

56c. McEwen, C. M. (1972). The soluble monoamine oxidase of human plasma and sera. *Adv. Biochem. Pharm. 5*: 151.
57. Bardsley, W. G., Crabbe, M. J. C., and Scott, I. V. (1974). The amine oxidase of human placenta and pregnancy plasma. *Biochem. J. 139*: 169.
58. Kapellier-Alder, R. (1965). Recent attempts at the purification and identification of human placental "histaminase." *Clin. Chim. Acta 11*: 191.
59. Smith, J. K. (1967). The purification and properties of placental histaminase. *Biochem. J. 103*: 110.
60. Paolucci, F., Cronenberger, L., Plan, R., and Pacheco, H. (1971). Purification and properties of diamine oxidase: Oxygen oxido-reductase from human placenta. *Biochimie 53*: 735.
61. Crabbe, M. J. C., Waight, R. D., Bardsley, W. G., Barker, R. W., Kelly, I. A., and Knowles, P. F. (1976). Human placental diamine oxidase. Improved purification and characterization of a copper- and manganese-containing amine oxidase with novel substrate specificity. *Biochem. J. 155*: 679.
62. Hata, A. (1976). Purification and properties of diamine oxidase from pig kidney and human placenta. *Bull. Tokyo Med. Den. Univ. 23*: 63.
63. Lin, C.-W., Kirley, S. D., and St. Pierre, M. (1981). Tumor and placental histaminase. I. Affinity chromatography purification and characterization of the placental enzyme. *Ocodevelopmental Biol. Med. 2*: 267.
64. Hölltä, E., Pulkkinen, P., Elfving, K., and Jänne, J. (1975). Oxidation of polyamines by diamine oxidase from human semenal plasma. *Biochem. J. 145*: 373.
65. Shindler, J. S., and Bardsley, W. G. (1976). Human kidney diamine oxidase: Inhibition studies. *Biochem. Pharmacol. 25*: 2689.
66. Bieganski, T., Kushe, J., Lorenz, W., Hesterberg, R., Stahlnecht, C.-D., and Feusser, K.-D. (1983). Distribution and properties of human intestinal diamine oxidase and its relevance for the histamine catabolism. *Biochim. Biophys. Acta 756*: 196.
67. Bambardieri, G., Milani, A., and Rossi, L. (1985). Copper-dependent amine oxidases; clinical aspects. In *Structure and Function of Amine Oxidases* (Mondovi, B., ed.), CRC Press, Boca Raton, FL, p. 195.
68. Lewinsohn, R. (1990). Benzylamine-oxidizing enzymes in human pregnancy fluids and pharyngeal aspirate of the newborn: nature and origin. *Biogenic Amines 7*: 651.
69. White-Tabor, C., Tabor, H., and Rosenthal, S. M. (1954). Purification of amine oxidase from beef plasma. *J. Biol. Chem. 208*: 645.
70a. Yamada, H., and Yasunobu, K. (1962). Monoamine oxidase. I. Purification, crystallization and properties of plasma monoamine oxidase. *J. Biol. Chem. 237*: 1511.
70b. Yamada, H., and Yasunobu, K. (1963). Monoamine oxidase. IV. Nature of the second prosthetic group of plasma amine oxidase. *J. Biol. Chem. 238*: 2669.
71. Achee, F. M., Chervenka, C. H., Smith, R. A., and Yasunobu, K. T. (1968). Monoamine oxidase. XII. The association and dissociation, and number of subunits of beef plasma amine oxidase. *Biochemistry 7*: 4329.
72. Watanabe, K., and Yasunobu, K. (1970). Carbohydrate content of bovine plasma amine oxidase and isolation of a carbohydrate-containing fragment attached to asparagine. *J. Biol. Chem. 245*: 4612.

73. Yasunobu, K., and Smith, R. A. (1971). Amine oxidase (beef plasma). *Methods Enzymol. 17B*: 698.
74. Watanabe, K., Smith, R. A., Inamasu, M., and Yasunobu, K. (1972). Recent investigations on the prosthetic group of beef plasma amine oxidase. *Adv. Biochem. Pyschopharmacol. 5*: 107.
75. Yamada, H., Gee, P., Ebata, M., and Yasunobu, K. T. (1964). Monoamine oxidase. VI. Physicochemical properties of plasma monoamine oxidase. *Biochim. Biophys. Acta 81*: 165.
76a. Turini, P., Sabatini, S. Befani, O., Chimenti, F., Casanova, C., Riccio, P. L., and Mondovi, B. (1982). Purification of bovine plasma amine oxidase. *Anal. Biochem. 125*: 294.
76b. Mondovi, B., Turini, P., Befani, O., and Sabatini, S. (1983). Purification of bovine plasma amine oxidase. *Methods Enzymol. 94*: 314.
77. Mondovi, B., Sabantini, S., and Befani, O. (1984). The active site of copper amine oxidases. *J. Mol. Catal. 23*: 325.
78. Rinaldi, A., Floris, G., Sabatini, S., Finazzi-Argo, A., Giartosio, A., Rotilio, G., and Mondovi, B. (1983). Reaction of beef plasma amine and lentil seedling Cu-amine oxidases with phenylhydrazine. *Biochem. Biophys. Res. Commun. 115*: 841.
79. Suzui, S., Sakarai, T., Nakahara, Manabe, T., and Okuyama, T. (1983). Effects of metal substitution on the chromophore of bovine serum amine oxidase. *Biochemistry 22*: 1630.
80. Janes, S. M., and Klinman, J. P. (1991). An investigation of bovine serum amine oxidase active site stoichiometry: Evidence for an aminotransferase mechanism involving two carbonyl cofactors per enzyme dimer. *Biochemistry 30*: 4599.
81. Nakano, G., Harada, M., and Nagatsu, T. (1974). Purification and properties of an amine oxidase in bovine dental pulp and its comparison with serum amine oxidase. *Biochim. Biophys. Acta 341*: 366.
82. Megrabyan, Z. B., Gulkhandanyan, C. V., and Nalbandy, R. M. (1981). Soluble benzylamine oxidase from aorta. *Biokhimiya 46*: 612.
83. Fernandez de Arriba, A., Balsa, D., Tipton, K. F., and Unzeta, M. (1990). Monoamine oxidase and semicarbazide-sensitive amine oxidase activities in bovine eye. *J. Neural Transm.* [Suppl.] *32*: 327.
84. Lizcano, J. M., Balsa, D., Tipton, K. F., and Unaeta, M. (1990). Amine oxidase activities in bovine lung. *J. Neural Transm.* [Suppl] *32*: 341.
85a. Mondovi, B., Rotilio, G., Finazzi, A., and Scioscia-Santoro, A. (1964). Purification of pig kidney diamine oxidase and its identity with histaminase. *Biochem. J. 91*: 408.
85b. Mondovi, B., Rotilio, G., Costa, M. T., and Finazzi-Argo, A. 1971. Diamine oxidase (pig kidney). *Methods Enzymol. 17B*: 735.
86. Mondovi, B., Rotilio, G., Costa, M. T., Finazzi-Argo, A., Chiancone, E., Hansen, R. E., and Beinert, H. (1967). Diamine oxidase from pig kidney. Improved purification and properties. *J. Biol. Chem. 242*: 1160.
87. Yamada, H., Kumagai, H., Kawasaki, H., Matsui, H., and Ogata, K. (1967). Crystallization and properties of diamine oxidase from pig kidney. *Biochem. Biophys. Res. Commun. 29*: 723.

88. Swedin, B., Mossle, I., and Jörnvall, H. (1976). Studies of pig kidney diamine oxidase. *Acta Chem. Scand. B30*: 970.
89. Kluetz, M. D., and Schmidt, P. G. (1977). Diamine oxidase: Molecular weight and subunit analysis. *Biochem. Biophys. Res. Commun. 76*: 40.
90a. Buffoni, F., and Blaschko, H. (1964). Benzylamine oxidase and histaminase: purification and crystallization of an enzyme from pig plasma. *Proc. Roy. Soc. B. 161*: 153.
90b. Buffoni, F., and Blaschko, H. (1971). Amine oxidase (pig plasma). *Methods Enzymol. 17B*: 682.
91. Buffoni, F., and Igesti, I. (1975). Active-site titration of pig plasma benzylamine oxidase with phenylhydrazine. *Biochem. J. 145*: 369.
92. Lindstrom, A., and Petterson, G. (1973). Active-site titration of pig plasma benzylamine oxidase with hydrazine derivatives. *Eur. J. Biochem. 34*: 564.
93. Lindstrom, A., and Petterson, G. (1978). Active-site titration of pig plasma benzylamine oxidase. *Eur. J. Biochem. 83*: 131.
94. Barker, R., Boden, N., Cayley, G., Charlton, S. C., Henson, R., Holmes, M. C., Kelly, I. D., and Knowles, P. F. (1979). Properties of cupric ions in benzylamine oxidase from pig plasma as studied by magnetic-resonance and kinetic methods. *Biochem. J. 177*: 289.
95. Falk, M. C. (1983). Stoichiometry of phenylhydrazine inactivation of pig plasma amine oxidase. *Biochemistry 23*: 3740.
96. Falk, M. C., Staton, A. J., and Williams, T. J. (1983). Heterogeneity of pig plasma amine oxidase: Molecular and catalytic properties of chromatographically isolated forms. *Biochemistry 22*: 3746.
97a. Hysmith, R. M., and Boor, P. J. (1987). In vitro expression of benzylamine oxidase activity in cultured porcine smooth muscle cells. *J. Cardiovasc. Pharmacol. 9*: 668.
97b. Hysmith, R. M., and Boor, P. J. (1988). Purification of benzylamine oxidase from cultured porcine aortic smooth muscle cells. *Biochem. Cell Biol. 66*: 821.
98. Norquist, A., and Oreland, L. (1988). Amine oxidases in dental pulp—A study with special regard to a semicarbazide-sensitive amine oxidase with catalytic potency toward seratonin. *Pharmacol. Res. Commun. 20*, [Suppl. IV]: 169.
99. Corda, M., Pellegrini, M., and Rinaldi, A. (1984). Diamine oxidase from horse kidney: ionic strength dependence and activity. *Ital. J. Biochem. 33*: 303.
100. Rucker, R. B., and Goettlich-Riemann, W. (1972). Purification and properties of sheep plasma amine oxidase. *Enzymologia 43*: 33.
101a. McEwen, C. M., Cullen, K. T., and Sober, A. J. (1966). Rabbit serum monoamine oxidase: purification and characterization. *J. Biol. Chem. 241*: 4544.
101b. McEwen, C. M. (1971). Monoamine oxidase (rabbit serum). *Methods Enzymol. 17B*: 686.
102. Buffoni, F., Banchelli, G., Bertocci, B., and Raimondi, L. (1989). Effects of pyridoxamine on semicarbazide-sensitive amine oxidase activity of rabbit lung and heart. *J. Pharm. Pharmacol. 41*: 469.
103. Argento-Cerù, M. P., Sartori, C., and Autori, F. (1973). Diamine oxidase in rabbit liver. 1. Assay conditions and subcellular localization. *Eur. J. Biochem. 34*: 369.

104. Sartori, C., Bargagli, A. M., and Argento-Cerù, M. P. (1983). Subcellular localization of diamine oxidase in rabbit kidney cortex. *Biochim. Biophys. Acta 758*: 58.
105. Barrand, M. A., and Callingham, B. A. (1984). Solubilization and some properties of a semicarbazide-sensitive amine oxidase in brown adipose tissue of rat. *Biochem. J. 222*: 467.
106. Raimondi, L., Pirisino, R., Ignesti, G., Capecchi, S., Banchelli, G., and Buffoni, F. (1991). Semicarbazide-sensitive amine oxidase (SSAO) activity of rat white adipose tissue. *Biochem. Pharmacol. 41*: 467.
107. Ignesti, G., Banchelli, G., Falai, C., Pirisino, R., Raimondi, L., and Buffoni, F. (1983). Characterization of the amine oxidase activities of liver microsomes of different vertebrate and invertebrate species. *Biochemistry 32*: 3653.
108. Coquil, J. F., Goridis, C., Mack, G., and Neff, N. H. (1973). Monoamine oxidase in rat arteries: Evidence for different forms and selective location. *Br. J. Pharm. 48*: 590.
109a. Andree, T. H., and Clarke, D. E. (1982). Characteristics of rat skull benzylamine oxidase. *Proc. Soc. Exp. Biol. Med. 171*: 298.
109b. Clarke, D. E., Lyles, G. A., and Callingham, B. A. (1982). A comparison of cardiac and vascular clorgyline-resistant amine oxidase and monoamine oxidase. *Biochem. Pharmacol. 31*: 27.
109c. Arai, Y., Toyoshima, Y., and Kinemuchi, H. (1986). Studies of monoamine oxidase and semicarbazide-sensitive amine oxidase. I. Inhibition by a selective monoamine oxidase-B inhibitor, MD 78236. *Japan. J. Pharmacol. 41*: 183.
110. Lyles, G. A., and Singh, I. (1985). Vascular smooth muscle cells: A major source of the semicarbazide-sensitive amine oxidase of the rat aorta. *J. Pharm. Pharmacol. 37*: 637.
111. Backus, B., and Kim, K. S. (1970). Subcellular distribution of diamine oxidase. *Comp. Gen. Physiol. 1*: 196.
112. Almeida, A. P., Flye, W., Deveraux, D., Horakova, Z., and Beaven, M. A. (1980). Distribution of histaminase (diamine oxidase) in blood of various species. *Comp. Biochem. Physiol. 67C*: 187.
113. Hansson, R., and Thysell, H. (1971). Heparin-induced increase of diamine oxidase in lymph and blood of rat, guinea pig and rabbit. *Acta Physiol. Scand. 81*: 208.
114. Hampton, J. K., and Parmelee, M. L. (1969). Plasma diamine oxidase activity in humans and marmosets. *Comp. Biochem. Physiol. 30*: 367.
115. Matsuyima, M., and Otake, S. (1986). Purification and properties of diamine oxidase in the pyloric ceca of yellowtail *Seriola quinqueradiata*. *Bull. Japan. Soc. Sci. Fish 52*: 1617.
116. Matsuyima, M., and Otake, S. (1983). On the diamine oxidase in the pyloric ceca and the intestine of common mackerel *Scomber japonicus*. *Bull. Japan. Soc. Sci. Fish 49*: 1695.
117a. Holstein, B. (1975). Intestinal diamine oxidase of some teleostean fishes. *Comp. Biochem. Physiol. 50B*: 291.
117b. Matsuyima, M., and Otake, S. (1984). Comparison of the characteristics of the diamine oxidase in the digestive organs of several teleostean fish. *Bull. Japan. Soc. Sci. Fish 50*: 1261.

117c. Matsuyima, M., and Otake, S. (1985). Effects of feeding conditions on the diamine oxidase activity of several teleostean fishes. *Bull. Japan. Soc. Sci. Fish 51*: 272.
118. Kumazawa, T., Seno, H., and Suzuki, O. (1989). Semicarbazide-sensitive amine oxidase activity in catfish tissue. *Comp. Biochem. Physiol. 92B*: 347.
119. Cogoni, A., Farci, R., Cau, A., Melis, A., Medda, R., and Floris, G. (1990). Mullet plasma: Serum amine oxidase and ceruloplasmin: Purification and properties. *Comp. Biochem. Physiol. 95C*: 297.
120. Mondovi, B., and Riccio, P. (1985). Animal intracellular amine oxidases. In *Structure and Function of Amine Oxidases* (Mondovi, B., ed.), CRC Press, Boca Raton, FL, p. 63.
121a. Callingham, B. A., and Barrand, M. A. (1987). Some properties of semicarbazide-sensitive amine oxidases. *J. Neural. Transm.* [Suppl.] *23*: 37.
121b. Callingham, B. A., Holt. A., and Elliot, J. (1990). Some aspects of the pharmacology of semicarbazide-sensitive amine oxidases. *J. Neural. Transm.* [Suppl.] *32*: 279.
122. Weyler, W., Hsu, Y.-P., and Breakfield, X.O. (1990). Biochemistry and genetics of monoamine oxidase. *Pharmac. Ther. 47*: 391.
123. Tabor, C. W., and Tabor, H. (1984). Polyamines. *Ann. Rev. Biochem. 53*: 749.
124. Morgan, D. M. L. (1985). Polyamine oxidases. *Biochem. Soc. Trans. 13*: 322.
125. Boadle, M. C., and Blaschko, H. (1968). Cockroach amine oxidase: Classification and substrate specificity. *Comp. Biochem. Physiol. 25*: 129.
126. Pegg, A. E. (1986). Recent advances in the biochemistry of polyamines in eukaryotes. *Biochem. J. 234*: 249.
127. Seiler, N. (1987). Inhibition of enzymes oxidizing polyamines. In *Inhibition of Polyamine Metabolism* (McCann, P. P., Pegg, A. E., and Sjoerdsma, A., eds.), Academic Press, New York, p. 79.
128. Seiler, N. (1990). Polyamine metabolism. *Digestion* [Suppl. 2] *46*: 319.
129. Smith, T. A. (1985). The di- and polyamine oxidases of higher plants. *Biochem. Trans. 13*: 319.
130. Federico, R., Cona, A., Angelini, R., Schininà, M. E., and Giartosio, A. (1990). Characterization of maize polyamine oxidase. *Phytochem. 29*: 2411.
131. Smith, T. A. (1985). Polyamines. *Ann. Rev. Plant. Physiol. 36*: 117.
132. Haywood, G. W., and Large, P. J. (1984). Partial purification of a peroxisomal polyamine oxidase from *Candida boidinii* and its role in growth on spermidine as sole nitrogen source. *J. Gen. Microbiol. 130*: 1123.
133. Yamada, H. (1971). Putrescine oxidase (*Micrococcus rubens*). *Methods Enzymol. 17B*: 726.
134. Yamada, H., Kumagai, H., and Uwajima, T. (1971). Tyramine oxidase (*Sacina lutea*). *Methods Enzymol. 17B*: 722.
135. Parrott, S., Jones, S., and Cooper, R. A. (1987). 2-Phenylethylamine catabolism by *Escherichia coli* K12. *J. Gen. Microbiol. 133*: 347.
136a. Horiuchi, T., and Kurokawa, T. (1991). Purification and properties of fructosylamine oxidase from *Asperigillus* sp. 1005, *Agric. Biol. Chem. 55*: 333.
136b. Horiuchi, T., Kurokawa, T., and Saito, N. (1989). Purification and properties of fructosyl-amino acid oxidase from *Cornynebacterium* sp. 2-4-1. *Agric. Biol. Chem. 53*: 103.

136c. Watanabe, N., Ohtsuta, M., Takahashi, S.-I., Sakano, Y., and Fujimoto, D. (1990). Enzymatic deglycation of fructosyl-lysine. *Agric. Biol. Chem. 54*: 1063.

137. Mondovi, B., ed. (1985). *Structure and Function of Amine Oxidases*, CRC Press, Boca Raton, FL.

138. Pegg, A. E., and McCann, P. P. (1988). Polyamine metabolism and function in mammalian cells and protozoans. *ISI Atlas of Science: Biochemistry 1*: 11.

139. Danzin, C., and Mamont, P. S., (1987). Polyamine inhibition in vivo and in organ repair. In *Inhibition of Polyamine Metabolism* (McCann, P. P., Pegg, A. E., and Sjoerdsma, A., eds.), Academic Press, New York, p. 141.

140. Heby, O., and Perrson, L., (1990). Molecular genetics of polyamine synthesis in eukaryotic cells. *TIBS 15*: 153.

141a. Pierce, G. B., Gramzinski, R. A., and Parchment, R. E. (1990). Amine oxidases, programmed cell death and tissue renewal. *Phil. Trans. Roy. Soc. Lond. B 327*: 67.

141b. Coffino, P., and Poznanski, A. (1991). Killer polyamines? *J. Cell. Biochem. 45*: 54.

142. Marton, L. J., and Morris, D. R. (1987). Molecular and cellular functions of the polyamines. In *Inhibition of Polyamine Metabolism* (McCann, P. P., Pegg, A. E., and Sjoerdsma, A., eds.), Academic Press, New York, p. 79.

143. Piacentini, M., Melino, G., Cerù, M. P., Farrace, G., Piredda, L., Autuori, F., and Finazzi-Argo, A. (1988). Correlation between hypusine biosynthesis and spermidine level in human neuroblast cell induced to differentiate by retinoic acid and α-difluoromethylornithine. In *Perspectives in Polyamine Research* (Perrin, A., Scalabrino, G., Sessa, A., and Ferioli, M. E., eds.), Wichtig Editori, Milano, p. 119.

144. Albruzzesse, A., Isernia, T., Liguori, V., and Beninati, S. (1988). Polyamine dependent posttranslational modification of protein and cell proliferation. In *Perspectives in Polyamine Research* (Perrin, A., Scalabrino, G., Sessa, A., and Ferioli, M. E., eds.), Wichtig, Editori, Milano, p. 79.

145. Monti, M. G., Moruzzi, M. S., and Mezzetti, G. (1988). Effects of polyamines on protein kinase C directed phosphorylation of cytosolic proteins. In *Perspectives in Polyamine Research* (Perrin, A., Scalabrino, G., Sessa, A., and Ferioli, M. E., eds.), Wichtig Editori, Milano, p. 41.

146. Hegazy, M. G., Schlender, K. K., Reimann, E. M., and DiSalvo, J. (1988). Modulation of glycogen synthase kinase activity of skeletal and smooth muscle casein I by spermine. *Biochem. Biophys. Res. Commun. 156*: 653.

147. Piacentini, M., Martinet, N., Beninati, S., and Folk, J. E. (1986). Free and protein-conjugated polyamine in mouse epidermal cells: Effects of high calcium and retinoic acid. *J. Biol. Chem. 263*: 3790.

148. Clô, C., Tantini, B., Pignatti, C., Marmiroli, S., and Caldarera, C. M. (1988). Do polyamines mediate cellular desensitization to hormonal stimuli coupled to regulation of cyclic AMP level? In *Perspectives in Polyamine Research* (Perrin, A., Scalabrino, G., Sessa, A., and Ferioli, M. E., eds.), Wichtig Editori, Milano, p. 45.

149a. Corcoran, A. E., and Smyth, P. P. A. (1988). Polyamines as intracellular messengers in the human thyroid *Biochem. Soc. Trans. 16*: 397.

149b. Welsh, N., and Sjöholm, Å. (1988). Polyamine and insulin production in isolated pancreatic islets. *Biochem. J. 252*; 701.

149c. Moore, J. J., Lundgren, D. W., Moore, R. M., and Anderson, B. (1988). Polyamines control human chorionic gonadotropin production in the JEG-3 choriocarcinoma cell. *J. Biol. Chem. 263*: 12765.

150. González-Bosch, C., Miralles, V. J., Hernandez-Yago, J. and Grisolia, S. (1987). Spermidine and spermine stimulate the transport of the precursor of ornithine carbamoyltransferase into rat liver mitochondria. *Biochem. Biophys. Res. Commun. 149*: 21.

151a. Caron, P. C., Cote, L., and Kremzner, C. T. (1988). Putrescine, a source of γ-aminobutyric acid in the adrenal gland of the rat. *Biochem. J. 251*: 559.

151b. Fussi, F., Salvoldi, F., and Curti, M. (1988). Identification of *N*-carboxyethyl-γ-aminobutyric acid in brain: A key substance for a postulated pathway from spermidine to GABA? In *Perspectives in Polyamine Research* (Perrin, A., Scalabrino, G., Sessa, A., and Ferioli, M. E., eds.), Wichtig Editori, Milano, p. 37.

152. Perrin, A., Sessa, A., and Desiderio, M. A. (1985). Diamine oxidase in regenerating and hypertrophic tissues. In *Structure and Function of Amine Oxidases* (Mondovi, B., ed.), CRC Press, Boca Raton, FL, p. 179.

153. Seiler, N., Bolkenius, F. N., and Knödgen, B. (1985). The influence of catabolic reactions on polyamine excretion. *Biochem. J. 225*: 219.

154. McCann, P. P., Pegg, A. E., and Sjoerdsma, A., eds. (1987). *Inhibition of Polyamine Metabolism*, Academic Press, New York, Chapters 1,2,5–11,16.

155. Agostinelli, E., Riccio, P., Mucigrosso, J., Turini, P., and Mondovi, B. (1988). Amine oxidases as biological regulators. In *Perspectives in Polyamine Research* (Perrin, A., Scalabrino, G., Sessa, A., and Ferioli, M. E., eds.), Wichtig Editori, Milano, p. 11.

156. Bachrach, U. (1985). Copper amine oxidases and amines as regulators of cellular processes. *Structure and Function of Amine Oxidases* (Mondavi, B., ed.), CRC Press, Boca Raton, FL, p. 5.

157. Baylin, S. B., and Luk, G. D. (1985). Diamine oxidase activity in human tumors: clinical and biologic significance. In *Structure and Function of Amine Oxidases* (Mondovi, B., ed.), CRC Press, Boca Raton, FL, p. 187.

158a. Sunkara, P. S., Baylin, S. B., and Luk, C. D. (1987). Inhibition of polyamine biosynthesis: Cellular and in vivo effects on tumor proliferation. In *Inhibition of Polyamine Metabolism* (McCann, P. P., Pegg, A. E., and Sjoerdsma, A., eds.), Academic Press, New York, p. 121.

158b. Verma, A. K., and Bontwell, R. K. (1987). Inhibition of carcinogenesis by inhibitors of putrescine biosynthesis. In *Inhibition of Polyamine Metabolism* (McCann, P. P., Pegg, A. E., and Sjoerdsma, A., eds.), Academic Press, New York, p. 249.

158c. Barlow, P. J., Barlow, J. L. R., and Sjoerdsma, A. (1987). Clinical aspects of inhibition of ornithine decarboxylase with emphasis on therapeutic trials of eflornithine (DFMO) on cancer and protozoan disease. In *Inhibition of Polyamine Metabolism* (McCann, P. P., Pegg, A. E., and Sjoerdsma, A., eds.), Academic Press, New York, p. 345.

159a. Heby, O., Luk, G. D., and Schindler, J. (1987). Polyamine synthesis inhibitors act as both inducers and suppressors of cell differentiation. In *Inhibition of Polyamine Metabolism* (McCann, P. P., Pegg, A. E., and Sjoerdsma, A., eds.), Academic Press, New York, p. 165.

159b. Fozard, J. R. (1987). The contragestational effects of ornithine decarboxylase inhibition. In *Inhibition of Polyamine Metabolism* (McCann, P. P., Pegg, A. E., and Sjoerdsma, A., eds.), Academic Press, New York, p. 187.
160. Porter, C. W., and Jänne, J. (1987). Modulation of antineoplastic drug action by inhibitors of polyamine biosynthesis. In *Inhibition of Polyamine Metabolism* (McCann, P. P., Pegg, A. E., and Sjoerdsma, A., eds.), Academic Press, New York, p. 203.
161. Kallio, A., and Jänne, J. (1983). Role of diamine oxidase during the treatment of tumor-bearing mice with combinations of polyamine anti-metabolites. *Biochem. J. 212*: 895.
162. Máslinka, C., Bieganski, T., Fogel, W. A., and Kilter, M. E. (1985). Diamine oxidase in developing tissues. In *Structure and Function of Amine Oxidases* (Mondovi B., ed.), CRC Press, Boca Raton, FL, p. 153.
163. Bachrach, U., Ash, I., and Rahamim, E. (1988). Cytotoxicity of amine- and diamine-oxidases microinjected into cultured cells. In *Perspectives in Polyamine Research* (Perrin, A., Scalabrino, G., Sessa, A., and Ferioli, M. E., eds.), Wichtig Editori, Milano, p. 105.
164. Sutter, J., Hesterberg, R., Lorenz, W., Ennis, M., Stahlnecht, C.-D., and Kusche, J. (1985). Pathophysiological functions of diamine oxidase: An evaluation in animal studies using a probabalistic model with several types of causal relationships. In *Structure and Function of Amine Oxidases* (Mondovi, B., ed.), CRC Press, Boca Raton, FL, p. 167.
165. Buffoni, F. (1985). Lysyl oxidase (characterization and clinical importance). In *Structure and Function of Amine Oxidases* (Mondovi, B., ed.), CRC Press, Boca Raton, FL, p. 77.
166. Bardsley, W. G. (1985). Inhibitors of copper amine oxidases. In *Structure and Function of Amine Oxidases* (Mondovi, B., ed.), CRC Press, Boca Raton, FL, p. 135.
167a. Tipton, K. F., Dortert, P., and Benedetti, M. S., eds. (1984). *Monoamine Oxidase and Disease: Prospects for Therapy with Reversible Inhibitors*, Academic Press, New York.
167b. Singer, T. P., van Korff, R. W., and Murphy, D. L., eds. (1979). *Monoamine Oxidase: Structure, Function, and Altered Functions*, Academic Press, New York.
167c. Usdin, E., Weiner, N., and Youdim, M. B. H., eds. (1981). *Function and Regulation of Monoamine Enzymes: Basic and Clinical Aspects*, MacMillan Publishers Ltd., London.
168a. Singer, T. P., Castagnoli, Jr., N., Ramsay, R. R., and Trevor, A. J. (1987). Biochemical events in the development of parkinsonism induced by 1-methyl-4-phenyl-1,2,3,6-tetrahydropyridine. *J. Neurochem. 49*: 1.
168b. Singer, T. P., Trevor, A. T., and Castagnoli, Jr., N. (1986). Biochemistry of the neurotoxic action of MPTP: or how a faulty batch of 'designer drug' led to parkinsonism in drug abusers. *TIBS 12*: 266.
169. Bhatti, A. R., Burdon, J., William, A. C., Pall, H. S., and Ramsden, D. B. (1988). Synthesis, nuclear magnetic resonance spectroscopy, mass spectroscopy, and plasma amine oxidase inhibitory properties of analogues of 1-methyl-4-phenyl-1,2,3,6-tetrahydropyridine. *J. Neuralchem. 50*: 1097.

170. Devlin, A. J., Bhatti, A. R., Williams, A. C., and Ramsden, D. B. (1990). Inhibition of human benzylamine oxidase (BzAO) by analogues of 1-methyl-4-phenyl-1,2,3,6-tetrahydropyridine (MPTP). *Toxicol. Lett. 54*: 135.
171. Callingham, B. A., Holt, A., and Elliot, J. (1991). Properties and functions of the semicarbazide-sensitive amine oxidases. *Biochem. Soc. Trans. 19*: 228.
172. Boor, P. J., Hysmith, R. M., and Sauduji, R. (1990). A role for a new vascular enzyme in the metabolism of xenobiotic amines. *Circulation Res. 66*: 249.
173. Le Rudulier, D., and Goas, G. (1977). La diamine oxydase dans les jeunes plantes de *Glycine max. Phytochemistry 16*: 509.
174. Macholán, L., and Minár, J. (1974). The depression of synthesis of peak diamine oxidase due to light and the verification of its participation in growth processes using competitive inhibitors. *Biologia Plantarum* (Praha), *16*: 86.
175. Federico, R., and Angelini, R. (1988). Distribution of polyamines and their related catabolic enzymes in etiolated and light-grown leguminosae seedlings. *Planta 173*: 317.
176. Federico, R., Angelini, R., Cesta, A., and Pini, C. (1985). Determination of diamine oxidase in lentil seedlings by enzymatic activity and immunoreactivity. *Plant Physiol. 79*: 62.
177. Lobenstein-Verbeek, C. L., Jogejan, J. A., Frank, J., and Duine, J. A. (1984). Bovine serum amine oxidase: a mammalian enzyme having covalently bond PQQ as prosthetic group. *FEBS Lett. 170*: 305.
178a. Joshi, B. H., and Prakash, V. (1982). Naturally occurring pea diamine oxidase inhibitors obtained from *Sorghum vulgare* during germination, *Experientia 38*: 315.
178b. Sindu, P. K., and Desai, H. V. (1980). Inhibitors of diamine oxidase in cotyledons of groundnut seedlings. *Phytochemistry 19*: 317.
179. Angelini, R., Federico, R., and Mancinelli, A. (1988). Phytochrome-mediated control of diamine oxidase levels in the epicotyl of etiolated lentil (*Lens culinaris* Medicus) seedlings. *Plant Physiol. 88*: 1207.
180. Federico, R., Angelini, R., Argento-Cerù, M. P., and Manes, F. (1985). Immunohistochemical demonstration of lentil diamine oxidase. *Cell. Mol. Biol. 31*: 171.
181. Federico, R., and Angelini, R. (1986). Occurrence of diamine oxidase in the apoplast of pea epicotyls. *Planta 167*: 300.
182. Raimondi, L., Pirisino, R., Ignesti, G., Conforti, L., Banchelli, G., and Buffoni, F. (1990). Cultured preadipocytes produce a semicarbazide-sensitive amine oxidase (SSAO) activity, *J. Neural Transm.* [Suppl.] *32*: 331.
183. Tajima, S., Kanazawa, T., Takeuchi, E., and Yamamoto, Y. (1985). Characterization of a urate-degrading diamine oxidase-peroxidase enzyme system in soybean radicles. *Plant Cell Physiol. 26*: 787.
184. Goodwin, T. W., and Mercer, E. I. (1983). *Introduction to Plant Biochemistry*, 2nd ed., Pergamon Press, Oxford & New York, p. 335.
185. Rinaldi, A., Floris, G., and Giartosio, A. (1985). Plant amine oxidases. In *Structure and Function of Amine Oxidases* (Mondovi, B., ed.), CRC Press, Boca Raton, FL, p. 51.
186. Argento-Cerù, M. P., and Autuori, F. (1985). Localization of diamine oxidase in animal tissue. In *Structure and Function of Amine Oxidases* (Mondavi, B., ed.), CRC Press, Boca Raton, FL, p. 89.

187. Huneau, J. F., Tome, D., and Wal, J. M. (1989). Histamine content, diamine oxidase and histidine decarboxylase activities along the intestinal tract of the rat. *Agents and Actions 28*: 231.
188. D'Agostino, L., Pignata, S., Daniele, B., Ventriglia, R., Ferrari, G., Ferraro, C., Spaguolo, S., Lucchelli, P., and Mazzacca, G. (1989). Release of diamine oxidase into plasma by glycosaminoglycans in rat. *Biochim. Biophys. Acta 993*: 228.
189. Argento-Cerù, M. P., Bargagli, A. M., and Sartori, C. (1984). Topological location of diamine oxidase in the transverse plane of rabbit liver microsomal membranes. *Eur. J. Biochem. 138*: 645.
190. Ryder, T. A., MacKenzie, M. L., Pryse-Davies, J., Glover, V., Lewinsohn, R., and Sandler, M. (1979). A coupled peroxidatic oxidation technique for the histochemical localization of monoamine oxidase A and B and benzylamine oxidase. *Histochemistry 62*: 93.
191. Lewinsohn, R. (1981). Amine oxidase in human blood vessels and nonvascular smooth muscle. *J. Pharm. Pharmacol. 33*: 569.
192. Blicharski, J. R. D., and Lyles, G. A. (1990). Semicarbazide-sensitive amine oxidase activity in rat aortic cultured smooth muscle cells. *J. Neural Transm.* [Suppl] *32*: 337.
193. Precious, E., and Lyles, G. A. (1988). Properties of a semicarbazide-sensitive amine oxidase in human umbilical artery. *J. Pharm. Pharmacol. 40*: 627.
194. Buffoni, F., Marino, P., and Pirisino, R. (1976). Partial purification of pig aorta amine oxidases. *Ital. J. Biochem. 25*: 191.
195. Alper, C. A. (1990). The plasma proteins. In *Hematology* (Williams, W. J., Beutler, E., Erslev, A. J., and Lichtman, M. A., eds.), McGraw-Hill Publishing Co., New York, p. 1616.
196. Roth, J. A., and Gillis, C. N. (1975). Multiple forms of amine oxidase in perfused rabbit lung. *J. Pharmacol. Exp. Therapeutics 194*: 537.
197. Rao, S. B., Rao, K. S. P., and Mehendale, H. M. (1986). Absence of diamine oxidase activity from rabbit and rat lungs. *Biochem. J. 234*: 733.
198. Slocum, R. D., and Furey, M. J. (1991). Electron-microscopic cytochemical localization of diamine and polyamine oxidases in pea and maize tissues. *Planta 183*: 443.
199. Ruzicka, V., and Minár, J. (1974). The significance of the cotyledons for the formation of diamine oxidase in pea plants. *Biologia Plantarum* (Praha) *16*: 215.
200. Srivastava, S. K., and Prakash, V. (1977). Purification and properties of pea cotyledon and embryo diamine oxidase. *Phytochemistry 16*: 189.
201. Srivastava, S. K., Prakash, V., and Naik, B. I. (1977). Regulation of diamine oxidase activity in germinating pea seeds. *Phytochemistry 16*: 185.
202. Prakash, V., and Sindhu, R. K. (1979). Effect of phytic acid on diamine oxidase activity in germinating pea seeds. *Experientia 35*: 855.
203. Srivastava, S. K., Naik, B. I., and Raj, A. D. S. (1982). Effects of phenolics on plant amine oxidases. *Phytochemistry 21*: 1519.
204. Hirasawa, E. (1988). Diamine oxidase in cotyledons of *Piseum sativum* develops as a result of the supply of oxygen through embryonic axis during germination. *Plant Physiol. 88*: 441.
205. Srivastava, S. K., Raj, A. D. S., and Naik, B. I. (1981). Polyamine metabolism during aging and senescence of pea leaves. *Ind. J. Exp. Biol. 19*: 437.

206. Paprskárová, L., and Minár, J. (1976). The effects of potassium-deficiency on diamine oxidase activity in pea. *Biologia Plantrum* (Praha) *18*: 99.
207a. Delhaize, E., Dilworth, M. J., and Webb, J. (1986). The effects of copper nutrition and developmental state on the biosynthesis of diamine oxidase in clover leaves. *Plant Physiol. 82*: 1126.
207b. Delhaize, E., Loneragan, J. F., and Webb, J. (1985). Development of three copper metalloproteins in clover leaves. *Plant Physiol. 78*: 4.
208. Malmstrom, B. G., Andreasson, L. E., and Reinhammar, B. (1975). Copper-containing oxidases and superoxide dismutase. In *The Enzymes* (Boyer, P. D., ed.), Academic Press, New York, p. 507.
209. Yasunobu, K. T., Ishizaki, H., and Minamiura, N. (1976). The molecular, mechanistic and immunological properties of amine oxidases. *Molec. Cell. Biochem. 13*: 3.
210. Mondovi, B., and Avigiano, L. (1987). Amine oxidases. *Life Chem. Rep. 5*: 187.
211. Oda, O., Manube, T., and Okuyama, T. (1981). Oxidation of ϵ-amino groups of lysyl peptides by bovine serum amine oxidase. *J. Biochem. 86*: 1317.
212. Banchelli, G., Buffoni, F., Elliot, J., and Callingham, B. A. (1990). A study of the biochemical pharmacology of 3,5-ethoxy-4-aminomethylpyridine (B24), a novel amine oxidase inhibitor with selectivity for tissue bound semicarbazide-sensitive amine oxidase enzymes. *Neurochem. Int. 17*: 215.
213. Elliot, J., Callingham, B. A., and Barrand, M. A. (1989). In-vivo effects of (E)-2-(3′4′-dimethoxyphenyl)-3-fluoroallylamine (MDL 72145) on amine oxidase activities in the rat. Selective inhibition of semicarbazide-sensitive amine oxidase in vascular and brown adipose tissue. *J. Pharm. Pharmacol. 41*: 37.
214. Hiraoka, A., Ohtaka, J., Koike, S., Tsuboi, Y., Tsuchikawa, S., and Miura, I. (1988). Inhibition of copper-containing amine oxidases by oximes. *Chem. Pharm. Bull. 36*: 3027.
215. McIntire, W. S., Dooley, D. M., McGuirl, M. A., Cote, C. E., and Bates, J. L. (1990). Methylamine oxidase from *Arthrobacter* P1 as a prototype of eukaryotic plasma amine oxidase and diamine oxidase. *J. Neural Transm.* [Suppl.] *32*: 315.
216. Wang, T.-M., Achee, F. M., and Yasunobu, K. T. (1968). Amine oxidase. XI. Beef plasma amine oxidase, a sulfhydryl protein. *Arch. Biochem. Biophys. 128*: 106.
217. Suva, R. H., and Abeles, R. H. (1978). Studies on the mechanism of action of plasma amine oxidase. *Biochemistry 17*: 3538.
218. Floris, G., Giartosio, A., and Rinaldi, A. (1983). Essential sulfhydryl groups in diamine oxidase from *Euphorbia characias* latex. *Arch. Biochem. Biophys. 220*: 623.
219. Tsurushiin, S., Hiramatsu, A., Inamasu, M., and Yasunobu, K. T. (1975). The essential histidine residues of bovine plasma amine oxidase. *Biochim. Biophys. Acta 400*: 451.
220. Floris, G., and Fadda, M. B. (1978). Evidence for arginine residues in diamine oxidase of porcine kidney. *Boll. Soc. Ital. Biol. Sper. 54*: 1735.
221. Rauch, N., and Rauch, R. J. (1973). Isozymes of amine oxidase in human plasma and other tissue. *Experientia 29*: 215.
222. Pionette, J.-M. (1974). Analytical-band centrifugation of the active form of pig kidney diamine oxidase. *Biochem. Biophys. Res. Commun. 58*: 495.

223. Dooley, D. M., McGuirl, M. A., Cote, C. E., Knowles, P. F., Singh, I., Spiller, M., Brown, R. D., and Koenig, S. H. (1991). Coordination chemistry of copper-containing amine oxidases: Nuclear magnetic relaxation dispersion studies of copper binding, solvent-water exchange, substrate and inhibitor binding, and protein aggregation. *J. Am. Chem. Soc. 113*: 754.
224. Denney, R. M., Waguespack, A., and Varde, A. (1988). Isolation of a pH-dependent, inhibitory monoclonal antibody to human placental diamine oxidase. *Pharmacol. Res. Commun.* [Suppl. IV] *20*: 185.
225. Oratore, A., Banchelli, G., Buffoni, F., Sabatini, S., Mondovi, B., and Finazzi-Argo, A. (1981). Reaction between mammalian amine oxidases and their antibodies. *Biochem. Biophys. Res. Commun. 98*: 1002.
226. Oratore, A., Guerriere, P., Ballini, Mondovi, B., and Finazzi-Argo, A. (1979). Preparation and properties of antibodies against pig kidney diamine oxidase. *FEBS Lett. 104*: 154.
227. Shah, M., and Ali, R. (1989). Immunochemical reactivity of native and modified preparations of pig kidney diamine oxidase. *J. Biochem. 106*: 616.
228. Takano, K., Suzuki, T., and Yasuda, K. (1970). Localization of diamine oxidase and D-amino acid oxidase in kidney, demonstrated by means of immunohistochemical methods. *Acta Histochem. Cytochem. 3*: 105.
229. Hamilton, G. A. (1981). Oxidases with monocopper reactive sites. In *Copper Proteins* (Spiro, T. G., ed.), Wiley-Interscience, New York, p. 205.
230. Buffoni, F. (1966). Histaminase and related amine oxidases. *Pharmacol. Rev. 18*: 1163.
231. Knowles, P. F., and Yadav, D. S. (1984). Amine oxidases. In *Copper Proteins and Copper Enzymes* (Lontie, R., ed.), CRC Press, Boca Raton, FL, p. 103.
232. Buffoni, F., and Della Corte, L. (1972). Pig plasma amine oxidase. *Adv. Biochem. Psychopharmacol. 5*: 133.
232a. Shanbhag, V. M., and Martell, A. E. (1991). Oxidative deamination of amino acids by molecular oxygen with pyridoxal derivatives and metal ions as catalysts. *J. Am. Chem. Soc. 113*: 6479.
233. Salisbury, S. A., Forrest, H. S., Cruse, W. B. T., and Kennard, O. (1979). A novel coenzyme from bacterial primary alcohol dehydrogenases. *Nature 280*: 843.
234. Moog, R. S., McGuirl, M. A., Cote, C. E., and Dooley, D. M. (1986). Evidence for methoxatin (pyrroloquinoline quinone) as the cofactor in bovine plasma amine oxidase from resonance Raman spectroscopy. *Proc. Natl. Acad. Sci. USA 83*: 8435.
235. van der Meer, R. A., Mulder, A. C., Jongejan, J. A., and Duine, J. A. (1989). Determination of PQQ in quinoproteins with covalently bound cofactor and in PQQ-derivatives. *FEBS Lett. 254*: 99.
236. Knowles, P. F., Pandeya, K. B., Ruis, F. X., Spencer, C. M., Moog, R. S., McGuirl, M. A., and Dooley, D. M. (1987). The organic cofactor in plasma amine oxidase; evidence for pyrroloquinoline quinone and against pyridoxal phosphate. *Biochem. J. 241*: 603.
237. van der Meer, R. A., Jongejan, J. A., Frank, J., and Duine, J. A. (1986). Hydrazone formation of 2,4-dinitrophenylhydrazine with pyrroloquinoline quinone in porcine kidney diamine oxidase. *FEBS Lett. 206*: 111.

238. Glatz, Z., Kovár, J., Macholán, L., and Pec, P. (1987). Pea (*Pisum sativum*) diamine oxidase contains pyrroloquinoline quinone as a cofactor. *Biochem. J. 242*: 603.
239. Citro, G., Verdina, A., Galati, R., Floris, G. Sabatini, S., and Finazzi-Argò, A. (1989). Production of antibodies against the coenzyme pyrroloquinoline quinone. *FEBS Lett. 247*: 201.
240. Buffoni, F., and Cambi, S. A. (1990). A method for the isolation and identification of pyridoxal phosphate in proteins. *Anal. Biochem. 187*: 44.
241. Buffoni, F. (1990). Nature of the organic cofactor of pig plasma benzylamine oxidase. *Biochim. Biophys. Acta 1040*: 77.
242. *Enzyme Nomenclature*. (1979). Academic Press, New York, pp. 122–125.
243a. Clark, K., Penner-Hahn, J. E., Whittaker, M. M., and Whittaker, J. W. (1990). Oxidation-state assignments for galactose oxidase complexes from x-ray absorption spectroscopy. Evidence for Cu(I) in the active site. *J. Am. Chem. Soc. 112*: 6433.
243b. Whittaker, M. M., and Whittaker, J. W. (1990). A tyrosine-derived free radical in apogalactose oxidase. *J. Biol. Chem. 265*: 9610.
243c. Whittaker, M. M., DeVito, V. L., Asher, S. A., and Whittaker, J. W. (1989). Resonance Raman evidence for tyrosine involvement in the radical site of galactose oxidase. *J. Biol. Chem. 264*: 7104.
244. Lindstrom, A., Olsson, B., and Pettersson, G. (1974). Transient kinetics of benzaldehyde formation during the catalytic action of pig plasma benzylamine oxidase. *Eur. J. Biochem. 42*: 377.
245. Lindstrom, A., Olsson, B., Pettersson, G., an Syzmanska, J. (1974). Kinetics of the interaction between pig plasma benzylamine oxidase and various monoamines. *Eur. J. Biochem. 47*: 99.
246. Collison, D., Knowles, P. F., Mabbs, F. E., Rius, F. X., Singh, I., Dooley, D. M., Cote, C. E., and McGuirl, M. (1989). Studies on the active site of pig plasma amine oxidase. *Biochem. J. 264*: 663.
247. Pettersson, G. (1985). Plasma amine oxidase. In *Structure and Function of Amine Oxidases* (Mondovi, B., ed.), CRC Press, Boca Raton, FL, p. 105.
248. Finazzi-Argo, A. (1985). Optical spectroscopic properties of copper-containing amine oxidases. In *Structure and Function of Amine Oxidases* (Mondovi, B., ed.), CRC Press, Boca Raton, FL, p. 121.
249. Dooley, D. M., McIntire, W. S., McGuirl, M. A., Cote, C. E., and Bates, J. L. (1990). Characterization of the active site of *Arthrobacter* P1 methylamine oxidase: Evidence for copper-quinone interactions. *J. Am. Chem. Soc. 112*: 2782.
250. Dooley, D. M., and Cote, C. E. (1985). Copper(II) coordination chemistry in bovine plasma amine oxidase: Azide and thiocyanate binding. *Inorg. Chem. 24*: 3996.
251. Dooley, D. M., McGuirl, M. A., Peisach, J., and McCracken, J. (1987). The generation of an organic free radical in substrate-reduced pig kidney diamine oxidase-cyanide. *FEBS Lett. 214*: 274.
252. Dooley, D. M., McGuirl, M. A., Brown, D. E., Turowski, P. N., McIntire, W. S., and Knowles, P. F. (1991). A Cu(I)-semiquinone state in substrate reduced amine oxidases. *Nature 349*: 262.
253. Dooley, D. M., and Cote, C. E. (1984). Inactivation of beef plasma amine oxidase by sulfide. *J. Biol. Chem. 259*: 2923.

254. Dooley, D. M. (1987). The coordination chemistry of copper-containing metalloproteins. *Life Chem. Rep. 5*: 91.
255. Dooley, D. M., and McGuirl, M. A. (1986). Spectroscopic studies of pig kidney diamine oxidase-anion complexes. *Inorg. Chim. Acta 123*: 231.
256. Dooley, D. M., Cote, C. E., and Golnik, K. C. (1984). Inhibition of copper-containing amine oxidases by Cu(II) complexes and anions. *J. Molec. Catal. 23*: 243.
257. Morpurgo, L., Befani, O., Sabatini, S., Mondovi, B., Artico, M., Corelli, F., Massa, S., Stefancich, G., and Avigliano, L. (1988). Spectroscopic studies of the reaction between bovine serum amine oxidase (copper-containing) and some hydrazides and hydrazines. *Biochem. J. 256*: 565.
258. Brown, D. E., McGuirl, M. A., Dooley, D. M., Janes, S. M., Mu, D., and Klinman, J. P. (1991). The organic functional group in copper-containing amine oxidases. *J. Biol. Chem. 266*: 4049.
259. Lamkin, M. S., Williams, T. J., and Falk, M. C. (1988). Excitation energy transfer study of the spatial relationship between the carbonyl and metal cofactors in pig plasma amine oxidase. *Arch. Biochem. Biophys. 261*: 72.
260. Cogini, A., Farci, R., Rinaldi, A., and Floris, G. (1989). Amine oxidase from *Lathyrus cicera* and *Phaseolus vulgarus*: Purification and properties. *Prep. Biochem. 19*: 95.
261. Finazzi-Argo, A., Rinaldi, A., Floris, G., and Rotilio, G. (1984). A free-radical intermediate in the reduction of plant Cu-amine oxidase. *FEBS Lett. 176*: 378.
262. Yamada, H., Adachi, O., and Ogata, K. (1965). Amine oxidases of microorganisms. III. Properties of amine oxidase of *Aspergillus niger*. *Agric. Biol. Chem. 29*: 864.
263. Lindstrom, A., Olsson, B., and Pettersson, G. (1973). Kinetics of the interaction between pig plasma benzylamine oxidase and substrate. *Eur. J. Biochem. 35*: 70.
264. Yamada, H., and Yasunobu, K. (1962). Monoamine oxidase. II. Copper, one of the prosthetic groups of plasma monoamine oxidase. *J. Biol. Chem. 237*: 3077.
265. Mann, P. J. G. (1960). Further purification and properties of the amine oxidase of pea seedlings. *Biochem. J. 76*: 44P.
266. Morpugo, L., Agostinelli, E., Befani, O., and Mondovi, B. (1987). Reaction of bovine serum amine oxidase with *N,N*-diethyldithiocarbamate. Selective removal of one copper ion. *Biochem. J. 248*: 865.
267. Suzuki, S., Sakurai, T., and Nakahara, A. (1986). Roles of the two copper ions in bovine serum amine oxidase. *Biochemistry 25*: 338.
268. Williams-Smith, D. L., Bray, R. C., Barber, M. J., Tsopanakis, A. D., and Vincent, S. P. (1977). Changes in apparent pH on freezing aqueous buffer solutions and their relevance to biochemical electron-paramagnetic-resonance spectroscopy. *Biochem. J. 167*: 593.
269. Yamada, H., Yasunobu, K., Yamano, T., and Mason, H. S. (1963). Copper in plasma amine oxidase. *Nature 198*: 1092.
270. Boas, J. F. (1984). Electron paramagnetic resonance of copper proteins. In *Copper Proteins and Copper Enzymes* (Lontie, R., ed.), CRC Press, Boca Raton, FL, p. 5.
271. Suzuki, S., Sakurai, T., Nakahara, A., Oda, O., Manabe, T., and Okuyama, T. (1980). Spectroscopic aspects of copper binding site in bovine serum amine oxidase. *FEBS Lett. 116*: 17.

272. Buffoni, F., Della Corte, I., and Knowles, P. F. (1968). The nature of copper in pig plasma benzylamine oxidase. *Biochem. J. 106*: 575.
273. Yamada, H., Adachi, O., and Yamano, T. (1969). Electron paramagnetic resonance spectra of amine oxidase from *Aspergillus niger*. *Biochim. Biophys. Acta 191*: 751.
274. Bardsley, W. G., Childs, R. E., and Crabbe, M. J. C. (1974). Inhibition of enzymes by metal ion-chelating reagents. The action of copper-chelators on diamine oxidase. *Biochem. J. 137*: 61.
275. Baker, G., Knowles, P. F., Pandeya, K. B., and Rayner, J. B. (1986). Electron nuclear double resonance (ENDOR) spectroscopy of amine oxidase from pig plasma. *Biochem. J. 237*: 609.
276. McCracken, J., Peisach, J., and Dooley, D. M. (1987). Cu(II) coordination of amine oxidases: Pulsed EPR studies of histidine imidazole, water and eoxgenous ligand coordination. *J. Am. Chem. Soc. 109*: 4064.
277. Zeidan, H., Watanabe, K., Piette, L. H., and Yasunobu, K. T. (1980). Electron spin resonance studies of plasma amine oxidase. Probing of the environment about the substrate-liberated sulfhydryl groups in the active site. *J. Biol. Chem. 255*: 7621.
278. Greenaway, F. T., O'Gara, C. Y., Marchena, J. M., Poku, J. W., Urtiaga, J. G., and Zou, Y. (1991). EPR studies of spin-labeled bovine plasma amine oxidase: The nature of the substrate-binding site. *Arch. Biochem. Biophys. 285*: 291.
279. Kluetz, M. D., and Schmidt, P. G. (1977). Proton relaxation studies of the hog kidney diamine oxidase center. *Biochemistry 16*: 5191.
280. Williams, T. J., and Falk, M. C. (1986). Spatial relationship between the copper and carbonyl cofactors in the active site of pig plasma amine oxidase. *J. Biol. Chem. 261*: 15949.
281. Scott, R. A., and Dooley, D. M. (1985). X-ray absorption spectroscopy of the Cu(II) sites in bovine plasma amine oxidase. *J. Am. Chem. Soc. 107*: 4348.
282. Knowles, P. F., Strange, R. W., Blackburn, N. J., and Hasnain, S. S. (1989). EXAFS studies on pig plasma amine oxidase. A detailed structure analysis using the curved wave multiple scattering approach. *J. Am. Chem. Soc. 111*: 102.
283. Suzuki, S., Sakurai, T., Nakahara, A., Oda, O., Manabe, T., and Okuyama, T. (1981). Preparation and characterization of cobalt(II)-substituted bovine serum amine oxidase. *J. Biochem.* (Tokyo) *90*: 905.
284. Inamasu, M., Yasunobu, K. T., and Koenig, W. A. (1974). Cofactor investigation of bovine plasma amine oxidase. *J. Biol. Chem. 249*: 5265.
285. Ameyama, M., Hayashi, U., Matsushita, K., Shinagawa, E., and Adachi, O. (1984). Microbial production of pyrroloquinoline quinone. *Agric. Biol. Chem. 48*: 561.
286. Oi, S., Inamasu, M., and Yasunobu, K. T. (1970). Mechanistic studies of beef plasma amine oxidases. *Biochemistry 9*: 3378.
287. Taylor, C. E., Taylor, R. S., Rasmussen, C., and Knowles, P. F. (1972). A catalytic mechanism for enzyme benzylamine oxidase from pig plasma. *Biochem. J. 130*: 713.
288. Bardsley, W. G., and Ashford, J. S. (1972). Inhibition of pig kidney diamine oxidase by substrate analogues. *Biochem. J. 128*: 253.
289. Bardsley, W. G., Crabbe, M. J. C., and Shindler, J. S. (1973). Kinetics of the diamine oxidase reaction. *Biochem. J. 131*: 459.

290. McEwen, C. M. (1965). Human plasma amine oxidase. II Kinetic studies. *J. Biol. Chem. 240*: 2011.
291. Kelley, I. D., Knowles, P. F., Yadav, K. D. S., Bardsley, W. G., Leff, P., and Waight, R. D. (1981). Steady-state kinetic studies on benzylamine oxidase from pig plasma. *Eur. J. Biochem. 114*: 133.
292. Buffoni, F., Della Corte, L., and Ignesti, C. (1972). Pig plasma benzylamine oxidase (histaminase): Kinetic studies. *Pharmacol. Res. Commun. 4*: 99.
293. Lindstrom, A., Olsson, B., Olsson, J., and Pettersson, G. (1976). Effects of pH on the transient reduction of pig plasma benzylamine oxidase by benzylamine derivatives. *Eur. J. Biochem. 64*: 321.
294. Farnum, M. F., Palcic, M., and Klinman, J. P. (1986). pH dependence of deuterium isotope effects and tritium exchange in the bovine plasma amine oxidase reaction: A role for single base catalysis in amine oxidation and exchange. *Biochemistry 25*: 1898.
295. Berg, K. A., and Abeles, R. H. (1980). Mechanism of action of plasma amine oxidase: Products released under anaerobic conditions. *Biochemistry 19*: 3186.
296. Lindstrom, A., and Pettersson, G. (1978). The order of substrate addition and product release during the catalytic action of pig plasma benzylamine oxidase. *Eur. J. Biochem. 84*: 479.
297. Ruis, F. X., Knowles, P. F., and Pettersson, G. (1984). The kinetics of ammonia release during the catalytic cycle of pig plasma amine oxidase. *Biochem. J. 220*: 962.
298. Hartmann, C., and Klinman, J. P. (1987). Reductive trapping of substrate to bovine plasma amine oxidase. *J. Biol. Chem. 262*: 962.
299. Palcic, M. M., and Klinman, J. P. (1983). Isotopic probes yield microscopic constants: Separation of binding energy from catalytic efficiency in the bovine plasma amine oxidase reaction. *Biochemistry 22*: 5957.
300. Hartmann, C., and Klinman, J. P. (1990). Reductive trapping of substrate to methylamine oxidase from *Arthrobacter* P1. *FEBS Lett. 261*: 441.
301. Olsson, B., Olsson, J., and Pettersson, G. (1976). Stopped-flow spectrophotometric characterization of enzymic reaction intermediates in the anaerobic reduction of pig plasma benzylamine oxidase by amine substrates. *Eur. J. Biochem. 71*: 375.
302. Hartmann, C. (1988). Mechanistic studies on bovine plasma amine oxidase. Ph.D. thesis, University of California, Berkeley.
303. Hevey, R. C., Babson, J., Maycock, A. L., and Abeles, R. H. (1973). Highly specific enzyme inhibitors: inhibition of plasma amine oxidase. *J. Am. Chem. Soc. 95*; 6125.
304. Maycock, A. L., Suva, R. H., and Abeles, R. H. (1975). Novel inactivators of plasma amine oxidase. *J. Am. Chem. Soc. 97*: 5613.
305. Neumann, R., Hevey, R. C., and Abeles, R. H. (1975). The action of plasma amine oxidase on β-haloamines. *J. Biol. Chem. 250*: 6362.
306. Sleath, P. R., Noar, J. B., Eberlein, G. A., and Bruice, T. C. (1985). Synthesis of 7,7-didecarboxymethoxatin (4,5-dihydro-4,5-dioxo-1*H*-pyrrolo-[2,3-*f*]quinoline-2-carboxylic acid) and comparison of its chemical properties with those of methoxatin and analogous *o*-quinones. Model studies directed towards the action of PQQ requiring bacterial oxidoreductases and mammalian plasma amine oxidase. *J. Am. Chem. Soc. 107*: 3328.

307. Olsson, B., Olsson, J., and Pettersson, G. (1976). Kinetic isotope effects on the catalytic activity of pig plasma benzylamine oxidase. *Eur. J. Biochem. 64*: 327.
308. Hartmann, C., and Klinman, J. P. (1991). Structure-function studies of substrate oxidation by bovine serum amine oxidases: Relationship to cofactor structure and mechanism. *Biochemistry 30*: 4605.
309. Lovenberg, W., and Beaven, M. A. (1971). The release of tritium upon deamination of 3,4-dihydroxy[2-^{3}H]phenylethylamine by plasma amine oxidase. *Biochim. Biophys. Acta 251*: 452.
310a. Summers, M. C., Markovic, R., and Klinman, J. P. (1979). Stereochemistry and kinetic isotope effects in the bovine plasma amine oxidase catalyzed oxidation of dopamine. *Biochemistry 18*: 1969.
310b. Battersby, A. R., Stauton, J., Klinman, J. P., and Summers, M. C. (1979). Stereochemistry of oxidation of benzylamine by amine oxidase from beef plasma. *FEBS Lett. 99*: 297.
311. Farnum, M. F., and Klinman, J. P. (1986). Stereochemical probes of bovine plasma amine oxidase: Evidence for mirror image processing and a *syn* abstraction of hydrogens from C-1 and C-2 dopamine. *Biochemistry 25*: 6028.
312. Coleman, A. A., Hindsgaul, O., and Palcic, M. M. (1989). Stereochemistry of copper amine oxidase reactions. *J. Biol. Chem. 264*: 19500.
313. Coleman, A. A., Scaman, C. H., Kang, Y. J., and Palcic, M. M. (1991). Stereochemical trends in copper amine oxidase reactions. *J. Biol. Chem. 266*: 6795.
314. Hanson, K. R., and Rose, I. A. (1975). Interpretation of enzyme reaction stereospecificity. *Acc. Chem. Res. 8*: 1.
315. Yu, P. H. (1988). Three types of stereospecificity and the kinetic dueterium isotope effect in the oxidative deamination of dopamine as catalyzed by different amine oxidases. *Biochem. Cell. Biol. 66*: 853.
316. Battersby, A. R., Stauton, J., and Summers, M. C. (1976). Studies of enzyme mediated reactions. Part VI: Stereochemical course of the dehydrogenation of stereospecifically labeled benzylamines by the amine oxidase from pea seedling. *J. Chem. Soc. Perkin Trans. I*: 1052.
317. Battersby, A. R., Stauton, J., Summers, M. C., and Southgate, R. (1979). Studies of enzyme mediated reactions. Part IX: Stereochemistry of oxidative ring cleavage adjacent to nitrogen during biosynthesis of chelidonine. *J. Chem. Soc. Perkin Trans. I*: 45.
318. Gerdes, H. J., and Leistner, E. (1979). Stereochemistry of reactions catalyzed by L-lysine decarboxylase and diamine oxidase. *Phytochemistry 18*: 771.
319. Allen, J. K., Barrows, K. D., and Jones, A. J. (1979). Biosynthesis of echinulin. The stereochemistry of aromatic isoprenylation. *J. Chem. Soc. Chem. Commun.*: 280.
320. Battersby, A. R., Murphy, R., and Staunton, J. (1982). Studies of enzyme-mediated reactions. Part XIV. Stereochemical course of the formation of cadaverine by decarboxylation of (2S)-lysine with lysyl decarboxylase (E.C. 4.1.1.18) from *Bacillus cadaveris*. *J. Chem. Soc. Perkins Trans. I*: 449.
321. Battersby, A. R., Staunton, J., and Tippett, J. (1982). Studies of enzyme-mediated reactions. Part XV. Stereochemical course of the formation of γ-aminobutyric acid

(GABA) by decarboxylation of (2S)-glutamic acid with glutamate decarboxylase from *Escherichia coli*. *J. Chem. Soc. Perkins Trans. I*: 455.

322. Saunders, W. H. (1985). Calculation of isotope effects in elimination reactions. New experimental criteria for tunneling in slow proton transfers. *J. Am. Chem. Soc. 107*: 164.
323. Swain, C. G., Stivers, E. C., Reuwer, J. F., and Schaad, L. J. (1958). Use of hydrogen isotope effects to identify the attacking nucleophile in the enolization of ketone catalyzed by acetic acid. *J. Am. Chem. Soc. 80*: 5885.
324. Grant, K. L., and Klinman, J. P. (1989). Evidence that both protium and deuterium undergo significant tunneling in the reaction catalyzed by bovine serum amine oxidase. *Biochemistry 28*: 6597.
325. Lindstrom, A., Olsson, B., and Pettersson, G. (1974). Effect of azide on some spectral and kinetic properties of pig plasma benzylamine oxidase. *Eur. J. Biochem. 48*: 237.
326. Morpurgo, L., Agostinelli, E., Muccigrosso, J., Martini, F., Mondovi, B., and Avigiano, L. (1989). Benzylhydrazine as a pseudo-substrate of bovine serum amine oxidase. *Biochem. J. 260*: 19.
327. Grant, J., Kelly, I., Knowles, P. F., Olsson, J., and Pettersson, G. (1978). Changes in the copper centres of benzylamine oxidase from pig plasma during the catalytic cycle. *Biochem. Biophys. Res. Commun. 83*: 1216.
328. Mondovi, B., Rotilio, G., Finazzi-Argo, A., Vallogini, M. P., Malmström, B. G., and Antinini, E. (1969). Copper reduction by substrate in diamine oxidase. *FEBS Lett. 2*: 182.
329. Mondovi, B., Rotilio, G., Finazzi-Argo, A., and Antonini, E. 91971). Amine oxidases: A new class of copper oxidases. In *Magnetic Resonance in Biological Research* (Franconi, G., ed.), Gordon and Breach, London, p. 233.
330. Yadav K. P. S., and Knowles, P. F. (1981). A catalytic mechanism for benzylamine oxidase from pig plasma. Stopped-flow kinetic studies. *Eur. J. Biochem. 114*: 139.
331. Rotilio, G., Calabrese, L., Finazzi-Argo, A., and Mondovi, B. (1970). Indirect evidence for the production of superoxide anion radicals by pig plasma diamine oxidase. *Biochim. Biophys. Acta 198*: 618.
332. Younes, M., and Weser, U. (1978). Involvement of superoxide in the catalytic cycle of diamine oxidase. *Biochim. Biophys. Acta 265*: 644.

7

Lysyl Oxidase

Herbert M. Kagan and Philip C. Trackman

Boston University School of Medicine, Boston, Massachusetts

I. INTRODUCTION

Elastin and collagen are examples of proteins whose formation and maturation are critically dependent upon posttranslational modifications. Thus, collagen is post-translationally processed by several different catalysts to cleave signal peptides, hydroxylate proline and lysine residues, glycosylate hydroxylysines, and to oxidize lysine residues. The latter modification, catalyzed by lysyl oxidase (EC 1.4.3.13), generates peptidyl α-aminoadipic-δ-semialdehyde in collagen as well as elastin. The aldehyde residue can spontaneously condense with neighboring aldehydes or ε-amino groups to form inter- or intrachain covalent cross-linkages, two of which are shown (Fig. 1). Continuous, intermolecular condensations of this kind will then convert soluble monomers of elastin or collagen into insoluble fibers in the extracellular matrix. Thus, lysyl oxidase can potentially regulate the development and repair of the matrix in connective tissues. This chapter will summarize findings appearing since a previous detailed review on the properties and regulation of this catalyst [1].

II. PURIFICATION AND MOLECULAR WEIGHT

Prior to the finding that the bulk of tissue lysyl oxidase was rendered soluble by 4–6 M urea extraction buffers [2], the purification of this enzyme had been

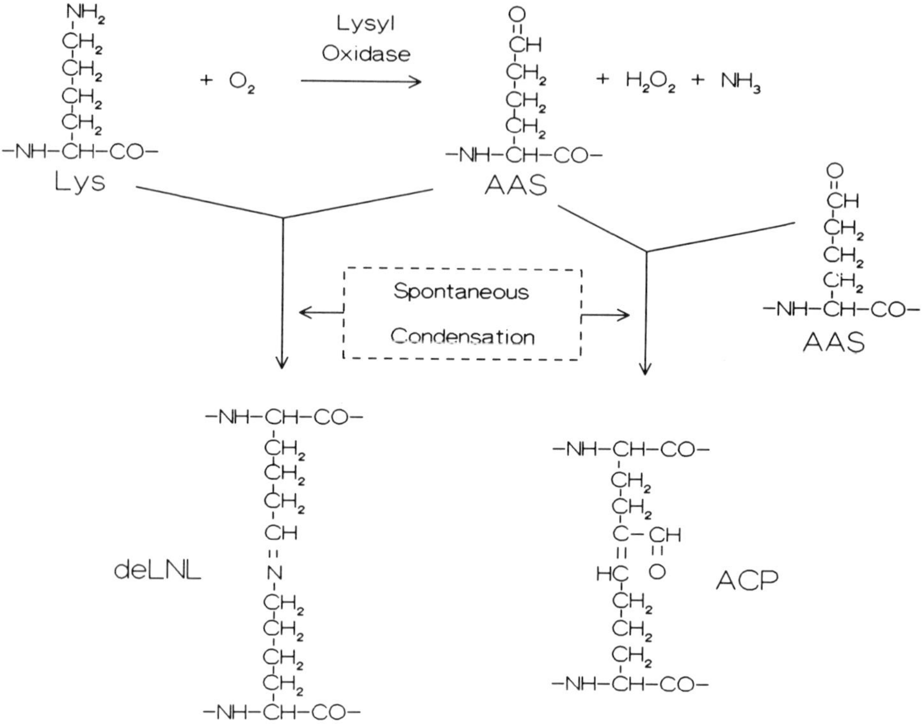

Figure 1 Role of lysyl oxidase in cross-linkage formation. AAS, α-aminoadipic-δ-semialdehyde; deLNL, dehydrolysinonorleucine, a Schiff base cross-link; ACP, aldol condensation produce cross-link. More complex cross-links can arise from the bifunctional ACP and deLNL cross-links in elastin and collagen.

a difficult task, which often yielded preparations significantly contaminated by other proteins [1]. Varying molecular weights had also been reported with values ranging from approximately 30 kDa to greater than 100 kDa [1]. Urea-extractable enzyme has since been highly purified from bovine aorta and lung [3,4], human placenta [5], and rat lung [6] yielding M_r values of 30–32 kDa in SDS. Lysyl oxidase species identified by Western blotting of urea extracts of various human tissues also have molecular weights approximating 30 kDa [7]. A recently modified version of current purification methods induced polymerization of the enzyme by removing urea from the urea-extracted preparation. The polymerized enzyme was thereby retained at the top of a gel filtration column. After washing with urea-free buffer, column-bound enzyme was disaggregated and eluted with urea-supplemented buffer, resulting in a high degree of purification [8].

III. COFACTORS

Early descriptions of the inhibition of elastin and collagen cross-linking in vivo by copper-deficient diets or by the administration of compounds directed at carbonyl functions led to hypotheses that the catalyst of this process contained both copper and carbonyl cofactors (reviewed in Ref. 1). The properties of these cofactors have recently been addressed.

A. Copper

Studies with purified preparations of lysyl oxidase have confirmed that copper is essential to the expression of enzyme activity [9,10]. The highly purified enzymes of chick bone [9] and bovine aorta [10] contained 1–1.1 g-atom of Cu per 32 kDa, while the catalytically inert metal-free bovine aortic enzyme fully regains activity upon reconstitution with 1 ± 0.1 g-atom of Cu^{2+} per 32,000 Da monomer [10]. Activity was not restored by divalent iron, cobalt, nickel, cadmium, mercury, or zinc, however [10]. The Cu EPR spectrum of the bovine aortic enzyme is consistent with a single Cu(II) species bound to the enzyme in a tetragonally distorted, octahedrally coordinated field containing nitrogenous ligands [10] and is similar in that sense to other mammalian copper-dependent amine oxidases.

B. Carbonyl Cofactor

A syndrome of connective tissue defects reflecting impaired cross-link formation can be induced by administration of lysyl oxidase inhibitors such as isonicotinic acid hydrazide and semicarbazide [1], which derivatize carbonyl functions. The resulting inhibition of collagen insolubilization in vivo can be significantly reversed by pyridoxal treatment while pyridoxine-deficient diets result in cross-linkage anomalies in chicks and in cell culture systems [11]. Such results, as well as spectral analyses of chick aorta lysyl oxidase [12], led to the hypothesis that pyridoxal phosphate (PLP) was the carbonyl cofactor of lysyl oxidase. However, various PLP-specific analyses of the highly purified bovine aorta enzyme were completely negative for this cofactor [13]. It has been noted that PLP deficiency results in the accumulation of homocysteine in vivo due to the lack of function of PLP-dependent cystathionine synthase [14]. Since homocysteine is an aminothiol, it can trap the peptidyl aldehyde product of lysyl oxidase, thus preventing its conversion to mature cross-linkages. Moreover, PLP deficiency can exert a generalized inhibitory effect on protein synthesis [15]. Such effects might underlie the observed relationships between PLP deficiency and the reduction in cross-linkage content and lysyl oxidase levels.

The progress of research on the carbonyl cofactor of lysyl oxidase has paralleled similar efforts concerned with other copper-dependent amine oxidases. The reports that plasma amine oxidase appeared to contain pyrroloquinoline quinone (4,5-dihydro-4,5-dioxo-1H-pyrrolo[2,3-f]-quinoline-2,7,9-tricarboxylic acid; methoxatin; PQQ) (see Fig. 2) [16,17] were of considerable interest to those involved in lysyl oxidase research. This *o*-quinone had previously been implicated as the carbonyl cofactor in various bacterial redox enzymes. PQQ is an effective redox catalyst, which can be derivatized by carbonyl reagents known to inhibit lysyl oxidase. Moreover, the three carboxyl groups of PQQ provide potential sites for covalent linkages with protein side chains and thus could account for the apparently covalent linkage of the carbonyl cofactor to lysyl oxidase [13]. Indeed, strong similarities were found between the resonance Raman spectra of the phenylhydrazones of bovine plasma amine oxidase and an active site peptide isolated from bovine aortic lysyl oxidase [18], while it was also reported that PQQ was found in a proteolytic digest of human placental lysyl oxidase [19]. Consistent with these results, PQQ-deficient rats develop growth anomalies and lathrytic symptoms, while the levels of lysyl oxidase protein and activity were reduced [20]. Feeding of PQQ to these deficient animals tended to reverse these effects [20] in support of a vitamin role and a cofactor function for this *o*-quinone in lysyl oxidase.

A more recent report contradicted the presence of PQQ in bovine plasma amine

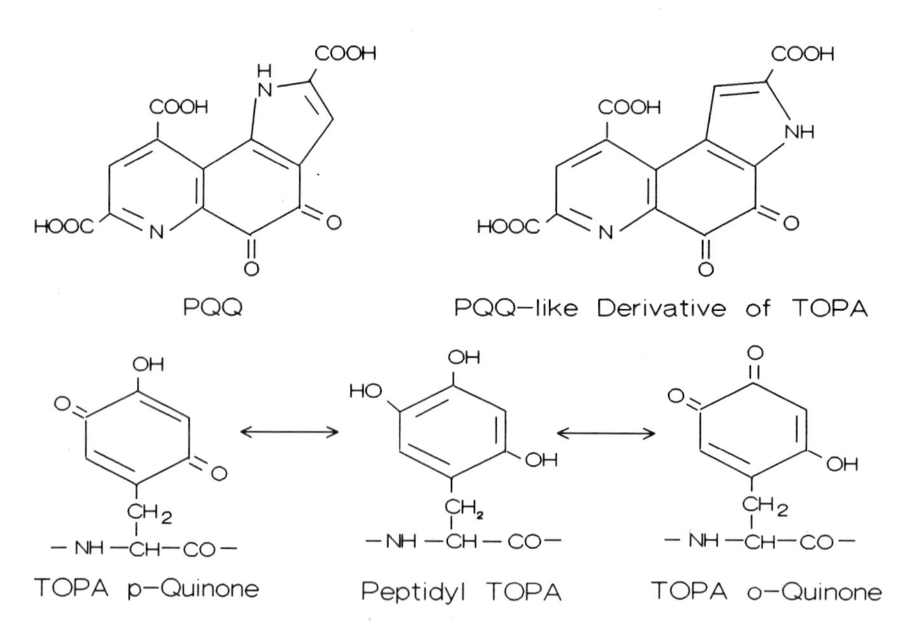

Figure 2 Possible carbonyl cofactors and related compounds. The *o*- and *p*-quinones of TOPA can arise by spontaneous oxidation of TOPA in the presence of oxygen.

oxidase [21], the enzyme that had served as the prototypic mammalian copper-dependent amine oxidase quinoprotein and as a model for the cofactor identification in lysyl oxidase. Thus, mass spectroscopic and NMR analyses of an active site peptide of bovine plasma amine oxidase indicated that a quinone form of a trihydroxyphenylalanine residue (TOPA; 6-hydroxydopa) (Fig. 2) in the -X-position in the sequence: -Leu-Asn-X-Asp-Tyr- was the site of attack by carbonyl reagents and therefore was the carbonyl cofactor in this enzyme [21]. Indeed, the UV-visible spectral characteristics of the TOPA and PQQ phenylhydrazones are similar, while TOPA quinones with free α-amino and α-carboxylate functions released in digests of TOPA-containing proteins can spontaneously convert to a compound structurally very similar to PQQ (Fig. 2). Although lysyl oxidase is not nearly as abundant as plasma amine oxidase and it is thus more difficult to obtain sufficient quantities of active site peptides for such thorough physical-chemical analyses, it remains of critical importance to characterize the properties of its carbonyl cofactor completely enough to allow its unequivocal identification. Such efforts are in progress.

IV. MECHANISM OF ACTION

Steady-state kinetic analyses [22] indicated that lysyl oxidase follows a ping-pong kinetic course with the order of substrate binding and product release as shown in Figure 3. A chemical mechanism following this kinetic pattern that assumes a PQQ-like, *o*-quinone cofactor is shown in Figure 4. It is likely that the reaction begins with Schiff base formation between the ϵ-amino group of a substrate lysine residue and a carbonyl of the organic cofactor forming an enzyme-substrate adduct (Fig. 4, I). Kinetic isotope effects seen with α-deuterated amine substrates are consistent with the subsequent, rate-contributing abstraction of an α-proton from the bound amine [23] to form a transient carbanion (Fig. 4, II-A), chemical support for which has been obtained [23]. Passage of an electron pair into the carbonyl cofactor and tautomerization would yield canonical form II-B from which the aldehyde product can be released by hydrolysis. The would generate a two-electron reduced, aminated form of the carbonyl cofactor, i.e., the amino-

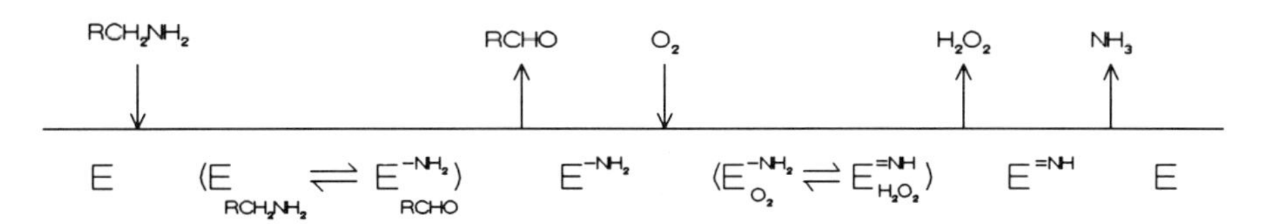

Figure 3 Order of substrate binding and product release in lysyl oxidase catalysis. Possible enzyme intermediates are symbolically represented below the line.

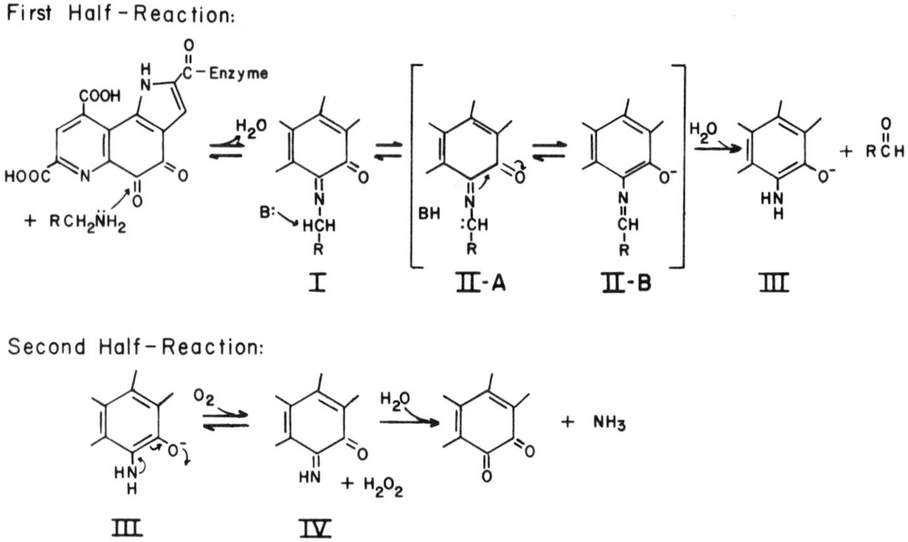

First Half-Reaction:

Second Half-Reaction:

Figure 4 Proposed mechanism of action of lysyl oxidase.

phenol of a PQQ- (Fig. 4, III) or TOPA-like residue. In the absence of oxygen, one mole of aldehyde is produced and released per mole of active site, consistent with the kinetic prediction for the independence of the first half-reaction of the oxygen substrate [22]. The order of substrate binding and product release in the second half-reaction (Fig. 4) was predicted by product inhibition kinetic studies [22]. In toto, the second half-reaction accounts for the reoxidation of the cofactor by oxygen, which is reduced to hydrogen peroxide in the process. Once the aminophenol is oxidized to the iminoquinone (Fig. 4, IV), ammonia can be released by hydrolysis to regenerate the oxidized enzyme for another round of catalysis.

The role of copper in this mechanism has yet to be ascertained. However, recent studies indicate that the metal-free apoenzyme cannot catalyze the first half-reaction in the presence or absence of oxygen [10]. Moreover, the absorption spectrum of the apoenzyme is not perturbed by the aerobic or anaerobic addition of an alkylamine substrate, whereas anaerobic addition of *n*-butylamine to the native holoenzyme induces a spectral change consistent with the formation of an aminophenol [10]. Thus, the presence of the metal ion is critical at least to the first half-reaction. It is possible that copper stabilizes a catalytically competent conformation and/or may align the carbonyl cofactor and the amine substrate for further processing.

In view of its redox activity, it might be expected that enzyme-bound Cu(II) might also function in the reoxidation of the substrate-reduced carbonyl cofactor, initially accepting one electron to yield a Cu(I)-semiquinone pair. In turn, Cu(I) could then transfer an electron to molecular oxygen, forming superoxide (O_2^-) and regenerating Cu(II). Another copper-mediated one-electron transfer from the semiquinone would then complete the reoxidation of the carbonyl cofactor and reduce a superoxide intermediate to peroxide (O_2^{2-}), a final product of the amine oxidases. Evidence for such events in the amine oxidases has been lacking until recently, when it was reported that the ESR spectrum determined at 22°C of various copper-dependent amine oxidases changed from that expected of protein-bound Cu(II) to spectra consistent with a Cu(I)-semiquinone complex upon anaerobic reduction by substrate in the absence of oxygen [24]. Such analyses remain to be carried out with lysyl oxidase.

V. INHIBITORS

The search for specific and highly potent inhibitors of lysyl oxidase remains an important goal in view of the potential of such compounds as antifibrotic agents. Clearly, this catalyst would seem to be a reasonable chemotherapeutic target for such agents in view of its accessibility in the extracellular compartment and since excess collagen molecules are degraded if they are not insolubilized by cross-linking into accreting collagen fibers. A variety of in vivo studies support the feasibility of this approach. For example, administration of BAPN has been shown to limit the fibrotic response following certain surgical procedures and to limit collagen deposition in models of lung fibrosis [1].

As insights into the mechanism of action of lysyl oxidase have been gained, various molecular candidates have been tested as active site-directed inhibitors, some of which behave as mechanism-based, irreversible enzyme inhibitors. For example, β-haloethylamines are irreversible inhibitors of lysyl oxidase. Like true substrates, these agents appear to undergo α-proton abstraction. Tautomerization of the resulting carbanion can lead to departure of the β-halide, thus generating a Michael acceptor subject to covalent derivatization by an enzyme nucleophile (Fig. 5) [25]. The potencies of these agents are similar to that for BAPN (K_I = 5 μM) with K_I in the 6- to 10-μM range. Irreversible inhibition also develops with BAPN [1], possibly because of enzyme processing of the bound inhibitor to a ketenimine, which can then be covalently derivatized by an enzyme nucleophile (Fig. 5). Similarly, benzylamines substituted in the *para* position (Fig. 5) with electron-withdrawing functions behave as ground-state inhibitors of lysyl oxidase, presumably by stabilizing the carbanion resulting from α-proton abstraction, delocalizing the carbanion electrons in a fashion that inhibits tautomerization of the carbanion to a form from which the aldehyde product can be hydrolyzed [26].

Although lysyl oxidase can productively oxidize distal diamines such as 1,4-diaminobutane, recent studies reveal that vicinal 1,2-diamines such as ethylene-diamine are irreversible inhibitors of this enzyme [27]. It was hypothesized that

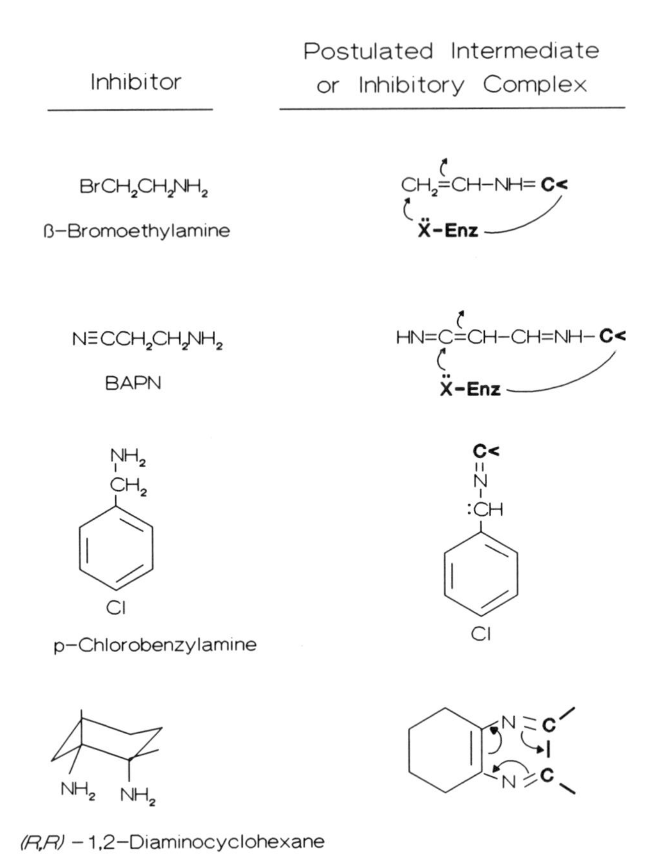

Figure 5 Representative inhibitors of lysyl oxidase. Symbols in bold print represent enzyme residues, including one or both carbonyls of the carbonyl cofactor and a possible enzyme nucleophile.

the inhibition involves initial *bis* Schiff base formation between the two amino functions of the diamine and the two carbonyls of the *o*-quinone cofactor. Consistent with the expected steric preferences of such a condensation, (*R*,*R*)-1,2-diaminocyclohexane proved to be the most potent irreversible inhibitor of lysyl oxidase found thus far ($K_I = 0.4\ \mu M$), while the *S*,*S* isomer of this compound was hardly inhibitory ($IC_{50} > 10$ mM) [27]. Chemical and kinetic analyses of the inhibition supported a mechanism whereby a *bis* Schiff base adduct appears to be oxidatively processed without turnover to form a fully conjugated six-membered ring composed of two carbons of the *o*-quinone cofactor and two carbons and two nitrogens of the diamine (Fig. 5) [27]. The proposed mechanism is consistent with the expected reactivity and steric preferences of an *o*-quinone such as PQQ or TOPA at the active site.

VI. BIOSYNTHESIS, PROCESSING, AND SECRETION

Insights into the nature of biosynthetic forms of lysyl oxidase and the amino acid sequence of the enzyme have recently been gained by the sequencing of lysyl oxidase cDNA isolated from a rat aortic cDNA library [28], a summary representation of which is shown in Figure 6 and the full coding region sequence of which is shown in Figure 7. This 2678 bp cDNA consisted of 286 bp of partial 5′- and 1159 bp of partial 3′-untranslated sequences flanking a 1233 bp region coding for a 411 amino acid, 46.6 kDa protein species. A putative signal peptide sequence exists at the N-terminal region of this species consistent with the secretory fate of lysyl oxidase. Proteolytic removal of the signal peptide by cleavage between Cys21 and Ala22 would yield a 44.5 kDa protein at the C-terminal region of which sequences corresponding to those of peptides isolated from the 32 kDa active bovine aortic enzyme are found.

The likelihood that such a larger species represents a lysyl oxidase precursor was supported by the specific immunoprecipitation of a 46–48 kDa cell-free

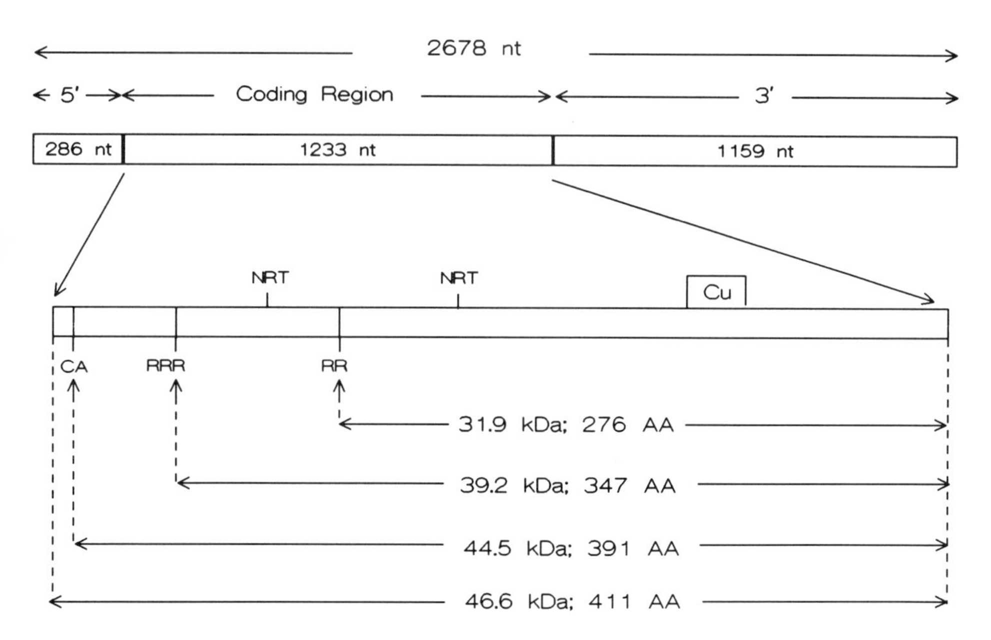

Figure 6 Lysyl oxidase cDNA and predicted protein species. Possible proteolytic processing sites, potential sites of *N*-glycosylation and localization of a putative copper-binding site are illustrated. nt, nucleotides; AA, amino acids; NRT (Asn-Arg-Thr), potential *N*-glycosylation site; RR (Arg-Arg), potential proteolytic site; CA (Cys-Ala), potential signal peptidase cleavage site.

```
ATG CGT TTC GCC TGG ACC GTG CTC TTT CTG GGA CAG CTG CAG TTC TGT CCC CTT CTC CGC
Met Arg Phe Ala Trp Thr Val Leu Phe Leu Gly Gln Leu Gln Phe Cys Pro Leu Leu Arg
                                    10                                       20
TGC GCC CCG CAG GCC CCG CGC GAG CCT CCC GCC GCC CCC GGT GCC TGG CGC CAG ACA ATC
Cys Ala Pro Gln Ala Pro Arg Glu Pro Pro Ala Ala Pro Gly Ala Trp Arg Gln Thr Ile
                                    30                                       40
CAA TGG GAG AAC AAC GGG CAG GTG TTC AGT CTG TTG AGC CTG GGG GCG CAG TAC CAG CCT
Gln Trp Glu Asn Asn Gly Gln Val Phe Ser Leu Leu Ser Leu Gly Ala Gln Tyr Gln Pro
                                    50                   *                   60
CAG CGA CGC CGC GAC TCC AGC GCC ACT GCC CCG AGA GCC GAC GGC AAC GCT GCA GCA CAG
Gln Arg Arg Arg Asp Ser Ser Ala Thr Ala Pro Arg Ala Asp Gly Asn Ala Ala Ala Gln
                                    70                                       80
CCA CGC ACG CCC ATT CTG CTG CTG CGT GAC AAC CGC ACT GCC TCT GCC CGT GCG AGG ACT
Pro Arg Thr Pro Ile Leu Leu Leu Arg Asp Asn Arg Thr Ala Ser Ala Arg Ala Arg Thr
                                    90                                      100
CCA AGC CCA TCT GGG GTC GCC GCG GGT CGT CCC CGG CCC GCA GCC CGC CAC TGG TTC CAA
Pro Ser Pro Ser Gly Val Ala Ala Gly Arg Pro Arg Pro Ala Ala Arg His Trp Phe Gln
                                    110                          **         120
GTT GGT TTC TCG CCG TCG GGG GCC GGC GAT GGA GCC TCA AGG CGC GCA GCG AAC CGG ACT
Val Gly Phe Ser Pro Ser Gly Ala Gly Asp Gly Ala Ser Arg Arg Ala Ala Asn Arg Thr
                                    130                                     140
GCG TCG CCA CAG CCT CCG CAG CTC AGT AAT CTG AGG CCA CCC AGC CAC GTA GAT CGC ATG
Ala Ser Pro Gln Pro Pro Gln Leu Ser Asn Leu Arg Pro Pro Ser His Val Asp Arg Met
 *                                  150                                     160
GTG GGC GAC GAC CCC TAC AAT CCC TAC AAG TAC TCC GAC GAC AAC CCC TAT TAT AAC TAC
Val Gly Asp Asp Pro Tyr Asn Pro Tyr Lys Tyr Ser Asp Asp Asn Pro Tyr Tyr Asn Tyr
                         **         170                                     180
TAT GAC ACT TAT GAG AGA CCC CGG TCC GGG AGC AGG CAC CGA CCT GGA TAT GGC ACC GGT
Tyr Asp Thr Tyr Glu Arg Pro Arg Ser Gly Ser Arg His Arg Pro Gly Tyr Gly Thr Gly
                                    190                                     200
TAC TTC CAG TAC GGT CTC CCG GAC CTG GTA CCC GAT CCC TAC TAC ATC CAG GCA TCC ACG
Tyr Phe Gln Tyr Gly Leu Pro Asp Leu Val Pro Asp Pro Tyr Tyr Ile Gln Ala Ser Thr
                                    210                                     220
TAC GTA CAA AAG ATG TCT ATG TAC CTG AGA TGC GCT GCG GAA GAA AAC TGC CTG GCC AGT
Tyr Val Gln Lys Met Ser Met Tyr Asn Leu Arg Cys Ala Ala Glu Glu Asn Cys Leu Ala
                                    230                                     240
AAC TCA GCA TAT AGG GCG GAT GTC AGA GAC TAT GAC CAC AGG GTA CTG CTA CGA TTT CCT
Ser Ser Ala Tyr Arg Ala Asp Val Arg Asp Tyr Asp His Arg Val Leu Leu Arg Phe Pro
                                    250                                     260
CAG AGA GTG AAA AAC CAA GGG ACG TCT GAC TTC TTA CCA AGC CGC CCC CGC TAC TCC TGG
Gln Arg Val Lys Asn Gln Gly Thr Ser Asp Phe Leu Pro Ser Arg Pro Arg Tyr Ser Trp
                                    270                                     280
GAG TGG CAC AGC TGC CAC CAA CAT TAC CAC AGC ATG GAT GAA TTC AGC CAC TAC GAC CTG
Glu Trp His Ser Cys His Gln His Tyr His Ser Met Asp Glu Phe Ser His Tyr Asp Leu
                                    290                                     300
CTG GAT GCC AGC ACA CAG AGG AGA GTG GCC GAG GGC CAC AAA GCA AGC TTC TGT CTG GAG
Leu Asp Ala Ser Thr Gln Arg Arg Val Ala Glu Gly His Lys Ala Ser Phe Cys Leu Glu
                                    310                                     320
GAC ACT TCC TGT GAT TAT GGG TAC CAC AGA CGA TTT GCC TGT ACT GCA CAC ACA CAG GGG
Asp Thr Ser Cys Asp Tyr Gly Tyr His Arg Arg Phe Ala Cys Thr Ala His Thr Gln Gly
                                    330                                     340
TTG AGT CCC GGA TGT TAT GAT ACT TAT GCA GCA GAC ATA GAC TGC CAG TGG ATT GAT ATT
Leu Ser Pro Gly Cys Tyr Asp Thr Tyr Ala Ala Asp Ile Asp Cys Gln Trp Ile Asp Ile
                                    350                                     360
ACA GAT GTA CAA CCC GGA AAT TAC ATT CTA AAG GTC AGT GTA AAC CCC AGC TAC CTG GTG
Thr Asp Val Gln Pro Gly Asn Tyr Ile Leu Lys Val Ser Val Asn Pro Ser Tyr Leu Val
                                    370                                     380
CCT GAA TCA GAC TAC AGT AAC AAT GTC GTA CGC TGT GAA ATT CGC TAC ACA GGA CAT CAC
Pro Glu Ser Asp Tyr Ser Asn Asn Val Val Arg Cys Glu Ile Arg Tyr Thr Gly His His
                                    390                                     400
GCC TAT GCC TCA GGC TGC ACC ATT TCA CCG TAT TAGAAAGAAGCT...............
Ala Tyr Ala Ser Gly Cys Thr Ile Ser Pro Tyr END
                                    410
```

Figure 7 Coding region cDNA and predicted amino acid sequence of rat aorta lysyl oxidase. The sequence shown differs with respect to the addition of six nucleotides and resulting changes in the amino acid sequence from that reported in Ref. 28. These nucleotides are marked by asterisks and are in bold print in the codons corresponding to Gly75, Ala137, Val 161, and Pro187.

translation product of bovine smooth muscle cell mRNA by rabbit antilysyl oxidase in this report [28] and by the finding of a 48 kDa cell-free translation product of mRNA isolated from fibrotic rat liver [29]. Moreover, monoclonal antihuman lysyl oxidase reacted with a 48 kDa, enzymatically inactive peak resolved by gel exclusion chromatography of extracts of human placenta [30]. The predicted amino acid sequence between residues 311 and 400 is consistent with known metal-binding sequences in certain copper metalloproteins and occurs within the region accounting for the functional 32 kDa enzyme. The putative precursor also contains Arg-Arg-Arg and Arg-Arg sequences at residues 62–64 and at 134–135, respectively. These polybasic sequences are potential proteolytic processing sites whereby cleavage at residues 62–64 would yield a 39.2 kDa species, while cleavage at 134–135 would yield a 31.9 kDa enzyme (Fig. 6). Using antibody raised against the purified 32 kDa bovine aortic enzyme, studies in progress in our laboratory have identified protein bands at 46–48, 38-, and 32 kDa specifically immunoprecipitated from fractions of rat aorta smooth muscle cells and rat lung fibroblasts pulse-labeled in culture, consistent with these predictions. Two Asn-Arg-Thr sequences also appear in the predicted sequence (Fig. 6), indicating the potential for *N*-glycosylation at both Asn residues. Our current studies also support the conclusion that the 46 kDa precursor is *N*-glycosylated.

Several reports have noted that lysyl oxidase is a secreted protein, as reviewed [1]. More recently, Kuivaniemi et al. [31] found that the secretion of enzyme by human skin fibroblasts was inhibited by monensin or nigericin, ionophores known to inhibit secretion of proteins that are processed through the Golgi pathway [31]. These workers also noted that matrix-bound enzyme activity decreased in parallel with that in the medium in monensin-treated cultures, suggesting that this bound form may be in equilibrium with the soluble catalyst in the medium.

VII. LOCALIZATION IN TISSUES AND CELLS

Enzyme was localized at the electron microscope level in the extracellular matrix of rat lung and aorta employing antilysyl oxidase and second antibody complexed with colloidal gold [32]. The electron dense particles were seen in association with the microfibrillar network surrounding elastic fibers and within the amorphous elastic fibers themselves. However, using the same technique, others have reported relatively sparse association of gold deposits on matrix microfibrils but distinct deposits on and within elastin and collagen fibers of human placenta, skin, and aorta [33]. The microfibrillar network appears in the extracellular space prior to the formation of elastic fibers and then appears to serve as a scaffolding within which amorphous elastin is deposited [34]. These structures persist as a filamentous network largely on the perimeter and, to a lesser degree, within the developing elastic fiber. The apparent association of lysyl oxidase with this microfibrillar material implies that the enzyme may function at such sites to cross-link new tropoelastin units to the radially accreting elastic fiber. Possible mechanisms by which the appropriate intermolecular associations are achieved to properly align tropoelastin, lysyl oxidase, and the microfibrillar scaffolding in extracellular space are yet to be discerned. As an additional component to be considered in such interactions, Fornieri et al. [35] reported that administration of lysyl oxidase inhibitors to developing chicks and rats resulted in the appearance of unusual, lateral aggregates of aortic elastic fibers, which were permeated by glycosaminoglycans. It was postulated that polyanionic glycosaminoglycans might pair with charge ϵ-amino groups of tropoelastin, thus contributing to appropriate intermolecular relationships prior to the cross-linking process. In view of the evidence that lysyl oxidase functions as an extracellular enzyme, it is of interest that lysyl oxidase was localized along cytoskelatal structures within cultured human fibroblasts by fluorescence immunohistochemistry [36]. The apparent predominance of the staining reaction at such loci warrants further investigation into the possible biological significance of this observation.

VIII. LYSYL OXIDASE IN PATHOLOGICAL PROCESSES

A. Dietary Factors and Environmental Toxins

The critical role played by lysyl oxidase in the development and homeostasis of the lung is reflected in the disruption of lung structure in chick and rat models of dietary copper deficiency [37,38]. Markedly decreased levels of lysyl oxidase seen in these models were consistent with the accompanying reduction in insoluble lung elastin, dilation of the airways, and increased level of soluble collagen [37,38]. Tinker et al. demonstrated that copper-deficient diets result in both a decrease in cross-linking activity and an increase in elastin degradation in chick aorta, while restoration of copper restores normal cross-linking and imparts resistance to elastolysis [39]. Thus, the activity of lysyl oxidase appears essential to the accumulation as well as the prevention of loss of this connective tissue protein.

Induction of lung fibrosis by exposure to agents such as bleomycin and cadmium is accompanied by large increases in lung lysyl oxidase activity [1]. The elevated enzyme activity in cadmium-exposed rat lung was recently shown to reflect increased enzyme protein [6]. The newly formed, cadmium-induced enzyme resolved into two major and additional minor peaks of activity upon elution from DEAE-Trisacryl by a salt gradient in 6 M urea [6], thus exhibiting the ionic heterogeneity characteristic of the highly purified enzymes of various bovine, chick, and human tissues [1]. While the basis of this phenomenon is unknown, this result suggests that it may occur in coincidence with enzyme synthesis.

In contrast to the stimulation of lysyl oxidase in the lung by cadmium exposure, bone lysyl oxidase activity decreases in cadmium- fed chicks apparently because of the replacement of copper in the enzyme by cadmium [40]. It is likely that the opposite effects on lysyl oxidase activities seen in lung and bone reflect the marked increase in metallothionein that occurs in the lung model, whereas this metal scavenging protein apparently is not synthesized in response to cadmium in the bone [41].

The activation of lung connective tissue protein formation in response to toxic agents could stem from factors released upon cell injury or from fibrogenic factors introduced into the lung or other connective tissues by infiltrating macrophages and neutrophils. One such mediator has been implicated in the development of fibrosis in silicotic rat lungs [42] in which lysyl oxidase and other enzymes of collagen biosynthesis are elevated [43]. Thus, a 16 kDa protein isolated from silicotic rat lungs and apparently derived from macrophages was found to stimulate collagen synthesis in granulation tissue cells at a concentration of 0.1 nM. Its effect on lysyl oxidase expression does not appear to have been determined. The first evidence for the regulation of lysyl oxidase at the genetic level in a fibrotic model stems from the three-fold increase in functional lysyl oxidase mRNA activity seen in cell-free translations of fibrotic rat liver mRNA [29]. Additional insights into the molecular levels of regulation should be forthcoming in view of the recent description of the sequence of lysyl oxidase cDNA [28].

B. Genetic Disease

As previously reviewed, levels of lysyl oxidase are markedly decreased in certain genetically related disease states, including Menkes' disease and X-linked cutis laxa [1]. A recent report describes an unusually severe case of Menkes' disease in which skin fibroblast lysyl oxidase levels were only 6–12% of controls [44], a level that clearly can account for the severe connective tissue pathology associated with this disease. The enzyme deficiency in both of these diseases appears to be secondary to disturbances in copper metabolism, however. Other studies implying deficiencies of lysyl oxidase in osteogenesis imperfecta and Marfan syndrome have recently been contradicted [45]. Aortic lysyl oxidase activity is decreased 30–50% in brown Norway rats, which are genetically susceptible to disruption of the internal elastic lamina [46]. Coupled with the apparent increase in aortic elastolytic activity in this rat, the decrease in lysyl oxidase could contribute to the manifestations of this syndrome. It should be noted, however, that decreases of lysyl oxidase activity of these magnitudes may not result in deficient cross-linking in view of the observation that the 64% reduction of enzyme activity in the skin of glucocorticoid-treated rats was accompanied by an apparent increase in collagen cross-linking [47].

IX. BIOLOGICAL REGULATION OF LYSYL OXIDASE

Lysyl oxidase activity responds to changes in levels of certain hormones, as illustrated by the estrogen-induced increase in mouse cervical lysyl oxidase activity [48]. The addition of 10–100 nM levels of testosterone to the medium of cultured aortic smooth muscle cells increased cell layer lysyl oxidase activity 5.5-fold [49]. This effect is of interest in view of the greater tendency of males to develop arterial lesions that can become fibrotic with the progression of atherosclerosis.

It has been commonly observed that the expression of lysyl oxidase by cells in culture is most active when proliferation has stopped or slowed, consistent with the most active stage of matrix production. It is of particular interest in this regard that lysyl oxidase activity as well as rates of collagen synthesis were markedly reduced in malignantly transformed human cell lines [50]. Another intriguing aspect of the relationship between lysyl oxidase and tumorigenesis has recently emerged. Thus, the normal expression of the *rrg* transcript in mouse NIH 3T3 cells is markedly down-regulated by c-H-*ras*-induced transformation of these cells to the tumorigenic RS485 cell line [51]. High levels of *rrg* expression were restored by interferon-induced reversion of RS485 cells to the nontumorigenic PR4 cell line. The specific importance of *rrg* to the reversion was emphasized by the fact that the normally persistent revertants developed the transformed phenotype when stably transfected with *rrg* antisense cDNA [51]. Since the appearance of this report, it has been learned that *rrg* is 92% homologous at the cDNA sequence level and 98% homologous at the predicted protein sequence level to lysyl oxidase [52]. It is not known at this writing whether the *rrg* protein is catalytically active or if lysyl oxidase catalysis is involved in the reversion of the tumorigenic state. However, these intriguing results do indicate that a lysyl oxidase–like protein may play a critical role in the maintenance of the nontumorigenic state in these fibroblast lines.

X. SUMMARY AND PROSPECTS

Progress has been made on several aspects of the enzymology and biology of lysyl oxidase. While the precise chemical identity of the organic cofactor requires further investigation, this moiety exhibits the reactivity of an *o*-quinone and thus is consistent with a PQQ- or TOPA-like residue. The catalytic mechanism of this enzyme clearly parallels those proposed for other mammalian copper-dependent amine oxidases. As further details of the active site structures of these proteins become available, structural homologies in active site regions may also be found, although it should be noted that the sequence of the TOPA-bearing active site peptide of plasma amine oxidase is not seen in the lysyl oxidase protein sequence. Recent insights into the secretory mechanism and the biosynthetic precursor make it likely that detailed information about the biosynthesis and processing of this important enzyme will soon be forthcoming. In addition, evidence pointing to an unexpected intracellular localization and the recent results implying a relationship between lysyl oxidase and tumorigenesis certainly raise important new prospects for future research.

ACKNOWLEDGMENTS

The work reviewed here is supported by NIH Grants R37 AR 18880, HL 13262, and HL 19717.

REFERENCES

1. Kagan, H. M. (1986). Characterization and regulation of lysyl oxidase. In *Biology of Extracellular Matrix: A Series Regulation of Matrix Accumulation*, Vol. 1. (R. P. Mecham, ed.), Academic Press, Orlando, FL, p. 321.
2. Narayanan, A. S., Siegel, R. C., and Martin, G. R. (1974). Stability and purification of lysyl oxidase. *Arch. Biochem. Biophys. 162*: 231.
3. Kagan, H. M., Sullivan, K. A., Olsson, T. A., and Cronlund, A. L. (1979). Purification and properties of four species of lysyl oxidase from bovine aorta. *Biochem. J. 177*: 203.
4. Cronlund, A. L., and Kagan, H. M. (1986). Comparison of lysyl oxidase of bovine aorta and lung. *Connect. Tiss. Res. 15*: 173.
5. Kuivaniemi, H., Savolainen, E. R., and Kivirikko, K. I. (1984). Human placental lysyl oxidase. Purification, partial characterization and preparation of two specific antisera to the enzyme. *J. Biol. Chem. 259*: 6996.
6. Almassian, B. J., Trackman, P. C., Iguchi, H., Boak, A., Calvaresi, D., and Kagan, H. M. (1990). Induction of lung lysyl oxidase activity and lysyl oxidase protein by exposure of rats to cadmium chloride: properties of the induced enzyme. *Connect. Tissue Res. 25*: 197.
7. Kuivaniemi, H. (1985). Partial characterization of lysyl oxidase from several human tissues. *Biochem. J. 230*: 639.
8. Shackleton, D. R., and Hulmes, D. J. S. (1990). Purification of lysyl oxidase from piglet skin by selective interaction with Sephacryl S-200. *Biochem. J. 266*: 917.
9. Iguchi, H., and Sano, S. (1985). Cadmium- or zinc-binding to bone lysyl oxidase and copper replacement. *Connect. Tiss. Res. 14*: 129.
10. Gacheru, S. N., Trackman, P. C., Shah, M. A., O'Gara, C. Y., Spacciapoli, P., Greenaway, F. T., and Kagan, H. M. (1990). Structural and catalytic properties of copper in lysyl oxidase. *J. Biol. Chem. 265*: 19022.
11. Levene, C. I., O'Shea, M. P., and Carrington, C. J. (1988). Protein lysine 6-oxidase (lysyl oxidase) cofactor; methoxatin (PQQ) or pyridoxal? *Int. J. Biochem. 20*: 1451.
12. Bird, T. A., and Levene, C. I. (1982). Lysyl oxidase: Evidence that pyridoxal phosphate is a cofactor. *Biochem. Biophys. Res., Commun. 108*: 1172.
13. Williamson, P. R., Kittler, J. M., Thanassi, J. W., and Kagan, H. M. (1986). Reactivity of a functional carbonyl moiety in bovine lysyl oxidase. Evidence against pyridoxal 5′-phosphate. *Biochem. J. 235*: 597.
14. Williamson, P. R., Moog, R. S., Dooley, D. M., and Kagan, H. M. (1986). Evidence for pyrroloquinoline quinone as the carbonyl cofactor in lysyl oxidase by absorption and resonance Raman spectroscopy. *J. Biol. Chem. 261*: 16302.
16. Ameyama, M., Hayashi, M., Matsushita, K., Shinagawa, E., and Adachi, O. (1984). Microbial production of pyrroloquinoline quinone. *Agric. Biol. Chem. 48*: 561.

17. Lobenstein-Verbeek, C. L., Jonjegan, J. A., Frank, J., and Duine, J. A. (1984). Bovine serum amine oxidase: A mammalian enzyme having covalently bound PQQ as a prosthetic group. *FEBS Lett. 170*: 305.
20. Killgore, J., Smidt, C., Duich, L., Romero-Chapman, N., Tinker, N., Reiser, K., Melko, M. Hyde, D., and Rucker, R. B. (1989). Nutritional importance of pyrroloquinoline quinone. *Science 245*: 850.
21. Janes, S. M., Mu, D., Wemmer, D., Smith, A. J., Kaue, S., Maltby, S,. Burlingame, A. L., and Klinman, J. P. (1990). A new redox coactor in eukaryotic enzymes: 6-Hydroxydopa at the active site of bovine serum amine oxidase. *Science 248*: 981.
22. Williamson, P. R., and Kagan, H. M. (1986). Reaction pathway of bovine aortic lysyl oxidase. *J. Biol. Chem. 261*: 9477.
23. Williamson, P. R., and Kagan, H. M. (1987). α-Proton abstraction and carbanion formation in the mechanism of action of lysyl oxidase. *J. Biol. Chem. 262*: 8196.
24. Dooley, D. M., McGuirl, M. A., Brown, D. E., Turowski, P. N., McIntire, W. S., and Knowles, P. F. (1991). A Cu(I) semiquinone state in substrate-reduced amine oxidases. *Nature 349*: 262.
25. Tang, S. S., Simpson, D. E., and Kagan, H. M. (1984). β-Substituted ethylamine derivatives as suicide inhibitors of lysyl oxidase. *J. Biol. Chem. 259*: 975.
26. Williamson, P. R., and Kagan, H. M. (1987). Electronegativity of aromatic amines as a basis for the development of ground state inhibitors of lysyl oxidase. *J. Biol. Chem. 262*: 14520.
27. Gacheru, S. N., Trackman, P. C., Calaman, S. D., Greenaway, F. T., and Kagan, H. M. (1989). Vicinal diamines as pyrroloquinoline quinone-directed irreversible inhibitors of lysyl oxidase. *J. Biol. Chem. 264*: 12963.
28. Trackman, P. C., Pratt, A. M., Wolanski, A., Tang, S. S., Offner, G. D., Troxler, R. T., and Kagan, H. M. (1990). Cloning of rat aorta lysyl oxidase cDNA: Complete codons and predicted amino acide sequence. *Biochemistry 29*: 4863.
29. Wakasaki, H., and Ooshima, A. (1990). Synthesis of lysyl oxidase in experimental hepatic fibrosis. *Biochem. Biophys. Res. Commun. 166*: 1201.
30. Burbelo, P. D., Monckeberg, A., and Chichester, C. O. (1986). Monoclonal antibodies to human lysyl oxidase. *Coll. Rel. Res. 6*: 153.
31. Kuivaniemi, H., Ala-Kokko, L., and Kivirikko, K. I. (1986). Secretion of lysyl oxidase by cultured human skin fibroblasts and effects of monensin, nigericin, tunicamycin and colchicine. *Biochim. Biophys. Acta 883*: 326.
32. Kagan, H. M. Vaccaro, C. A., Bronson, R. E., Tang, S. S., and Brody, J. S. (1986). Ultrastructural immunolocalization of lysyl oxidase in vascular connective tissue. *J. Cell Biol. 103*: 1121.
33. Baccarani-Contri, M., Vincenzi, D., Quaglino, D., Mori, G., and Pasquali-Ronchetti, I. (1990). Localization of human placenta lysyl oxidase on human placenta, skin and aorta by immunoelectron-microscopy. *Matrix 9*: 428.
34. Cleary, E. G. (1987). The microfibrillar component of elastic fibers. Morphology and biochemistry. In *Connective Tissue Disease. Molecular Pathology of the Extracellular Matrix* (Uitto, J., and Perejda, A. J., eds.), Marcel Dekker, New York, P. 55.
35. Fornieri, C., Baccarani-Contri, M., Quaglino, D., and Pasquali-Ronchetti, I. (1987). Lysyl oxidase activity and elastin/glycosaminoglycan interactions in growing chick and rat aortas. *J. Cell Biol. 105*: 1463.

36. Wakasaki, H., and Ooshima, A. (1990). Immunohistochemical localization of lysyl oxidase with monoclonal antibodies. *Lab Invest. 63*: 377.
37. Harris, E. D. (1986). Biochemical defect in chick lung resulting from copper deficiency. *J. Nutr. 116*: 252.
38. Dubick, M. A., Keen, C. L., and Rucker, R. B. (1985). Elastin metabolism during perinatal lung development in the copper-deficient rat. *Exp. Lung. Res. 8*: 237.
39. Tinker, D., Romero-Chapman, N., Reiser, K., Hyde, D., and Rucker, R. (1990). Elastin metabolism during recovery from impaired crosslink formation. *Arch. Biochem. Bioph. 278*: 326.
40. Iguchi, H., and Sano, S. (1985). Cadmium- zinc-binding to bone lysyl oxidase and copper replacement. *Connect. Tiss. Res. 14*: 129.
41. Hirano, S., Tsukamoto, N., and Suzuki, K. T. (1990). Biochemical changes in the rat lung and liver following intra-tracheal instillation of cadmium oxide. *Toxicol. Lett. 50*: 97.
42. Aalto, M., Kulonen, E., and Pikkarainen, J. (1989). Isolation of silica-dependent protein from rat lung with special reference to development of fibrosis. *Br. J. Exp. Path. 70*: 167.
43. Poole, A., Myllyla, R., Wagner, J. C., and Brown, R. C. (1985). Collagen biosynthesis enzymes in lung tissue and serum of rats with experimental silicosis. *Br. J. Exp. Pathol. 66*: 567.
44. Royce, P. M., and Steinman, B. (1990). Markedly reduced activity of lysyl oxidase in skin and aorta from a patient with Menkes' disease showing unusually severe connective tissue manifestations. *Pediat. Res. 28*: 136.
45. Royce, P. M., and Steinmann, B. (1988). Lysyl oxidase in osteogenesis imperfecta and Marfan's syndrome. *Coll. Rel. Res. 8*: 183.
46. Osborne-Pellegrin, M. J., Farjanei, J., and Hornebeck, W. (1990). Role of elastase and lysyl oxidase activity in spontaneous rupture of internal elastic lamina in rats. *Arteriosclerosis 10*: 1136.
47. Counts, D. F., Shull, S., and Cutroneo, K. (1986). Skin lysyl oxidase is not rate limiting for collagen crosslinking in the glucocorticoid-treated rat. *Connect. Tiss. Res. 14*: 237.
48. Ozasa, H., Tominaga, T., and Takeda, T. (1986). Evidence of estrogen-like effect of dehydroepiandrosterone on lysyl oxidase activity in the mouse cervix. *Acta Obstet. Gynecol. Scand. 65*: 543.
49. Bronson, R. E., Calaman, S. D., Traish, A. M., and Kagan, H. M. (1987). Stimulation of lysyl oxidase activity by testosterone and characterization of androgen receptors in cultured calf aorta smooth muscle cells. *Biochem. J. 244*: 317.
50. Kuivaniemi, H., Korhonen, R. M., Vaheri, A., and Kivirriko, K. I. (1986). Deficient production of lysyl oxidase in cultures of malignantly transformed human cells. *FEBS Lett. 195*: 261.
51. Contente, S., Kenyon, K., Rimoldi, D., and Friedman, R. F. (1990). Expression of gene *rrg* is associated with reversion of NIH 3T3 transformed by LTR-c-H-*ras*. *Science 249*: 796.
52. Kenyon, K., Contente, S., Trackman, P. C., Tang, J., Kagan, H. M., and Friedman, R. F. (1991). A suppressor of the phenotypic expression of *ras* encodes lysyl oxidase. *Science 253*: 802.

III

Biosynthesis, Structure, and Function of Quinoproteins

8

Regulation and Expression of Bacterial Quinoproteins

Kerstin Laufer and Mary E. Lidstrom

California Institute of Technology, Pasadena, California

I. INTRODUCTION

Bacterial quinoproteins are found in a variety of bacteria that represent a wide range of metabolic types. In each case these quinoproteins are involved in metabolic processes that are regulated by the host bacterium. For three of these proteins—methanol dehydrogenase (MDH), methylamine dehydrogenase (MADH), and glucose dehydrogenase (GDH)—information is available concerning regulatory parameters. However, until recently nothing was known about the mechanisms involved in this regulation or the level at which the regulation occurred. With the development of sophisticated genetic capabilities in many of the bacteria containing these proteins, it is now possible to carry out detailed analyses of the genetic basis of regulation. In this chapter, we will discuss the current evidence concerning regulation and expression of MDH, MADH, and GDH.

II. METHANOL DEHYDROGENASE

A. Regulatory Parameters

Pyrroloquinoline quinone (PQQ)–containing MDH has been found in all gram-negative methane- or methanol-utilizing bacteria thus far examined (for a review see Ref. 1). Contrary to preliminary findings, evidence is now accumulating that the enzyme consists of two subunits in an $\alpha_2\beta_2$ structure [2–5]. Enzyme activity has been located in the periplasm or loosely attached to the outer surface of the inner membrane [1]. Although regulatory phenomena of MDH have been studied extensively in methylotrophs over the last 10 years, not much information concerning the molecular basis for these observations is yet available. From the data so far accumulated it appears that the serine or RuMP pathway methylotrophs follow different regulatory strategies for MDH synthesis than the autotrophic methylotrophs. In the former strains, in which MDH is a part of the catabolic chain of the organisms, enzyme production occurs during growth on multicarbon compounds but can be increased under certain conditions. In these cases, regulation via both substrate induction and derepression at low growth rates has been suggested. In the autotrophic bacteria MDH is more tightly controlled and appears to have two regulatory mechanisms: catabolite repression and product (formaldehyde) induction. The presence of PQQ is apparently not a prerequisite for MDH protein production. However, enzyme activity might be controlled via functions involved in insertion of the cofactor into the apoenzyme. A protein that modulates substrate affinity of MDH has also been suggested to function in this purpose. Further factors found to influence enzyme synthesis and/or activity are the concentration of oxygen and calcium and possibly a low molecular compound involved in electron transfer to cytochromes. The observed stimulatory effect of ammonia on in vitro enzyme activity is not thought to be relevant in vivo. Each of these regulatory effects is discussed in more detail below.

1. Effects on Enzyme Production

Growth substrates. In serine and RuMP pathway methylotrophs, MDH serves to generate both ATP and formaldehyde. The formaldehyde produced is either assimilated into cell carbon or oxidized to CO_2 (for recent reviews see Refs. 6,7). In these organisms the enzyme is regulated separately from other dissimilatory enzymes [8–10]. In serine-pathway methylotrophs, both levels of MDH protein and activity are increased upon addition of methanol, and the presence of excess amounts of multicarbon compounds does not interfere with this effect [8–16]. In the RuMP pathway organism *Methylophilus methylotrophus*, both protein and activity were shown to be increased during growth on methanol as compared to methylamine [17]. Although these observations indicate a regulation of enzyme synthesis by substrate induction, further studies have revealed that protein production is not triggered by methanol alone. Overexpression of MDH activity was reported during growth of *Methylobacterium* species and a restricted RuMP pathway methylotroph on multicarbon compounds that were poor growth substrates [9,18] or for *Methylobacterium* species, *M. methylotrophus* and the restricted RuMP pathway organism at growth-limiting concentrations of methanol [9,10,18–20]. These observations show that MDH is also regulated by derepression in response to growth rate (see next section). When methanol was presented in excess, enzyme production was suppressed [9,10,17–20] probably due to accumulation of intermediates of methanol metabolism [10,17]. Activity of MDH was also significant during growth of these strains on methylamine [15,17,21,22].

In the autotrophic methylotrophs *Paracoccus denitrificans* and *Xanthobacter flavus* H4-14, MDH is used only for the generation of energy (for recent reviews see Refs. 5–7). When *P. denitrificans* was grown on multicarbon compounds at growth-limiting concentrations, methanol oxidation still operated as an auxilliary energy-generating system [23,24]. In the presence of excess concentrations of multicarbon compounds, however, levels of MDH protein were very low and enzyme activity was not detected in either strain [8,24–26]. From these findings it was concluded that in autotrophic methylotrophs MDH protein is kept at low levels by catabolite repression [24,26].

High levels of MDH were observed in *P. denitrificans* during growth on methanol, methylamine, choline, and formaldehyde [24]. As opposed to nonautotrophic methylotrophs, however, no direct effect of methanol on enzyme production could be detected. Surprisingly, maximal protein synthesis was found in the presence of formaldehyde. This C_1 compound might also serve as the signal for induced MDH production during methanol, methylamine, and choline degradation, where it occurs as an intermediate. The formation of high levels of MDH protein in autotrophs is therefore most likely achieved by product induction [24]. In nonautotrophic methylotrophs formaldehyde might be responsible for increased MDH levels in the presence of methylamine by a similar mechanism [5].

Oxygen. Oxygen, the terminal electron acceptor of the methanol oxidase system, has been found to influence MDH protein production in two methylotrophs. In *M. methylotrophus* protein synthesis and activity have been shown to be tuned to the standing concentration of oxygen [10,20]. This allows the cells to maintain the respiration rate needed to fulfill the energy demands of the cell under the respective growth rate. A similar observation was made with *P. denitrificans* grown on choline at different dissolved oxygen concentrations [24].

Growth rate and other physiological parameters. Studies with bacteria growing under more defined conditions in continuous culture have revealed that synthesis of MDH is also influenced by certain physiological factors, such as growth rate or the energetic state of the cell.

In RuMP and serine pathway methylotrophs, it was found that MDH production responds to the growth rate in a reverse, nonlinear relation [9,10,19,20]. High enzyme activities were observed with growth-limiting concentrations of either methanol or multicarbon compounds, or with substrates allowing for poor growth [8–10,18–20]. At high growth rates, MDH levels were decreased, and in the RuMP pathway strains this was true regardless of whether the limiting factor was methanol or nitrogen. It is therefore possible that in these strains the ratio of methanol to total protein is the key regulatory parameter under these conditions

[10,20]. These findings suggest that in both RuMP and serine pathway methylotrophs MDH is regulated by derepression at low growth rates. However, induction by methanol also seems to occur. In *M. methylotrophus*, cells grown in a chemostat on methylamine contained only one third of the MDH activity of cells grown at the same growth rate on methanol plus methylamine [17]. Likewise, in serine pathway methylotrophs, cells grown in batch cultures at similar growth rates had higher MDH activity on methanol than on other substrates [9]. The molecular basis for this phenomenon has, however, not yet been resolved.

In autotrophic methylotrophs the energetic state of the cells seems to be responsible for the fine-tuning of MDH protein levels during catabolite repression [24]. Immunologically detectable protein levels but no enzyme activity could be found when growth occurred under carbon-limiting conditions or on substrates allowing for only poor growth. However, with energized cells, i.e., in the presence of multicarbon substrates supporting rapid growth, no MDH protein could be detected. It has been suggested that a similar mechanism determines the degree of MDH derepression in serine- or RuMP pathway methylotrophs at different growth rates [24].

2. Effects on Enzyme Activity

PQQ. Active MDH requires the cofactor PQQ attached to the apoenzyme in a tight, noncovalent binding [1,7,27,28]. It is therefore conceivable that PQQ might be involved in regulation of apoenzyme synthesis and/or enzyme activity.

A loose correlation exists between the formation of both cofactor and apoenzyme [29]. Overproduction and subsequent excretion of large amounts of PQQ into the surrounding medium have been reported for *Hyphomicrobium* and *Methylobacterium* strains as well as for *Pseudomonas* species producing MDH or PQQ containing alcohol dehydrogenase [28–31]. In *M. organophilum* XX, methanol was shown to have no effect on internal PQQ levels [29].

On the other hand, synthesis of MDH protein is not strictly dependent on the presence of PQQ. Mutants of *M. organophilum* DSM 760 and *M. extorquens* AM1 lacking the ability to synthesize PQQ were found to still contain significant amounts of MDH apoprotein. These mutants were able to grow on methanol if the medium was supplemented with PQQ [29–33]. MDH activity was, however, not restored to full wild-type levels, indicating that the internal PQQ concentration might have an effect on MDH synthesis [28,30].

The second regulatory mechanism involving PQQ that affects in vivo methanol oxidation is the supply of cells with active holoenzyme. Three mutants of *M. extorquens* AM1 have been described to contain wild-type levels of MDH protein, which exhibited an altered UV-visible spectrum in the wavelengths attributed to absorbance by PQQ [34,35]. These enzymes proved to be incapable of oxidizing methanol. Therefore, it was suggested that functions impaired in these mutants might be involved in PQQ insertion into the apoenzyme or modification of the apoenzyme after PQQ insertion [33,35]. If this hypothesis is correct, such functions could be involved in modifying enzyme activity by regulating the supply of intact cofactor-apoenzyme complexes [33,35]. A similar mechanism has been suggested to control MDH activity in *P. denitrificans* and *X. flavus* H4-14 [24]. Several observations showed that enzyme activity does not always correlate with the levels of MDH protein, indicating that enzyme synthesis and activity must be independently regulated [24].

Calcium. Binding of PQQ to its respective apoenzyme requires calcium ions, as could easily be demonstrated in vitro for glucose and alcohol dehydrogenase [36,37]. Recently it has been reported that the calcium concentration is also an essential parameter for growth and MDH activity in *Methylobacillus glycogenes*. This divalent cation was shown to be attached to the enzyme and could be replaced by strontium, yielding a 10-fold increase in enzyme activity [4,38]. The role of calcium in regulating MDH activity is unknown.

Small molecules (effectors) and protein factors. In general, MDH does not seem to be regulated by small molecules (effectors). Ammonium stimulates activity in vitro (for a review see Ref. 1) and could therefore be considered an effector. However, this stimulation is probably an artefact due to the high pH and artificial electron acceptors used for the in vitro assay. At high protein concentrations, activities measured with cytochromes as electron acceptors proved to be independent of ammonium [39,40]. The rates of methanol oxidation in these systems, however, were extremely low. Experimental evidence is available, suggesting that the dependence of MDH activity on ammonia in vitro might be due to prolonged exposure of the purified enzyme to oxygen [41,42]. In this respect it has been suggested that an additional low molecular weight component might be involved in electron transport to cytochromes [42]. This factor is presumably easily oxidized and thus inactivated under aerobic conditions in the presence of cytochromes.

In addition, data have been presented indicating that a protein factor designated M-protein (modifier-protein) might be involved in controlling MDH activity by modifying its affinity for substrates and products [1,43]. Under in vitro conditions greater utilization of less favorable substrates such as higher or polyalcohols was observed in the presence of purified M-protein. The reverse reaction (reduction of the respective aldehydes) was at the same time repressed by a decrease in affinity.

B. Genetic Organization and Expression

The system mediating conversion of methanol to formaldehyde in methylotrophic bacteria is designated the Mox system. This name was formerly also used to refer to methanol oxidation in methylotrophic yeasts, where methanol is converted to formaldehyde via alcohol oxidase, a non-PQQ, H_2O_2-producing enzyme. The functions involved in this mechanism are now known as the Aox system (for a recent review see Ref. 44). The Mox system of methylotrophic bacteria seems to be complex, since approximately 20 different complementation groups have as yet been identified (for recent reviews see Refs. 5,33). On a molecular basis some unique features have been reported for *mox* genes, as codon usage and promoter structures are different from those known for *Escherichia coli*. The MDH structural genes are both organized in one gene cluster, and short signal peptides could be derived from the DNA sequences of both genes.

1. Genes

Genes involved in methanol metabolism have been detected by intra- and interspecific complementation analyses of mutants impaired in growth on methanol [22,32,33,45–47]. Complementation has been carried out by conjugal mobilization, since transformation does not occur at significant frequencies (for a review see Ref. 48) and electroporation has only recently been made applicable in methylotrophs [49]. At least 20 different complementation classes have been distinguished in *Methylobacterium* species (for a recent review see Ref. 45). In *M. extorquens* AM1 these genes are organized in five separate clusters, and in some of these groups phenotypic characterization and expression studies have allowed the determination of gene functions (Table 1) [2,33–35,46,48]. The structural genes for the α and β subunits of MDH, *moxF* and *moxI*, are located in a single gene cluster together with two additional genes, *moxJ* and *moxG*, arranged in the order *moxFJGI* [2,3,34,35]. *MoxG* encodes cyt c_L, the electron acceptor of MDH, and

Table 1 Order and Function of Identified *Mox* Genes in *Methylobacterium extorquens* AM1[a]

Gene order	Gene designation	Proposed function
(Q E)	*moxQ, moxE*	Unknown
FJGI AKL B	*moxF, moxI*	MDH structural genes
	moxG	Cytochrome c_L structural gene
	moxJ	Unknown
	moxA, moxK, moxL	Modification or assembly of MDH complex
	moxB	Regulation
PCTVOM	*moxC, moxO, moxP, moxT, moxV*	PQQ synthesis and/or transport
(ND)	*moxD, moxM, moxN*	Unknown
(HU)	*moxH, moxU*	PQQ synthesis and/or transport
R S	*moxR, moxS*	Unknown

[a]The data are taken from inter- and intraspecific complementation analyses with mutants from *M. extorquens* AM1 and *M. organophilum* XX [2,33–35,46]. Parentheses indicate that the order of the genes is still unknown.

moxJ is the structural gene for a 30 kD protein of unknown function, the expression of which seems to be co-regulated with *moxF* and *moxI* [2,34,35]. A similar organization of the MDH structural genes has been reported for *M. organophilum* XX [22,46]. *P. denitrificans* [5], and the methanotrophs *Methylosporovibrio methanica* and *Methylomonas* sp. A4 [45,46]. For *M. methylotrophus* five classes of methanol oxidation mutants were found by complementation analysis and biochemical characterization, but the functions of each were not determined [50].

The *moxF* gene has been cloned from several methylotrophs, including *Methylobacterium* strains [20,30,34], *P. denitrificans* [51], and five methanotrophs [45–47,52]. The sequence is known for *M. extorquens* AM1 [53], *M. organophilum* XX [54], and *P. denitrificans* [51], and a comparison of these data indicates that MDH is highly conserved among different methylotrophs: a similarity of 96% has been found for the sequences of the *Methylobacterium* species and of 82% when these were compared to *P. denitrificans*. These results were even higher at the level of the deduced amino acid sequences. Partial sequences of the N-terminus of *moxF* from the methanotrophs *Methylomonas albus* BG8 and *Methylomonas* sp. A4 also showed a high degree of similarity (75–80%) with the *Methylobacterium* genes (for a recent review see Ref. 45). Hybridization studies have suggested that *moxF* shows strong conservation in all gram-negative methylotrophs that use methane or methanol [46,47]. Interesting enough, the DNA sequence of PQQ-containing alcohol dehydrogenase from *Acetobacter aceti* [55] and *Acetobacter polyoxogenes* [56] have 30% similarity to MDH, the highest match being found at the N-terminus. Due to these findings it has been speculated that both dehydrogenases might be derived from the same ancestral gene [55]. The gene for the MDH small subunit, *moxI*, has been sequenced for *M. extorquens* AM1 [3] and recently also for *P. denitrificans* [5]. A comparison of the nucleotide sequences yielded an equally high similarity as found with *moxF* [5]. Although MDHs of autotrophic and nonautotrophic bacteria seem to be very similar on the nucleotide level, considerable differences apparently exist in the amino acid sequences [25,57].

The proximity of *moxF* and *moxJ* might suggest a functional relationship, but the significance of the *moxJ* gene product for MDH has not yet been clarified. A transposon insertion mutant of *M. extorquens* AM1 in *moxJ* has no MDH, but polarity effects on downstream genes could not be ruled out [58]. Nevertheless, it should be mentioned here that *moxJ* has been identified and sequenced for *M. extorquens* AM1 [53] and that it shows considerable identity with ORF2 of *P. denitrificans*, also located immediately downstream of *moxF* in this organism [59].

The codon usage in all hitherto characterized methylotrophs with a high GC content (65–69%) shows a strong bias in the third position for G or C (for a recent review see Ref. 45). This feature has previously been observed in bacteria with a high GC content [60] and might partially account for the difficulties reported for *mox* gene expression in *E. coli*. *Methylobacillus flagellatum*, a low % GC methylotroph did not show this codon preference [45].

2. Transcription Regulation

The *moxF* gene has been shown to be transcriptionally regulated during growth on methanol in two *Methylobacterium* species. In *M. organophilum* XX the levels of MDH mRNA were lower in methanol-grown cells than during growth on methylamine [54]. In *M. extorquens* AM1 mRNA levels were 5 to 10-fold higher in methanol-grown cells than in succinate grown cells [61]. However, the mechanism of this transcriptional regulation has not yet been resolved.

In *M. extorquens* AM1 all four *moxFJGI* genes are transcribed in the same direction [2]. Tn5 insertion analysis has suggested that the whole cluster is expressed coordinately, but this has not yet been confirmed by transcript studies [34]. A promoter upstream of *moxF* has recently been identified [11]. In *M. organophilum* XX, however, *moxF* seems to be transcribed in a single unit [46,54]. For the *moxFJGI* cluster in *P. denitrificans*, only one promoter upstream of *moxF* was indicated, and it has been assumed that in this organism all four genes are co-expressed [5,59]. The intergenic regions all contain standard ribosome-binding site sequences [53], and immediately downstream of the ORFs for all three known *moxF* genes, strong potential stem and loop structures could be detected [33,51,53,54]. The role of these structures in expression is not known.

The transcriptional start sites for *moxF* have been documented in *M. organophilum* XX [54] and *M. extorquens* AM1 [53,61,62], both approximately 170 bp upstream of the translation start. A comparative analysis of the upstream regions revealed identical sequences in the -10 and -35 positions of these organisms (Table 2), possibly representing the promoter sequence for these genes. Sequences of high similarity have also been found upstream of *moxF* in the methanotrophs *M. albus* BG8 and *Methylomonas* sp. A4 (Table 2) and in two other methylotrophic genes not related to the *mox* system [45]. From these data a putative consensus sequence has been proposed which is different from known *E. coli* promoters (Table 2). The autotrophic methylotroph *P. denitrificans* contains no comparable promoter sequences in the 170 bp stretch of known sequence upstream of *moxF*. Instead, a GC-rich sequence 84 bp upstream from *moxF* has been suggested as a possible promoter, in analogy to similar sequences found directly upstream of transcriptional start sites for genes of *Pseudomonas putida* and *Rhodopseudomonas sphaeroides* (for a review see Ref. 63) (Table 2). However, the transcriptional start site has not yet been reported for *moxF* of *P. denitrificans*, and so any promoter assignments are still speculative.

These findings initially suggested substantial differences in regulation of MDH mRNA production for autotrophic- and nonautotrophic methylotrophs. However, sequences with partial homology to the GC-rich region of *P. denitrificans* have

Table 2 Putative Regulatory Sequences of *moxF* Gene Transcription in Autotrophic and Nonautotrophic Methylotrophs

Organism	Sequence upstream of *moxF*	Distance to start site (bp): Transcription	Distance to start site (bp): Translation
Paracoccus (consensus)	g t g GCGGC GGCcc cggac ucgug gcc t c t g t g g		
Paracoccus denitrificans	GCT GCGGC GGTTC GGGAG TGGAG GCC	no data	84
Methylobacterium organophilum XX	GTC GCGCC GGCCG TCGCG GGGAT AGA	265	435
Methylobacterium extorquens AM1	GGT GCGGC GCATC GTCTC CAAGT GCG	282	444
	CAC GCGGC GGCCG CCTTT TGCGG CAC	463	625
Methylobacterium (consensus)	AAAGACA –(16–18)– TAGAAA	5–7	
Methylobacterium organophilum XX	AAAGACA –(18)– TAGAAA	5	175
Methylobacterium extorquens AM1	AAAGACA –(18)– TAGAAA	6	168
Methylomonas albus BG8	AAAGGAA –(16)– CGGAAA	no data	212[b]
Methylomonas sp. A4	AAATCCA –(17)– AAGAAA	no data	85[c]
Escherichia coli (consensus)	TTGACA –(15–19)– TATAAT	5–9	

[a]Nucleotides in the upstream sequences matching the respective consensus sequences are underlined. Sequence data and transcription start sites from Refs. 45, 51, 53, 54, 61, 63.

[b]Kuhn and Lidstrom, unpublished.

[c]Waechter-Brulla and Lidstrom, unpublished.

also been reported at large distances upstream of *moxF* in *M. organophilum* XX (435 bp) and *M. extorquens* AM1 (444 bp and 625 bp) (Table 2) [45,63]. It is possible that in all three strains these GC-rich regions are involved in transcriptional regulation. Regulatory regions located at comparable distances upstream of the translation start site have been described in other gram-negative organisms [64–67].

Experimental proof for the function of regions upstream of *moxF* as a methanol-responsive promoter has been provided in *M. extorquens* AM1 with β-galactosidase as a reporter enzyme. In this case it was demonstrated that a DNA fragment covering a region 1300 bp upstream of *moxF* exhibits promoter activity [11]. This activity was found to respond to the growth substrate and was increased in cells grown on methanol as compared to succinate, but only when the transcriptional fusion was inserted into the chromosome. When the fusion was carried on a plasmid, expression was constitutive. These *mox* specific promoters seem to allow independent regulation of *mox* functions, as there is evidence that expression of other genes in methylotrophs requires the equivalent of a σ^{70} RNA polymerase subunit [45,63,67,68].

C. Secretion and Posttranslational Processing

Little experimental data are available concerning secretion of MDH into the periplasm and possible further modifications leading to formation of active enzyme. From the DNA sequence upstream of *moxF*, *moxI*, and *moxJ*, signal peptides of 27–32, 22, and 25 amino acids, respectively, could be derived [3,51,53,54]. The signal sequences for *moxF* are almost identical for both *Methylobacterium* species but show significant differences when compared to *P. denitrificans*. However, all periplasmic *mox* polypeptides for which sequence data are available contain signal sequences with structures that are similar to these found in secreted polypeptides of other gram-negative bacteria [62], including a common Ala-X-Ala sequence before the cleavage site [51,53,62]. The transport of premature MDH into the periplasm most likely occurs cotranslationally since no higher molecular weight polypeptides could be detected with pulse-chase labeling experiments [27].

From the finding that in vitro reconstitution of apo-MDH and PQQ could not be achieved under conditions sufficient for the reconstitution of PQQ-linked glucose dehydrogenase, it was further concluded that additional protein factors might be involved in posttranslational MDH activation [27]. This suggestion was supported by the detection of *mox* mutants (*moxA*, *moxK*, and *moxL*) which showed wild-type levels of MDH without activity and with abnormal PQQ spectra. The finding that these mutants were not capable of growth on methanol in the presence of PQQ could be explained by a lack of protein factors responsible for correct insertion of PQQ into the apoenzyme [34,35].

III. METHYLAMINE DEHYDROGENASE

A. Regulatory Parameters

Methylamine dehydrogenase is found in a variety of gram-negative methylotrophic bacteria that utilize methylamine as a sole source of carbon and nitrogen, where it functions to convert methylamine to formaldehyde and ammonia [1]. The bacteria include those that are obligate methylotrophs containing the ribulose monophosphate pathway, such as *Methylobacillus* [69], *Methylophilus* [70–72], or "*Methylomonas*" [73,74] strains, facultative methylotrophs containing the serine pathway (some *Methylobacterium* strains [74,75]), and autotrophic methylotrophs that use the Calvin-Benson cycle for CO_2 fixation, such as *Paracoccus denitrificans* [76] and *Thiobacillus versutus* [71]. Regulation of MADH and its electron acceptor, amicyanin, in these strains is summarized in Table 3.

1. *Effects on Enzyme Production*

The bacteria that contain MADH are capable of using alternate carbon, energy, and nitrogen sources. In all cases, MADH activity has been shown to be present in cells grown on methylamine as the sole carbon and energy source but at reduced or nondetectable levels in cells grown on other substrates such as methanol, succinate, or glucose [15,69,71,73,75,77–79]. This is in contrast to MDH in these strains, which is present at high levels in cells grown on either methanol or methylamine (see Sec. II). The electron acceptor for MADH is amicyanin, a blue copper protein [74]. Amicyanin has been shown to exhibit a pattern of regulation similar to that of MADH in *P. denitrificans* [76], organism 4025, an obligate methylotroph [80], *Methylobacillus flagellatum* KT [81], "*Methylomonas*" J [74], an uncharacterized *Methylobacterium* species [74] and *M. extorquens* AM1 [74]. Amicyanin has so far not been detected in the *Methylophilus* species [70,82].

In *M. extorquens* AM1, amicyanin levels did not change appreciably in response to the copper concentration (0–4 μM) in the growth medium [74], but in both organism 4025 and *M. flagellatum* KT, amicyanin levels were strongly regulated by copper concentration, showing optima at 10–12 μM copper [80,81]. No evidence is available that copper concentration regulates MADH activity in any of these strains.

MADH is also present in cells grown on other methylated amines. *M. extorquens* AM1 grows on ethylamine and propylamine using MADH, and MADH is present in cells grown on these substrates at levels of 20% and 40%, respectively, of cells grown on methylamine [83; L. Chistoserdova and M. Lidstrom, unpublished data]. During growth on trimethylamine, *M. methylotrophus* and *Methylophilus* W3A1 also express MADH [70,72,79], which is required for utilization of the methylamine produced as an intermediate during growth on trimethylamine. Likewise, *Methylophilus* W3A1 contains MADH activity during growth with dimethylamine [79].

Table 3 Summary of Methylamine Dehydrogenase and Amicyanin Regulation

Class	Organism	Induction		Repression		Inhibition
		MADH	Amicyanin	MADH	Amicyanin	
Facultative serine pathway	*Methylobacterium extorquens* AM1	MA+ EA+ PA+ M− Cu− S−	MA+ M− Cu+	M+[a] S+		
Facultative autotroph	*Paracoccus denitrificans*	MA+ M−	MA+ M− S−	M−[b] S+	S+	TMA+
Obligate RuMP pathway	*Methylobacillus flagellatum* KT	MA+ M−	MA+ M− Cu+			
	Organism 4025	MA+ M−	MA+ M− Cu+			
	Methylophilus methylotrophus	TMA+ MA+ M−		M+		
	Methylophilus W3A1	TMA+ DMA+ MA+ M− G−				TMA+

+, effect is observed; −, no effect observed; MA, methylamine; M, methanol; EA, ethylamine; PA, propylamine; S, succinate; Cu, copper; TMA, trimethylamine; DMA, dimethylamine; G, glucose.
[a]Partial.
[b]Data not given, may be partial.

The evidence described above suggests that MADH and amicyanin are induced in response to the presence of methylamine and other monomethylated primary amines. A few studies have also been carried out concerning the regulation during growth with multiple carbon sources. In *P. denitrificans*, cells grown in the presence of methanol plus methylamine with ammonia as the nitrogen source were stated to express MADH activity [5], but levels were not mentioned. Amicyanin and MADH were not detectable in cells grown with methylamine plus succinate [76,84]. In *M. extorquens* AM1 cells grown with equimolar levels of methanol and methylamine, MADH activity is decreased to 65% of that in methylamine-grown cells [L. Chistoserdova and M. Lidstrom, unpublished]. Cells grown with equimolar levels of methylamine plus succinate contained only 12% of the MADH activity found in methylamine-grown cells [L. Chistoserdova and M. Lidstrom, unpublished]. These data suggest that in the facultative methylotrophs, MADH is regulated by a catabolite repression type mechanism. The enzyme is only mildly repressed by methanol but is affected more severely by growth on multicarbon compounds. However, in the obligate methylotroph, *M. methylotrophus*, MADH is not detectable in cells grown on methanol plus methylamine with ammonia [17], suggesting that strong catabolite repression by methanol occurs.

Additional information exists suggesting that when methylamine is used as the sole nitrogen source in the presence of an alternative carbon source, MADH is repressed and a second methylamine oxidation system is induced. In *M. methylotrophus* grown with methanol plus methylamine (no other nitrogen source), methylamine oxidation was detected at high levels in whole cells, but MADH activity was too low to account for the growth rate [17]. In cells grown in continuous culture with methanol plus methylamine (no other nitrogen source), the MADH activity was dependent upon the C:N ratio used, at a ratio of 2.3, a 12.5-fold decrease in MADH activity was observed, and at a ratio of 15.9, MADH activity was not detectable. However, methylamine was oxidized in both cases, indicating that two different enzymes might be present. This suggestive evidence is supported by the isolation of mutants in both *M. flagellatum* KT [85] and *M. extorquens* AM1 [A. Chistoserdov, L. Chistoserdova, and M. Lidstrom, unpublished] that are unable to grow on methylamine as a carbon and energy source but grow normally on methylamine as a nitrogen source. These data suggest that an alternative methylamine oxidation system may be present in these organisms, regulated by nitrogen source availability, and provides further evidence that MADH is repressed by carbon sources other than methylated amines.

2. Effects on Enzyme Activity

Very little information is available on effectors of MADH activity. In *P. denitrificans* and *Methylophilus* W3A1, trimethylamine inhibits activity of purified MADH and methylamine-dependent respiration [86]. This is a mixed-type inhibition (trimethylamine is suggested to bind to both the free enzyme and the enzyme-substrate complex), with apparent K_i values of 1.1 and 4.7 mM, respectively. *Methylophilus* W3A1 grows on both trimethylamine and methylamine, but trimethylamine is a better substrate in terms of energy generation. It has been postulated that the inhibition of MADH by trimethylamine allows the cell to selectively utilize the more efficient substrate when it is available in high concentration [86]. However, *P. denitrificans* does not grow on trimethylamine, and here the role of this inhibition is not clear. Since MADH is a periplasmic enzyme [1,87], it would not be surprising to find that it is not subject to regulation by effectors. Most signals for carbon, energy, or nitrogen metabolism that would be candidates for effectors should not be present in the periplasm.

B. Genetic Organization and Expression

1. Methylamine Utilization Genes

Genetic techniques are now available in most of the bacteria that contain MADH [48,63], and as a result, studies of the organization and expression of genes involved in methylamine utilization are beginning to appear. We have proposed calling these genes *mau*, for *m*ethyl*a*mine *u*tilization [88]. *Mau* genes have been cloned from two bacteria: *M. extorquens* AM1 and *P. denitrificans*. In *P. denitrificans*, the gene for amicyanin was cloned by using an oligonucleotide probe based on the known amino acid sequence [84]. Sequence analysis of a 0.65 kb *Eco*RI fragment containing this gene revealed the C-terminal region of the gene for the MADH small subunit directly upstream and the N-terminal region of an additional open reading frame for an unknown polypeptide downstream. In *M. extorquens* AM1, genes for both amicyanin (*mauC*) and the MADH small subunit (*mauA*) have been cloned by a similar approach and sequenced [88,89]. As in *P. denitrificans*, these two genes are adjacent on the chromosome. The first few amino acids of a potential open reading frame are present downstream of the amicyanin gene, and these are identical to the amino acids in the open reading frame discovered downstream of the amicyanin gene in *P. denitrificans*. A comparison of the amicyanin sequences shows 53% similarity [89]. For the two amicyanin genes, the proteins are predicted to contain leader sequences, as expected for periplasmic polypeptides.

The *mau* gene cluster in *M. extorquens* AM1 has been examined in more detail. The polypeptides present on a 5.2 kb DNA fragment containing the two known *mau* genes have been studied using an *E. coli* T7 expression system [83]. This analysis has shown the region to contain five genes, all transcribed in the same direction. Immunoblot analysis has demonstrated that the first gene encodes the MADH large subunit (*mauB*). The next two genes encode polypeptides of unknown function of approximately 20 and 23 kDa (*mauE* and *mauD* genes), and the next two genes encode the MADH small subunit and amicyanin. As noted above, it is likely that at least one more gene is present immediately downstream of *mauC*. The gene for azurin, a second blue copper protein found in *M. extorquens* AM1 whose function is not known [74], is apparently not present within the large clone containing the known *mau* genes [83]. The genes of unknown function that are present in this gene cluster might be involved in electron transport, regulation, secretion, or the posttranslational processing that must occur to generate the MADH cofactor, tryptophan tryptophylquinone (TTQ) [90]. At this early stage in these studies it is not known how many *mau* genes will be identified, but it seems clear that this system is complex, and it is probably conserved between *M. extorquens* AM1 and *P. denitrificans*.

2. Transcriptional Regulation

Since the first *mau* genes have been cloned only recently, very little is known concerning transcriptional regulation of this system. It is generally assumed that either transcriptional or translational regulation occurs, since the MADH polypeptides are known to be strongly induced by methylamine in *P. denitrificans* [5] and *M. extorquens* AM1 [83; A. Chistoserdov and M. Lidstrom, unpublished data]. However, at this point no promoters have been cloned or studied, no data are available concerning transcripts, and no studies have been carried out involving transcript stability or processing. The sequence data for the *M. extorquens* AM1 *mau* region show the presence of two putative hairpin structures flanking *mauC* [89]. These regions and two other unsequenced regions flanking *mauB* have been shown to contain transcriptional terminator activity in *E. coli* [83]. It is not known whether these putative hairpin/terminator structures function in transcriptional regulation or transcriptional termination in *M. extorquens* AM1, but their position flanking the first and fourth genes in the cluster raises the possibility that these closely grouped *mau* genes may not be cotranscribed.

C. Secretion and Posttranslational Processing

As noted above, both MADH and amicyanin are periplasmic proteins and must be secreted. The amicyanin polypeptides from both *M. extorquens* AM1 and *P. denitrificans* appear to contain leader sequences similar in structure to those found in other gram-negative bacteria [84,89], and therefore would be expected to be involved in a normal secretion pathway. In support of this idea, the amicyanin from *M. extorquens* AM1 appears to be processed in *E. coli* [83]. However, secretion of the MADH is more complicated. The MADH small subunit (MauA) contains a cofactor (TTQ) that is derived from two tryptophan residues in the polypeptide chain by a proposed posttranslational modification that must include cross-linking and oxidative reactions [90]. The temporal and spatial sequence of TTQ synthesis is not clear. If MauA is synthesized in the cytoplasm, then modified, then secreted, it might require special secretion components in the membrane due to the cross-linked structure. Alternatively, if TTQ synthesis occurs after secretion of MauA, it might then require reduced cofactors (e.g. NADH) not available in the periplasm and therefore could involve membrane-bound enzymes. The enzyme(s) involved in secretion of MauA or in TTQ synthesis have not been identified, but it is possible that one or more of the unknown genes present in the *mau* gene cluster described above are involved in these specialized functions.

More evidence suggesting that MauA may require a special secretion system has been generated during studies of *mauA*. The N-terminal sequence of *mauA* revealed an unusual leader sequence for this polypeptide [91]. This sequence is quite long (57 amino acids) and in addition has an unusual structure. The N-terminal region is hydrophilic and is predicted to produce a soluble alpha-helix structure [91]. This is followed by a normal hydrophobic core, but surprisingly, this is followed by a region of high net positive charge, just preceding the peptidase cleavage site. In *E. coli*, the addition of net positive charges to the region of leader sequences preceding the cleavage site results in marked slowing of secretion and leader peptide processing [92]. Evidence from T7 expression experiments and studies with *mauA-phoA* (alkaline phosphatase) fusions show that MauA is not processed by the *E. coli* secretion system to a significant extent, but *mauA-phoA* fusions are expressed and processed in *M. extorquens* AM1 [83,91]. Since these T7 expression experiments suggest that MauB (the MADH large subunit) is processed quite well in *E. coli* [83], it seems likely that *M. extorquens* AM1 contains special secretion components that allow secretion of MauA.

IV. GLUCOSE DEHYDROGENASE

Quinoprotein glucose dehydrogenase (GDH) is found in a variety of gram-negative bacteria, where it functions to convert glucose to glucono-1,4-lactone, which is converted to gluconate either enzymatically or nonenzymatically [93]. In most cases it is bound to the outer surface of the cytoplasmic membrane [94], but one soluble periplasmic form is known [95,96]. In some bacteria, such as some *Klebsiella pneumoniae*, *Rhodobacter sphaeroides*, *Pseudomonas*, *Acetobacter*, and *Gluconobacter* strains, the enzyme is present in active form with its cofactor, PQQ [97,98]. However, in other strains, such as *Acinetobacter lwoffi* [98], *Agrobacterium radiobacter* [99], *Agrobacterium tumefaciens*, *Azotobacter vinelandii*, *Rhizobium leguminosarum* [36], *E. coli* [100], and *Salmonella typhimurium* [101], the apoenzyme is synthesized but PQQ is not. In these cases, active holoenzyme is reconstituted in vivo when PQQ is added to growth medium, and so PQQ has been suggested to act as a vitamin [102]. A recent report suggests that *E. coli* contains cryptic genes for PQQ synthesis, since mutants were isolated that were capable of synthesizing PQQ [103]. *Acinetobacter calcoaceticus* contains two different GDHs, one membrane-bound (GDH-A) and one soluble (GDH-B), apparently located in the periplasm [94,104,105]. Curiously, this organism, like most *Acinetobacter* strains containing GDH, is unable to grow on glucose or other aldoses oxidized by GDH or on gluconate [106].

A. Regulatory Parameters

1. Effects on Enzyme Production

In most strains, the quinoprotein GDHs are not regulated by the presence of glucose or other aldoses, but are modulated by other parameters [93] (Table 4). One exception is *Pseudomonas aeruginosa*, in which GDH activity is induced under aerobic conditions by glucose, gluconate, and glycerol [107]. In some instances, such as in *Pseudomonas cepacia* [108] and *A. calcoaceticus* [109], the changes are moderate, while in others, such as *K. pneumoniae* [110] and *E. coli* [100], more marked differences in GDH activities are observed. These studies are complicated by the fact that in the strains that are able to synthesize PQQ, synthesis of the apoprotein and PQQ is regulated separately. In order to understand how the system is controlled, it is necessary to assess each independently. So far, apoprotein and PQQ levels have not been measured directly by immunoblotting and PQQ assays, but have been inferred from activity assays in the presence and absence of PQQ. Therefore, the level of PQQ under different growth conditions is not always clear. However, in the cases in which activity in the presence and absence of PQQ has been tested, it is clear that GDH activity is regulated by physiological parameters either directly through apoprotein levels or indirectly through PQQ synthesis (see below).

GDH activity has been shown to be regulated by the presence of oxygen, by growth rate, and by the type of limiting nutrient, but the mechanism apparently differs with different strains. In all strains tested so far, GDH is not present in cells grown anaerobically or with limiting oxygen [93]. In *K. pneumoniae*, this regulation has been shown to occur by decreased availability of both PQQ and the apoprotein [111]. However, in *P. aeruginosa* and *R. sphaeroides*, the apoprotein is produced at high levels under both aerobic and anaerobic conditions, but PQQ is apparently not present during anaerobic growth, since addition of PQQ to anaerobically grown cells completely restores GDH activity to the levels found in aerobically grown cells [98]. In *A. calcoaceticus*, the lack of activity is apparently not due to lack of apoprotein or PQQ, but rather to some factor involving O_2 limitation, as addition of O_2 to O_2-limited cultures resulted in immediate restoration of GDH activity [111].

Growth rate has been shown to affect GDH activity in *A. calcoaceticus*. A 20-fold decrease in activity was observed in carbon-limited cells as the dilution rate was increased from 0.05 to 1.0 [112]. In this case, addition of PQQ did not increase activities, suggesting that the apoprotein and possibly PQQ were the targets of regulation. However, in *K. pneumoniae* under potassium limitation, and in *P. aeruginosa* under carbon limitation, GDH activity showed the opposite response to growth rate [97,113]. In these cases, only activity was measured.

In *K. pneumoniae*, GDH activities were lower during carbon, ammonia, and sulfate limitation than during phosphate or potassium limitation [111]. Studies of

Table 4 Regulation of Quinoprotein Glucose Dehydrogenase

Organism	GDH component	Response to aldoses	Presence during anaerobic growth	Response to increasing growth rate	Limiting nutrient (GDH levels)				
					NH_4^+	SO_4^{2-}	PO_4^{3-}	K^+	Carbon
Pseudomonas aeruginosa	apo		+						
	PQQ		−						
	holo	I	−	increases[b]					
Pseudomonas cepacia	holo	C							
Acinetobacter calcoaceticus	apo		+[a]	decreases[b]					
	PQQ		+	?					
	holo	C	−	decreases[b]					hi[d]
Klebsiella pneumoniae	apo		lo		hi	hi	hi	hi	lo
	PQQ		?		lo	lo	med	med	?
	holo	C	−	increases[c]	lo	lo	med	med	lo
Escherichia coli	apo	C				hi	hi		lo
Rhodobacter sphaeroides	apo		+						
	PQQ		−						
	holo	C	−						
Acinetobacter lwoffi	apo	C	+	no change[b,d]					hi
Agrobacterium radiobacter	apo	C		no change[b,d]	hi				hi

apo, apoprotein; PQQ, pyrroloquinoline quinone; holo, holoenzyme; I, inducible; C, constitutive; +, present; −, absent; lo, low; hi, high; med, medium.
[a]O_2-limited growth; activity is limited by lack of O_2.
[b]Carbon-limited growth.
[c]Potassium-limited growth.
[d]With respect to batch culture.

activities in the presence and absence of PQQ suggested that carbon-limited cells contained low levels of apoprotein and possibly also PQQ, while ammonia and sulfate limited cells contained higher levels of apoprotein, but low levels of PQQ. Therefore, in this organism, PQQ and apoprotein appear to be regulated separately, and one or both can be targets for regulation. A similar patter of regulation of the apoprotein was observed in *E. coli*, which is unable to synthesize PQQ [100]. However, in *A. lwoffi* and *A. radiobacter*, which also produce apoprotein but apparently cannot synthesize PQQ, the apoprotein was produced constitutively during growth in batch and continuous culture under a variety of conditions [98,114].

In both *K. pneumoniae* and *A. calcoaceticus*, it has been hypothesized that the regulation of GDH activity by growth rate and limited nutrient reflects response to the cellular energy status [111,112]. In both organisms, culture conditions characterized by a heavy demand for respiratory energy (e.g., low growth rate, growth in the presence of uncouplers) result in increased levels of GDH [111,112]. It has been shown that the oxidation of glucose by GDH contributes to the membrane potential [102]. Therefore, it has been suggested that under conditions of heavy energy demand, the synthesis of GDH would be beneficial to the cell and would account for the pattern of regulation observed [93,112]. In *E. coli*, however, a GDH mutant showed no change in growth rate with respect to the wild-type in minimal or rich media and PQQ also had no effect on growth rate [115]. This result may not be contradictory to the energy status hypothesis, though, since the effect of this mutation has not been tested under growth-limiting conditions.

Another possible regulatory parameter for apoprotein synthesis is the level of PQQ. However, the available evidence suggests that apoprotein synthesis is insensitive to PQQ availability. As shown in Table 4, production of apoprotein and PQQ are only loosely coordinated. In *E. coli*, which only synthesizes apoprotein, no change in apoprotein levels was observed during culture in the presence or absence of PQQ [93], and in other strains, apoprotein is produced at high levels in the absence of PQQ [36,98,99–101]. This is also true in *Pseudomonas testosteroni*, in which a PQQ-linked alcohol dehydrogenase apoprotein is produced in the absence of PQQ [116].

2. Effects on Enzyme Activity

PQQ. Since several factors affect PQQ availability, GDH activity can change depending upon each of these factors. Phosphate may affect PQQ entry into the periplasm. It has been suggested that PQQ enters *E. coli* cells through the PhoE porin, which is only expressed under conditions of low phosphate [117]. Therefore, the level of phosphate may influence how much PQQ is available to the apoprotein. In addition, it has been shown that amino acids can lower the effective concentration of PQQ in the medium, apparently by forming biologically inactive oxazoles [29]. Therefore, the use of amino acid autotrophs may complicate studies of PQQ effects on GDH activity. Another important parameter is the available Ca^{+2} and Mg^{+2} concentrations, since it has been shown that divalent cations are necessary for apoprotein-PQQ assembly [36,118,119]. Assembly can occur at low divalent metal concentrations, but it requires much higher concentrations of PQQ.

An additional mechanism that could influence PQQ availability is regulation of PQQ synthesis. As noted previously, PQQ synthesis is regulated by oxygen levels and other parameters (Table 4). However, the evidence available so far suggests that PQQ levels do not respond to the amount of PQQ in the medium or in the cell or to the amount of apoprotein present [29]. In *A. calcoaceticus* GDH mutants, PQQ is produced in the absence of apoprotein [105]. In *A. calcoaceticus* and *P. putida*, PQQ can be underproduced, and only about 25% of the GDH apoprotein in cell extracts contains PQQ under some growth conditions [29]. It has been suggested that the lack of response of PQQ synthesis to external PQQ levels reflects the lack of an uptake mechanism [29].

Small Molecules (Effectors). Little information is available concerning effectors of GDH. Like the other quinoproteins discussed here, it is unlikely that candidate indicators of catabolic activity would be present in the periplasmic. However, pH of the growth medium can have a strong effect on measured activities. In *K. pneumoniae* a low pH (5.5–6.5) is optimal both in vitro and in vivo [111], while in *A. calcoaceticus* a broader pH optima (5.0–8.2) has been observed [120].

B. Genetic Organization and Expression

1. Genes

The genes for the GDH apoprotein have been cloned and sequenced from *A. calcoaceticus* [104,105,121] and *E. coli* [115]. In *A. calcoaceticus*, the structural genes for both GDH-A and GDH-B have been isolated. The GDH-A gene (*gdhA*) was cloned by complementation of a GDH mutant with a clone bank constructed in a broad-host range cosmid vector [105], and the resultant clone complemented all 15 independently isolated GDH mutants. The *gdhA* gene was localized to a 3 kb complementing fragment, and expression experiments showed the presence of a protein of approximately 83,000 [105]. This region was sequenced, revealing the presence of an open reading frame predicted to encode a protein of 86,900 [121]. This protein contained at the N-terminal region five hydrophobic stretches predicted to be transmembrane segments, in keeping with the membrane location of GDH-A [121]. A clone containing *gdhA* was able to complement a GDH mutant of *E. coli* [105].

The *gdhB* gene was cloned using an oligonucleotide based on N-terminal amino acid sequence of GDH-B [104]. Sequencing in the region identified by the oligonucleotide revealed an open reading frame predicted to encode a protein of 52,772, which included a 24-amino-acid leader sequence, as expected for a periplasmic protein [104]. No significant sequence similarity was found between *gdhA* and *gdhB* [104], and the isolated clones show no overlap in restriction enzyme patterns.

Insertion mutants were constructed in *gdhA* and *gdhB*, and a double mutant was also constructed [104]. No in vivo activity of GDH-B has been detected in the past and consistent with this finding, the *gdhA* mutant showed no activity on MacConkey plates, while the *gdhB* mutant showed normal activity. However, GDH-B is detectable in vitro with artificial electron acceptors, and has a different substrate specificity than GDH-A [122]. Analysis of the in vitro activity of the three mutants confirmed that GDH-B oxidizes maltose and lactose and GDH-A does not [104]. The role of GDH-B in this organism is not known, but it has been suggested that *A. calcoaceticus* lacks a component of the electron transport chain for GDH-B that might be provided externally in its natural habitat [104].

E. coli has two systems for utilizing glucose: one involving GDH and a second involving the PTS system [93,115]. The GDH structural gene (*gcd*) has been cloned from *E. coli* by complementation of a PTS GDH double mutant ($pts^- gcd^-$) [115]. This complementation was difficult, as $pts^- gcd^+$ mutants do not grow well on glucose plus PQQ as single colonies. However, transformants were identified by selecting for the vector marker and then testing individual colonies for growth on glucose plus PQQ. The gene was localized to a 3 kb subfragment and was sequenced [115]. Like *gdhA* from *A. calcoaceticus*, *gcd* from *E. coli* encoded a protein of 87,064 that contained five putative membrane-spanning segments in the N-terminal portion. Strong similarity was observed between the two proteins at the amino acid level. When the sequences of several quinoproteins were compared, a region of 47 amino acids was identified that contained 40–50% similarity, and six amino acids were found that were conserved in all six quinoproteins examined. It was proposed that this region might represent a PQQ-binding domain [115]. The gene was mapped to 3.1 min on the *E. coli* chromosome.

2. *Transcriptional Regulation*

At this writing, little is known concerning transcriptional regulation of GDH genes. Promoters have not been identified, transcripts have not been studied, and transcriptional start sites have not yet been mapped. Although genes involved in PQQ synthesis have been cloned from *A. calcoaceticus* [123,124] and *K. pneumoniae* [125], little is known about their transcriptional regulation.

C. Secretion and Posttranslational Processing

As noted in the previous section, GDH-B appears to be synthesized with a standard signal sequence, and evidence has been presented that it is processed by the *E. coli* secretion system [104]. The other GDH enzymes appear to be periplasmic proteins anchored in the membrane [115,121]. In both types of enzymes, it appears that PQQ assembles spontaneously with apoprotein in the presence of Ca^{+2} or Mg^{+2}

both in vitro [118,119] and in vivo [36], and therefore it is unlikely that special assembly or processing functions are involved, as suggested for the MDH system (see section on MDH).

V. SUMMARY

The preceding portions of this review have documented a variety of regulatory strategies that are employed by bacteria for controlling the quinoproteins, MDH, MADH, and GDH. In some cases they appear to be regulated by derepression, in other cases by induction, and in some cases by a combination of the two. Some examples of catabolite repression are known, but in other instances the enzymes are not subject to this type of control. In cells containing only apoproteins, expression is regulated in some strains, while in others it is constitutive, with little change in activity over a wide range of growth conditions. Not only are there differences between these quinoproteins, marked variations are also found for a specific quinoprotein when diverse strains containing them are compared. Likewise, various genetic organization schemes are present, although more data are needed before complete comparisons can be made.

Although in most cases information is available concerning parameters that affect enzyme activity of these quinoproteins in cells, very little is known concerning the molecular basis of these effects. The cloning and characterization of structural genes for each of these quinoproteins and the development of genetic techniques for the bacteria containing these enzymes should now open the way for detailed mechanistic studies for all three quinoproteins.

REFERENCES

1. Anthony, C. (1986). Bacterial oxidation of methane and methanol. *Adv. Microb. Physiol. 27*: 113.
2. Anderson, D., and Lidstrom, M. (1988). The *moxFG* region encodes four polypeptides in the methanol- oxidizing bacterium *Methylobacterium* sp. strain AM1. *J. Bacteriol. 170*: 2254.
3. Nunn, D., Day, D., and Anthony, C. (1989). The second subunit of methanol dehydrogenase of *Methylobacterium extorquens* AM1. *Biochem. J. 260*: 857.
4. Adachi, O., Matsushita, K., Shinagawa, E., and Ameyama, M. (1990). Purification and properties of methanol dehydrogenase and aldehyde dehydrogenase from *Methylobacillus glycogenes*. *Agr. Biol. Chem. 54*: 3123.
5. Harms, N., and Van Spanning, R. (1991). C1 metabolism in *Paracoccus denitrificans*: Genetics of *Paracoccus denitrificans*. *J. Bioenerg. Biomembr. 23*: 187.
6. Anthony, C. (1991). Assimilation of carbon by methylotrophs. In *Biology of Methylotrophs* (I. Goldberg and J. S. Rokem, eds.), Butterworth-Heinemann, Boston, P. 79.

7. Rokem, J. S., and Goldberg, I. (1991). Oxidation pathways in methylotrophs. In *Biology of Methylotrophs* (I. Goldberg and J. S. Rokem, eds.), Butterworth-Heinemann, Boston, p. 111.
8. Weaver, C., and Lidstrom, M. (1985). Methanol dissimilation in *Xanthobacter* H4-14: Activities, induction and comparison to *Pseudomonas* AM1 and *Paracoccus denitrificans*. *J. Gen. Microbiol. 131*: 2183.
9. Roitsch, T., and Stolp, H. (1986). Synthesis of dissimilatory enzymes of serine type methylotrophs under different growth conditions. *Arch. Microbiol. 144*: 245.
10. Jones, C. W., Greenwood, J. A., Burton, S. M., Santos, H., and Turner, D. L. (1987). Environmental regulation of methanol and formaldehyde metabolism by *Methylophilus methylotrophus*. *J. Gen. Microbiol. 133*: 1511.
11. Morris, C. J., and Lidstrom, M. E. (1992). A methanol-inducible promoter from *Methylobacterium extorquens* AM1. *J. Bacteriol. 174*: (in press).
12. McNerney, T., and O'Connor, M. (1980). Regulation of enzymes associated with C-1 metabolism in three facultative methylotrophs. *App. Environ. Microbiol. 40*: 370.
13. O'Connor, M. (1981). Regulation and genetics in facultative methylotrophic bacteria. In *Microbiol Growth on C_1 Compounds* (H. Dalton, ed.), Heyden & Son, London, p. 294.
14. Dunstan, P. M., Anthony, C., and Drabble, W. T. (1972). Microbial metabolism of C_1 and C_2 compounds: The role of glyoxylate, glycollate and acetate in the growth of *Pseudomonas* AM1 on ethanol and on C_1 compounds. *Biochem. J. 128*: 107.
15. Marison, I. W., and Attwood, M. M. (1982). A possible alternative mechanism for the oxidation of formaldehyde to formate. *J. Gen. Microbiol. 128*: 1441.
16. O'Connor, M. L., and Hanson, R. S. (1977). Enzyme regulation in *Methylobacterium organophilum*. *J. Gen. Microbiol. 102*: 327.
17. Dawson, A., Southgate, G., and Goodwin, P. (1990). Regulation of methanol and methylamine dehydrogenase in *Methylophilus methylotrophus*. *FEMS Microbiol. Lett. 68*: 93.
18. Roitsch, T., and Stolp, H. (1985). Overproduction of methanol dehydrogenase in glucose grown cells of a restricted RuMP type methylotroph. *Arch. Microbiol. 142*: 34.
19. Girio, M. F., and Attwood, M. M. (1991). Metabolic aspects of the growth of a pink-pigmented facultative methylotroph in carbon-limited chemostat culture. *Appl. Microbiol. Biotechnol. 35*: 77.
20. Greenwood, J., and Jones, C. (1986). Environmental regulation of the methanol oxidase system of *Methylophilus methylotrophus*. *J. Gen. Microbiol. 132*: 1247.
21. Stone, S., and Goodwin, P. (1989). Characterization and complementation of mutants of *Methylobacterium* AM1 which are defective in C-1 assimilation. *J. Gen. Microbiol. 135*: 227.
22. Machlin, S., Tam, P., Bastien, C., and Hanson, R. (1987). Genetic and physical analyses of *Methylobacterium organophilum* XX genes encoding methanol oxidation. *J. Bacteriol. 170*: 141.
23. Van Versefeld, H. W., and Stouthamer, A. H. (1979). Growth yields and the efficiency of oxidative phosphorylation of *Paracoccus denitrificans* during two (carbon) substrate-limited growth. *Arch. Microbiol. 121*: 213.

24. De Vries, G., Harms, N., Maurer, K., Papendrecht, A., and Stouthamer, A. (1988). Physiological regulation of *Paracoccus denitrificans* methanol dehydrogenase synthesis and activity. *J. Bacteriol. 170*: 3731.
25. Harms, N., De Vries, G., Maurer, K., Veltkamp, E., and Stouthamer, A. (1985). Isolation and characterization of *Paracoccus denitrificans* mutants with defects in the metabolism of one-carbon compounds. *J. Bacteriol. 164*: 1064.
26. Meijer, W. G., Croes, L. M., Jenni, B., Lehmicke, L. G., Lidstrom, M. E., and Dijkhuizen, L. L. (1990). Characterization of *Xanthobacter* strains H4-14 and 24a and enzyme profiles after growth under autotrophic and heterotrophic conditions. *Arch. Microbiol. 153*: 360.
27. Davidson, V. L., Neher, J. W., and Cecchini, G. (1985). The biosynthesis and assembly of methanol dehydrogenase in bacterium W3A1. *J. Biol. Chem. 260*: 9642.
28. Biville, F., Mazodier, P., Gasser, F., Van Kleef, M., and Duine, J. (1988). Physiological properties of a pyrroloquinoline quinone mutant of *Methylobacterium organophilum. FEMS Microbiol. Lett. 52*: 53.
29. Van Kleef, M., and Duine, J. (1989). Factors relevant in bacterial pyrroloquinoline quinone production. *Appl. Environ. Microbiol. 55*: 1209.
30. Mazodier, P., Biville, F., Turlin, E., and Gasser, F. (1988). Localization of a pyrroloquinoline quinone biosynthesis gene near the methanol dehydrogenase structural gene in *Methylobacterium organophilum* DSM 760. *J. Gen. Microbiol. 134*: 2513.
31. Biville, F., Mazodier, P., Turlin, E., and Gasser, F., (1989). Mutants of *Methylobacterium organophilum* unable to synthesize PQQ. *Ant. v. Leeuw. 56*: 103.
32. Biville, F., Turlin, E., and Gasser, F. (1989). Cloning and genetic analysis of six pyrroloquinoline quinone biosynthesis genes in *Methylobacterium organophilum* DSM 760. *J. Gen. Microbiol. 135*: 2917.
33. Lidstrom, M. E. (1990). Genetics of carbon metabolism in methylotrophic bacteria. *FEMS Microbiol. Lett. 87*: 431.
34. Nunn, D., and Lidstrom, M. (1986). Isolation and complementation analysis of 10 methanol oxidation mutant classes and identification of the methanol dehydrogenase structural gene of *Methylobacterium* sp. strain AM1. *J. Bacteriol. 166*: 581.
35. Nunn, D., and Lidstrom, M. (1986). Phenotypic characterization of 10 methanol oxidation mutant classes in *Methylobacterium* sp. strain AM1. *J. Bacteriol. 166*: 592.
36. Van Schie, B., De Mooy, O., Linton, J., Van Dijken, J., and Kuenen, J. (1987). PQQ-dependent production of gluconic acid by *Agrobacterium* and *Rhizobium* species. *J. Gen. Microbiol. 133*: 867.
37. Poels, P. A., Groen, B. W., and Duine, J. A. (1987). $NAD(P)^+$-independent aldehyde dehydrogenase from *Pseudomonas testosteroni*: A novel type of molybdenum containing hydroxylase. *Eur. J. Biochem. 166*: 575.
38. Adachi, O., Matsushita, K., Shinagawa, E., and Ameyama, M. (1990). Calcium in quinoprotein methanol dehydrogenase can be replaced by strontium. *Agric. Biol. Chem. 54*: 2833.
39. Dijkstra, M., Frank, J., and Duine, J. (1989). Studies on electron transfer from methanol dehydrogenase to cytochrome c_L, both purified from *Hyphomicrobium* X. *Biochem. J. 257*: 87.

40. Mukai, K., Fukumori, Y., and Yamanaka, T. (1990). The methanol-oxidizing system of *Methylobacterium extorquens* AM1 reconstituted with purified constituents. *J. Biochem. 107*: 714.
41. Duine, J. A., Frank J., Jzn., and De Ruiter, L. G. (1979). Isolation of a methanol dehydrogenase with a functional coupling to cytochrome c. *J. Gen. Microbiol. 115*: 523.
42. Dijkstra, M., Frank, J., Jzn., and Duine, J. A. (1988). Methanol oxidation under physiological conditions using methanol dehydrogenase and a factor isolated from *Hyphomicrobium* X. *FEBS Lett. 227*: 198.
43. Page, M. D., and Anthony, C. (1986). Regulation of formaldehyde oxidation by the methanol dehydrogenase modifier protein of *Methylophilus methylotrophus* and *Pseudomonas* AM1. *J. Gen. Microbiol. 132*: 1553.
44. Van der Klei, I. J., Harder, W., and Veenhuis, M. (1991). Biosynthesis and assembly of alcohol oxidase, a peroxisomal matrix protein in methylotrophic yeasts: A review. *Yeast 7*: 195.
45. Lidstrom, M. E., and Tsygankov, Y. D. (1991). Molecular genetics of methylotrophic bacteria. In *Biology of Methylotrophs* (I. Goldberg and J. S. Rokem, eds.), Butterworth-Heinemann, Boston, p. 273.
46. Bastien, C., Machlin, S., Zhang, Y., Donaldson, K., and Hanson, R. (1989). Organization of genes required for the oxidation of methanol to formaldehyde in three Type II methylotrophs. *Appl. Env. Microbiol. 55*: 3124.
47. Stephens, R., Haygood, M., and Lidstrom, M. (1988). Identification of putative methanol dehydrogenase (*moxF*) structural genes in methylotrophs and cloning of *moxF* genes from *Methylococcus capsulatus* Bath and *Methylomonas albus* BG8. *J. Bacteriol. 170*: 2063.
48. Lidstrom, M. E., and Stirling, D. I. (1990). Methylotrophs: Genetics and commercial applications. *Annu. Rev. Microbiol. 44*: 27.
49. Ueda, S., Matsumoto, S., Shimizu, S., and Yamane, T. (1991). Transformation of a methylotrophic bacterium, *Methylobacterium extorquens*, with a broad-host-range plasmid by electroporation. *Appl. Environ. Microbiol. 57*(4): 924.
50. Dawson, A., and Goodwin, P. (1990). Investigation of mutants of *Methylophilus methylotrophus* which are defective in methanol oxidation. *J. Gen. Microbiol. 136*: 1373.
51. Harms, N., DeVries, G., Maurer, K., Hoogendijk, J., and Stouthamer, A. (1987). Isolation and nucleotide sequence of the methanol dehydrogenase structural gene from *Paracoccus denitrificans*. *J. Bacteriol. 169*: 3969.
52. Al-Taho, N., Cornish, A., and Warner, P. (1990). Molecular cloning of the methanol dehydrogenase structural gene from *Methylosinus trichosporium* OB3b. *Curr. Microbiol. 20*: 153.
53. Anderson, D., Morris, C., Nunn, D., Anthony, C., and Lidstrom, M. (1990). Nucleotide sequence of the *Methylobacterium extorquens* AM1 *moxF* and *moxJ* genes involved in methanol oxidation. *Gene 90*: 173.
54. Machlin, S., and Hanson, R. (1988). Nucleotide sequence and transcriptional start site of the *Methylobacterium organophilum* XX methanol dehydrogenase structural gene. *J. Bacteriol. 170*: 4739.
55. Inoue, T., Sunagawa, M., Mori, A., Imai, C., Fukuda, M., Takagi, M., and Yano,

K. (1989). Cloning and sequencing of the gene encoding the 72-kilodalton dehydrogenase from *Acetobacter aceti*. *J. Bacteriol. 171*: 3115.

56. Tamaki, T., Fukaya, M., Takemura, H., Tayama, K., Okumura, K., Kawamura, Y., Nishiyama, M., Horinouchi, S., and Beppu, T. (1991). Cloning and sequencing of the gene cluster encoding two subunits of membrane-bound alcohol dehydrogenase from *Acetobacter polyoxogenes*. *Biochim. Biophys. Acta 1088*: 292.
57. Bamforth, C. W., and Quayle, J. R. (1978). Anaerobic and aerobic growth of *Paracoccus denitrificans* on methanol. *Arch. Microbiol. 119*: 91.
58. Lee, K., Stone, S., Goodwin, P., and Holloway, B. (1991). Characterization of transposon insertion mutants of *Methylobacterium extorquens* AM1 (*Methylobacterium* strain AM1) which are defective in methanol oxidation. *J. Gen. Microbiol. 137*: 895.
59. Harms, N., VanSpanning, R., Oltmann, L., and Stouthamer, A. (1989) Regulation of methanol dehydrogenase synthesis in *Paracoccus denitrificans*. *Ant. v. Leeuw. 56*: 47.
60. West, S., and Iglewski, B. (1988). Codon usage in *Pseudomonas aeruginosa*. *Nucl. Acid Res. 16*: 9323.
61. Anderson, D. J. (1988). Characterization of a methanol oxidation gene cluster in the facultative methylotroph *Methylobacterium* sp. strain AM1. Ph.D. thesis, University of Washington, Seattle, pp. 47–92.
62. Lidstrom, M. E. (1991). The genetics and molecular biology of methanol-utilizing bacteria. In *Methane and Methanol Utilizers* (J. C. Murrell and H. Dalton, eds.), Plenum, New York, in press.
63. DeVries, G., Kues, U., and Stahl, U. (1990). Physiology and genetics of methylotrophic bacteria. *FEMS Microbiol. Rev. 75*: 57.
64. Sawers, G., and Böck, A. (1989). Novel transcriptional control of the pyruvate formate-lyase gene: Upstream regulatory sequences and multiple promoters regulate anaerobic expression. *J. Bacteriol. 171*: 2485.
65. Sawers, G., Wagner, A. F. V., and Böck, A. (1989). Transcription initiation at multiple promoters of the *pfl* gene by $E\sigma^{70}$-dependent transcription in vitro and heterologous expression in *Pseudomonas putida in vivo*. *J. Bacteriol. 171*: 4930.
66. Valentin-Hansen, P., Aiba, H., and Schümperli, D. (1982). The structure of tandem regulatory regions in the *deo* operon of *Escherichia coli* K12. *EMBO J. 1*: 317.
67. Chistoserdov, A., Eremashvili, M., Mashko, S., Lapidus, A., Skvortsova, M., and Sterkin, V. (1987). Expression of human interferon F gene in obligate methylotroph *Methylobacillus flagellatum* KT and *Pseudomonas putida* (in Russian). *Molec. Genet. Microbiol. Virol. 8*: 36.
68. Byrom, D. (1984). Host-vector systems for *Methylophilus methylotrophus*. In *Microbial Growth on C_1 Compounds* (H. Dalton, ed.), Heyden & Son, London, p. 221.
69. Kirukhin, M. Y., Chistoserdov, A. Y., and Tsygankov, Y. D. (1990). Methylamine dehydrogenase from *Methylobacillus flagellatum*. *Meth. Enzymol. 188*: 247.
70. Burton, S. M., Byrom, D., Carver, M., Jones, G. D. D., and Jones, C. W. (1983). The oxidation of methylated amines by methylotrophic bacterium *Methylophilus methylotrophus*. *FEMS Microbiol. Lett. 17*: 185.
71. Haywood, G. W., Janschke, N. S., Large, P. J., and Wallis, J. M. (1982). Properties

and subunit structure of methylamine dehydrogenase from *Thiobacillus* A2 and *Methylophilus methylotrophus*. *FEMS Microbiol. Lett. 15*: 79.
72. Kenny, W. C., and McIntire, W. (1983). Characterization of methylamine dehydrogenase from bacterium W3A1: interaction with reductant and amino-containing compounds. *Biochemistry 22*: 3858.
73. Mehta, R. J. (1977). Methylamine dehydrogenase from the obligate methylotroph *Methylomonas methylovora*. *Can. J. Microbiol. 23*: 402.
74. Tobari, J. (1984). Blue copper proteins in electron transport in methylotrophic bacteria. *Microbial Growth on C1 Compounds* (R. L. Crawford and R. S. Hanson, eds.), ASM Press, Washington, DC, pp. 106–112.
75. Eady, R. R., and Large, P. J. (1968). Purification and properties of an amine dehydrogenase from *Pseudomonas* AM1 and its role in growth on methylamine. *Biochem. J. 106*: 245.
76. Husain, M., and Davidson, V. L. (1985). An inducible periplasmic blue copper protein from *Paracoccus denitrificans*: Purification, properties, and physiological role. *J. Biol. Chem. 260*: 14626.
77. Troyan, O. S., Netrusov, A. I., Skirdov, I. V., and Kondratyeva, E. N. (1978). Oxidation of monocarbon compounds by bacteria of different genera. *Mikrobiologiya 14*: 202 (in Russian)
78. Large, P. J., and Haywood, G. W. (1981). *Methylophilus methylotrophus* grows on methylated amines. *FEMS Microbiol. Lett. 11*: 207.
79. Davidson, V. L. (1985). Regulation by carbon source of enzyme expression and slime production in Bacterium W3A1. *J. Bacteriol. 164*: 941.
80. Auton, K., and Anthony, C. (1989). The 'methylamine oxidase' system of an obligate methylotroph. *Biochem. J. 260*: 75.
81. Dinarieva, T., and Netrusov, A. (1989). Cupredoxines of obligate methylotroph. *FEBS Lett. 259*: 47.
82. Chandrasekar, R., and Klapper, M. H. (1986). Methylamine dehydrogenase and cytochrome c552 from the bacterium W3A1. *J. Biol. Chem. 261*: 3616.
83. Chistoserdov, A. Y., Tsygankov, Y. D., and Lidstrom, M. E. (1991). Genetic organization of methylamine utilization genes from *Methylobacterium extorquens* AM1. *J. Bacteriol. 173*: 5901.
84. Van Spanning, R., Wansell, C., Reijnders, W., Oltmann, L., and Stouthamer, A. (1990). Mutagenesis of the gene encoding amicyanin of *Paracoccus denitrificans* and the resultant effect on methylamine oxidation. *FEBS Lett. 275*: 217.
85. Gak, E. R., Chistoserdov. A. Y, and Tsygankov, Y. D. (1989). Mutants of *Methylobacterium flagellatum* defective in catabolism of methylamine. *Proceedings of the 6th International Symposium Microbial Growth C1 Compounds*, Göttingen, Germany, 417.
86. Davidson, V. L., and Kumar, M. A. (1990). Inhibition by trimethylamine of methylamine oxidation by *Paracoccus denitrificans* and bacterium W3A1. *Biochim. Biophys. Acta 1016*: 339.
87. Kasprazak, A., and Steenkamp, D. (1983). Localization of the major dehydrogenases in two methylotrophs by radiochemical labeling. *J. Bacteriol. 156*: 348.
88. Chistoserdov, A., Tsygankov, Y., and Lidstrom, M. (1990). Cloning and sequencing

of the structural gene for the small subunit of methylamine dehydrogenase from *Methylobacterium extorquens* AM1: Evidence for two tryptophan residues involved in the active center. *Biochem. Biophys. Res. Commun. 172*: 211.

89. Chistoserdov, A,. Tsygankov, Y., and Lidstrom, M. (1991). Nucleotide sequence of the amicyanin gene from *Methylobacterium extorquens* AM1. *DNA Sequence, 2*: 53.
90. McIntire, W., Wemmer, D., Chistoserdov, A., and Lidstrom, M. (1991). A new cofactor in a prokaryotic enzyme: Tryptophan tryptophylquinone as the redox prosthetic group in methylamine dehydrogenase. *Science 252*: 817.
91. Chistoserdov, A. Y., and Lidstrom, M. E. (1991). The small subunit of methylamine dehydrogenase from *Methylobacterium extorquens* AM1 has an unusual leader sequence. *J. Bacteriol. 173*: 5909.
92. Boyd, D., and Beckwith, J. (1990). The role of charged amino acids in the localization of secreted and membrane proteins. *Cell 62*: 1031.
93. Neijssel, O. M., Hommes, R. W. J., Postma, P. W., and Tempest, D. W. (1989). Physiological significance and bioenergetic aspects of glucose dehydrogenase. *Ant. v. Leeuw. J. Microbiol. 56*: 51.
94. Matsushita, K., Shinagawa, E., Inoue, T., Adachi, O., and Ameyama, M. (1986). Immunological evidence for two types of PQQ-dependent D-glucose dehydrogenases in bacterial membranes and the location of the enzyme in *Escherichia coli*. *FEMS Micro. Lett. 37*: 141.
95. Hauge, J. G. (1960). Purification and properties of glucose dehydrogenase and cytochrome *b* from *Bacterium anitratum*. *Biochim. Biophys. Acta 45*: 250.
96. Hauge, J. G. (1964). Glucose dehydrogenase of *Bacterium anitratum*: An enzyme with a novel prosthetic group. *J. Biol. Chem. 239*: 3630.
97. Neijssel, O. M., Tempest, D. W., Postma, P. W., Duine, J. A., and Frank, J. J. (1983). Glucose metabolism by K^+-limited *Klebsiella aerogenes*: Evidence for the involvement of a quinoprotein glucose dehydrogenase. *FEMS Microbiol. Lett. 20*: 35.
98. Van Shie, B. J., Van Dijken, J. P., and Kuenen, J. G. (1984). Non-coordinated synthesis of glucose dehydrogenase and its prosthetic group PQQ in *Acinetobacter* and *Pseudomonas* species. *FEMS Microbiol. Lett. 24*: 133.
99. Ameyama, M., Sinagawa, E., Matsushita, K., and Adachi, O. (1985). Growth stimulating activity for microorganisms in naturally occurring substances and partial characterization of the substance for the activity as PQQ. *Agric. Biolog. Chem. 49*: 699.
100. Hommes, R. W. J., Postma, P. W., Neijssel, O. M. Tempest, D. W., Dokter, P., and Duine J. A. (1984). Evidence of a quinoprotein glucose dehydrogenase apoenzyme in several strains of *Escherichia coli*. *FEMS Microbiol. Lett. 24*: 329.
101. Hommes, R. W. J., Loenen, W. A. M., Neijssel, O. M., and Postma, P. W. (1986). Galactose metabolism in *gal* mutants of *Salmonella typhimurium* and *Escherichia coli*. *FEMS Microbiol. Lett. 36*: 187.
102. Van Schie, B., Hellingwerf, K., Van Dijken, J., Elferink, M., Van Dijl, J., Kuenen, J., and Konings, W. (1985). Energy transduction by electron transfer via a pyrroloquinoline quinone-dependent glucose dehydrogenase in *Escherichia coli*, *Pseudomonas aeruginosa*, and *Acinetobacter calcoaceticus* (var. lwoffi). *J. Bacteriol. 163*: 493.

103. Biville, F., Turlin, E., and Gasser, F. (1991). Mutants of *Escherichia coli* producing pyrroloquinoline quinone. *J. Gen. Microbiol. 137*: 1775.
104. Cleton-Jansen, A.-M., Goosen, N., Vink, K., and Van De Putte, P. (1989). Cloning, characterization and DNA sequencing of the gene encoding the Mr 50000 quinoprotein glucose dehydrogenase from *Acinetobacter calcoaceticus. Molec. Gen. Genet. 217*: 430.
105. Cleton-Jansen, A.-M., Goosen, N., Wenzel, T. J., and Van De Putte, P. (1988). Cloning of the gene encoding quinoprotein glucose dehydrogenase from *Acinetobacter calcoaceticus*: Evidence for the presence of a second enzyme. *J. Bacteriol. 170*: 2121.
106. Juni, E. (1978). Genetics and physiology of *Acinetobacter. Ann. Rev. Microbiol. 32*: 349.
107. Midgley, M., and Dawes, E. A. (1973). The regulation of glucose and methylglucoside in *Pseudomonas aeruginosa. Biochem. J. 132*: 141.
108. Berka, R. T., Allenza, P., and Lessie, T. G. (1984). Hyperinduction of enzymes of the phosphorylative pathway of glucose dissimilation in *Pseudomonas cepacia. Curr. Microbiol. 11*: 143.
109. DeBont, J. A. M., Dokter, P., Van Shie, B. J., Van Dijken, J. P., Frank, J. J., and Kuenen, J. G. (1984). Role of quinoprotein glucose dehydrogenase in gluconic acid production by *Acinetobacter calcoaceticus. Ant. v. Leeuw. J. Microbiol. 50*: 76.
110. Hommes, R., Van Hell, P. W., Postma, P., Neijssel, O., and Tempest, D. (1985). The functional significance of glucose dehydrogenase in *Klebsiella aerogenes. Arch. Microbiol. 143*: 163.
111. Hommes, R., Herman, P., Postma, P., Tempest, D., and Neijssel, O. (1989). The separate roles of PQQ and apo-enzyme synthesis in the regulation of glucose dehydrogenase activity in *Klebsiella pneumoniae. Arch. Microbiol. 151*: 257.
112. Van Shie, B. J., Van Dijken, J. P., and Kuenen, J. G. (1988). Effects of growth rate and oxygen tension on glucose dehydrogenase activity in *Acinetobacter calcoaceticus* LMD 79.41. *Ant. v. Leeuw. J. Microbiol. 55*: 53.
113. Ng, F. M. W., and Dawes, E. A. (1973). Chemostat studies on the regulation of glucose metabolism in *Pseudomonas aeruginosa* by citrate. *Biochem. J. 132*: 129.
114. Linton, J. D., Woodard, S., and Gouldney, D. G. (1987). The consequence of stimulating glucose dehydrogenase activity by the addition of PQQ on metabolite production by *Agrobacterium radiobacter* NCIB 11883. *Appl. Microbiol. Biotech. 25*: 357.
115. Cleton-Jansen, A.-M. Goosen, N., Fayet, O., and Van De Putte, P. (1990). Cloning, mapping, and sequencing of the gene encoding *Escherichia coli* quinoprotein glucose dehydrogenase. *J. Bacteriol. 172*: 6308.
116. Groen, B. W., Van Kleef, M. A. G., and Duine, J. A. (1986). Quinohaemoprotein alcohol dehydrogenase apoenzyme from *Pseudomonas testosteroni. Biochem. J. 2343*: 611.
117. Adamowicz, M., Conway, T., and Nickerson, K. W. (1991). Nutritional complementation of oxidative glucose metabolism in *Escherichia coli* via pyrroloquinoline quinone-dependent glucose dehydrogenase and the Entner-Doudoroff pathway. *Appl. Environ. Microbiol. 57*: 2012.

118. Duine, J. A., and Jongejan, J. A. (1989). Quinoproteins, enzymes with pyrroloquinoline quinone as cofactor. *Ann. Rev. Biochem. 58*: 403.
119. Geiger, O., and Görisch, H. (1989). Reversible thermal inactivation of the quinoprotein glucose dehydrogenase from *Acinetobacter calcoaceticus*- Ca^{2+} ions are necessary for reactivation. *Biochem. J. 261*: 415.
120. Van Shie, B. J., Rouwenhorst, R. J., De Bont, J. a. M., Van Dijken, J. P., and Kuenen, J. G. (1987). An in vivo analysis of the energetics of aldose oxidation by *Acinetobacter calcoaceticus. Appl. Microbiol. Biotech. 26*: 560.
121. Cleton-Jansen, A.-M., Goosen, N., Odle, G., and Van De Putte, P. (1988). Nucleotide sequence of the gene coding for quinoprotein glucose dehydrogenase from *Acinetobacter calcoaceticus. Nucl. Acids Res. 16*: 6228.
122. Dokter, P., Frank, J., and Duine, J. A. (1986). Purification and characterization of quinoprotein glucose dehydrogenase from *Acinetobacter calcoaceticus* LMD 79.41. *Biochem. J. 239*: 163.
123. Goosen, N. Vermaas, D. A., and Van De Putte, P. (1987). Cloning of the genes involved in synthesis of coenzyme pyrroloquinoline-quinone from *Acinetobacter calcoaceticus. J. Bacteriol. 169*: 303.
124. Goosen, N., Horsman, H., Huinen, R., and Van De Putte, P. (1989). *Acinetobacter calcoaceticus* genes involved in biosynthesis of the coenzyme pyrrolo-quinoline-quinone: nucleotide sequence and expression in *Escherichia coli* K-12. *J. Bacteriol. 171*: 447.
125. Meulenberg, J., Loenen, W., Sellink, E., and Postma, P. (1990). A general method for the transfer of plasmid-borne mutant alleles to the chromosome of *Klebsiella pneumoniae* using bacteriophage λ: transfer of pqq genes. *Mol. Gen. Genet. 220*: 481.

9

The Role of Quinoproteins in Bacterial Energy Transduction

Christopher Anthony

University of Southampton, Southampton, England

I. INTRODUCTION

For the purpose of this chapter, quinoproteins are considered to be proteins with either pyrroloquinoline quinone (PQQ) or a similar quinone structure as prosthetic group and so will include methylamine dehydrogenase as well as the quinoprotein dehydrogenases responsible for oxidation of glucose, methanol, and ethanol.

In order to be used for ATP production, bacterial dehydrogenases must be coupled to electron transport chains. These must be arranged so as to produce a protonmotive force across the inner cytoplasmic membrane, containing an ATP synthase driven by the protonmotive force. Bacterial electron transport chains vary considerably in their composition. The most common variations concern the presence or absence of *c*-type cytochromes, the type of terminal oxidase that is present, and whether or not an anaerobic nitrate reductase is operational. The electron transport chains differ from one organism to another and are also affected by the growth medium. Figure 1 provides an outline of a number of typical electron transport chains, which will provide the framework for consideration of energy transduction by way of the quinoprotein dehydrogenases.

In all systems studied it has been found that soluble cytochromes *c* and blue copper proteins are located exclusively in the periplasm, and membrane-bound cytochrome *c* is most commonly part of a cytochrome bc_1 complex. This complex mediates the oxidation of membrane-bound ubiquinol by the periplasmic cytochrome *c*, which is then oxidized by a membrane-bound oxidase.

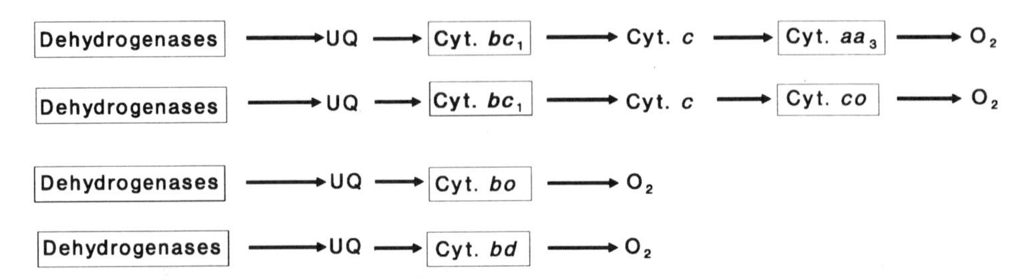

Figure 1 Typical aerobic bacterial electron transport chains. Membrane components are enclosed in boxes. Ubiquinone (UQ) is soluble in the membrane. The most common dehydrogenases are those for NADH and succinate or lactate. Cyt. aa_3 is the classical *a*-type oxidase, cyt. *bo* and cyt. *co* are *o*-type oxidases, and cyt. *bd* is the *d*-type oxidase.

The oxidases fall into two main functional categories: those that oxidize the periplasmic cytochrome *c*, and those that oxidize ubiquinol without the mediation of a cytochrome bc_1 complex [1]. As shown in Figure 1, those that oxidize cytochrome *c* include cytochrome aa_3 and cytochrome *co*. The latter is an *o*-type oxidase, and so it has an O_2-reactive *b*-type cytochrome (cytochrome *o*). The second type of *o*-type oxidase is cytochrome *bo*. This oxidase and cytochrome *bd* differ from other oxidases in being able directly to oxidize ubiquinol. The physiological importance of the various types of oxidases depends on the amount of energy able to be harnessed by the electron transport chains of which they are a part. For example, when the oxidase uses ubiquinol as substrate during oxidation of NADH, the maximum ATP yield (P/O ratio) is likely to be about 70% of the P/O ratio for electron transport chains containing cytochrome *c* and its appropriate oxidases. A second consideration with respect to the various oxidases is their differing affinities for oxygen. This is particularly important when a single organism is able to synthesize more than one type of oxidase. For example, in enteric bacteria the affinity of cytochrome *bd* for oxygen is about 100 times greater than that of cytochrome *bo*; and the cytochrome *bd* is induced instead of the cytochrome *bo* under conditions where oxygen is scarce or when oxygen must be removed in order to protect oxygen-sensitive enzymes [2,3].

For each substrate considered below, the location of the quinoprotein dehydrogenase and the nature of the electron transport components will be considered together with evidence for the sequence of components. When available, evidence for the involvement of the quinoprotein in establishing a protonmotive force will be discussed.

II. METHANOL OXIDATION

The quinoprotein methanol dehydrogenase (MDH) occurs in methylotrophic bacteria that grow on methane or methanol. The majority of these are aerobic, although there are some important exceptions, including bacteria that are able to grow anaerobically on methanol with nitrate as an alternative electron acceptor to oxygen (*Hyphomicrobia* and *Paracoccus denitrificans*). MDH reacts directly with a specific cytochrome called cytochrome c_L [4], and these proteins together constitute a high proportion of the periplasmic proteins in bacteria growing on methanol. In *Methylophilus methylotrophus*, for example, it has been estimated that the concentration of these proteins is about 0.5 mM [5]. How these proteins are involved in electron transport and energy transduction will now be considered (see Ref. 4 for an extensive review including all references on this subject prior to 1985).

A. The Cytochromes and Oxidases Involved in Methanol Oxidation

It is usually emphasized in discussions of electron transfer systems of methylotrophs that they contain large amounts of *c*-type cytochromes. This is not surprising because between 50 and 90% of the electrons passing to the oxidase in methylotrophs arise from MDH, and these bypass the lower redox potential parts of the electron transport chain, providing electrons exclusively at the level of the *c*-type cytochromes [4,6].

In addition to the specific cytochrome c_L, methylotrophs have at least one other cytochrome *c*, which when first described was called cytochrome c_H. Its high isoelectric point distinguishes it from the lower isoelectric point of the acidic cytochrome c_L [6]. For convenience I shall continue to use this nomenclature, but it should be borne in mind that in some methylotrophs the cytochrome c_H is exceptional in being acidic, and the cytochrome is named according to its absorption maximum (e.g., cytochrome *c*-550 of *P. denitrificans* [7]). The properties of the two main types of *c*-type cytochromes in a wide range of different methylotrophs have been fully described elsewhere [6]. Cytochrome c_H is a typical small Class I *c*-type cytochrome. The amino acid sequences of these cytochromes from *P. denitrificans* and *M. methylotrophus* are typical of Class I cytochromes *c* and the crystal structure of the *Paracoccus* protein confirms this conclusion [8]. In other aerobic bacteria such cytochromes mediate the oxidation of the cytochrome bc_1 complex by the terminal cytochrome oxidase (Fig. 1), and confirmation of this role in methylotrophs comes from work with the purified oxidases.

As indicated in Figure 1, the major oxidases found in bacteria that use cytochrome *c* are cytochrome aa_3 and cytochrome *co*, and as expected, these are the oxidases that function in methylotrophs (Table 1). The most important common feature of these oxidases is that the preferred substrate is the typical Class I *c*-type (cytochrome c_H), the rate with this substrate being usually at least 50 times greater than with cytochrome c_L.

Table 1 The Oxidases of Methylotrophs During Methanol Oxidation

Organism	Type of methylotroph	Type of oxidase	Main substrate	Alternative substrate	Ref.
M. methylotrophus	Obligate	Cyt. *co*	Cyt. c_H	—	9, 10
		Cyt. aa_3	[Cyt. c_H]	—	9, 10
Organism 4025	Obligate	Cyt. *co*	Cyt. c_H	Azurin	11, 12
M. extorquens AM1	Facultative	Cyt. aa_3	Cyt. c_H	—	13–15
A. methanolicus	Acidophilic	Cyt. *co*	Cyt. c_H	—	16
P. denitrificans	Facultative autotroph	Cyt. aa_3	Cyt. c_{550}	[Cyt. c_{553i}]	17–19

When the substrate is given in parentheses, there is no in vitro evidence. Cytochrome c_{553i} has also been called cytochrome c_{co}.

B. Electron Transport Chains Involving Methanol Dehydrogenase

It is well established that cytochrome c_L is able to react rapidly with cytochrome c_H, and so the simplest "methanol oxidase" electron transport chain that can be envisaged is one containing these two cytochromes plus MDH and an oxidase (Fig. 2). The absolute importance of cytochrome c_L in this scheme is demonstrated by the properties of mutants lacking cytochrome *c*, which lose their ability to grow on methanol [19–21]. However, because methylotrophs typically contain more than two periplasmic *c*-type cytochromes and also one or two blue copper proteins, it is difficult to rule out more complex or alternative pathways involving these proteins. Certainly, if these proteins are able to interact, then a few electrons might wander down insignificant less well-trodden paths. The main routes will be indicated by the specificity of some of the interactions, by the higher concentrations of some potential intermediates (cytochromes c_L and c_H), by induction of some intermediates during growth on methanol, and by the demonstration that loss of the particular component leads to loss of ability to oxidize methanol. Some of the variations in the pathways proposed in different methylotrophs are illustrated in Figure 2 and are summarized below. A critical point with respect to energy transduction from MDH is that all these electron transport chains are similar to those operating in the oxidation of inorganic substrates [22] in bypassing the low potential ubiquinol/cytochrome *b* parts of the chain.

1. Methylobacterium extorquens *AM1*

In this pink facultative methylotroph the electron transport chain terminates in the sole oxidase, cytochrome aa_3, and usually involves cytochrome c_L and cytochrome c_H (Fig. 2), as confirmed in vitro using the pure proteins [13,14]. Mutants lacking all *c*-type cytochromes were unable to grow on methanol but, remarkably,

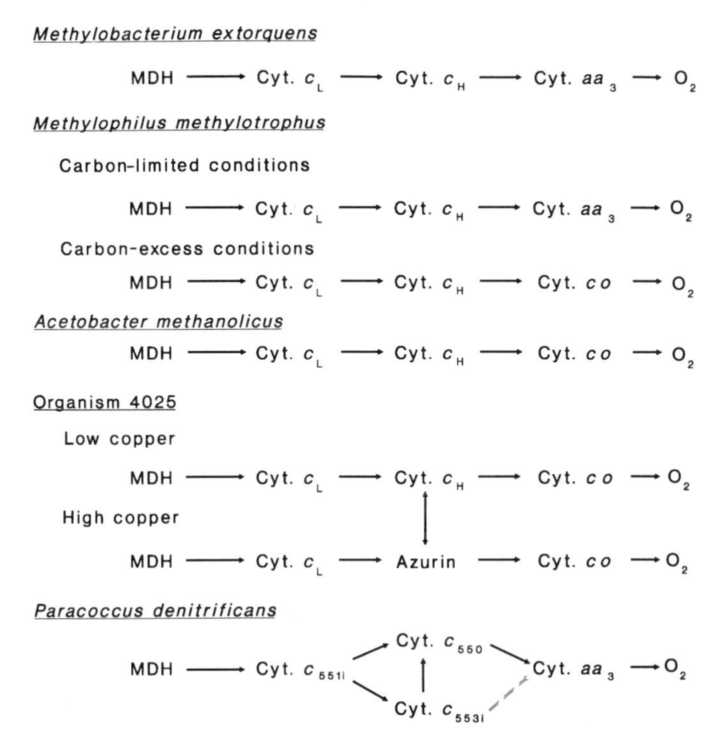

Figure 2 Electron transport in the "methanol oxidase" systems of methylotrophs. In organism 4025 under many conditions, cytochrome c_H and azurin are both present, and under these conditions both these electron acceptors may operate in electron transport. In *P. denitrificans* cytochrome *c*-551 is entirely analogous in structure and function to cytochrome c_L and is also called cytochrome *c*-552 [7,19]. Cytochrome *c*-550 is the best substrate for the oxidase (see Table 1). Cytochrome *c*-553i (also called cytochrome c_{co}) is induced during growth on methanol and methylamine, and mutants grow less well on methanol without it. It has been suggested that it possibly replaces *c*-550 in a mutant lacking this cytochrome [19].

were able to grow on other substrates such as succinate, leading to the suggestion that *c*-type cytochromes are not essential for the oxidation of succinate or NADH in these bacteria [23,24]. A similar observation was later made using cytochrome *c* mutants of *P. denitrificans* [25,26]. This conclusion is supported by the observation that in some growth conditions the stoichiometry of proton translocation in *M. extorquens* is consistent with no involvement of cytochrome *c* [27,28]. This was probably the first example of cytochrome *c*-independent electron flow from the *b*-type cytochromes to cytochrome aa_3 that is now becoming of more general interest [29].

2. Methylophilus methylotrophus

This obligate methylotroph has two possible pathways depending on its growth conditions (Fig. 2) [29]. Under conditions of methanol limitation the predominant oxidase is cytochrome aa_3, but under oxygen limitation the *o*-type oxidase, cytochrome *co*, is the sole oxidase. The reason for the alternative oxidases is not obvious. Induction of cytochrome *co* also occurs under nitrogen limitation when oxygen is in excess. The response thus appears to be to the carbon concentration in the growth medium and not to differing oxygen concentrations. The *o*-type oxidase was the first to be purified and clearly shown to be cytochrome *co*, containing two subunits of *c*-type cytochrome and two subunits of a CO-reactive cytochrome *b* (cytochrome *o*) [10]. The electron transport chain shown for methanol-excess conditions (Fig. 2) has been confirmed by its complete reconstitution using the pure proteins [10]. An unexpected observation when analyzing the membrane cytochromes was that the only *c*-type cytochromes in the membrane were the cytochrome *c* component of cytochrome *co* and membrane-bound cytochrome c_L. There was no separate cytochrome c_1 that would be expected if there is a typical cytochrome bc_1 [30]. It appears that cytochrome c_1 may be replaced by cytochrome c_L in these bacteria and a similar situation might also occur in *P. denitrificans* [31].

3. Acetobacter methanolicus

Because of its cytochrome complement, it is probable that the electron transport chain for methanol oxidation in this acidophilic methylotroph is similar to that in *M. methylotrophus* grown under carbon-excess conditions (Fig. 2) [32]. The oxidase of this organism is remarkable in that it appears to be a typical cytochrome *co*, but the cytochrome *c* component is lost on purification and the oxidase is then only able to oxidize those cytochromes with which it is able to reconstitute an active cytochrome *co* complex [16].

4. *Organism 4025*

The electron transport chain of this obligate methylotroph is very similar to that in *A. methanolicus* and in *M. methylotrophus* during growth under carbon-excess conditions (Fig. 2). It uses cytochrome *co* exclusively and is unusual in requiring high concentrations of copper for growth [33,34]. In some growth conditions the concentration of azurin is relatively high, and as was shown in vitro using purified proteins, it can replace cytochrome c_H as a mediator between cytochrome c_L and the oxidase [11,12].

5. Paracoccus denitrificans

P. denitrificans is an unusual methylotroph in being a facultative autotroph, fixing its carbon at the level of CO_2 by the Calvin cycle [35]. There is also a great deal known about its energetics, and it is becoming very important in the molecular biology of energetic systems [19]. There has been some debate over the cytochromes involved in methanol oxidation in this organism. This is discussed at some length here because it illustrates some of the problems of determining paths of electron flow in the periplasm of bacteria [29,36].

The first evidence relating to methanol oxidation in these bacteria was the description of a CO-reactive cytochrome *c* induced during growth on methanol. This was assumed to be involved specifically in methanol oxidation and was called cytochrome c_{co} [37]. Three cytochromes *c* (all acidic) were subsequently purified from the periplasm of *P. denitrificans* and characterized by Davidson and colleagues [18,38,39]. One was the well-known cytochrome *c*-550 (15 kDa), and the other two were induced during growth on methanol. These were designated cytochrome *c*-551i (22 kDa) and cytochrome *c*-553i (30 kDa), and it was proposed that this CO-binding cytochrome *c*-553i is the electron acceptor for MDH [18]. It was also suggested, however [40,41], that the 22 kDa cytochrome *c*-551i (also called cytochrome *c*-552) is not restricted to growth on C_1-compounds but is the electron donor to cytochrome aa_3.

Some of this confusion has been sorted out by our investigations using the pure cytochromes and MDH [7]. We showed that cytochrome *c*-553i (or cytochrome c_{co}) is not the electron acceptor for MDH; the sole electron acceptor is the periplasmic 22 kDa cytochrome (cytochrome *c*-551i or *c*-552). The previous assumption that the cytochrome *c*-553i is the MDH electron acceptor was based on the observation that it was one of the two *c*-type cytochromes induced during growth on methanol, and on the assumption that the CO reactivity is in some way diagnostic of the methanol-linked cytochrome *c* of methylotrophs. As has been emphasized previously [4,35], there never has been any evidence that this feature is important to the physiological function of these cytochromes. This work raises the question of the function of the cytochrome *c*-553i (or cytochrome c_{co}) that is also induced during growth on methanol. It was shown [18] that spheroplasts oxidize this cytochrome, but only at low rates compared with cytochrome *c*-550. This suggests that the sequence of electron transport intermediates in *P. denitrificans* is as follows:

$$\text{MDH} \rightarrow \text{cyt. } c\text{-551i} \rightarrow \text{cyt. } c\text{-553i} \rightarrow c\text{-550} \rightarrow \text{oxidase} \quad (1)$$

The importance of cytochrome *c*-553i in this sequence is supported by its induction during growth on methanol and by the diminished growth rate on methanol of a mutant lacking it [19]. By contrast, a mutant lacking cytochrome *c*-550 was capable of normal growth on methanol; in this mutant the cytochrome *c*-553i was presumably acting as donor to the oxidase [42]. This raises an important general point about evaluating alternative pathways from properties of mutants. If two proteins are both able to fulfill a particular function in a pathway, then loss of one of them will permit the other to be the sole operator. It does not prove it to be the normal sole operator in the wild-type organism. In this particular case the final conclusion must be that the alternative pathways indicated in Figure 2 may operate and that the predominant path will be determined by the relative amounts of the proteins under a particular growth condition, and by the kinetic constants (affinities and rates of reaction) of the proteins concerned.

It should be noted that although cytochrome *c*-550 is probably normally involved in methanol oxidation, there is considerable evidence that it is not involved in the oxidation of NADH and succinate, and that electron transport during oxidation of these substrates involves direct electron transfer between the cytochrome bc_1 complex and the oxidase [17,26], as shown for *M. extorquens* (above). It has been suggested that membrane-bound cytochrome *c*-552 is essential for this process, and that this cytochrome is identical to the periplasmic cytochrome *c*552 [40,41], which has subsequently has been shown to be the electron acceptor for MDH (also called cytochrome *c*-551i) [7]. This would raise a problem with respect to regulation of synthesis of this cytochrome in *P. denitrificans* because the periplasmic form is induced only during growth on methanol (not succinate), but it is present on membranes at a high level when cells are grown on succinate [31,40].

C. Anaerobic Electron Transport from Methanol Dehydrogenase

Figure 3 shows the pathway operating in *P. denitrificans* growing anaerobically on methanol with nitrate as electron acceptor. In this case it is proposed that nitrite is the terminal electron acceptor from the *c*-type cytochromes. The electron donor to nitrate reductase is ubiquinol [43], which is reduced by oxidation of NADH, which is produced during oxidation of formaldehyde and formate. The nitrite produced

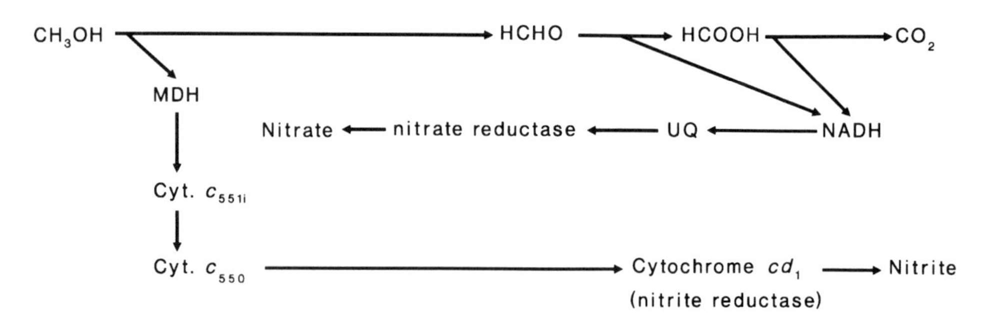

Figure 3 Electron transport during the anaerobic oxidation of methanol by *P. denitrificans*. This same scheme is also likely to operate during anaerobic growth in *Hyphomicrobia*. The produce of reduction of nitrate by nitrate reductase is nitrite, which is then used as terminal electron acceptor for the electron transport chain from methanol dehydrogenase.

by this process is used as terminal electron acceptor for *c*-type cytochromes that are reduced by MDH and oxidized by the nitrite reductase, which is a periplasmic cytochrome cd_1 [44].

D. Protein/Protein Interactions Involved in the "Methanol Oxidase" System

The soluble proteins of this electron transport chain may either operate as separate entities, forming short-lived bimolecular complexes during which electron transfer occurs, or they might form stable complexes of more than two proteins in a "wire" system. In such a system electrons would flow, for example, from MDH through cytochrome c_L to cytochrome c_H, the site of entry of electrons into cytochrome c_L being different from the site of exit. There is now strong evidence that such a wire system does not operate in periplasmic electron transport from MDH (see Chapter 2) [45,46]: for the MDH/cytochrome c_L interaction all the evidence is consistent with the proteins reversibly "docking" between lysyl residues on the α-subunit of MDH and carboxyl groups on the cytochrome. Furthermore, cross-linking experiments show that the site on cytochrome c_L that is involved in docking with MDH is the same site that is involved in docking with cytochrome c_H.

E. Proton Translocation and ATP Synthesis During Methanol Oxidation

By analogy with other proton-translocating electron transport chains that involve cytochrome *c*, it would be expected that up to three proton-translocating segments might operate during the oxidation of NADH, and only one during the oxidation of methanol. This has been confirmed in an extensive investigation of the bioenergetics of *M. methylotrophus* by Dawson and Jones [47,48] and of proton translocation in *M. extorquens* [27,28]. In these bacteria it appears that the oxidases are not proton-pumping, by contrast with the cytochrome aa_3 of *P. denitrificans* [49]. *M. methylotrophus* is able to sustain a protonmotive force of up to -165 mV during respiration with methanol as substrate. At pH 7.0, the growth pH, this is almost exclusively due to the membrane potential [50]. It was concluded that phosphorylation of ADP to ATP is coupled to translocation of about four protons and that the oxidation of methanol and NADH is accompanied by the synthesis of about 0.5 and 1.4–2.0 molecules of ATP, respectively [51]. This suggestion is consistent with the measured ATP production during oxidation of methanol and NADH by membrane vesicles isolated from *M. extorquens* [52]. All the information on electron transport components and on proton translocation and ATP synthesis is consistent with the arrangement shown in Figure 4 in which protons are released into the periplasm from reduced MDH when it passes electrons to cytochrome c_L, and protons are consumed on the inner surface of the periplasmic membrane.

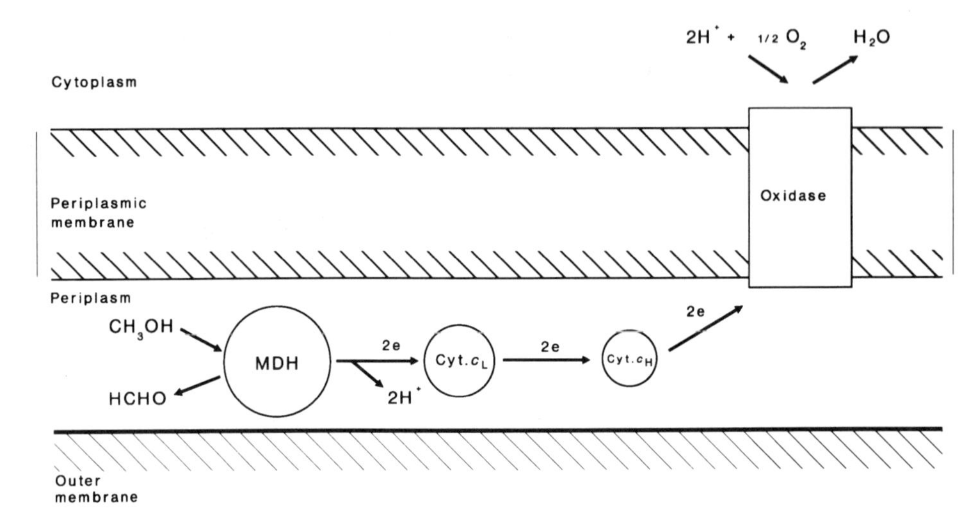

Figure 4 Proton translocation in the periplasmic "methanol oxidase" system. The size of the periplasmic proteins is roughly to scale and assumes that the distance between the two membranes is about 65 Å [36].

III. ELECTRON TRANSPORT BY WAY OF METHYLAMINE DEHYDROGENASE

A. The Electron Acceptor for Methylamine Dehydrogenase

The physiological electron acceptor for methylamine dehydrogenase (MADH) is usually the type I blue copper protein, induced during growth on methylamine, which was first discovered by Tobari in *M. extorquens* AM1 and called amicyanin [53,54]. It is one of the two blue copper proteins, or cupredoxins, found in many methylotrophs [55]. Its function was initially indicated by the observation that it is induced only during growth on methylamine, that the rates of electron transfer to amicyanin were very rapid, and that no significant electron transfer occurred with *c*-type cytochromes or azurin from the same bacteria. It was shown that amicyanin is located solely in the periplasmic space together with the dehydrogenase and *c*-type cytochromes [33,56] and that the rate of electron transfer from MADH to amicyanin is sufficient to account for the respiration rate in whole bacteria [12,33,57]. The conclusion that amicyanin is the electron acceptor from MADH is supported by work with at least five very different methylotrophs including the pink facultative methylotroph *Methylobacterium* [53,54], the obligate methylotrophs *Methylomonas* J [53,54,58] and organism 4025 [11,12,33,34], and the facultative autotrophs *Thiobacillus versutus* [59] and *P. denitrificans* [39,56, 60,62,63].

Amicyanin is not detectable, however, in all methylotrophs growing on methylamine by way of MADH. In bacteria genuinely lacking all amicyanin, the obvious alternative is one of the periplasmic *c*-type cytochromes. These may be involved (instead of amicyanin) in the oxidation of methylamine in trimethylamine-grown *M. methylotrophus* [64] and in the closely related organism W3A1 [65], in copper-deficient *M. extorquens* [66], and perhaps in copper-deficient organism 4025 [11] (Fig. 5). In support of this conclusion is the observation that MADH is sometimes able to react directly with the typical small cytochrome *c* of methylotrophs (usually called cytochrome c_H), but whether or not this has any physiological significance is not known. If amicyanin is present and able to react with cytochrome *c*, then addition of methylamine to whole bacteria will, of course, lead to rapid reduction of the cytochrome *c*, as was previously demonstrated in *Methylo-*

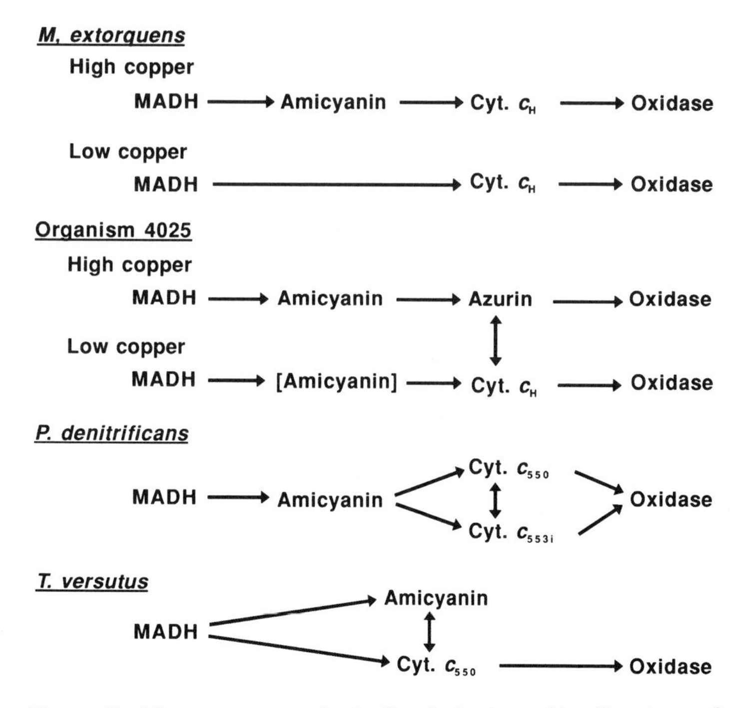

Figure 5 Electron transport in the "methylamine oxidase" systems of methylotrophs. Azurin may replace the typical Class I *c*-type cytochromes in other bacteria as well as organism 4025. In many conditions in the bacteria listed here, more than one route of electron transport may operate. In *P. denitrificans* cytochrome *c*-553i may be able to replace cytochrome *c*-550 completely in a mutant lacking that cytochrome.

bacterium [23]. Neither of the two oxidases described in methylotrophs (cytochrome aa_3 or cytochrome *co*) is able to oxidize amicyanin, although the second blue copper protein (azurin) may be oxidized [11]. As amicyanin is able to interact with azurin or periplasmic *c*-type cytochromes, some of which can act as electron donor to the oxidase, then the simplest electron transport chain will involve MADH, amicyanin, a periplasmic cytochrome *c* (or azurin), and a membrane oxidase (Fig. 5).

B. Electron Transport Chains Involving Methylamine Dehydrogenase

An illustration of the difficulty of coming to definitive conclusions with respect to electron transport systems involving amicyanin is given by our studies with organism 4025. This obligate methylotroph usually produces so much amicyanin during growth on methylamine that it appears blue in color [33,34]. We have shown that in oxygen-limited continuous culture, copper deficiency leads to lower growth densities of bacteria and lower respiration rates with methylamine and that blue copper proteins were not readily detectable [11]. It was impossible, however, to demonstrate that cytochrome c_H is an electron acceptor from MADH or to reconstitute from pure proteins a complete electron transport chain consisting of dehydrogenase, cytochrome c_H, and the oxidase (cytochrome *co*). We concluded that amicyanin may be the only electron acceptor for MADH but are surprised that such low levels of amicyanin may be sufficient to permit growth. When sufficient copper was present to achieve maximum growth, very large amounts of amicyanin and azurin were produced and complete electron transport chains could be reconstituted using MADH plus amicyanin and oxidase, with either cytochrome c_H or azurin as the intermediate electron carrier between amicyanin and the oxidase. This provides an excellent example of an organism in which blue copper proteins completely replace soluble cytochromes [11,12] (Fig. 5). In contract to organism 4025, it has been shown, using the pure proteins from *T. versutus* [57], that the rate constants are sufficiently high for either amicyanin or cytochrome *c*-550 to act as electron acceptor from MADH (Fig. 5).

As shown for studies of electron transport from methanol dehydrogenase, the results with mutants lacking the various components are extremely important, as illustrated by the demonstration that mutants of *P. denitrificans* lacking amicyanin fail to grow on methylamine although they are able to synthesize all the periplasmic *c*-type cytochromes [68]. It had been suggested that the cytochrome *c*-551i of *P. denitrificans*, which is induced together with cytochrome *c*-553i during growth on methanol or methylamine, might be involved in electron transport from methylamine [38,39,61]. However, is has now been conclusively demonstrated that cytochrome *c*-551i is equivalent to cytochrome c_L, the electron acceptor from methanol dehydrogenase, and mutants lacking it grow perfectly well on methylamine [7,19,20,21]. Remarkably, mutants lacking the expected donor to the oxidase (cytochrome *c*-550) are still able to grow (more slowly) on methylamine, suggesting that the other periplasmic *c*-type cytochromes, or azurin, are able to take over some of the role of cytochrome *c*-550 [42].

In summary, the most likely electron transport chains for the oxidation of methylamine in methylotrophic bacteria are as shown in Figure 5, in which amicyanin is usually the electron acceptor and in which electrons flow to the oxidase is by way of the typical Class I cytochrome *c* (*c*-550 or c_H) or azurin. The proton translocation measured during respiration with methylamine in *M. extorquens* [27] is consistent with the proposal that does not involve the mid-chain *b*-type cytochromes and that the protonmotive force is established as in methanol oxidation (Fig. 4), the yield of ATP being always one or less.

IV. GLUCOSE OXIDATION

In the direct route for oxidation of glucose in bacteria, this substrate is oxidized in the periplasm to gluconic acid by way of the quinoprotein glucose dehydrogenase (GDH). Although a great deal is known about the dehydrogenases, the picture of electron transport from them in order to obtain energy has been confused. This is mainly because there are two types of GDH in one of the main bacteria used in this work—*Acinetobacter calcoaceticus*. The membrane-bound form of the enzyme is sometimes referred to as GDH-A or m-GDH, and the soluble form of the enzyme is sometimes referred to as GDH-B or s-GDH.

A. The Primary Electron Acceptor from the Soluble GDH-B in *Acinetobacter calcoaceticus*

The soluble, periplasmic form (GDH-B) has only been described in *A. calcoaceticus*, and its structure shows very little similarity to GDH-A in the same organism or in any other organism having GDH (see Ref. 68 for discussion of primary sequences of these and other quinoproteins). GDH-B does not react directly with ubiquinone although it is able to react with short-chain homologs, and it has been concluded from this [69] that GDH-B does not couple to the electron transport chain at the level of ubiquinone. It has been suggested that a soluble cytochrome *b*-562 is the electron acceptor for GDH-B and that its electrons are then passed onto the ubiquinone [70,71]. This conclusion was based on the slow reduction of the cytochrome by the GDH-B and on the observation that the cytochrome and GDH-B were released together from the bacteria in various conditions that made the outer membrane permeable [71]. There is no evidence for electron transfer from the cytochrome *b*-562 to ubiquinone, and there is no other evidence to support the involvement of this system in glucose oxidation in these bacteria [72,73].

B. The Primary Electron Acceptor from Membrane-Bound GDH

The membrane-bound GDH is the enzyme responsible for glucose oxidation in a wide range of bacteria. In *A. calcoaceticus* it is sometimes referred to as GDH-A to distinguish it from the periplasmic GDH-B (see Ref. 8 for review of primary sequences of the glucose dehydrogenases). Studies in vitro have demonstrated that the primary electron acceptor of GDH is ubiquinone in *Pseudonomas* [74], *E. coli* [75], *Gluconobacter* [76], and *A. calcoaceticus*. This has been confirmed in vivo for *A. calcoaceticus* [72] and in reconstituted membrane systems in *E. coli* [75] and *Gluconobacter* [77].

C. The Electron Transport Chains Involved in Glucose Oxidation

As GDH interacts directly with membrane ubiquinone, the subsequent electron transport (Fig. 6) is determined by the nature of the oxidases and their substrates in the various bacteria that oxidize glucose, and these are all variations of the systems outlined in Figure 1.

In *A. calcoaceticus* the ubiquinone is oxidized directly by oxidase complexes (Fig. 6). In conditions of high aeration this is the *o*-type oxidase (cytochrome *bo*), and in oxygen-deficient conditions it is the *d*-type oxidase (cytochrome *bd*) [72,73]. The same type of oxidases also operate in *E. coli* [75]. In *Pseudomonas* the oxidase is the second type of *o*-type oxidase (cytochrome *co*) that uses *c*-type cytochromes as substrate [78] (Fig. 6).

The nature of the oxidase in acetic acid bacteria raises an interesting question. *Gluconobacter suboxydans* produces large amounts of membrane-bound cytochrome *c* during growth on sugars, although it was not stated whether or not soluble *c*-type cytochromes were produced [79,80]. In most bacteria containing cytochrome *c* the predominant oxidases are those that are able to oxidize this cytochrome (cytochrome aa_3 or cytochrome *co*). This organism appears to be very unusual in that it contains *c*-type cytochromes, and yet its predominant oxidase appears to be a typical, cyanide-sensitive, ubiquinol oxidase (cytochrome *bo*). It contains no *a*-type oxidase and there is no evidence that it contains a typical cytochrome *co* [81]. It appears that these bacteria have a cyanide-insensitive alternative oxidase that is induced at low pH together with cytochrome *c* and alcohol dehydrogenase (see below). It was concluded that "*Gluconobacter suboxydans* produces a KCN-insensitive and non-energy-generating respiratory bypass, which may be related to a cytochrome *c* associated with alcohol dehydrogenase, concomitant with a decrease in the extracellular pH during growth" [80] (see Fig. 6).

Figure 6 summarizes the various electron transport chains likely to be involved in glucose oxidation. These systems are arranged as in other bacteria that oxidize

Pseudomonas

GDH → UQ → Cyt.*bc*$_1$ → Cyt.*c* → Cyt.*co* → O_2

A. calcoaceticus and E. coli

GDH → UQ → Cyt.*bd* → O_2 [Low oxygen tension]

GDH → UQ → Cyt.*bo* → O_2 [High oxygen tension]

Gluconobacter

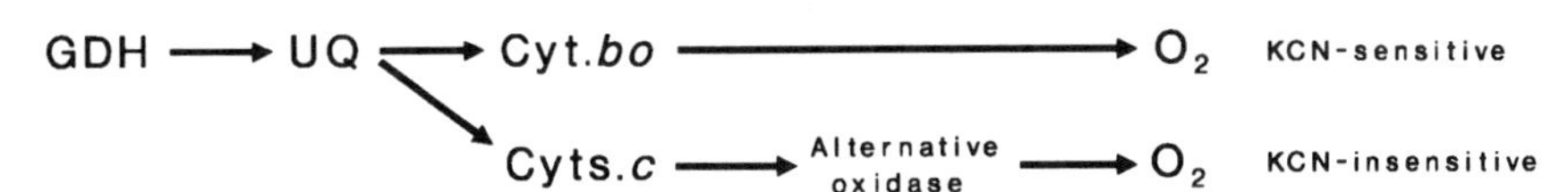

Soluble ADH

ADH → Periplasmic Cyts. *c* → Oxidase → O_2

Membrane ADH (e.g. acetic acid bacteria)

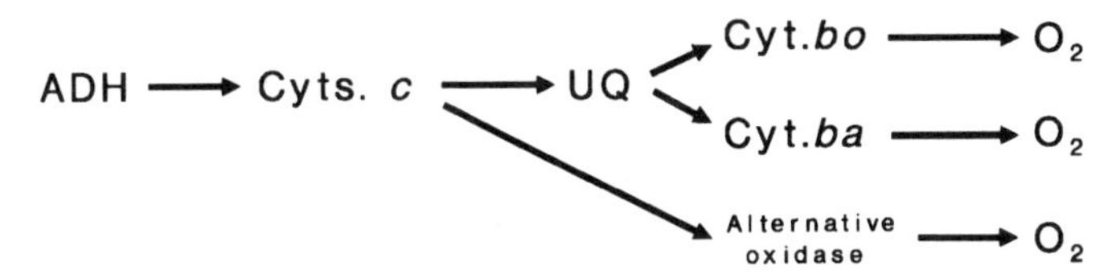

Figure 6 Electron transport in the oxidation of glucose and ethanol. The glucose dehydrogenase (GDH) is the typical integral membrane protein in which the binding site for glucose is on the periplasmic side of the membrane. Bacteria having a soluble alcohol dehydrogenase (ADH) include *P. aeruginosa*, *P. putida*, and *R. acidophila*, and *P. testosteronii*, in which the ADH is a soluble quinohemoprotein. The membrane ADH occurs in the acetic acid bacteria, but it should be noted that the amino acid sequences of the dehydrogenase subunits are not typical of integral membrane proteins [69]. These subunits are presumably bound to membrane components of the three-subunit ADHs. The ubiquinol oxidase in *Gluconobacter* is cytochrome *bo*, and in *Acetobacter aceti* it is cytochrome *ba* (previously called cytochrome a_1).

ubiquinol to establish a protonmotive force and hence make ATP. Such systems have been established in reconstituted glucose-oxidizing membrane systems for *E. coli* [75,82], *Pseudomonas* [82], *A. calcoaceticus* [82], and *Gluconobacter* [76]. This is consistent with previous observations that glucose oxidation by way of the membrane GDH is able to lead to increased growth yields [75,76,78,82–87].

V. ALCOHOL OXIDATION

The route for electron transport from ethanol depends on the type of dehydrogenase involved. Some of the alcohol dehydrogenases (ADHs) are soluble and similar in many ways to the typical MDH of methylotrophs; these include ADHs from *Pseudomonas aeruginosa* and *Pseudomonas putida* [88] and from *Rhodopseudomonas acidophila* [89]. There is also a second type of soluble ADH, which occurs in *Pseudomonas testosteronii* and contains cytochrome *c* [90]. No information is available on the location of the soluble alcohol dehydrogenases, but if their electron acceptors are *c*-type cytochromes as for MDH then they are likely to be periplasmic and energy transduction is likely to be very similar to that in methylotrophs (Figs. 3 and 4).

The ADHs from acetic acid bacteria differ from the soluble enzymes in being membrane-bound hemoquinoprotein complexes. These complexes (with three subunits) contain heme in the dehydrogenase component, and they also have a separate cytochrome *c* component [68,91–93]. The only oxidases described in these bacteria are cytochrome *bo* and cytochrome *ba* (previously cytochrome a_1), and both of these oxidize ubiquinol, whose redox potential is lower than that of most *c*-type cytochromes [81,94]. It appears unlikely that these bacteria will have an electron transport chain in which the *c*-type cytochromes are involved at a redox potential lower than ubiquinol, and it is possible that the oxidase involved in the oxidation of ethanol is the alternative, cyanide-resistant oxidase described above in *Gluconobacter*. This oxidase is induced at low pH together with the alcohol dehydrogenase and its associated cytochrome *c* [79,80]. An alternative possibility is that the membrane cytochrome *c* subunit of ADH is involved in electron transport to ubiquinone, as indicated by reconstitution experiments involving ADH, ubiquinone, and the *o*-type ubiquinol oxidase [95] (Fig. 6).

REFERENCES

1. Poole, R. K. (1988). Bacterial cytochrome oxidases. In *Bacterial Energy Transduction* (C. Anthony, ed.), Academic Press, London, p. 231.
2. Smith, A., Hill, S., and Anthony, C. (1990). The purification, characterization and role of the *d*-type cytochrome oxidase of *Klebsiella pneumoniae* during nitrogen fixation. *J. Gen. Microbiol. 136*: 171.

3. Hill, S., Viollet, S., Smith, A. T., and Anthony, C. (1990). Roles for enteric *d*-type cytochrome oxidase in nitrogen fixation and microaerobiosis. *J. Bacteriol. 172*: 2071.
4. Anthony, C. (1986). The bacterial oxidation of methane and methanol. *Adv. Microbial. Physiol. 27*: 113.
5. Beardmore-Gray, M., O'Keeffe, D. T., and Anthony, C. (1983). The methanol: cytochrome *c* oxidoreductase activity of methylotrophs. *J. Gen. Microbiol. 129*: 923.
6. Anthony, C. (1992). The *c*-type cytochromes of methylotrophic bacteria. *Biochim. Biophys. Acta 1099*: 1.
7. Long, A. R., and Anthony, C. (1991). Characterization of the periplasmic cytochromes *c* of *Paracoccus denitrificans*: Identification of the electron acceptor for methanol dehydrogenase, and description of a novel cytochrome *c* heterodimer. *J. Gen. Microbiol. 137*: 415.
8. Timkovich, R., and Dickerson, R. E. (1976). The structure of *Paracoccus denitrificans* cytochrome *c*-550. *J. Biol. Chem. 251*: 4033.
9. Cross, A. R., and Anthony, C. (1980). The electron transport chains of the obligate methylotroph. *Methylophilus methylotrophus. Biochem. J. 192*: 429.
10. Froud, S. J., and Anthony, C. (1984). The purification and characterization of the o-type oxidase from *Methylophilus methylotrophus*, and its reconstitution into a 'methanol oxidase' electron transport chain. *J. Gen. Microbiol. 130*: 2201.
11. Auton, K. A., and Anthony, C. (1989). The role of cytochromes and blue copper proteins in growth of an obligate methylotroph on methanol and methylamine. *J. Gen. Microbiol. 135*: 1923.
12. Auton, K. A., and Anthony C. (1989). The 'methylamine oxidase' system of an obligate methylotroph. *Biochem. J. 260*: 75.
13. Fukumori, Y., Nakayama, K., and Yamanaka, T. (1985). Cytochrome *c* oxidase of *Pseudomonas* AM1: Purification, and molecular and enzymatic properties. *J. Biochem. 98*: 493.
14. Mukai, K., Fukumori, Y., and Yamanaka, T. (1990). The methanol-oxidising system of *Methylobacterium extorquens* AM1 reconstituted with purified constituents. *J. Biochem. 107*: 714.
15. Sone, N., Sekimach, M., Fukumori, Y., and Yamanaka, T. (1987). Evidence against proton pump activity by cytochrome oxidase of *Pseudomonas* AM1. *J. Biochem. 102*: 481.
16. Chan, H. T. C., and Anthony, C. (1991). The *o*-type oxidase of the acidophilic methylotroph *Acetobacter methanolicus. J. Gen. Microbiol. 137*: 693.
17. Bolgiano, B., Smith, L., and Davies, H. C. (1988). Kinetics of the interaction of the cytochrome *c* oxidase of *Paracoccus denitrificans* with its own and bovine cytochrome *c*. *Biochim. Biophys. Acta 933*: 341.
18. Davidson, V. L., and Kumar, M. A. (1989). Cytochrome *c*-550 mediates electron transfer from inducible periplasmic c-type cytochromes to the cytoplasmic membrane of *Paracoccus denitrificans. FEBS Lett. 245*: 271.
19. Harms, N., and van Spanning, R. J. M. (1991). C-1 metabolism in *Paracoccus denitrificans*: Genetics of *Paracoccus denitrificans. J. Bioenerg. Biomembrane 23:* 187.
20. Nunn, D. N., and Lidstrom, M. E. (1986). Isolation and complementation analysis of

10 methanol oxidation mutant classes and identification of the methanol dehydrogenase structural gene of *Methylobacterium* sp. strain AM1. *J. Bact. 166*: 581.
21. Nunn, D. N., and Lidstrom, M. E. (1986). Phenotypic characterisation of 10 methanol oxidation mutant classes of *Methylobacterium* sp. strain AM1. *J. Bact. 166*: 591.
22. Wood, P. M. (1988). Chemolithotrophy. In *Bacterial Energy Transduction* (C. Anthony, ed.), Academic Press, London, pp. 183–230.
23. Anthony, C. (1975). The microbial metabolism of C1 compounds: The cytochromes of *Pseudomonas* AM1. *Biochem. J. 146*: 289.
24. Widdowson, D., and Anthony, C. (1975). The microbial metabolism of C1 compounds: The electron transport chain of *Pseudomonas* AM1. *Biochem. J. 152*: 349.
25. Willison, J. C., and John, P. (1979). Mutants of *Paracoccus denitrificans* deficient in *c*-type cytochromes. *J. Gen. Microbiol. 115*: 443.
26. Bolgiano, B., Smith, L., and Davies, H. C. (1989). Electron transport reactions in a cytochrome c deficient mutant of *Paracoccus denitrificans*. *Biochim. Biophys. Acta 973*: 227.
27. O'Keeffe, D. T., and Anthony, C. (1978). The microbial metabolism of C1 compounds. The stoichiometry of respiration-driven proton translocation in *Pseudomonas* AM1 and in a mutant lacking cytochrome *c*. *Biochem. J. 170*: 561.
28. Keevil, C. W., and Anthony, C. (1979). Effect of growth conditions on the involvement of cytochrome c in electron transport, proton translocation and ATP synthesis in the facultative methylotroph, *Pseudomonas* AM1. *Biochem. J. 182*: 71.
29. Ferguson, S. J., and Page, M. D. (1990). The functions and biosynthesis of the *c*-type cytochromes of bacterial respiratory chains with particular reference to autotrophs and the handling of C1-compounds. *FEMS Microbiol. Rev. 87*: 227.
30. Froud, S. J., and Anthony, C. (1984). The role of cytochrome *c* in membranes of *Methylophilus methylotrophus*. *J. Gen. Microbiol. 130*: 3319.
31. Berry, E. A., and Trumpower, B. L. (1985). Isolation of ubiquinol oxidase from *Paracoccus denitrificans* and resolution into cytochrome bc_1 and cytochrome caa_3 complexes. *J. Biol. Chem. 260*: 2458.
32. Elliott, E. J., and Anthony, C. (1988). The interaction between methanol dehydrogenase and cytochrome *c* in the acidophilic methylotroph *Acetobacter methanolicus*. *J. Gen. Microbiol. 134*: 369.
33. Lawton, S. A., and Anthony, C. (1985). The roles of cytochromes and blue copper proteins in the oxidation of methanol and methylamine in organism 4025, an obligate methylotroph. *J. Gen. Microbiol. 131*: 2165.
34. Lawton, S. A., and Anthony, C. (1985). The role of blue copper proteins in the oxidation of methylamine by an obligate methylotroph. *Biochem. J. 228*: 719.
35. Anthony, C. (1982). *The Biochemistry of Methylotrophs*. Academic Press, London.
36. Ferguson, S. J. (1988). Periplasmic electron transport reactions. In *Bacterial Energy Transduction* (C. Anthony, ed.), Academic Press, London, pp. 141–182.
37. Van Verseveld, H. W., and Stouthamer, A. H. (1978). Electron transport chain and coupled oxidative phosphorylation in methanol-grown *Paracoccus denitrificans*. *Arch. Microbiol. 118*: 13.
38. Husain, M., and Davidson, V. L. (1986). Characterisation of two inducible periplasmic *c*-type cytochromes from *Paracoccus denitrificans*. *J. Biol. Chem. 261*: 8577.

39. Gray, K. A., Knaff, D. B., Husain, M., and Davidson, V. L. (1986). Measurement of the oxidation-reduction potentials of amicyanin and *c*-type cytochromes from *Paracoccus denitrificans*. *FEBS Lett.* *207*: 239.
40. Bosma, G., Braster, M., Stouthamer, A. H., and van Verseveld, H. W. (1987). Isolation and characterisation of ubiquinol oxidase complexes from *Paracoccus denitrificans* cells cultured under various limiting growth conditions in the chemostat. *Eur. J. Biochem.* *165*: 657.
41. Bosma, G., Braster, M., Stouthamer, A. H., and van Verseveld, H. W. (1987). Subfractionation and characterisation of soluble *c*-type cytochromes from *Paracoccus denitrificans* cultured under various limiting conditions in the chemostat. *Eur. J. Biochem.* *165*: 665.
42. van Spanning, R. J. M., Wansell, C., Harms, N., Oltmann, L. F., and Stouthamer, A. H. (1990). Mutagenesis of the gene encoding cytochrome *c*-550 of *Paracoccus denitrificans* and analysis of the resultant physiological effects. *J. Bacteriol.* *172*: 986.
43. Ferguson, S. J. (1988). The redox reactions of the nitrogen and sulphur cycles. *Symp. Soc. Gen. Microbiol.* *42*: 1.
44. Zumft, W. G., Viebrock. A., and Korner, H. 91988). Biochemical and physiological aspects of denitrification. *Symp. Soc. Gen. Microbiol.* *42*: 245.
45. Chan, H. T. C., and Anthony, C. (1991). The interaction of methanol dehydrogenase and cytochrome c_L in an acidophilic methylotroph, *Acetobacter methanolicus*. *Biochem. J.* *280*: 139.
46. Cox, J. M., Day, D. J., and Anthony, C. (1991). The interaction of methanol dehydrogenase and its electron acceptor, cytochrome c_L, in the facultative methylotroph *Methylobacterium extorquens* AM1 and in the obligate methylotroph *Methylophilus methylotrophus*. *Biochim. Biophys. Acta* *1119*: 97.
47. Dawson, M. J., and Jones, C. W. (1981). Energy conservation in the terminal region of the respiratory chain of the methylotrophic bacterium *Methylophilus methylotrophus*. *Eur. J. Biochem.* *118*: 113.
48. Dawson, M. J., and Jones, C. W. (1981). Respiration-linked proton translocation in the obligate methylotroph *Methylophilus methylotrophus*. *Biochem. J.* *194*: 915.
49. Solioz, M., Carafoli, E., and Ludwig, B. (1982). The cytochrome *c* oxidase of *Paracoccus denitrificans* pumps protons in a reconstituted system. *J. Biol. Chem.* *257*: 1579.
50. Dawson, M. J., and Jones, C. W. (1982). The protonmotive force and phosphorylation potential developed by whole cells of the methylotrophic bacterium *Methylophilus methylotrophus*. *Arch. Microbiol.* *133*: 55.
51. Jones, C. W., Carver, M. A., Dawson, M. J., and Patchett, R. A. (1984). Respiration and ATP synthesis in *Methylophilus methylotrophus*. *Microbial Growth on C-1 Compounds*. (R. L. Crawford and R. S. Hanson, eds.), American Society for Microbiology, Washington, pp. 134–140.
52. Netrusov, I. A., and Anthony, C. (1979). Oxidative phosphorylation in membrane preparations of *Pseudomonas* AM1. *Biochem. J.* *178*: 353.
53. Tobari, J., and Harada, Y. (1981). Amicyanin: An electron acceptor of methylamine dehydrogenase. *Biochem. Biophys. Res. Comm.* *101*: 502.
54. Tobari, J. (1984). Blue copper proteins in electron transport in methylotrophic

bacteria. *Microbial Growth on C-1 Compounds*. (R. L. Crawford and R. S. Hanson, eds.), American Society for Microbiology, Washington, pp. 106–112.

55. Anthony, C. (1988). Quinoproteins and energy transduction. In *Bacterial Energy Transduction* (C. Anthony, ed.), Academic Press, London, pp. 293–316.
56. Husain, M., and Davidson, V. L. (1985). An inducible periplasmic blue copper protein from *Paracoccus denitrificans*. *J. Biol. Chem. 260*: 14626.
57. van Wielink, J. E., Frank, J., and Duine, J. A. (1989). Electron transport from methylamine dehydrogenase to oxygen in the gram-negative bacterium *Thiobacillus versutus*. In *PQQ and Quinoproteins*. (J. A. Jongejan and J. A. Duine, eds.), Kluwer Academic Publishers, Dordrecht, pp. 269–278.
58. Ambler, R., and Tobari, J. (1985). The primary structures of *Pseudomonas* AM1 amicyanin and pseudoazurin. *Biochem. J. 232*: 451.
59. van Houwelingen, T., Canter, G. W., Stobbelaar, G., Duine, J. A., Frank, J., and Tsugita, A. (1985). Isolation and characterization of a blue copper protein from *Thiobacillus versutus*. *Eur. J. Biochem. 153*: 75.
60. Husain, M., Davidson, V. L., and Smith, A. J. (1986). Properties of *Paracoccus denitrificans* amicyanin. *Biochemistry 25*: 2431.
61. Gray, K. A., Davidson, V. L., and Knaff, D. B. (1988). Complex formation between methylamine dehydrogenase and amicyanin from *paracoccus denitrificans*. *J. Biol. Chem. 263*: 13987.
62. Lim, L. W., Mathews, F. S., Husain, M., and Davidson, V. L. (1986). *J. Mol. Biol. 189*: 257.
63. Kumar, M. A., and Davidson, V. L. (1990). Chemical cross-linking study of complex formation between methylamine dehydrogenase and amicyanin from *Paracoccus denitrificans*. *Biochemistry 29*: 5299.
64. Burton, S. M., Byrom, D., Carver, M. A., Jones, G. D. D., and Jones, C. W. (1983). The oxidation of methylated amines by the methylotrophic bacterium *Methylophilus methylotrophus*. *FEMS Microbiol. Lett. 17*: 185.
65. Chandrasekar, R., and Klapper, M. H. (1986). Methylamine dehydrogenases and cytochrome *c*-552 from the bacterium W3A1. *J. Biol. Chem. 261*: 3616.
66. Fukumori, Y., and Yamanaka, T. (1987). The methylamine oxidising system of *Pseudomonas* AM1 reconstituted with purified components. *J. Biochem. 101*: 441.
67. van Spanning, R. J. M., Wansell, C. W., Reijnders, W. N. M., Oltmann, L. F., and Stouthamer, A. H. (1990). Mutagenesis of the gene encoding amicyanin of *Paracoccus denitrificans* and the resultant effect on methylamine oxidation. *FEBS Lett. 275*: 217.
68. Anthony, C. (1991). The structure of bacterial quinoprotein dehydrogenases. *Int J. Biochem. 24*: 29.
69. Matsushita, K., Shinagawa, E., Adachi, O., and Ameyama, M. (1989). Quinoprotein D-glucose dehydrogenase of the *Acinetobacter calcoaceticus* respiratory chain: Membrane-bound and soluble forms are different molecular species. *Biochemistry 28*: 6276.
70. Hauge, J. G. (1961). Mode of action of glucose dehydrogenase from *Bacterium anitratum*. *Arch. Biochem. Biophys. 94*: 308.
71. Dokter, P., Van Wielink, J. E., Geerlof, A., Oltmann, L. F., Stouthamer, A. H., and

Duine, J. A. (1990). Purification and partial characterization of the membrane-bound haem-containing proteins from *Acinetobacter calcoaceticus* LMD- 79.41. *J. Gen. Microbiol. 136*: 2457.

72. Beardmore-Gray, M., and Anthony, C. (1986). The oxidation of glucose by *Acinetobacter calcoaceticus*: Interaction of the quinoprotein glucose dehydrogenase with the electron transport chain. *J. Gen. Microbiol. 132*: 1257.
73. Dokter, P., van Wielink, J. E., van Kleef, M. A. G., and Duine, J. A. (1988). Cytochrome *b*-562 from *Acinetobacter calcoaceticus* L. M. D. 79.41. *Biochem. J. 254*: 131.
74. Matsushita, K., Ohno, Y., Shinagawa, E., Adachi, O., and Ameyama, M. (1982). Membrane-bound, electron transport-linked, D-glucose dehydrogenase of *Pseudomonas fluorescens*. Interaction of the purified enzyme with ubiquinone or phospholipid. *Agric. Biol. Chem. 46*: 1007.
75. Matsushita, K., Nonobe, M., Shinagawa, E., Adachi, O., and Ameyama, M. (1987). Reconstitution of pyrroloquinoline quinone-dependent D-glucose oxidase respiratory chain of *Escherichia coli* with cytochrome *o* oxidase. *J. Bacteriol. 169*: 205.
76. Matsushita, K., Shinagawa, E., Adachi, O., and Ameyama, M. (1989). Reactivity with ubiquinone of quinoprotein D-glucose dehydrogenase from *Gluconobacter suboxydans*. *J. Biochem. 105*: 633.
77. Matsushita, K., Shinagawa, E., Adachi, O., and Ameyama, M. (1989). Quinoprotein D-glucose dehydrogenases in *Acinetobacter calcoaceticus* LMD 79.41: Purification and characterization of the membrane-bound enzyme distinct from the soluble enzyme. In *PQQ and Quinoproteins* (J. A. Jongejan and J. A. Duine, eds.), Kluwer Academic Publishers, Dordrecht, pp. 69–78.
78. Matsushita, K., Shinagawa, E., Adachi, O., and Ameyama, M. (1982). O-type cytochrome oxidase in the membrane of aerobically-grown *Pseudomonas aeruginosa*. *FEBS Lett. 139*: 255.
79. Ameyama, M., Matsushita, K., Shinagawa, E., and Adachi, O. (1987). Sugar-oxidizing respiratory chain of *Gluconobacter suboxydans*. Evidence for a branched respiratory chain and characterization of respiratory chain-linked cytochromes. *Agric. Biol. Chem. 51*: 2943.
80. Matsushita, K., Nagatani, Y., Shinagawa, E., Adachi, O., and Ameyama, M. (1989). Effect of extracellular pH on the respiratory chain and energetics of *Gluconobacter suboxydans*. *Agric. Biol. Chem. 53*: 2895.
81. Matsushita, M., Shinagawa, E., Adachi, O., and Ameyama, M. (1987). Purification, characterization and reconstitution of cytochrome oxidase from *Gluconobacter suboxydans*. *Biochim. Biophys. Acta 894*: 304.
82. van Schie, B. J., Hellingwerf, K. J., van Dijken, J. P., Elferink, M. G. L., van Dijl, J. M., Kuenen, J. G., and Konings, W. L. (1985). Energy transduction by electron transfer by way of a pyrroloquinoline quinone-dependent glucose dehydrogenase in *Escherichia coli, Pseudomonas aeruginosa*, and *Acinetobacter calcoaceticus* (var. lwoffi). *J. Bacteriol. 163*: 493.
83. van Schie, B. J., de Mooy, O. H., Linton, J. D., van Dijken, J. P., and Kuenen, J. G. (1987). PQQ-dependent production of gluconic acid by *Acinetobacter*, *Agrobacterium* and *Rhizobium* species. *J. Gen. Microbiol. 133*: 867.

84. van Schie, B. J., Rouwenhorst, R. J., de Bont, J. A. M., van Dijken, J. P., and Kuenen, J. G. (1987). An *in vivo* analysis of the energetics of aldose oxidation by *Acinetobacter calcoaceticus*. *Appl. Microbiol. Biotechnol. 26*: 560.
85. van Schie, B. J., Pronk, J. T., Hellingwerf, K. J., van Dijken, J. P., and Kuenen, J. G. (1987). Glucose dehydrogenase-mediated solute transport and ATP synthesis in *Acinetobacter calcoaceticus*. *J. Gen. Microbiol. 133*: 3427.
86. Neijssel, O. M., Hommes, R. W. J., Postma, P. W., and Tempest, D. W. (1989) Physiological significance and bioenergetic aspects of glucose dehydrogenase. In *PQQ and Quinoproteins* (J. A Jongejan and J. A. Duine, eds.), Kluwer Academic Publishers, Dodrecht, pp. 57–68.
87. Mueller, R. H., and Babel, W. (1986). Glucose as an energy donor in acetate-growing *Acinetobacter calcoaceticus*. *Arch. Mirobiol. 144*: 62.
88. Gorisch, H., and Rupp, M. (1989) Quinoprotein ethanol dehydrogenase from *Pseudomonas*. In *PQQ and Quinoproteins* (J. A. Jongejan and J. A. Duine, eds.), Kluwer Academic Publishers, Dordrecht, pp. 23–34.
89. Bamforth, C. W., and Quayle, J. R. (1978). The dye-linked alcohol dehydrogenase of *Rhodopseudomonas acidophila*: Comparison with dye-linked methanol dehydrogenase. *Biochem. J. 169*: 677.
90. Groen, B. W., van Kleef, M. A. G., and Duine, J. A. (1986). Quinohaemoprotein alcohol dehydrogenase apoenzyme from *Pseudomonas testosteroni*. *Biochem. J. 234*: 611.
91. Ameyama, M., and Adachi, O. (1982). Alcohol dehydrogenase from acetic acid bacteria. *Met. Enzymol. 89*: 450.
92. Inoue, T., Sunagawa, M., Mori, A., Imai, C., Fukuda, M., Takagi, M., and Yano, K. (1989). Cloning and sequencing of the gene encoding the 72-kilodalton dehydrogenase subunit of alcohol dehydrogenase from *Acetobacter aceti*. *J. Bacteriol. 171*: 3115.
93. Inoue, T., Sunagawa, M., Mori, A., Imai, C., Fukuda, M., Takagi, M., and Yano, K. (1990). Possible functional domains in a quinoprotein alcohol dehydrogenase from *Acetobacter aceti*. *J. Ferment. Bioeng. 70*: 58.
94. Matsushita, K., Shinagawa, E., Adachi, O., and Ameyama, M. (1990). Cytochrome a_1 of *Acetobacter aceti* is a cytochrome *ba* functioning as ubiquinol oxidase. *Proc. Natl. Acad. Sci. USA 87*: 9863.
95. Shinagawa, E., Matsushita, K., Adachi, O., and Ameyama, M. (1990). Evidence for electron transfer via ubiquinone between quinoproteins for D-glucose dehydrogenase and alcohol dehydrogenase of *Gluconobacter suboxydans*. *J. Biochem. 107*: 863.

10

X-Ray Crystallographic Studies of Quinoproteins

F. Scott Mathews

Washington University School of Medicine, St. Louis, Missouri

Wim G. J. Hol

University of Groningen, Groningen, The Netherlands

I. INTRODUCTION

A number of redox active enzymes of both bacterial and mammalian origin were thought at one time to contain the cofactor pyrroloquinoline quinone (PQQ). This class of enzymes has collectively been called quinoproteins [1]. The X-ray crystal structures of three of these enzymes have now been determined and will be discussed in this chapter. In one of these cases, methanol dehydrogenase, the PQQ cofactor is noncovalently attached to the enzyme, and its identity as PQQ is firmly established [2,3]. In the other two cases, it is now known that the cofactors are not PQQ, but are in fact derived from side chains of gene-encoded amino acids through posttranslational modification [4–6]. In one of these, methylamine dehydrogenase, the cofactor is an orthoquinone [4] and therefore a rightful member of the quinoprotein class. The other, galactose oxidase, does not contain a quinone, but rather a redox center consisting of a tyrosine side chain, which is covalently linked to a cysteine side chain and coordinated to a copper atom [6]. However, it is included in this chapter as an honorary quinoprotein.

Although the three quinoproteins discussed in this chapter all have different cofactors, they have one unifying features of molecular architecture in common. They all contain a rare structural motif discovered approximately 8 years ago in the influenza virus neuraminidase [7]. This motif is a repeating β leaflet arranged with circular symmetry. The β leaflet is a four-stranded antiparallel β sheet simply connected as in the letter W. The symmetry of the leaflet arrangement is pseudo-sixfold, sevenfold, or eightfold. However, the structural and catalytic roles played by this structural element differ substantially among the three enzymes.

The structure of methylamine dehydrogenase is the most thoroughly studied of these quinoproteins, having been solved in two bacterial species [8–10]. In addition, the structure of its complex with the natural electron acceptor amicyanin, a blue copper protein, has been determined for one of these species [11]. However, the precision of these structures is somewhat limited since their amino acid sequences are known only in part, the balance being estimated from the X-ray data. The structure of galactose oxidase [6] is also well characterized, and its known amino acid sequence has been placed in the model with high precision. Finally, for methanol dehydrogenase, only an alpha carbon tracing is available at the present time [12], since the amino acid sequence is unknown and the structure is unrefined.

II. METHYLAMINE DEHYDROGENASE

A. Background

In methylotrophic bacteria, oxidation of methylamine is catalyzed by an enzyme, methylamine dehydrogenase (MADH), which is located in the periplasmic space [13]. This enzyme is a quinoprotein containing tryptophan tryptophylquinone (TTQ), an amino acid–derived cofactor discovered recently [4,5]. The reaction catalyzed is

$$CH_3NH_3^+ + H_2O \rightarrow NH_4^+ + CH_2O + 2H^+ + 2e^-$$

Subsequently, electrons are transferred to the membrane-bound terminal oxidase, cytochrome aa_3, via a series of several soluble electron carrier proteins [14,15]. In several methylotrophs, the first electron acceptor from MADH is amicyanin, a blue copper protein, of molecular weight 12.5 kDA [16]. In the case of *Paracoccus denitrificans*, a second sequential electron acceptor is cytochrome c_{551i}, a protein of MW 22 kDa [17]. Both of these electron carriers, along with MADH, are induced when the bacteria are grown on methylamine as the sole carbon source. A proposed sequence [18] of electron carriers is

$$\text{MADH} \rightarrow \text{amicyanin} \rightarrow \text{cyt } c_{551i} \rightarrow \text{cyt } c_{550} \rightarrow \text{cyt } aa_3$$

The structure of MADH from *Thiobacillus versutus* (TV-MADH) was the first to be solved, at a resolution of 2.25 Å [8,9]. Even though the sequence was not known, the polypeptide chain could be traced and a hypothetical amino acid sequence deduced from the shapes of side chains in the electron density map. Shortly afterwards, the structure of MADH from *P. denitrificans* (PD-MADH) was solved by molecular replacement using TV-MADH as a search molecule. The TV-

MADH sequence was used to model the structure since the sequence of PD-MADH was also unknown. Similarly, the structure of the binary complex between PD-MADH and amicyanin was solved by molecular replacement. The latter showed a similar structure for MADH and provided considerable information on the structure and mode of binding of amicyanin.

B. TV-MADH

1. X-Ray Analysis

TV-MADH is an H_2L_2 heterotetramer of molecular weight 121 kDa. Its amino acid sequence is unknown. The molecular weights for the two types of subunits are 47,500 for the H (heavy) and 12,900 for the L (light) [19], respectively. The protein crystallizes in space group P3121 with unit cell parameters a = b = 129.8 Å and c = 104.3 Å. There is one H and one L subunit per asymmetric unit. Thus, there is a molecular twofold axis coincident with one of the crystallographic twofold axes.

The structure of TV-MADH was solved by the MIR method combined with solvent flattening using three heavy atom derivatives at a resolution of 2.25 Å [8]. An "X-ray" sequence was determined for the protein during the course of refinement based on the shapes of the amino acid side chains. The structure has been refined by molecular dynamics and restrained least squares methods to a residual R-factor of 0.286 in the range 6.0 to 2.25 Å resolution [9].

2. Structure of the Protein

Quaternary structure. The H_2L_2 tetramer forms an approximate parallelopiped when projected along the molecular 2-fold axis, with dimensions 76 by 61 Å, as shown in Figure 1. In this orientation the molecule is about 45 Å thick. The tetramer can best be described as a dimer of HL dimers. The subunits of the heterotetramer are arranged about the molecular twofold axis so that there is virtually no interaction between L subunits and only minimal interaction between the H subunits, this interaction being limited to about two residues.

The interaction between H_1 and L_2 subunits (Fig. 1) is also rather limited, although more extensive than between the two H subunits. The interaction is mostly limited to the N-terminal region of the H subunit comprising the first 31 residues. The interactions between the H_1 and L_1 (and H_2, L_2) subunits are the most extensive, with many hydrophobic residues located at the interface. The active site crevice is close to this interface.

Structure of the light subunit. The L subunit of MADH comprises 121 residues, numbered 7–127, and consists entirely of β strands, turns, and irregular structures. It contains a total of 10 β strands. Six of these form a twisted, rather distorted antiparallel β sheet (Fig. 2). The order of strands is 8,3,10,7,6,9 from right to left (Fig. 2). Some of these strands are quite long, although only a few

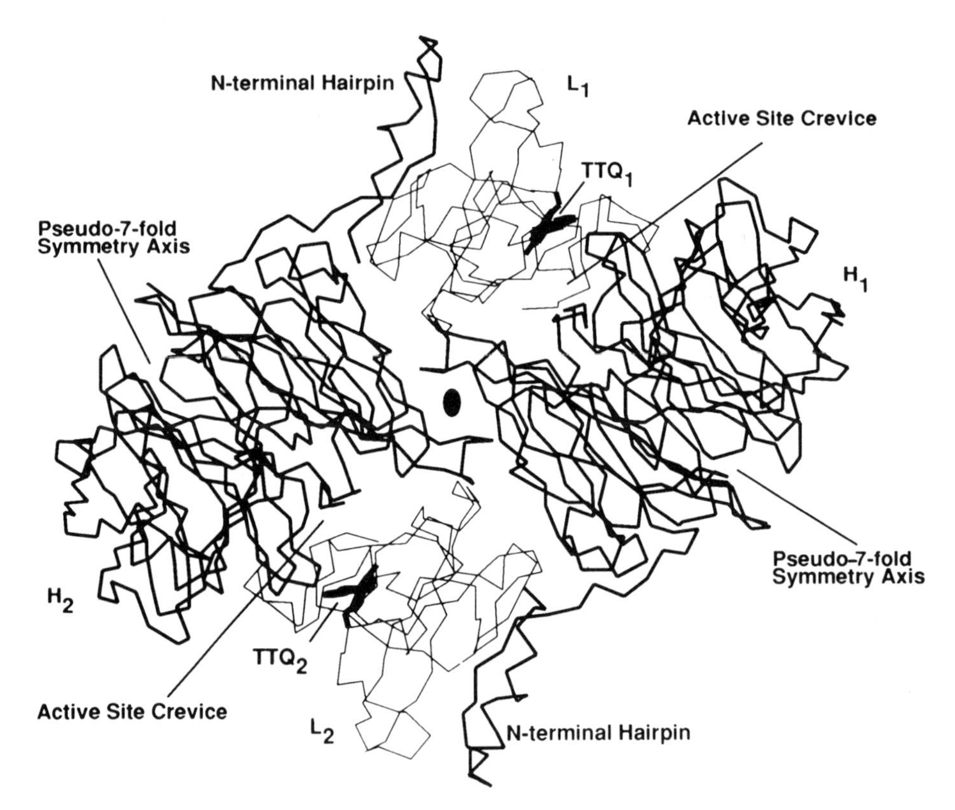

Figure 1 Schematic diagram of the H_2L_2 tetramer of methylamine dehydrogenase viewed along the molecular twofold axis.

residues of each are involved in the hydrogen bonding network of the sheet. The sheet is somewhat curled so that the end strand (β9) lies somewhat above the third from the end (β7). The TTQ cofactor is bound to the 5th and 6th strands of the sheet (β6 and β9). There are also two antiparallel β hairpins. Strands 1 and 2 form one such hairpin, although only about two pairs of residues connect the strands. The other is formed by strands β4 and β5 and is more regular. Both pairs of β hairpins are located on the edge of the molecule and are not involved with the central six-stranded sheet.

There are a large number of covalent linkages within the L subunit. There are six pairs of disulfide bonds, corresponding to the 12 cysteine residues deduced from the amino acid composition [20] and identified from the X-ray sequence [8]. These intrasubunit cross-links are probably responsible for the resistance of the subunit to heat denaturation [21]. In addition, there is a cross-link between the two

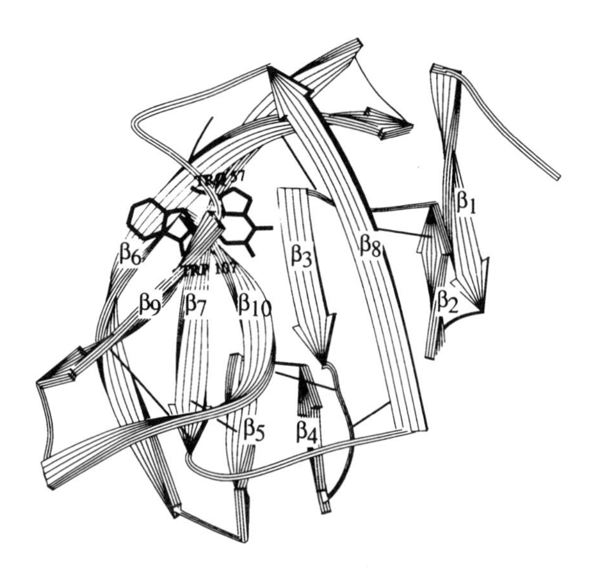

Figure 2 Ribbon diagram showing the light subunit of methylamine dehydrogenase. The position of TTQ is indicated as Trq 57 and Trp 107. The six disulfide cross-links are indicated by straight lines connecting β strands or loops.

Trp residues at positions 57 and 107, which form the TTQ cofactor [5]. These positions correspond to those located in the amino acid sequence of the L subunit from *Methylobacterium extorquens* determined chemically [22] and by DNA sequencing [23].

Structure of the heavy subunit. The heavy subunit of TV-MADH contains 370 residues and can be divided into two sections, as shown in Figure 3. The first section is very small, consisting of 31 residues that extend away from the rest of the subunit. The first 16 residues of this section form about 4½ turns of alpha helix, while the remaining 15 residues are in extended conformation and fold back along the side of the helix. For subunit H_1, this extension is packed against subunit L_2 (Fig. 1), i.e., the one not involved in the close HL dimer interaction.

The second section of the H subunit is contained in a single domain of 329 residues. This domain is in predominantly β structure and forms a cylinder made up of seven four-stranded antiparallel β sheets, each forming a simple up-down-up-down W-like β-leaflet structure. These leaflets are arranged around a central axis with pseudo sevenfold symmetry. In the H_2L_2 tetramer, these pseudo-sevenfold axes lie in a plane approximately perpendicular to the molecular twofold axis (see Fig. 1). The inside strand of each W is approximately parallel to the pseudo-sevenfold axis. For each of the W substructures, if the individual strands are labeled A–D, then strand A is located closest to the central axis and strand D is

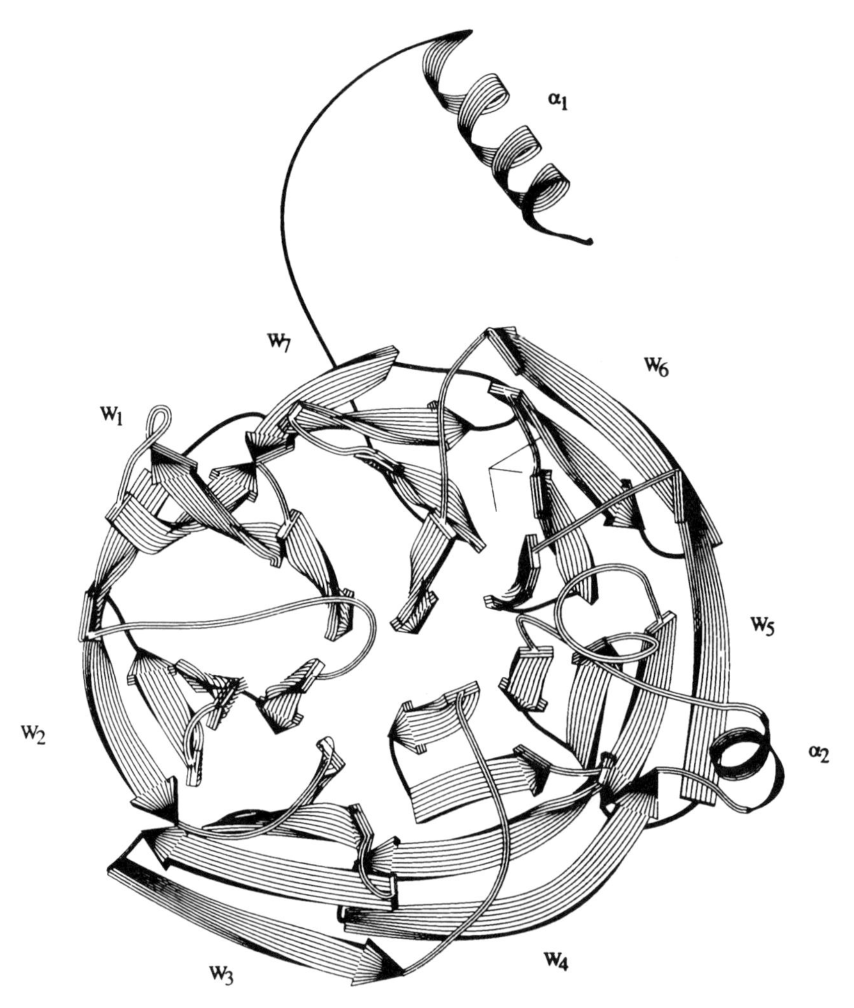

Figure 3 Ribbon drawing of the heavy subunit of methylamine dehydrogenase. The seven β leaflets are indicated by W_1–W_7. The N-terminal helix, α_1, and a second helix between W_4 and W_5 are also indicated. The pseudo sevenfold axis is approximately perpendicular to the diagram.

located on the outside of the domain. The one exception to the simple W motif is W7, which contains both chain termini. In W7, strand A is formed by the C-terminal segment, while strands B, C, and D are formed by the N-terminal part of the chain. Consecutive leaflets are connected by long loops, which link the D strand of one to the A strand of the next. The only helix in this domain is located between W4 and W5.

C. PD-MADH

1. *X-Ray Analysis*

PD-MADH crystallizes in space group $P2_12_12_1$ with unit cell parameters a = 151.8 Å, b = 135.8 Å, and c = 55.1 Å [24]. Its subunit molecular weight is 46,700 for the H subunit and 15,500 for the L subunit, for a total of M_r 124,400 for the tetramer. There is one heterotetramer per asymmetric unit, and the crystals diffract to at least 2.4 Å resolution. At the time of the X-ray analysis, none of the amino acid sequence was known. The structure was solved by molecular replacement using the structure of TV-MADH as a search molecule. The structure has been refined by molecular dynamics and restrained least squares methods to an R-factor of 0.246 in the range 10–2.8 Å resolution. Individual atomic thermal parameters were varied in the final few cycles.

2. *Structure of PD-MADH*

The present model of PD-MADH is very similar to that of TV-MADH. Examination of the amino acid side chains in the electron density maps of PD-MADH showed generally a very good match, although some differences in side-chain shape were evident. During refinement of the thermal parameters, the temperature factors for the first 18 residues of the H subunit rose to very high values compared with the remainder of the protein. Furthermore, the electron density corresponding to these residues was very weak. It was subsequently discovered by SDS PAGE [10] that the large subunit had lost approximately 2000 MW during the 3-week aging period required for crystal growth. As shown for TV-MADH in Figure 3, the first 31 residues of the H_1 and (H_2) subunit are extended away from the remainder of the H subunit toward the L_2 (and L_1) subunit and are likely to be susceptible to proteolytic degradation during aging.

Recently, the sequence of the first 25 [10] and the last 31 [25] amino acid residues of the L subunit of PD-MADH became available. In the latter segment, two insertions into the X-ray sequence were needed. This sequence data has now been incorporated into the final model.

D. Nature of the Cofactor

During the analysis of the TV-MADH structure [8], an attempt was made to fit a model of PQQ to the electron density at the expected position of the cofactor, based on the sequence of the enzyme from *M. extorquens* determined by chemical methods [22]. At that time the cofactor was thought to be a form of PQQ bound covalently at two positions along the polypeptide chain [26], although some chemical evidence against that assumption existed [27]. However, the electron density would not accommodate PQQ, even though the remainder of the electron density was quite clear. At this point alternate possibilities were tested, and two alternate models called pro-PQQ and TGA-proPQQ were discussed [28].

In the average 2.8 Å electron density map of PD-MADH, the expected density

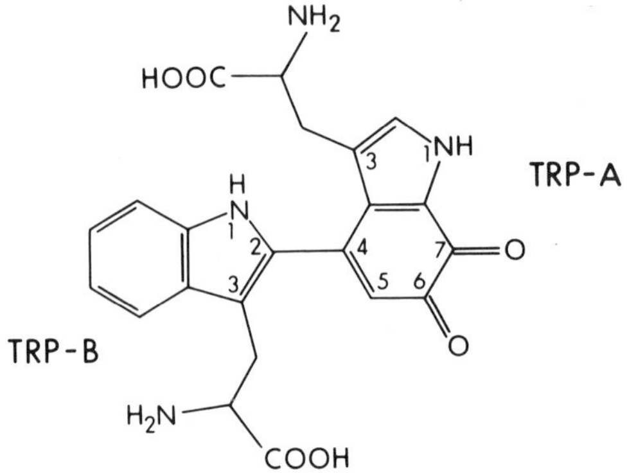

Figure 4 Tryptophan tryptophylquinone (TTQ). Trp-A and Trp-B correspond to positions 57 and 107 of the "X-ray sequence" of the light subunit of TV-MADH. (Reproduced by permission from Ref. 5.)

for PQQ, although somewhat larger than for TV-MADH, would not satisfactorily fit PQQ or the two pro-PQQ models discussed earlier. At this point, it was learned that the true cofactor for MADH was TTQ (Fig. 4), which is formed from two gene-encoded tryptophan side chains linked covalently and modified so that one of them becomes an *ortho*-quinone. Introduction of TTQ into the electron density maps of PD-MADH and TV-MADH [5] improved the fitting dramatically (Fig. 5). The electron density shows that the two Trp side chains are inclined by about 40–45° and strongly suggests that Trp 57 contains the quinone and that the atoms at positions 6 and 7 of the indole ring of residue 57 are modified. In the case of TV-MADH, the TTQ has been successfully refined using the observed X-ray data.

The environment of TTQ in TV-MADH is shown in Figure 6. It is clear that all potential hydrogen bond donors or acceptors on TTQ are satisfied by interactions with the L subunit. The two quinone oxygens at positions 6 and 7 are accessible to the active site channel (Fig. 1), which would be able to accommodate substrate. Also, atom O_6 is close to the binding site for phenylhydrazine, which can bind to TTQ covalently, and was previously identified by X-ray crystallography [8].

E. Complex of MADH with Amicyanin

1. Background

The natural electron acceptor for MADH in vivo is amicyanin, a small blue copper protein. Most of the physical, redox, and spectroscopic properties of amicyanin are very similar to those of other small blue copper proteins such as azurin and plastocyanin. Available amino acid sequence data indicate that amicyanins are a separate and unique class of copper proteins bearing closest similarity to plasto-

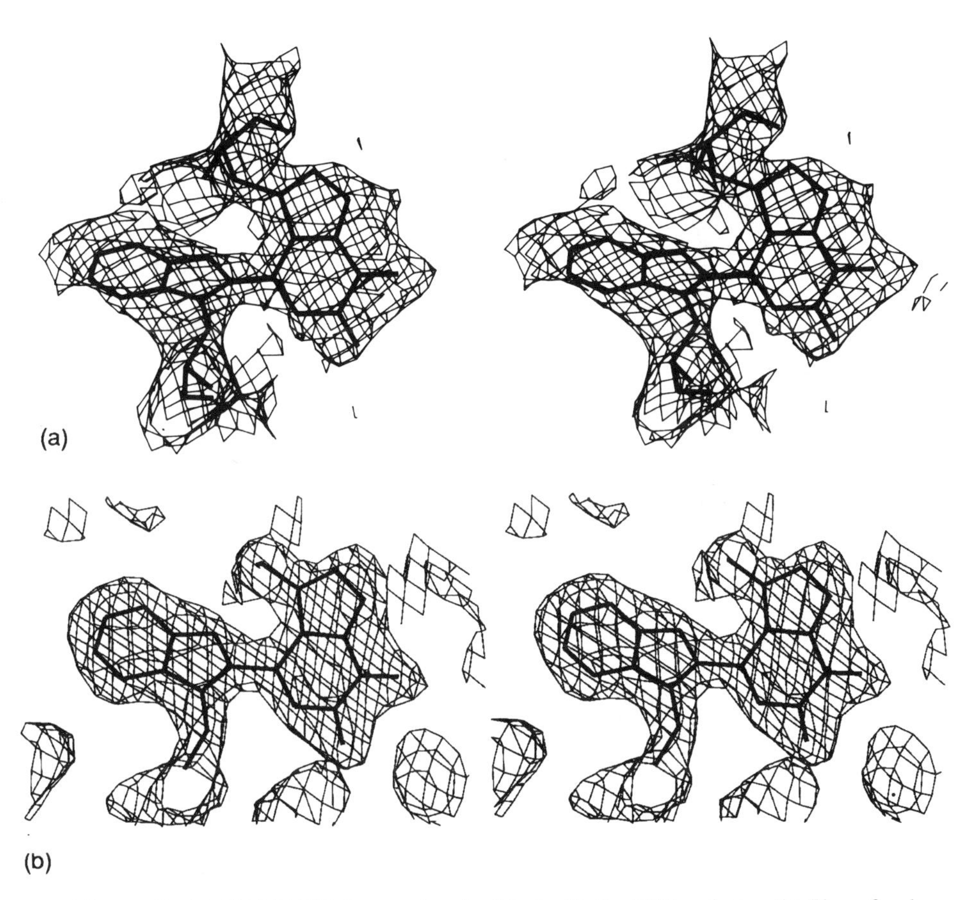

Figure 5 (a) PD-MADH electron density fitted with the TTQ cofactor. (b) The refined model and electron density of the TTQ cofactor in TV-MADH. (Reproduced by permission from Ref. 5.)

cyanin [29,30]. Resonance Raman spectroscopic studies [31] and two-dimensional NMR studies [32] support this hypothesis.

The specificity of interaction between amicyanin and MADH has been best demonstrated by studies of the isolated proteins from *P. denitrificans* Each of these proteins is induced in this bacterium only during growth on methylamine as a carbon source [16]. Furthermore, the amicyanin gene is located immediately downstream from that for the small subunit of MADH. Inactivation of the former by means of gene replacement resulted in complete loss of the ability to grow on methylamine [25].

Complex formation between MADH and amicyanin in solution is indicated by

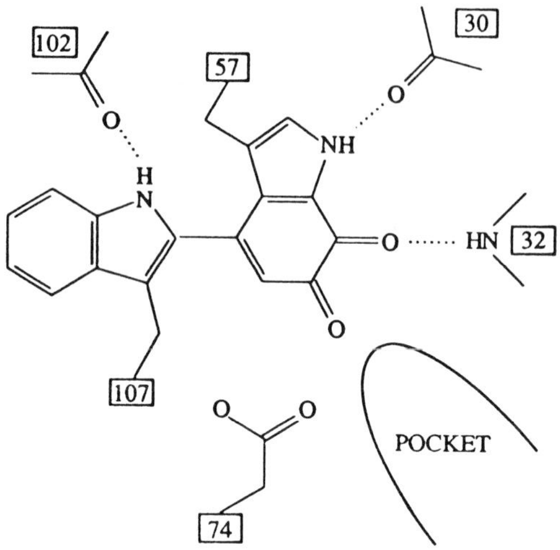

Figure 6 Schematic diagram of the hydrogen bonds between the TTQ cofactor and surrounding residues in TV-MADH. (Reproduced by permission from Ref. 5.)

spectral perturbation of TTQ and by a −73 mV shift in the redox potential of amicyanin upon mixing the two proteins [34]. Further evidence for complex formation is provided by steady-state kinetic analyses [35], chemical cross-linking studies [36], and resonance Raman spectroscopy [37]. These studies are summarized in Chapter 5.

2. *X-Ray Analysis*

Crystals of the complex between MADH and amicyanin, both isolated from *P. denitrificans*, were grown by vapor diffusion against 2.4 M Na/K phosphate buffer, pH 6.5 [24]. The complex is a heterohexamer composed of three types of subunits, the H and L subunits of MADH, of molecular weights 46.7 and 1.5 kDa, respectively, and amicyanin, of molecular weight 12.5 kDa. The crystals are tetragonal, space group $P4_12_12_1$, with cell parameters a = b = 124.6 Å, c = 247.3 Å. They contain one heterohexamer of M_r 149.4 kDa per asymmetric unit.

The structure of the complex was solved by the molecular replacement method using the structure of TV-MADH as the search probe [11]. The procedure was very similar to that employed to solve the PD-MADH structure itself. After the position and orientation of the MADH molecule were determined, its coordinates were refined for several cycles at 2.5 Å resolution. A difference electron density map revealed the outline of the amicyanin molecule in the complex after the density map had been averaged about molecular twofold axis. The chain of amicyanin was traced in the map. The model of the complex has subsequently been rebuilt after the complex was refined by restrained least squares methods to an R-factor of 0.299.

3. Structure of the Complex

Quaternary and component structures. The complex of MADH with amicyanin consists of a heterohexamer of the type $H_2L_2A_2$ as shown in Figure 7. The H_2L_2 unit corresponds to the heterotetramer of MADH described above. Each amicyanin molecule is in contact with both H and L subunits of a single HL dimer. The more extensive interactions are between amicyanin and the L subunit.

The structure of the MADH portion of the complex is virtually the same as that of the TV and PD-MADH molecules described above. Each H subunit consists of a disk-shaped domain composed of seven four-stranded antiparallel β sheets arranged in pseudo-sevenfold symmetry, preceded by a 31-residue extended arm. Each L subunit contains 10 β strands, which participate in three β structures, and is held together by six disulfide cross-links. Of particular importance is the fact that the electron density for the TTQ cofactor, which was not included in the calculation of structure factors for the complex, is consistent with the shape of TTQ. Also, the N-terminal arm of the H subunit appears to be intact, in contrast to that of the uncomplexed crystalline PD-MADH described above, which undergoes proteolytic degradation of the first 18 residues during crystallization.

The current model of amicyanin consists of 102 residues corresponding to

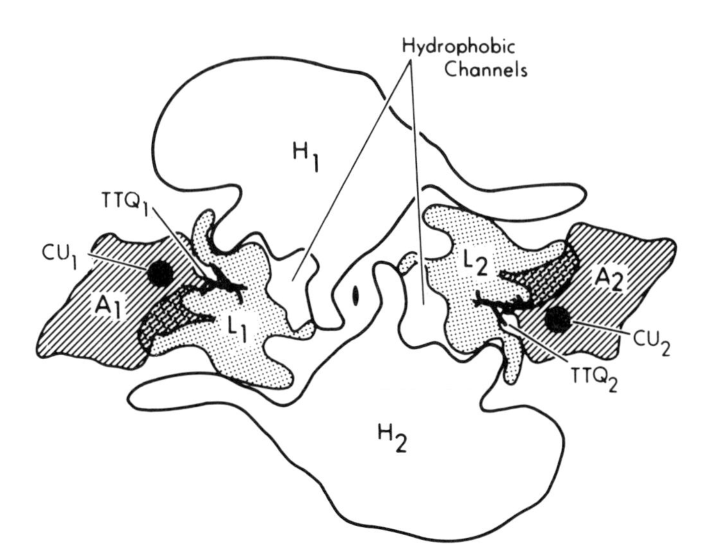

Figure 7 Schematic diagram of the $H_2L_2A_2$ heterohexamer of the MADH-amicyanin complex. The approximate positions of the TTQ cofactors and copper atoms are indicated.

positions 4–105 of the amino acid sequence. A ribbon diagram of the amicyanin backbone is shown in Figure 8. The backbone forms a nine-stranded β sandwich topologically similar to plastocyanin [38]. Their folding topologies are compared in Figure 9. The β sandwich of amicyanin consists of a four-stranded mixed β sheet (strands β1, β7, β4, β2, Fig. 8) facing a five-stranded mixed sheet (strands β3, β9, β8, β5, β6). The outside strand of the four-stranded sheet on the left of Figure 9a (strands 1, 7 4, 2) is absent in plastocyanin (Fig. 9b).

The interior of amicyanin is hydrophobic, while its surface is relatively polar. The copper atom is located close to one end of the molecule near the top in Figure 8. There are four ligands to the copper atom: His 53, Cys 92, His 95, and Met 98. This pattern of copper ligation is similar to that found in other blue copper proteins [38] as predicted from the amino acid sequence [25]. His 53 is located near the N-terminal end of β strand 5 while the remaining ligands are located on a loop connecting strands 8 and 9. The copper ligands are arranged as a distorted tetrahedron. Three of the ligands are buried within the protein, but a fourth, His

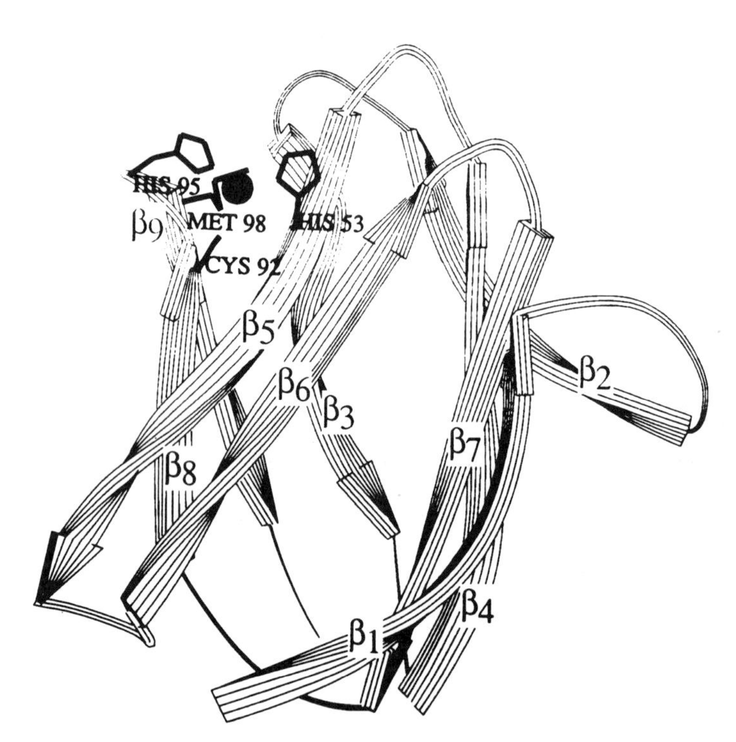

Figure 8 Ribbon diagram of amicyanin. The positions of the copper atom and its four ligands are indicated.

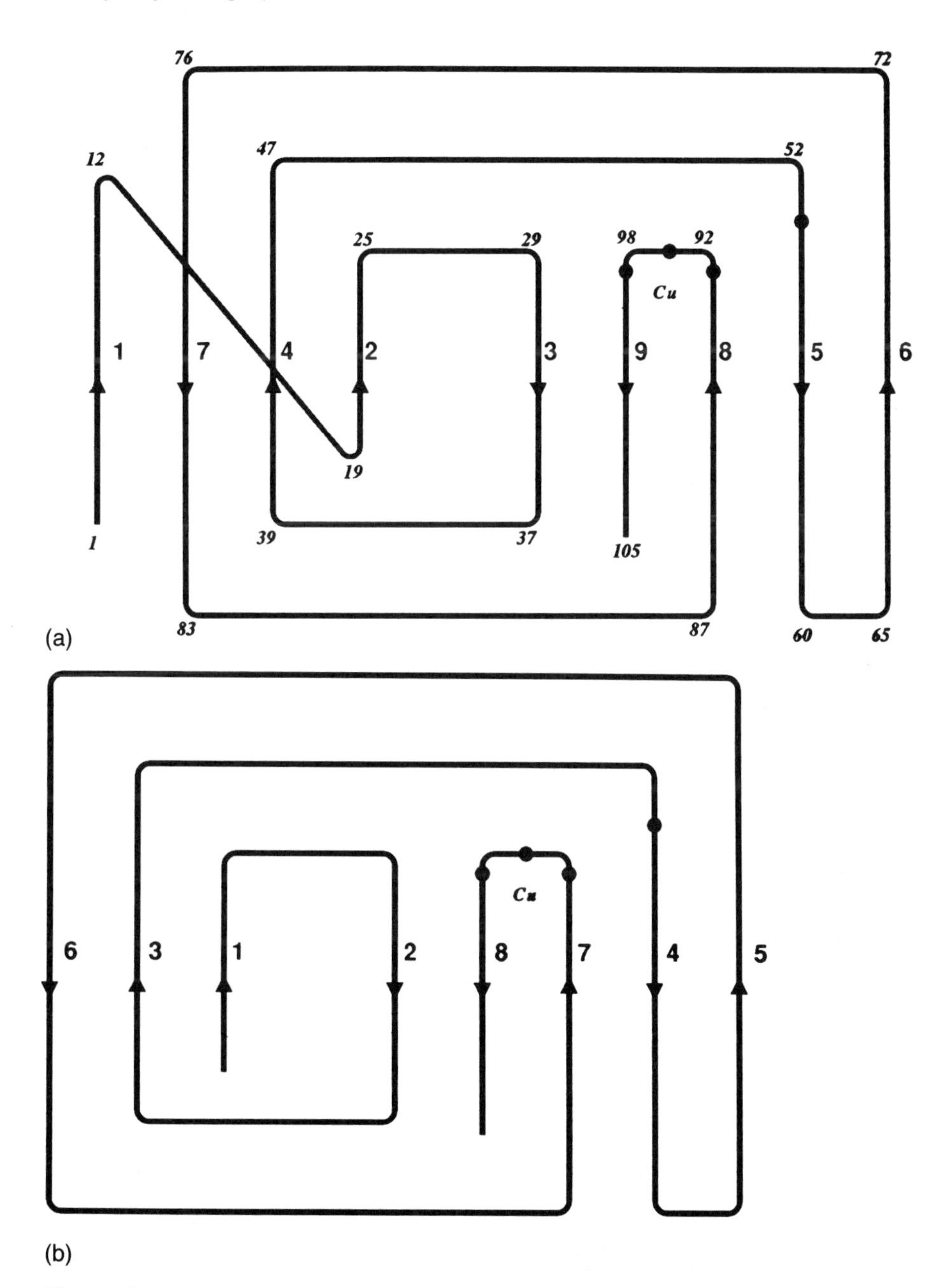

Figure 9 Topological diagrams of (a) amicyanin and (b) plastocyanin. Plastocyanin consists of an eight-stranded β sandwich, while amicyanin contains one additional β strand at the edge of one face of the sandwich.

95, lies on the protein surface, with its N^{δ} atom bound to the copper and its N^{ϵ} atom exposed to the surface. Surrounding His 95 on the protein surface is a cluster of hydrophobic residues.

Interaction between MADH and amicyanin. The major interaction between amicyanin and MADH occurs at a rather hydrophobic interface. The copper atom is located close to this interface. Approximately eight hydrophobic side chains surrounding His-95 of amicyanin interact with about 12 hydrophobic or uncharged hydrophilic side chains of MADH, according to the X-ray sequence from TV-MADH. Three of these are on the H subunit, and nine on the L subunit. Included in the latter is the Trp 107 moiety of the TTQ cofactor. The interface residues of amicyanin include three Met, three Pro, one Val, and one Phe. For the H subunit, two residues are tentatively identified as Ser and one as Phe. For the L subunit the composition of the interface appears to be more varied, but identification is uncertain since the complete sequence of MADH has not been reported.

The relative positions of the TTQ cofactor and the copper atom of amicyanin are shown in Figure 10. The TTQ is oriented so that the orthoquinone portion of Trp 57, 06 and 07, is pointed away from the copper atom, while the Trp 107 portion lies closest to the copper. The shortest distance between the redox centers in the two proteins, about 9.5 Å, is from the $C^{\eta 2}$ of Trp 107 to the copper atom. Thus, the active site of MADH, centered on the ortho-quinol of TTQ, is pointed away from the amicyanin-binding site, allowing substrate binding and product release without interfering with the electron transfer to the copper atom.

His 95 of amicyanin, one of the copper ligands, is located in the complex close to the TTQ of MADH at a distance of about 5.6 Å, from $C^{\delta 2}$ of the histidine to $C^{\eta 2}$ of Trp 107 of TTQ. Thus, His 95 of amicyanin is in a good position to mediate the electron transfer.

4. Discussion

A conserved feature among blue copper proteins is a histidine residue analogous to His 95 of amicyanin, which is surrounded by a surface patch of hydrophobic amino acid residues. It has long been speculated that for azurin and plastocyanin this region is involved in mediating electron transfer to copper. Support for this view is provided by studies of electron self-exchange rates [39] and by mutagenic studies [40] of azurin. The present data provides direct evidence that this is true.

That Trp 107 is exposed on the surface of MADH and lies close to His 95 of amicyanin implies that it may be involved in mediating electron transfer to this His from the *o*-quinone of Trp 57. Although the planes of the two tryptophan residues in TTQ are inclined by about 45°, electron transfer through Trp 107 to His 95 of amicyanin is likely to be much more efficient than through space electron transfer directly from Trp 57 [41].

Electrostatic interactions have long been thought to play a critical role in stabilizing protein-protein interactions between soluble redox partners. For amicyanin and MADH, current data as well as previous cross-linking studies [36]

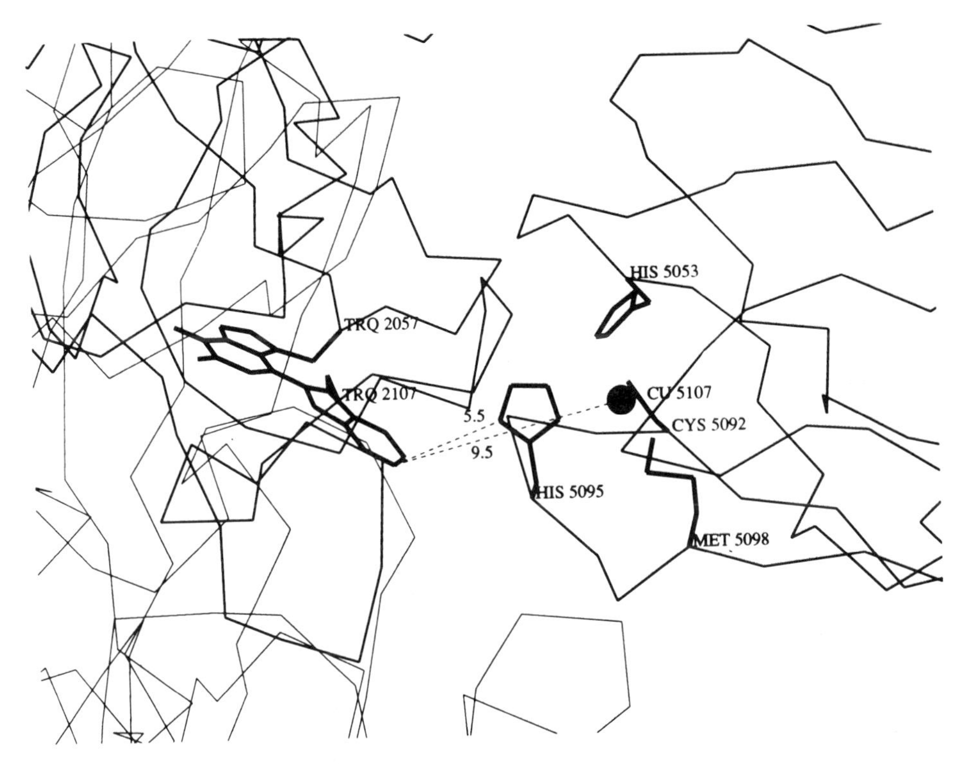

Figure 10 Interface between amicyanin (right) and plastocyanin (left). The L subunit is in thicker lines, and the H subunit is in thinner lines. The shortest distances from TTQ (Trq 2057 and Trq 2107) to the copper atom and to the histidine ligand (His 5095) are indicated in Å.

indicate that the interactions that stabilize this complex are primarily hydrophobic. However, it was previously observed that the complex-dependent changes in the absorption spectrum of TTQ and the redox potential of amicyanin [39], as well as the kinetic efficiency [35] of this complex, are sensitive to ionic strength. An important question, given these results, is which interactions between amicyanin and MADH can account for these observed ionic strength–dependent phenomena. Also, the underlying cause of the −73 mV redox potential shift of amicyanin upon complex formation remains to be determined.

III. GALACTOSE OXIDASE

A. Background

Galactose oxidase (GAO) is an extracellular enzyme secreted by a variety of fungi such as *Dactylus dendroides*, *Giberrella feyikuroi*, and *Fusarium graminearum*, which catalyzes stereospecifically the two-electron oxidation of primary alcohols to corresponding aldehydes [42–49] according to the following reaction:

$$RCH_2OH + O_2 \rightarrow RCHO + H_2O_2$$

The enzyme from *D. dendroides* has been most extensively characterized [42, 45–48]. It is a monomeric protein of M_r 68 kDa containing a unique mononuclear copper site. In its oxidized form the metal center exhibits extremely intense visible and near IR absorption features and is EPR silent [43]. It has been shown that these features relate to a free radical associated with a cupric site in the active form of the enzyme [47]. Consequently, galactose oxidase belongs to the group of enzymes exhibiting free radical active site chemistry—other examples being cytochrome *c* peroxidase [50] and ribonucleotide reductase [51]. The mononuclear copper site is in close proximity to a most unusual organic cofactor. This cofactor, once believed to be PQQ [52], has been shown by X-ray analysis to be a covalently linked tyrosine-cysteine group [6], which apparently stabilizes a free radical that resides on the tyrosine in the oxidized apoenzyme [49].

A well-refined crystal structure of the holoenzyme is available at a resolution of 1.7 Å at pH 4.5 [6]. The crystals, grown from ammonium sulfate as precipitant in 0.80 M acetate buffer have space group symmetry C_2 with unit cell parameters a = 98.0 Å, b = 89.4 Å, c = 86.7 Å, and β = 117.8°. Three heavy atom derivatives yielded an excellent initial electron density distribution. Using 1.7 Å resolution area detector data, a molecular model has been obtained consisting of 639 amino acid residues, 316 water molecules, two acetate ions, and a sodium ion, plus, of course, a cupric ion. It appears that not only is the side chain– derived cofactor of GAO unusual, but the molecular architecture is as well.

B. Description of the Structure

The polypeptide chain folds into three domains, with a high percentage of β structure and only a single alpha helix. In this respect the structure is quite similar to that of methylamine dehydrogenase. However, there are more striking similarities as well as numerous differences. The first of the three domains consists of residues 1–155 and has a β-sandwich structure. The second domain (Fig. 11) consists of residues 156–532 with a topology that is very similar to that of methylamine dehydrogenase: seven four-stranded antiparallel β sheets, or Ws, are arranged with pseudo-sevenfold symmetry like a flower with seven W petals. The third domain comprises residues 533–639 and contains seven β strands. The inorganic component of the redox center consists of a copper ion which is located near the pseudo-sevenfold axis from the second domain, on the opposite side from the third domain. Most remarkably, two of the β strands of the third domain pierce like a finger (comprising residues 574–589) through the central pore in the second domain, providing a ligand to the copper ion—a nitrogen in the imidazole ring of histidine 581.

Figure 11 Ribbon diagram of the second domain (approximately from residues 155 to 530) of galactose oxidase. The pseudo seven-fold symmetry axis is perpendicular to the diagram. The Met-Tyr-Cu redox center is located approximately in the center of the "doughnut" (see test). (This diagram was kindly supplied by Dr. Nobutoshi Ito.)

With respect to the organic part of the redox center, Ito et al. [6] discovered in their crystal structure an intriguing covalent bond between the C^{ϵ} atom of Tyr 272 and the S^{γ} atom of Cys 228 (Fig. 12). The C^{ϵ}-S^{γ} bond length was refined to 1.84 Å without stereochemical restraints. The C^{β} and S^{γ} atoms of the cysteine residue are virtually in the same plane as the phenol ring of Tyr 272, allowing for extensive conjugation in this system. The indole ring of Trp 290 is stacked parallel to the Tyr-

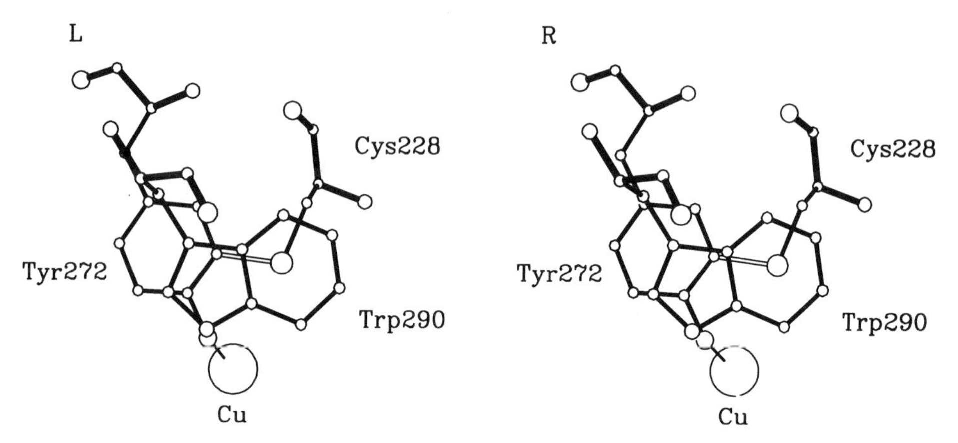

Figure 12 Stereo diagram showing the redox center of galactose oxidase. The S^{γ} of Cys 228 is covalently linked to the C^{ϵ} of Tyr 272, which is also a ligand of the copper atom. The plane of Trp 290 is located directly below and parallel to the plane formed by the Cu, the tyrosine ring, and the C^{β}-S^{γ} of the cysteine. (This diagram was kindly supplied by Dr. Nobutoshi Ito.)

Cys plane with the center of its benzene ring exactly above the sulfur atom with virtually equal distances of 3.84 ± 0.04 Å between the S^{γ} and the six ring atoms.

Tyr 272 is not only covalently linked via its C^{ϵ} to Cys 228, but its O^{η} is also one of the ligands of the copper ion [6]. Together with the side-chain nitrogens of His 496 and His 581 as well as an acetate ion, the O^{η} of Tyr 272 forms an almost perfect square with the Cu^{2+} in its center. Above this plane Tyr 495 donates another oxygen ligand, which becomes the apex of a square pyramid with the Cu^{2+} in the center of the basal plane of the pyramid. When crystals are transferred to PIPES buffer at pH 7.0, this perfect pyramid becomes somewhat distorted because a water molecule replaces the acetate ion at a slightly different position. This beautiful structure has shown another surprise in nature's reservoir of cofactors. Further studies unraveling the catalytic mechanism are awaited with great interest.

C. Discussion

When GAO is compared with MADH, one sees some striking similarities. The first one is that in both enzymes the organic cofactor, although very different in detail, is constructed from the side chains of two gene-encoded amino acids, which are covalently linked. It is most intriguing how this modification is carried out—either by one or more additional enzymes or perhaps by GAO or MADH itself through a self-catalytic process.

The second similarity concerns the topology of the sevenfold W-propellor, which is identical in the large subunit of MADH and in the large domain of GAO.

In each of these W-motifs the N-terminal strand runs in the center of the molecule approximately parallel to the pseudo-sevenfold symmetry axis. The functions of the W7 domains are very different, however, in the two enzymes. In GAO this domain contains residues that are deeply involved in cofactor formation and catalysis. In MADH the W7 domain, or subunit rather, is hardly involved in catalysis. It does not contribute to the cofactor, which is formed by tryptophans from the small subunit, and only marginally contributes to the active site pocket. Since the sequence of the large subunit of MADH is not yet known, we cannot determine the degree of sequence similarity between these two W7 propellor domains. A comparison of MADH, GAO, and neuraminidase at sequence and structural level would be most interesting. In a most surprising manner it appears that also the quinoprotein methanol dehydrogenase (MEDH) will have to be incorporated into this comparison, as will be described in the next section.

IV. METHANOL DEHYDROGENASE

A. Background

Methanol dehydrogenase is a bacterial quinoprotein that catalyzes the oxidation of methanol to formaldehyde [53]. Other primary alcohols as well as formaldehyde are also substrates [1]. The primary reaction catalyzed is

$$CH_3OH \rightarrow CH_2O + 2H^+ + 2e^-$$

The enzyme is located in the periplasmic space of many methlotrophic bacteria. It is an H_2L_2 tetramer with subunit molecular weights of approximately 60 kDa and 8 kDa, respectively [14]. It contains two molecules of PQQ per tetramer, bound noncovalently [2,3]. The primary electron acceptor in vivo is cytochrome c_L, an acidic protein of approximately 20 kDa molecular weight [14]. Electrons are subsequently transferred to a membrane-bound aa_3-type oxidase via a second carrier, cytochrome c_H, a basic protein similar to soluble mitochondrial or bacterial cytochromes. A more detailed discussion of the biochemical properties of MEDH is given in Chapter 2.

The amino acid sequences of MEDH from the heavy subunits of *P. denitrificans* [54] and *Methylobacterium organophilum* XX [55] have been obtained by gene-sequencing methods. Similarly, the sequences of both the heavy and light subunits of MEDH from *Methylobacterium extorquens* AM1 have been determined [56,57]. The amino acid sequence of MEDH from *Methylophilus methylotrophus* is not known.

B. X-Ray Analysis

MEDH from *M. methylotrophus* crystallizes in space group $P2_1$ with unit cell parameters a = 124.9 Å, b = 62.7 Å, c = 85.0 Å, and β = 93.4°. The crystals contain one full heterotetramer per asymmetric unit and diffract to beyond 2.4 Å resolution. Crystals of MEDH from bacterium W3A1 have also been prepared [58] which are isomorphous to the *M. methylotrophus* crystals.

An electron density map was calculated at 3.0 Å resolution using X-ray phases obtained by the multiple isomorphous replacement technique including anomalous scattering. The X-ray data were collected on an area detector. A total of three heavy atom derivatives were used. The quality of the map was improved by a solvent-leveling procedure, and the map was then averaged about the molecular twofold axis.

A preliminary chain tracing for MEDH was obtained from the 3.0 Å resolution electron density map plotted on transparent sheets. The amino acid sequences of the heavy subunits of MEDH from both *P. denitrificans* and *M. extorquens* AM1 and of the light subunit from *M. extorquens* AM1 were used as guides during the fitting. In this process, the aromatic side chains for Trp, Tyr, Phe, and His were used as markers to help identify bulky side chain density. Over 90% of these groups could be fitted to characteristic density. The fitting has been improved and verified by using the procedure "o" on a molecular graphics system [59]. The PQQ cofactor has not yet been located.

C. Structure of the Molecule

The heavy subunit of MEDH has been fitted with about 580 amino acid residues. In the light subunit, about 57 residues could be fitted. In the heterotetramer, two heavy subunits are in contact across a noncrystallographic twofold axis (Fig. 13). The interface is relatively small, with approximately 40 residues from each subunit, located mostly in two β strands, in contact. The light subunits make no contact with each other. Each heavy subunit consists of a single disk-shaped domain topologically similar to that of MADH and of GAO (Fig. 14). The light subunit consists of an irregularly folded extended chain of about 35 residues followed by a long helix, which together lie on the surface of the heavy subunit. Thus the two nonidentical subunits, H and L, make up a single domain.

The heavy subunit of MEDH is made up of mostly of β structure. The major structural feature of the heavy subunit consists of eight β leaflets, each composed of four antiparallel β strands, which form a W-like structure. These β leaflets are arranged about a pseudo-eightfold axis of symmetry, as shown in Figure 14. The axes of the two subunits of the heterohexamer are inclined by approximately 45° to the molecular twofold axis, as shown in Figure 13.

The arrangement of Ws in MEDH is topologically similar to that of MADH and of GAO except that the latter possesses pseudo-sevenfold symmetry rather than pseudo-eightfold symmetry. Like MADH and GAO, the strands of each leaflet are arranged in the pattern A, B, C and D, progressing outward from the central axis. The "top" of the subunit can be defined as the N-terminal end of the central barrel of A-strands [60]. These strands are oriented approximately parallel to the central

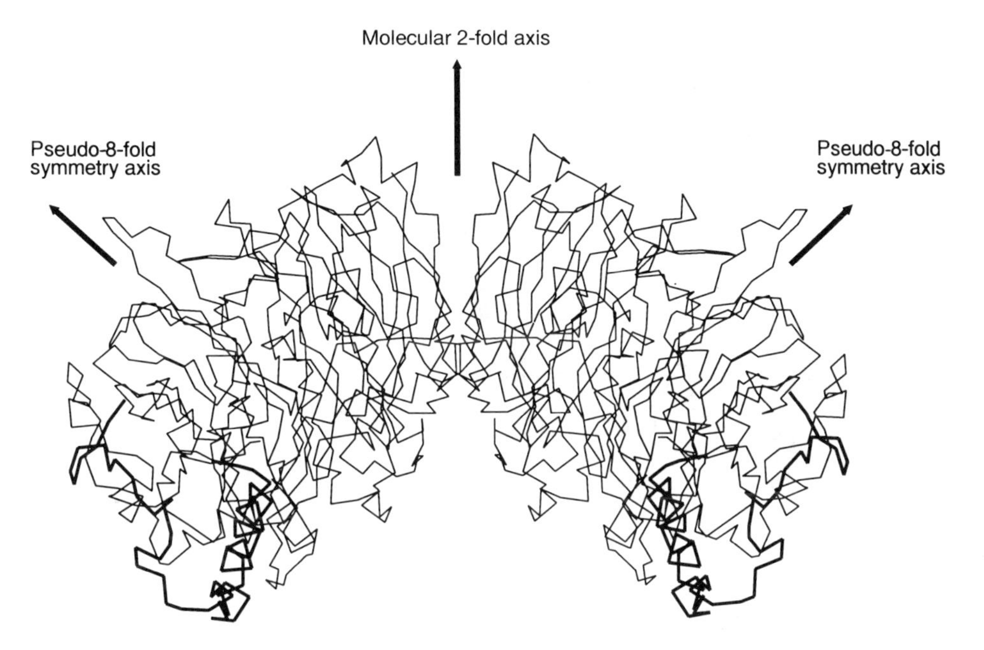

Figure 13 α-Carbon drawing of the MEDH dimer. The molecular twofold axis is vertical and the pseudo-eightfold symmetry axes are inclined by about 45° to the twofold axis. The heavy subunits are drawn in thin lines, and the light subunits are drawn in thick lines.

axis. Because of the twist of the β leaflets, the D strands are inclined by approximately 45° to the axis.

The N-terminal segment of the heavy subunit starts in the central channel, extends for about 40 residues along the bottom of the subunit, and then forms the outside strand (D) of leaflet W_8. It then goes on to form the inside strand (A) of leaflet W_1. Leaflet W_8 is completed by the inner three strands from the C-terminus. This construction is similar to that of the influenza neuraminidase [5] and of GAO but differs from that of MADH.

There are several excursions from the W structures, all located at the top of the subunit. The longest excursion occurs between strands B and C of W_6, which includes two antiparallel β strands located above the subunit. An additional excursion is located between W_5 and W_6. There are one or two short helices located in the segments between W_1 and W_2, between W_3 and W_4, and between strands B and C of W_8.

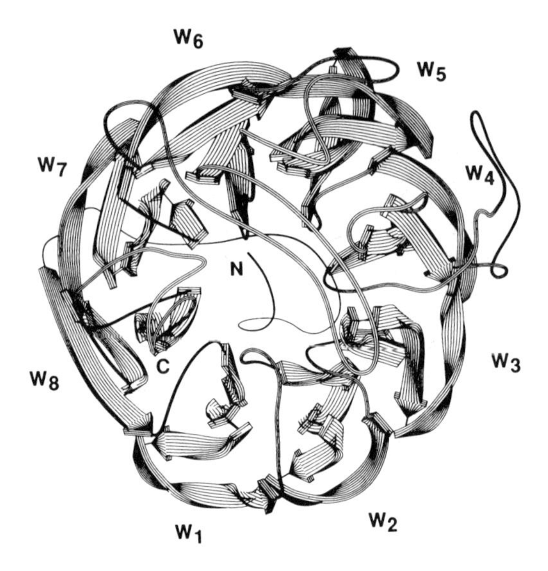

Figure 14 Ribbon diagram of the H subunit of MEDH. Those β sheets contained in the eight β leaflets (W_1–W_8) are indicated, as are the N- and C-termini. The pseudo-eightfold axis is approximately perpendicular to the diagram.

D. Discussion

At this stage of the crystallographic analysis of MEDH the PQQ cofactor has not been located. When the amino acid sequence becomes available, the structure can be fully refined and analysis of electron density difference maps may reveal the position of PQQ. It may also be possible to label the active site of MEDH by reaction with a "suicide substrate" such as cyclopropanol [61].

A likely place for the active site of MEDH is in the top end of the subunit. This region contains all the excursions from the basic superstructure of this motif, which would provide many amino acid side chains that could serve in catalysis and for binding the cofactor. In the analagous enzyme neuraminidase, this region contains the active site [6].

The light subunit of MEDH is very unusual. It does not fold into an independent globular domain, but is extended. Its tertiary structure, at least for the portion visible in the electron density map, consists entirely of interactions with the heavy subunit. The light subunit serves no obvious function at the moment. It is located mostly on the outside edge of the heavy subunit and has little contact with the presumed active site region in the "top" of the subunit. However, it may be involved in the catalytic process in some manner unknown at present. Further speculation on this point must wait until the structure analysis is completed.

V. CONCLUSION

As mentioned above, there are two themes that are common to methylamine dehydrogenase, galactose oxidase, and methanol dehydrogenase: an unusual cofactor as well as a domain or subunit consisting of a multiple W β sheet motif. The latter feature is, among the proteins with known three-dimensional structure, only shared with influenza virus neuraminidase.

The unusual cofactors in methylamine dehydrogenase and galactose oxidase are both constructed at some as yet unknown stage of biogenesis by formation of a covalent bond between two gene-encoded side chains. This is the only common feature, however, since in MADH two tryptophans are connected and in GAO a tyrosine and a cysteine residue are linked. Moreover, in MADH one of the tryptophans is modified such that it contains an ortho-quinone moiety, while in GAO a cupric ion liganded to the cofactor tyrosine is incorporated. In view of these differences, it is unlikely that a similar mechanism will be responsible for the generation of the TTQ and the Tyr-Cys-Cu redox centers. A most intriguing question is whether additional proteins are involved in cofactor synthesis or we are dealing with a self-catalytic process. The same question concerns the modified TOPA tyrosine in bovine serum amine oxidase (BSAO) [62]. The latter enzyme requires the modification of a single residue, while in MADH and GAO the two redox center side chains are approximately 50 residues apart. It should also be emphasized that the TTQ cofactor in MADH resides in the small subunit and *not* in the W7-subunit, while in GAO residues of the W7 domain form the cofactor. It appears therefore most likely that the generation of cofactors in MADH, GAO, and BSAO will proceed along quite, if not entirely, different pathways.

The W-motif has now been observed in neuraminidase, MADH, GAO, and MEDH, in chronological order of discovery. If current knowledge of protein structures has taught us anything, then we can expect a considerable proliferation of similar domains in protein structures still to be elucidated, although the W7/W8 propellor may not turn out to be as common as the TIM-barrel and its variants. Only future structure determinations will settle this point.

It should be pointed out that neuraminidase contains not only six Ws but also one more irregular motif between W4 and W5, which may have replaced or may have developed from a previous W motif. In this connection the analysis of the TV-MADH W7-domain is relevant. Vellieux et al. [8] suggested that the W7 can be thought of as being composed of a regular W3 plus a regular W4 half-domain. This raises the question whether a primodial W7 was once generated as such, leading later by divergent evolution to variants like W8 in MEDH and possibly W6 domains in neuraminidase, or that the W7 domain is the result of "W-loss" by a previous W8 domain or "W-gain" by an earlier W6 domain. Another possibility is that we are dealing with convergent evolution and that the W-motif by itself might rapidly fold because of its simple nature and relatively short loops so that in the

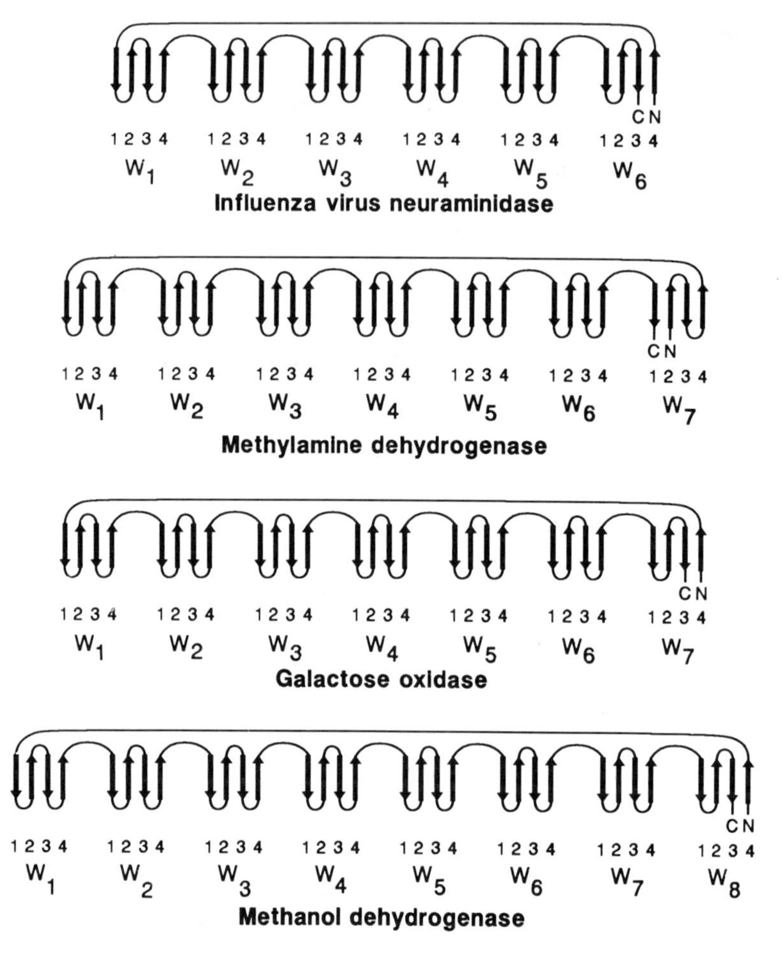

Figure 15 Schematic diagram of the folding topology of the six-, seven-, and eightfold β leaflets of influenza neuraminidase, methylamine dehydrogenase, galactose oxidase, and methanol dehydrogenase.

folding process such a fold would be likely to occur, although in many cases perhaps only as a transient phenomenon. Depending, then, on the nature and size of the side chains, combined with the degree of twist in each W, Ws can be arranged relatively easily with pseudo seven- and eightfold symmetry. Some evidence for this latter route lies in the fact that the positions of the active sites in the four W-containing enzymes have no similarity. Moreover, in GAO the central pore runs along the pseudo-symmetry axis (filled by a hairpin and solvent molecules), while in MADH and MEDH there is no central pore, and side chains are closely packed along much of the pseudo symmetry axis. But since sequences of the W domains in neither MADH nor MEDH are available yet, it is best to refrain here from further speculation. It should, finally, be pointed out that the last W (which can also be considered the first W!) of the four enzymes depicted in Figure 15 occurs in two variants. In neuraminidase, GAO, and MEDH this W is composed of one (outer) N-terminal strand plus three (inner) C-terminal strands. In both MADH structures the last W is composed of *one* (inner) C-terminal strand and three (outer) N-terminal strands. It is quite remarkable that these last Ws are so similar to the other Ws since, in complete contrast with the in-chain Ws, these Ws are composed of β strands that are not at all adjacent in sequence. Such observations make one wonder what really determines the three-dimensional structure of proteins: free energy of the native state, kinetics of the folding process, or assistance by chaperonins.

ACKNOWLEDGMENTS

The authors gratefully acknowledge the studies of Fred Vellieux, Eric Huizinga, and Kor Kalk leading to the three-dimensional structure of *Thiobacillus versutus* MADH, for which Hans Duine and coworkers at the Technical University in Delft generously provided protein material. They further acknowledge the efforts of Longyen Chen and Rosemary Durley on the structural analysis of MADH from *Paracoccus denitrificans* and its complex with amicyanin, of Zong-xiang Xia on the studies of MEDH, and of Victor Davidson in supplying material for these projects. The authors also wish to thank Nobutoshi Ito and Simon Phillips for helpful comments on galactose oxidase and Nobutoshi Ito for providing figures. Financial support has been provided by the Netherlands Foundation for Chemical Research (SON) and the Netherlands Organization for Scientific Research (NWO) and by grants from the USPHS and the National Science Foundation.

REFERENCES

1. Duine, J. A., and Jongejan, J. A. (1989). Quinoproteins, enzymes with pyrroloquinoline quinone as cofactor. *Annu. Rev. Biochem. 58*: 403.
2. Salisbury, S. A., Forrest, J. S., Cruse, W. B. T., and Kennard, O. (1979). A novel coenzyme from bacterial primary alcohol dehydrogenases. *Nature 280*: 843.
3. Duine, J. A., Frank, J., and Verwiel, P. E. J. (1980). Structure and activity of the prosthetic group of methanol dehydrogenase. *Eur. J. Biochem. 108*: 187.
4. McIntire, W. S., Wemmer, D. E., Chistoserdov, A., and Lidstrom, M.E. (1991). A new cofactor in a prokaryotic enzyme: Tryptophan tryptophylquinone as the redox prosthetic group in methylamine dehydrogenase. *Science 252*: 981.
5. Chen, L., Mathews, F. S., Davidson, V. L., Huizinga, E. G., Vellieux, F. M. D., Duine, J. A., and Hol, W. G. J. (1991). Crystallographic investigation of the

tryptophan-derived cofactor in the quinoprotein methylamine dehydrogenase. *FEBS Lett. 287*: 163.

6. Ito, N., Phillips, S. E. V., Stevens, C., Ogel, Z. B., McPherson, M. J., Keen, J. N., Yadov, K. D. S., and Knowles, P. F. (1991). Novel thioether bond revealed by a 1.7 Å crystal structure of galactose oxidase. *Nature 350*: 87.
7. Varghese, J. N., Laver, W. G., and Colman, P. W. (1983). Structure of the influenza virus glycoprotein antigen neuroaminidase at 2.9 Å resolution. *Nature 303*: 35.
8. Vellieux, F. M. D., Huitema, F., Groendijk, H., Kalk, K. H., Frank Jzn., J., Jongejan, J. A., Duine, J. A., Petratos, K., Drenth, J., and Hol, W. G. J. (1989). Structure of the quinoprotein methylamine dehydrogenase at 2.25 Å resolution. *EMBO J. 8*: 2171.
9. Vellieux, F. M. D., Kalk, K. H., Drenth, J., and Hol, W. G. J. (1990). Structure determination of quinoprotein methylamine dehydrogenase from *Thiobacillus versutus*. *Acta Cryst. B46*: 806.
10. Chen, L., Mathews, F. S., Davidson, V. L., Huizinga, E. G., Vellieux, F. M. D., and Hol, W. J. G. (1992). Three-dimensional structure of the quinoprotein methylamine dehydrogenase from *Paracoccus denitrificans* determined by molecular replacement at 2.8 Å resolution. *Proteins: Struct., Funct. and Genet.* (in press).
11. Chen, L., Durley, R., Poliks, B. J., Lim, L. W., Hamada, K., Chen, Z.-w., Mathews, F. S., Davidson, V. L., Satow, Y., Huizinga, E., Vellieux, F. M. D., and Hol, W. G. J. (1992). Interaction between a quinoprotein and a blue copper protein revealed by the crystal structure of an electron-transfer complex between methylamine dehydrogenase and amicyanin. *Biochem.* (in press).
12. Xia, Z.-x., Xiong, G.-P., Dai, W.-W., Davidson, V. L., White, S., and Mathews, F. S. (1992). Three dimensional structure of methanol dehydrogenase from methylophilus methylotrophus at 2.6 Å resolution. (submitted).
13. De Beer, R., Duine, J. A., Frank, J., and Large, P. J. (1980) The prosthetic group of methylamine dehydrogenase from *Pseudomonas* AM1. *Biochim. Biophys. Acta 622*: 370.
14. Anthony, C. (1989). Quinoproteins in C_1-dissimilation by bacteria. In *PQQ and Quinoproteins* (J. A. Jongejan and J. A. Duine, eds.), Kluwer Academic Publishers, Dordrecht, pp. 1–11.
15. van Wielink, J. E., Frank, J., and Duine, J. A. (1989). Electron transport from methylamine to oxygen in the gram-negative bacterium *Thiobacillus versutus*. In *PQQ and Quinoproteins* (J. A. Jongejan and J. A. Duine, eds.), Kluwer Academic Publishers, Dordrecht, pp. 269–278.
16. Husain, M., and Davidson, V. L. (1985). An inducible periplasmic blue copper protein from *Paracoccus denitrificans*. *J. Biol. Chem. 260*: 14626.
17. Husain, M., and Davidson, V. L. (1986). Characterization of two inducible periplasmic *c*-type cytochromes from *Paracoccus denitrificans*. *J. Biol. Chem. 261*: 8577.
18. Davidson, V. L., and Kumar, M. A. (1989). Cytochrome c-550 mediates electron transfer from inducible periplasmic c-type cytochromes to the cytoplasmic membrane of *Paracoccus denitrificans*. *FEBS Lett. 245*: 271.
19. Vellieux, F. M. D., Frank, J., Jzn., Swarte, M. B. A., Groendijk, H., Duine, J. A., Drenth, J., and Hol, W. G. J. (1986). Purification, crystallization and preliminary

x-ray investigation of quinoprotein methylamine dehydrogenase from *Thiobacillus versutus*. *Eur. J. Biochem. 154*: 383.
20. Huitema et al., unpublished.
21. Shirai, S., Matsumoto, T., and Tobari, J. (1978). Methylamine dehydrogenase of *Pseudomonas* AM1. *J. Biochem.* (Tokyo) *83*: 1599.
22. Ishii, Y., Hase, T., Fukumori, Y., Matsubara, H., and Tobari, J. (1983). Amino acid sequence of the light subunit of methylamine dehydrogenase from *Pseudomonas* AM1: Existence of two residues in the binding of the prosthetic group. *J. Biochem.* (Tokyo) *93*: 107.
23. Chistoserdov, A. Y., Tsygankov, Y. D., and Lidstrom, M. E. (1990). Cloning and sequencing of the structural gene for the small subunit of methylamine dehydrogenase from *Methylobacterium extorquens* AM1: Evidence for two tryptophan residues involved in the active center. *Biochem. Biophys. Res. Commun. 172*: 211.
24. Chen, L., Lim, L. W., Mathews, F. S., Davidson, V. L., and Husain, M. (1988). Preliminary x-ray crystallographic studies of methylamine dehydrogenase and methylamine dehydrogenase-amicyanin complexes from *Paracoccus denitrificans*. *J. Mol. Biol. 203*: 1137.
25. van Spanning, R. J. M., Wansell, C. W., Reijnders, W. N. M., Oltmann, L. F., and Stouthammer, A. H. (1990). Mutagenesis of the gene encoding amicyanin of *Paracoccus denitrificans* and the resulting effect on methylamine dehydrogenase. *FEBS Lett. 275*: 217.
26. van der Meer, R. A., Jongejan, J. A., and Duine, J. A. (1987). Phenylhydrazine as a probe for cofactor identification in amine oxidoreductases. Evidence for PQQ as the cofactor in methylamine dehydrogenase. *FEBS Lett. 221*: 299.
27. McIntire, W. S., and Stultz, J. T. (1986). On the structure and linkage of the covalent cofactor of methylamine dehydrogenase from the methylotrophic bacterium W3A1. *Biochem. Biophys. Res. Commun. 141*: 562.
28. Vellieux, F. M. D., and Hol, W. G. J. (1989). A new model for the pro-PQQ cofactor of quinoprotein methylamine dehydrogenase. *FEBS Lett. 255*: 460.
29. Ambler, R. P., and Tobari, J. (1985). The primary structure of *Pseudomonas* AM1 amicyanin and pseudoazurin. *Biochem. J. 232*: 451.
30. van Beeumen, J., van Bun, S., Canters, G. W., Lommen, A., and Chothis, C. (1991). The structural homology of amicyanin from *Thiobacillus versutus* to plant plastocyanins. *J. Biol. Chem. 266*: 4869.
31. Sharma, K. D., Loehr, T. M., Sanders-Loehr, J., Husain, M., and Davidson, V. L. (1988). Resonance Raman spectroscopy of amicyanin, a blue copper protein from *Paracoccus denitrificans*. *J. Biol. Chem. 263*: 3303.
32. Lommen, A., Canters, G. W., and van Beeumen, J. (1988). A ^{1}H-NMR study on the blue copper protein amicyanin from *Thiobacillus versutus*. *Eur. J. Biochem. 176*: 213.
33. Hussain, M., and Davidson, V. L. (1987). Purification and properties of methylamine dehydrogenase from *Paracoccus denitrificans*. *J. Bacteriol. 169*: 1712.
34. Gray, K. A., Davidson, V. L., and Knaff, D. B. (1988). Complex formation between methylamine dehydrogenase and amicyanin from *Paracoccus denitrificans*. *J. Biol. Chem. 263*: 13987.

35. Davidson, V. L., and Jones, L. H. (1991). Intermolecular electron transfer from quinoproteins and its relevance to biosensor technology. *Anal. Chim. Acta. 249*: 235.
36. Kumar, M. A., and Davidson, V. L. (1990). Chemical cross-linking study of complex formation between methylamine dehydrogenase and amicyanin from *Paracoccus denitrificans*. *Biochemistry 29*: 5299.
37. Backes, G., Davidson, V. L., Huitema, F., Duine, J. A., and Sanders-Loehr, J. (1991). Characterization of the tryptophan-derived quinone cofactor of methylamine dehydrogenase by resonance Raman spectroscopy. *Biochemistry 30*: 2901.
38. Guss, J. M., and Freeman, H. C. (1983). Structure of oxidized poplar plastocyanin at 1.6 Å resolution. *J. Mol. Biol. 169*: 521.
39. Groeneveld, C. M., Ouwerling, M. C., Erkelens, C., and Canters, G. W. (1988). NMR study of the protonation behavior of the histidine residues and the electron self exchange reaction of azurin from *Alcaligenes denitrificans*. *J. Mol. Biol. 200*: 189.
40. van der Kamp, M., Floris, R., Hali, F. C., and Canters, G. W. (1991). Site-directed mutagenesis reveals that the hydrophobic patch of azurin mediates electron transfer. *J. Am. Chem. Soc. 112*: 907.
41. Beratan, D. N., Onuchic, J. N., Betts, J. N., Bowler, B. E., and Gray, H. B. (1990). Electron tunneling pathways in ruthenated proteins. *J. Am. Chem. Soc. 112*: 7915.
42. Avigad, G., Amaral, D., Asensio, C., and Horecker, B. L. (1962). The D-galactose oxidase of *Polyporus circinatus*. *J. Biol. Chem. 237*: 2736.
43. Aiska, K., and Terada, O. (1982). Purification and properties of galactose oxidase from *Gibberalla fujikuroi*. *Agric. Biol. Chem. 46*: 1101.
44. Koroleva, O. V., Rabinovich, M. L., Buglova, T. T., and Yaropolov, A. I. (1983). Some properties of galactose oxidase from *Fusarium graminearum*. *Prik. Biochim. Mikrobiol. 19*: 632.
45. Amaral, D., Bernstein, L., Morse, D., and Horecker, B. L. (1963) Galactose oxidase of *Polyporus circinatus*: A copper enzyme. *J. Biol. Chem. 238*: 2281.
46. Hamilton, G. A., Adolf, P. K., de Jersey, J., DuBois, G. C., Dyrakacz, G. R., and Libby, R. D. (1978). Trivalent copper, superoxide and galactose oxidase. *J. Am. Chem. Soc. 100*: 1899.
47. Whittaker, M. M., and Whittaker, J. W. (1988). The active site of galactose oxidase. *J. Biol. Chem. 263*: 6074.
48. Whittaker, M. M., DeVito, V. I., Asher, S. A., and Whittaker, J. W. (1989). Resonance Raman evidence for tyrosine involvement in the radical site of galactose oxidase. *J. Biol. Chem. 264*: 7104.
49. Whittaker, M. M., and Whittaker, J. W. (1990). A tyrosine-derived free radical in apogalactose oxidase. *J. Biol. Chem. 265*: 9610.
50. Poulos, T. L., and Kraut, J. (1980). The stereochemistry of peroxidase catalysis. *J. Biol. Chem. 255*: 8199.
51. Nordlund, O., Sjöberg, B.-M., and Eklund, H. (1990). Three-dimensional structure of the free radical protein of ribonucleotide reductase. *Nature 345*: 593.
52. Van der Meer, R. A., Jongejan, J. A., and Duine, J. A. (1989). Pyrroloquinoline quinone as cofactor in galactose oxidase. *J. Biol. Chem. 264*: 7792.
53. Anthony, C., and Zatman, L. J. (1967). The microbiol oxidation of methanol. The

prosthetic group of the alcohol dehydrogenase of *Pseudomonas* spec. M27: A new oxidoreductase prosthetic group. *Biochem. J. 104*: 960.

54. Harms, N., de Vries, G. E., Maurer, K., Hoogendijk, J., and Stouthamer, A. H. (1987). Isolation and nucleotide sequence of the methanol dehydrogenase of the methanol dehydrogenase structural gene from *Paracoccus denitrificans. J. Bacteriol. 164*: 3969.
55. Machlin, S. M., and Hanson, R. H. (1988). Nucleotide sequence and transcriptional start site of the *Methylobacterium organophilum* XX methanol dehydrogenase structural gene. *J. Bacteriol. 170*: 4739.
56. Anderson, D. J., Morris, C. J., Nunn, D. N., Anthony, C., and Lidstrom, M. E. (1990). Nucleotide sequence of the Methylobacterium extorquens AM1 *moxF* and *moxJ* genes involved in methanol oxidation. *Gene 90*: 173.
57. Anderson, D. J., and Lidstrom, M. E. (1988). The *MoxFG* region encodes four polypeptides in the methanol-oxidizing bacterium *Methylobacterium* sp. strain AM1. *J. Bacteriol. 170*: 2254.
58. Xia, Z.-x., Hao, Z.-p., Davidson, V. L., and Mathews, F. S. (1989). Crystallization and preliminary x-ray crystallographic study of the quinoprotein methanol dehydrogenase from bacterium W3A1. *FEBS Lett. 258*: 175.
59. Jones, T. A., Zuo, J.-Y., Cowan, S. W., and Kjeldgaard, M. (1991). Improved methods for building protein models in electron density maps and the location of errors in these models. *Acta Cryst. A47*: 110.
60. Branden, C., and Tooze, J. (1991). *Introduction to Protein Structure*. Garland Publishing, Inc., New York.
61. Mincey, T., Bell, J. A., Mildvan, A. S., and Ables, R. H. (1981). Mechanism of action of methoxatin-dependent alcohol dehydrogenase. *Biochemistry 20*: 7502.
62. Janes, S. M., Mu, D., Wemmer, D., Smith, A. J., Kaur, S., Maltby, D., Burlingame, A. L., and Klinman, J. P. (1990). A new redox cofactor in eukaryotic enzymes: 6-Hydroxydopa at the active site of bovine serum amine oxidase. *Science 248*: 981.

11

Resonance Raman Spectroscopy of Quinoproteins

David M. Dooley and Doreen E. Brown

Amherst College, Amherst, Massachusetts

I. INTRODUCTION

Raman spectroscopy has emerged as an extremely powerful and informative technique for probing the structure and chemistry of biological molecules [1–3]. Although the physical mechanisms (or quantum mechanics) that underlie Raman scattering and infrared absorption are very different, the information contained in the spectra are similar: the frequencies and intensities of vibrational and rotational transitions. It must be recognized that different selection rules govern Raman scattering and infrared absorption, so that intensities of Raman and infrared transitions are not comparable. In molecules of sufficiently high symmetry, certain transitions may be observed in the infrared absorption spectrum but not in the Raman spectrum, and vice versa. Nevertheless, by considering the detailed and useful information available from a standard infrared spectrum of a molecule, one gets a good idea of the utility of a Raman spectrum as well. As will be discussed below, in many cases Raman spectroscopy offers significant advantages over infrared spectroscopy, particularly in the study of biological systems. Moreover, the continuing improvement and development of lasers and the advent of fourier transform instruments have greatly expanded the range of problems that can be productively studied by Raman spectroscopy.

A. The Resonance Raman Effect

The classical Raman effect (as reported by C. V. Raman in 1928) [4] arises from the inelastic scattering of a photon by a molecule. During an inelastic collision, some energy is transferred between the photon and molecule, but the total energy is, of course, conserved. Hence the energy lost or gained by the photon must exactly correspond to an energy change or transition of the molecule. Given the energy-level spacings of a typical molecule, the molecular transitions generally involved in Raman scattering are rotational and vibrational transitions. As an example, the Raman spectrum of acetonitrile is shown in Figure 1. Note that the energy scale (in wavenumbers) gives the *difference* in energy between the incident laser light and the scattered light. In this spectrum the scattered light is lower in energy than the incident light, indicating that some fraction of the incident photons

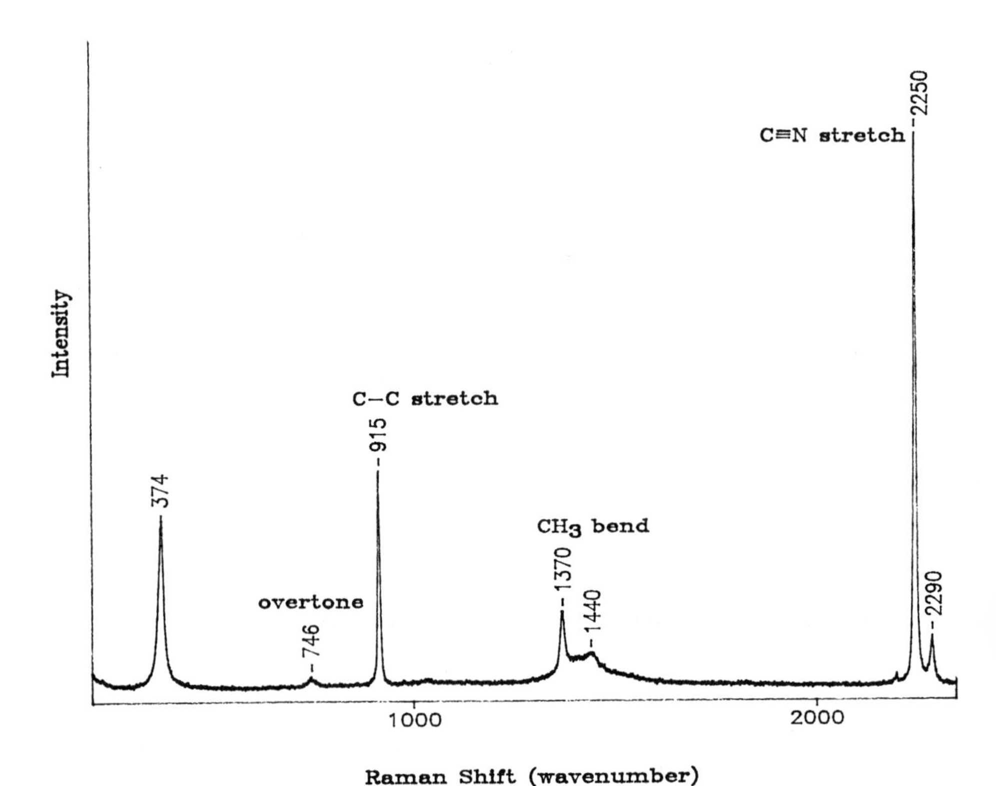

Figure 1 Raman spectrum of acetonitrile obtained with 514.5-nm excitation at room temperature. This is a single scan at 5 cm^{-1} resolution. The feature at 2290 cm^{-1} is probably a combination band (915 plus 1370 cm^{-1}).

have transferred energy to the molecule, exciting vibrational and rotational transitions. Therefore the energies of the bands observed in the Raman spectrum correspond to vibrational transitions of acetonitrile, since the rotational fine structure would not be resolved in a condensed phase, as is also the case for infrared spectra. Just as in an infrared spectrum it is possible to associate each band in a Raman spectrum to fundamental molecular vibrations, overtones, and combinations. Some assignments for acetonitrile are indicated in Figure 1. Normal Raman scattering is a relatively improbable event, that is, the vast majority of collisions between the incident photon beam and the molecule are elastic, with no change in the energy of the scattered photon. The discussion to this point has assumed that the energy of the incident light differs considerably from the energies of molecular electronic transitions. In the language of quantum mechanics, the interaction of the photons with the molecule does not produce an excited eigenstate of the molecular Hamiltonian. More naively, we can say that the incident photons are not absorbed. Under such conditions many of the basic features of Raman scattering may be adequately described by classical physics. Excellent discussions of the theoretical (ranging from introductory to advanced) and practical aspects of Raman spectroscopy are available in the literature [1–3,5,6].

Profound changes in the Raman spectrum of a molecule are observed as the incident light is selected or tuned to be closer to a molecular electronic transition. Most importantly, the intensities of certain bands, corresponding to particular vibrational modes of the molecule, may be enhanced by 2–6 orders of magnitude compared to the normal Raman scattering intensity. If the incident light falls within the envelope of an electronic absorption band, the light is "in resonance" with the electronic transition; the resulting Raman spectrum is termed a resonance Raman spectrum. The resonance enhancement phenomenon is a principal reason why resonance Raman spectroscopy has become such a valuable tool for the study of biological molecules. Some bands may be significantly enhanced as the exciting light approaches, but does not fall within, the absorption envelope. This is termed, appropriately enough, the preresonance Raman effect, and has also proven advantageous for biochemical applications. The focus here will be on applications of resonance Raman spectroscopy. Understanding resonance and preresonance enhancement of Raman bands has provided intriguing theoretical challenges, and several reasonably successful approaches were ultimately developed [5]. A common theme of these approaches is a description of resonance Raman spectroscopy as a two-photon process involving absorption followed by emission. Resonance Raman scattering is generally distinguished from fluorescence emission by virtue of the former's much shorter time scale. In this chapter one theoretical approach will be briefly described. The goal is to provide a framework for interpreting, in molecular terms, resonance Raman spectra.

Heller has developed an appealing, often intuitive, semiclassical approach to the "simple aspects" of resonance Raman scattering [7–9]. This approach examines Raman scattering in the time domain, which proves advantageous because the intrinsic time scale for Raman scattering is short compared to the period of molecular vibrations. Figure 2 pictorially presents Raman scattering as viewed by Heller's time-dependent approach [7]. Two potential surfaces are shown, each relevant to two vibrational degrees of freedom, x and y. On the bottom is shown a vibrational wave function χ on the ground electronic state potential surface. It is vertically displaced to the upper (excited) potential surface by the absorption of a photon, represented by the vertical arrow. This is simply another way of representing the familiar Franck-Condon approximation that electronic transitions occur on a time scale that is very short compared to that for nuclear motion. The wavefunction on the excited potential surface is

$$\phi = \mu\chi \tag{1}$$

where μ is the electronic transition moment and χ is a vibrational wave function for a particular normal mode in the electronic ground state. Since ϕ is not a stationary

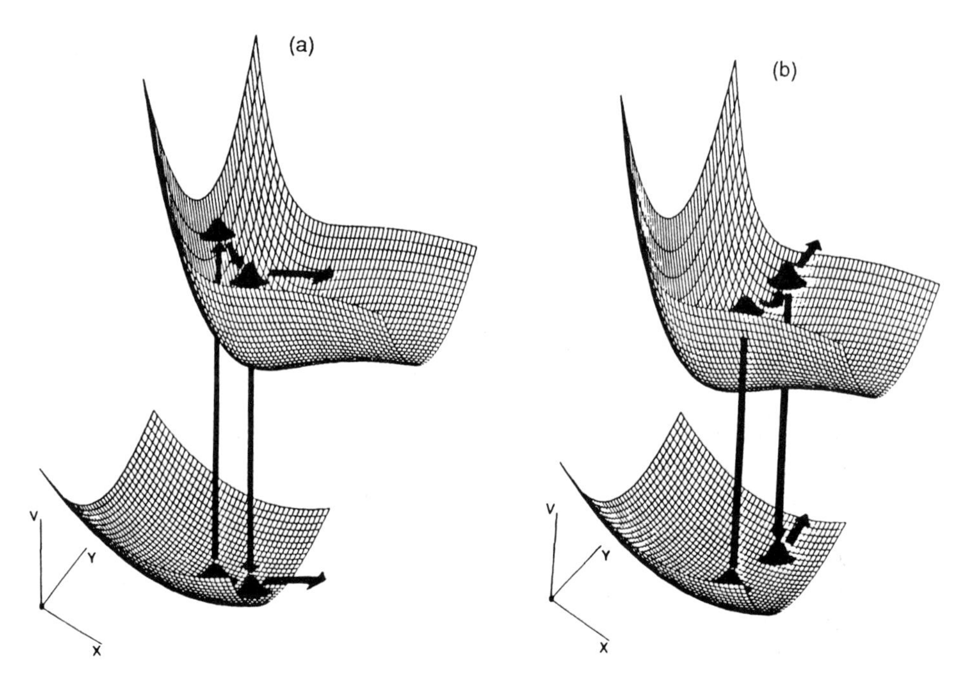

Figure 2 (a) Raman scattering from an upper potential surface, where the wave packet acquires motion in the x direction initially; the x-mode will be enhanced in the Raman as resonance is approached. (b) y modes enhanced due to y-motion initially. (Reproduced with permission from Ref. 7.)

state (i.e., it is not the appropriate vibrational wavefunction for the excited electronic state), it can be regarded as a displaced wave packet that will evolve with time before returning (vertically again, illustrated by the down arrow) to the ground state potential surface. Its initial motion for two possible cases is shown in Figure 2. The important point is that for short times the forces acting on the wave packet and its resultant motion are *classical* [7–9]. Put another way, the wave packet moves like a ball on a sloping driveway. If the force (the slope of the potential surface) in the excited state is along x, then the vibrational mode involving motion along x will be enhanced, and the y mode will not (Fig. 2a). If the force is along the y direction, then the y mode will be enhanced (Fig. 2b). Mathematically, the intensity of a particular normal mode is given by [7]:

$$I^{k}_{0n} = \left|\int_0^{\infty} e^{i\nu_{\text{laser}}t-\Gamma t}\langle\phi_n|\phi(t)\rangle dt\right|^2 + \text{nonresonant term} \tag{2}$$

Here $\phi(t)$ represents the time evolution of $\phi = \mu\chi_0$ and $\phi_n = \mu\chi_n$; χ_0 is the ground vibrational wavefunction for normal mode k, and χ_n is an excited vibrational wavefunction. Γ is a damping term, and ν_{laser} is the frequency of the incident laser light.

Although the equation appears formidable, it is easy to extract the physical significance for resonance Raman scattering. For example, if we set $n = 1$, the equation corresponds to scattered photons transferring sufficient energy to the molecule to excite the fundamental (i.e., $0 \rightarrow 1$) transition of the kth normal mode of vibration. The intensity of the Raman band is determined by the overlap $\langle\phi_n|\phi(t)\rangle$. Since the overlap is zero at $t = 0$ (the χs are required to be orthogonal), resonance Raman bands have nonzero intensities only to the extent that $\phi(t)$ evolves with time in response to the force exerted on it by the excited state potential surface. In other words, the steepness of the surface where the wave packet finds itself after the photon is absorbed largely determines the intensity. Like the ball on a steep driveway, the wave packet is displaced and gains momentum, and this causes it to overlap the first excited state ($n = 1$) of either the x or y normal mode when it returns to the lower potential surface (Fig. 2). The horizontal arrows in Figure 2 indicate the momentum of $\phi(t)$ at the time of its return to the ground electronic state. The damping factor Γ also affects the intensity of a Raman band. Γ represents the decay of $\phi(t)$ due to vibrational degrees of freedom other than those included by ϕ and ϕ_n [7–9]. For complex molecules absorption of a photon generally produces displacements in several normal modes of different frequencies, which results in a large effective Γ. As Γ increases, the amount of time that ϕ spends on the upper surface decreases, thereby magnifying the influence of the initial motion of ϕ, as illustrated in Figure 3. In the short-time regime, the following simple formula gives the intensity ratio for two normal modes k and l [7,8]:

$$\frac{I^{k}_{01}}{I^{l}_{01}} = \left(\frac{\nu_l}{\nu_k}\right)\left[\frac{V'_k}{V'_l}\right]^2 \tag{3}$$

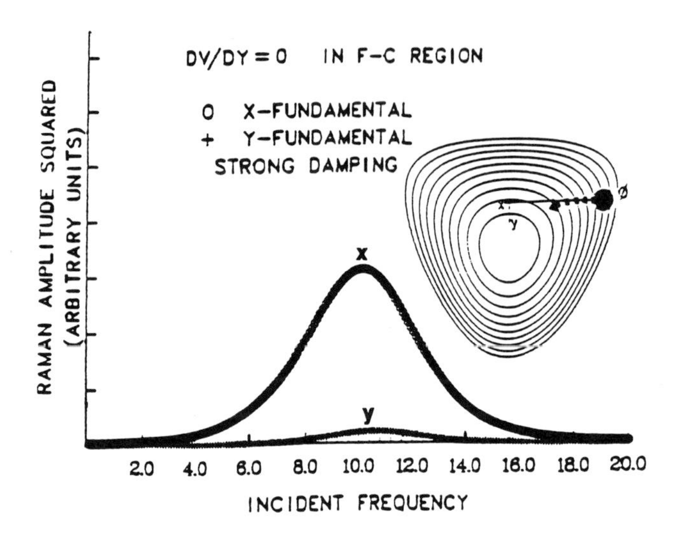

Figure 3 Potential surface, wave-packet ϕ, and the early motion of $\phi(t)$ is shown as an insert. For a large damping constant Γ, the motion of $\phi(t)$ results in the displayed x-mode and y-mode fundamental intensities ($0 \to 1$) below, in, and above resonance. (Adapted with permission from Ref. 7.)

Thus the relative intensities of resonance Raman bands corresponding to different normal modes are directly related to the slopes ($V'_{k,l}$) of the potential surface along the normal mode directions. If the ground and excited potential surfaces are both harmonic, and in the special case where certain simplifying assumptions are accurate, the intensity ratio can be related to excited-state displacements [8]:

$$\frac{I_{01}^k}{I_{01}^l} = \left(\frac{\nu_k^3 \Delta_k^2}{\nu_l^3 \Delta_l^2}\right) \tag{4}$$

Here $\Delta_{k,l}$ is the displacement of the nuclei from their equilibrium configuration *in the excited state*, projected on the ground state normal modes k and l.

These theoretical considerations teach some important lessons regarding the interpretation of resonance Raman spectra. One must be aware of the intrinsic relationship between the intensities of resonance Raman bands and the nature of the excited state produced by absorption of the photon. To a first approximation, those normal modes involving nuclei that are significantly displaced from their equilibrium positions in the excited state (Eq. 3) and where changes in the bonding/antibonding character of the associated molecular orbitals have occurred are likely to be enhanced the most. This follows from the expectation that if the bonding among certain nuclei has been perturbed in the excited state, then the equilibrium positions of those nuclei will differ significantly from their positions in the ground state. Often the slope of the upper potential surface in the Franck-Condon region will be large in such a case. If not, the mode will not be strongly resonance enhanced; displacement of the equilibrium nuclear positions in the excited state relative to the ground state is a necessary, but not sufficient, condition for resonance enhancement. Certain modes may be strongly enhanced and readily observed whereas others *may go undetected because they are only weakly enhanced, or not enhanced at all*, and thus are lost in the experimental background.

Regardless of the theoretical perspective one uses, there is an intimate connection between the nature of the electronic transition with which the laser is in resonance and the identity of the normal modes that are enhanced. Thus the observed resonance Raman spectrum provides information about the excited state involved, and, conversely, the resonance Raman spectrum can be rationalized (perhaps even predicted in part) if there is independent information already available on the properties of the excited state. Tsuboi and Hirakawa and coworkers have formulated two general rules related to these points [10]:

1. If a Raman band becomes much stronger when the exciting frequency is brought closer to the frequency of an electronic band, then the equilibrium conformation of the molecule is distorted along the normal coordinate for the Raman band on going from the ground to the excited electronic states.
2. For $\pi \rightarrow \pi^*$ transitions in conjugated double-bond systems, Raman bands arising from asymmetric stretching vibrations, rather than those from totally symmetric stretching vibrations, become dominant when the exciting frequency is brought close to that of the longest wavelength absorption band.

Rule 2 follows from the observation that low energy electronic transitions in conjugated molecules often involve changes in the nodal pattern of the molecular orbitals such that some bonds are strengthened and others weakened in the excited state; the asymmetric change in the equilibrium positions of the nuclei would then correspond to an asymmetric stretching vibration of the molecule.

B. Advantages and Limitations of Resonance Raman Spectroscopy in the Study of Biological Chromophores

We are now in a position to understand both the great utility of resonance Raman spectroscopy and its existing limitations in elucidating the structures and properties of quinoproteins.

1. Resonance Raman spectroscopy can be used to probe intact enzymes, subunits, isolated peptides, and both free and bound cofactors, as long as the compounds absorb laser light. Hence chromophores can be studied as a function of their environment, e.g., the effects of binding a chromophore to a protein may be probed directly. In favorable circumstances information is obtainable on specific

molecular interactions between a probe molecule, inhibitor, or substrate analog and the binding (active) site. Thorough reviews of such applications are available [1–3]. From Eqs. 1 and 2 we see that the resonance Raman intensity depends on μ^2 (*if* μ is independent of nuclear coordinates, it can be factored out of the overlap in Eq. 2). The more familiar electronic extinction coefficient ϵ is directly proportional to μ, so we can see that the resonance Raman intensity will be proportional to the square of the absorption intensity. This is one reason why porphyrins and other intensely absorbing species give such beautiful resonance Raman spectra.

2. Resonance Raman spectroscopy is information rich; the vibrational spectrum of the chromophore, in whole or in part, is selectively obtained. In the nonresonance Raman or infrared spectra of proteins or other macromolecules, it is extremely difficult, if not impossible, to assign bands to particular groups. Absorption of the incident photon confers enormous selectivity through resonance enhancement. The literally thousands of other, nonenhanced vibrational modes in a medium-sized protein are generally not observed under the signal-acquisition conditions employed. Of course the resonance Raman spectra may themselves be complex and difficult to interpret. Isotopic substitution, chemical perturbation, and theoretical calculations have all proven very useful in making specific assignments.

3. Resonance Raman spectroscopy is a sensitive technique so that relatively small amounts of sample are required; in many cases about 1–10 nmol will suffice. The concentration, and therefore the amount, of sample needed depends upon the extinction coefficient of the chromophore at the incident photon frequency. If the sample is photosensitive, more material is generally needed.

4. The primary limitations of resonance Raman spectroscopy are related to the requirement that the incident photon be absorbed. The molecular excited state produced by absorption of light can emit a photon following vibrational relaxation or internal conversion (fluorescence or phosphorescence), dissipate the photon energy via radiationless processes, or react. Fluorescence may be much more intense than even the resonance Raman effect and thus swamp the signal of interest. Various strategies have been developed to circumvent this problem, but in some cases it has proven insurmountable. Radiationless deactivation is basically a process by which the photon energy is dissipated into the chromophore's surroundings as heat, which may cause local thermal denaturation of a sensitive sample. Photochemical reactions may or may not be a problem depending upon the quantum efficiency of the specific reaction.

C. Experimental Aspects

A wide variety of sources may be consulted for all the necessary details [1–3,6,11]. Some specific points for the study of quinoproteins will be briefly summarized here. We have found that high quality resonance Raman spectra can be obtained on quinoproteins and certain derivatives when the sample absorbance is approximately in the range of 3–7 for a 1-cm pathlength. It is difficult to achieve satisfactory signal-to-noise for less absorbing samples, and self-absorption of the Raman scattered light by the sample is a problem for more absorbing samples. Although such absorbancies may require fairly concentrated samples, it is often possible to use only small volumes, on the order of 20–50 μl. Data are readily collected on such volumes by placing the sample in a small-diameter capillary tube. If necessary the sample can be cooled by flowing cold N_2 gas around the tube or by mounting it on a cold-finger. Photochemical degradation has not been a problem in experiments on quinoproteins, with one notable exception discussed below.

II. VIBRATIONAL PROPERTIES OF QUINONES AND TRYPTOPHAN

A. General Aspects

The vibrational properties of quinones have been extensively studied and reviewed [12]. Some especially pertinent features of the vibrational spectra of *p*-benzoquinone [13–19] will be summarized here. Selected data on tryptophan [20–22] will also be presented; these are relevant owing to the recent discovery that tryptophan tryptophanylquinone (TTQ) is the cofactor in methylamine dehydrogenase [23]. Table 1 summarizes relevant vibrational frequencies for *p*-benzoquinone and tryptophan; the vibrational spectra of these compounds provide some useful guidelines for interpreting the spectra of quinoproteins. Although *p*-benzoquinone has been studied via vibrational spectroscopy more thoroughly than any other quinone, considerable uncertainty has surrounded the assignments of certain key vibrational modes, including the carbonyl stretching mode(s). Recently Becker has reexamined the infrared and Raman data for *p*-benzoquinone and formulated assignments that are internally consistent and apparently compatible with all the existing data [14]. These assignments are given in Table 1. Of particular importance is the assignment of the in-phase C═O stretch as two bands (owing to Fermi resonance coupling effects) at 1665 and 1684 cm^{-1}. Fermi resonance can arise in polyatomic molecules when two vibrational levels belonging to different modes (or combinations of modes) are *accidently degenerate* [24]. This accidental degeneracy can cause a mixing of the vibrational modes and a splitting of the energies. Isotope effects were instrumental in the assignment of the *p*-benzoquinone vibrational spectrum. Note, for instance, that the bands assigned as primarily $(C{=}O)_{\text{in-phase}}$ shift by −11 and −14 cm^{-1} upon ^{18}O substitution (Table 1). The C═O bend is found at 483 cm^{-1} (−7 cm^{-1} shift for ^{18}O substitution). The region from 900 to 1600 cm^{-1} is dominated by C—H bending, C═C stretching, and ring modes. Certain of these modes, especially the ring

Table 1 Selected Vibrational Frequencies and Assignments for *p*-Benzoquinone and Tryptophan. Frequencies are Taken from Raman and Resonance Raman Spectra.

Frequencies (cm^{-1})			Assignment
p-benzoquinone[a]			
^{16}O		^{18}O	
447 (vs)		434 (vs)	ring bending (υ_6)
483 (w)		476 (w)	C=O bending (υ_{11})
1147 (vs)		1147 (vs)	C—H bending (υ_4)
1616 (w)		1595 (s)	C=C stretching (υ_3)
1665 (vs)		1654 (vs)	C=C stretching (υ_2)
1684 (m)		1670 (m)	C=O stretching (υ_2) Fermi resonance
p-benzosemiquinone[b]			
$Q^{\cdot}$	$Q^{\dot{-}}$	$(HQ^{\cdot +})$	
1613 (vs)	1620 (vs)	1644 (vs)	C=C stretching (υ_{8a})
1511 (s)	1435 (m)	1466 (w)	C—O stretching (υ_{7a})
469 (m)	481 (m)	474 (w)	ring deformation (υ_{6a})
Tryptophan[c]			
N-H		N-D	
762 (s)		758 (s)	in-phase ring breathing
882 (s)		865 (w)	ring deformation (+) N-H(D) in-plane deformation
1016 (s)		1016 (s)	out-of-phase ring breathing
1344 (m)		1340 (m)	ring deformation (Fermi resonance)[d]
1363 (m,sh)		1355 (m)	
1436 (m)		1385 (s)	pyrrole ring stretching (+) N-H(D) in-plane deformation
1555 (s)		1553 (s)	benzenelike ring stretching

vs, very strong; s, strong; m, medium; w, weak.
[a]Frequencies and assignments from Ref. 14.
[b]From Ref. 19.
[c]From Ref. 20.
[d]From Ref. 21.

bending deformation and the totally symmetric C=C stretch (υ_6 and υ_3, respectively) (Table 1) are remarkably sensitive to ^{18}O substitution; this is believed to reflect a change in composition of the normal mode induced by isotopic substitution [14].

Semiquinone forms of *p*-benzoquinone have also been investigated in some detail [17–19]. Protonation of the oxygen atoms has little effect on the principal $(C—H)_{bending}$ and $(C=C)_{stretching}$ vibrations but does significantly perturb the $C—O)_{stretching}$ vibration. The number of bands is quite variable in the resonance Raman spectra of the semiquinones and the hydroquinone radical cation. Strong enhancement of the $(C—H)_{bending}$, $(C{=}C)_{stretching}$ and $(C—O)_{stretching}$ modes have been taken as evidence for a strong interaction between the OH p(π) electron pair and the phenoxy radical π-electron system [19].

For comparison to methylamine dehydrogenase, selected tryptophan vibrational data are also included in Table 1. Exchange of the pyrrole N-H by deuterium produces sizable shifts in bands associated with modes possessing substantial N-H deformation character [20]. The doublet at 1340 and 1360 cm^{-1} is characteristic of tryptophan and sensitive to its microenvironment [21].

B. Pyrroloquinoline Quinone (PQQ)

The Raman spectrum of PQQ in neutral aqueous solution is shown in Figure 4. Under these conditions PQQ is partially hydrated at the C-5 position [25]. The band at 1666 cm^{-1} may be assigned to the carbonyl stretching vibration of the C-4

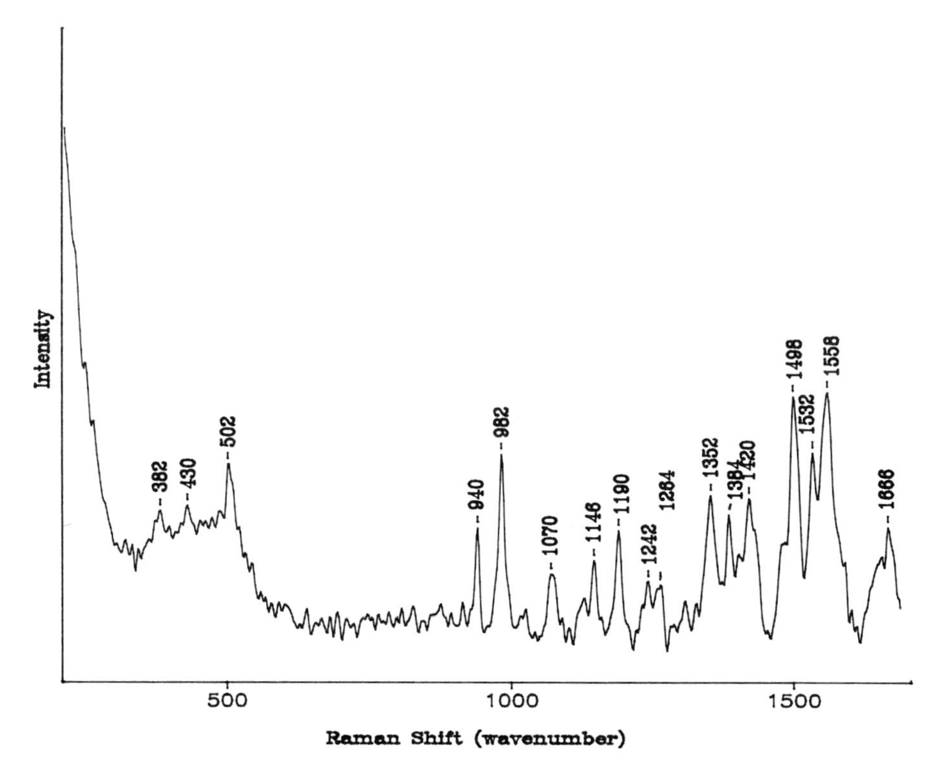

Figure 4 Raman spectrum of PQQ in phosphate buffer at pH 7.0. This spectrum was obtained at room temperature using 514.5-nm excitation with a resolution of 5 cm^{-1}.

carbonyl group, whereas the bands at 1558, 1532, and 1498 cm^{-1} might be related to the prominent pyridine bands at ~1600 cm^{-1} and in the 1400–1500 cm^{-1} region. A portion of the resonance Raman spectrum [26] (obtained with 350.7 nm excitation) of PQQ is displayed in Figure 5. Clearly, bands at ~1550, ~1415, and 977 cm^{-1} are enhanced relative to other PQQ modes. These modes then are probably related to the distortions in the excited state associated with a $\pi \rightarrow \pi^*$ transition of PQQ. The relative enhancements may be rationalized in part by reference to Tsuboi and Hirakawa's second rule (see Sec. I.A) since the pyridine bands in the 1400–1600 cm^{-1} region are related to the e_{1u} and e_{2g} modes of benzene, which lower its sixfold symmetry. However, the ~980 cm^{-1} band also appears to be resonance enhanced, and it might be related to the totally symmetric ring-breathing mode of pyridine at 992 cm^{-1}.

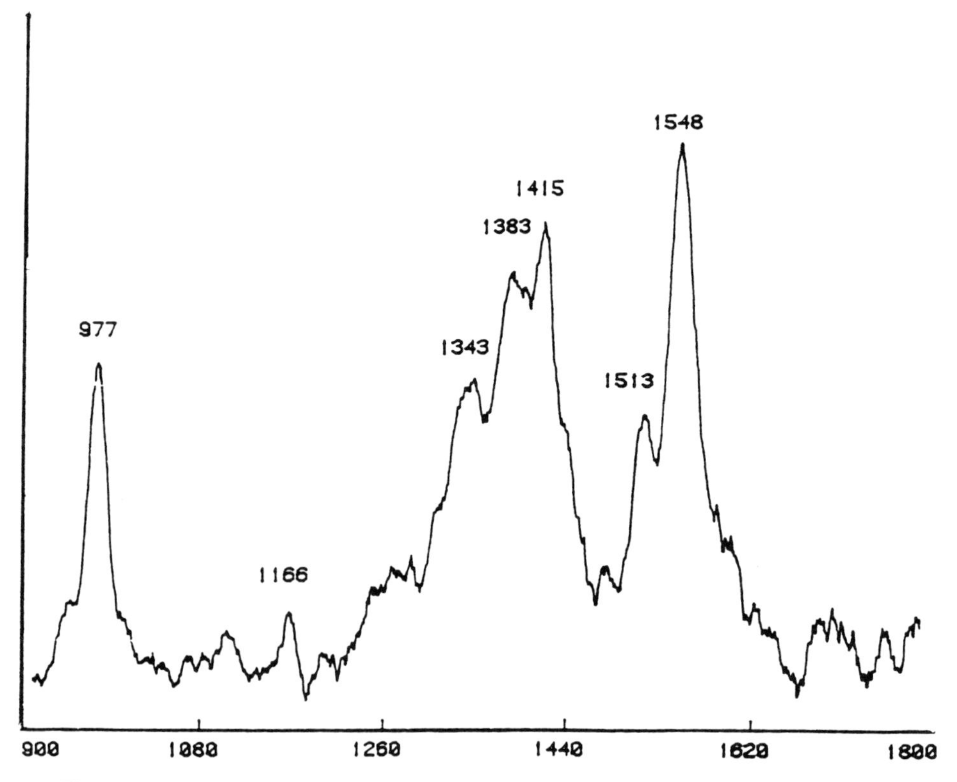

Figure 5 Resonance Raman spectrum of PQQ in Tris buffer at pH 7.5 plus 30% DMF. The exciting wavelength was 350.7 nm and the temperature was 90 K. (Figure courtesy of Dr. J. Sanders-Loehr.)

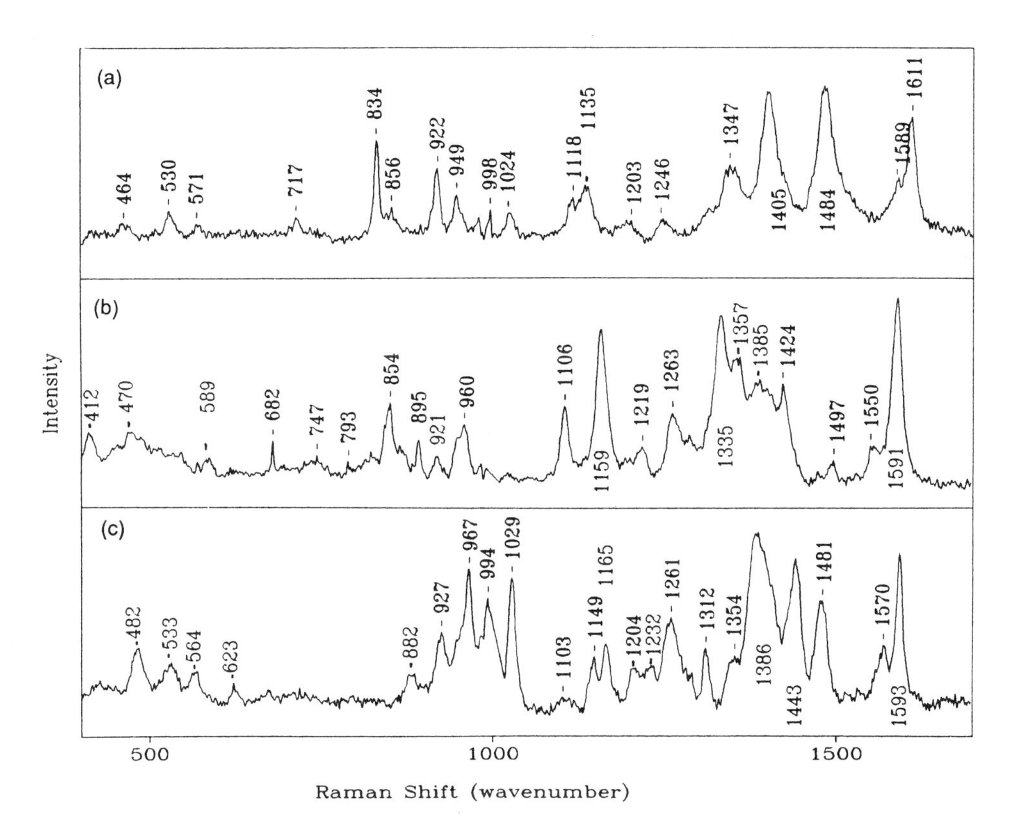

Figure 6 Resonance Raman spectra of three hydrazine derivatives of PQQ. All three spectra were obtained at room temperature and pH 7.0 using 457.9-nm excitation. (a) 2,4-dinitrophenylhydrazone; (b) *p*-nitrophenylhydrazone; (c) 2-hydrazinopyridine derivative.

Resonance Raman spectra of three hydrazine derivatives of PQQ are shown in Figure 6. Each spectrum was obtained on a 10–20 μl sample containing a few nanomoles of material. Additional experiments have shown that a wide variety of quinones and other carbonyl compounds may be distinguished via resonance Raman spectroscopy of their hydrazine derivatives, especially if two derivatives are examined (J. L. Bates, D. E. Brown, and D. M. Dooley, unpublished observations). Given the sensitivity and wide applicability of the technique, the identification (or discrimination) of quinonoid and other carbonyl compounds by resonance Raman spectroscopy of their hydrazine derivatives is an attractive analytical method.

III. RESONANCE RAMAN SPECTROSCOPY OF QUINOPROTEINS

Three different applications of resonance Raman spectroscopy to the study of quinoproteins have been developed so far. These are: (1) as an aid to the identification of the quinone "cofactor," (2) for comparisons of quinoproteins from different sources or organisms, (3) as a probe of the active-site structure of quinoproteins, and variations in the structure induced by dissociation, binding, or other reactions.

A. Amine Oxidases

This widespread and diverse class of quinoproteins has been thoroughly reviewed in Chapter 6 by McIntire and Hartmann. Following the initial reports of PQQ in amine oxidases in 1984 [27,28], it appeared to us that examination of amine oxidase-hydrazine derivatives by resonance Raman spectroscopy would be worthwhile for two important reasons. First, the vibrational spectrum would provide independent information on the identity of the carbonyl group. Second, both the intact enzyme and any carbonyl derivative released from the protein could be examined. This would allow a check of the integrity of the cofactor's structure following the isolation procedure.

The spectrum of the bovine plasma amine oxidase-2,4-dinitrophenylhydrazone is shown in Figure 7. As noted previously [29], the observed frequencies closely correspond to the frequencies of the PQQ-DNP (Fig. 6), although there are significant variations in the relative intensities between the two spectra. In contrast, the resonance Raman spectrum of the pyridoxal-DNP derivative is distinctly different [29]. On this basis we concluded that the cofactor in bovine plasma amine oxidase could not be a pyridoxal compound and was probably a PQQ derivative or a similar quinone. We subsequently discovered, by comparing the resonance Raman spectra of the 2-hydrazinopyridine derivatives (Fig. 8), that the similarity between the PQQ-DNP and amine oxidase-DNP derivatives was somewhat fortuitous. These results implied that, although the amine oxidase cofactor had quinonoid structure with some similarities with PQQ, it was not identical to PQQ. In 1990 Klinman and coworkers elegantly and convincingly demonstrated the quinone "cofactor" in bovine plasma amine oxidase was trihydroxyphenylalanine (topa) quinone [30]. The strategy of Klinman and coworkers was to isolate phenylhydrazine and *p*-nitrophenylhydrazine derivatives of the active-site carbonyl group in sufficient quantities for definitive structural characterization by NMR and mass spectroscopic techniques. The high yields obtained ruled out the possibility that the topa quinone–containing peptide was derived from a contaminant. Comparisons of the resonance Raman spectra of the phenylhydrazine and *p*-nitrophenylhydrazine derivatives of intact bovine plasma amine oxidase with the corresponding derivatives of the isolated peptides (Fig. 9)

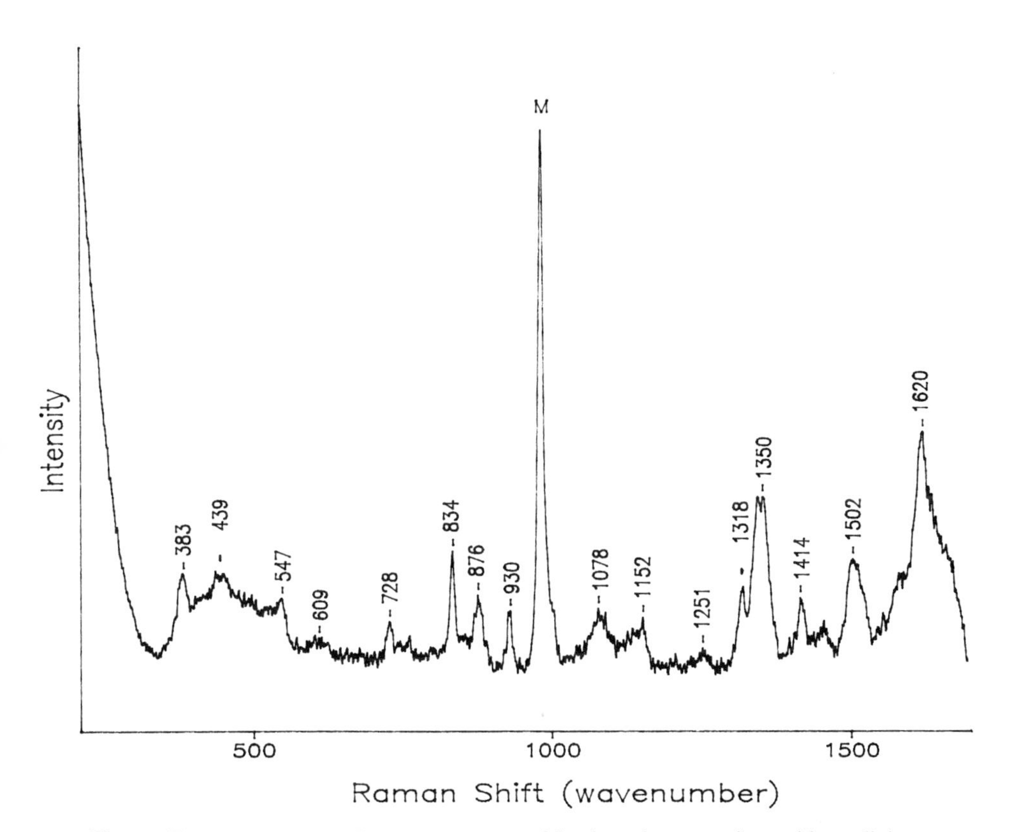

Figure 7 The resonance Raman spectrum of bovine plasma amine oxidase–dinitrophenylhydrazone. Experimental conditions as in Figure 6. The band marked (M) is from Na_2SO_4, added to the enzyme solution as a wavelength marker.

established that the topa quinone structure was not produced during the labeling and isolation of the peptides [31]. Figure 10 shows the resonance Raman spectra of the peptide derivatives compared to synthetic topa-hydantoin derivatives. The *identity* of the phenylhydrazone spectra, and the near identity of the *p*-nitrophenylhydrazone spectra, provide strong supporting evidence for the topa structure. In addition, the resonance Raman spectra of the *p*-nitrophenylhydrazones of PQQ and pyridoxal are sufficiently different from the spectra of the isolated peptides so that these compounds may be unequivocally ruled out as the cofactor in amine oxidases. In collaboration with Klinman and coworkers, we have recently obtained the resonance Raman spectra of several additional peptides isolated from a variety of amine oxidases, including the enzyme from the yeast *Hansenula polymorpha* (Fig. 11) [32]. Clearly topa quinone is the cofactor in this enzyme as well. Aligning

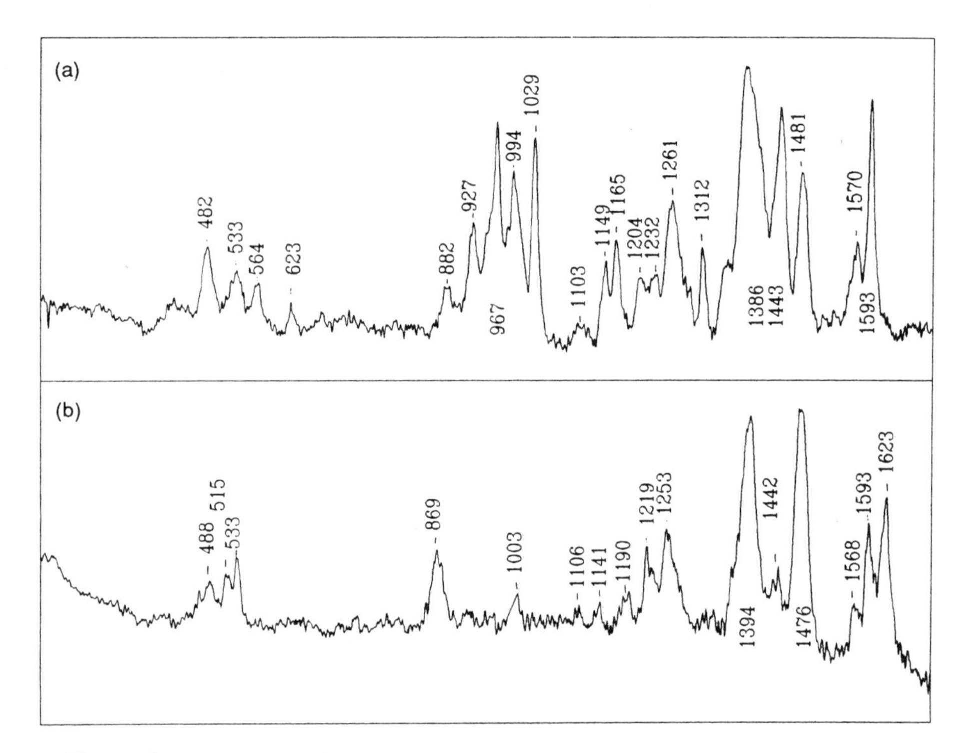

Figure 8 Comparison of the resonance Raman spectra of the 2-hydrazinopyridine derivatives of (a) PQQ and (b) *Arthrobacter* P1 methylamine oxidase. The spectra were measured under identical conditions (as in Figure 6).

the cDNA-derived protein sequence with that of the isolated topa-containing peptide demonstrates that the tyrosine codon (UAC) corresponds to topa in the native enzyme [32].

Phenylhydrazones of intact amine oxidases display closely similar resonance Raman spectra [31,33], whereas the spectra of the isolated active-site peptides are identical. A plausible interpretation is that the microenvironment around the active site varies somewhat from one enzyme to another, which is consistent with the variations in specificity and reactivity among amine oxidases. There is one notable exception among peptides isolated from amine oxidases—the lysyl amine oxidase phenylhydrazone. Although the lysyl oxidase spectrum has many similarities to the spectra of the other amine oxidases examined to date, there are also some noticeable differences [34]. This suggests that there might be some structural differences between the lysyl oxidase cofactor and topa quinone, or perhaps that some modification of the lysyl oxidase cofactor has occurred during its isolation

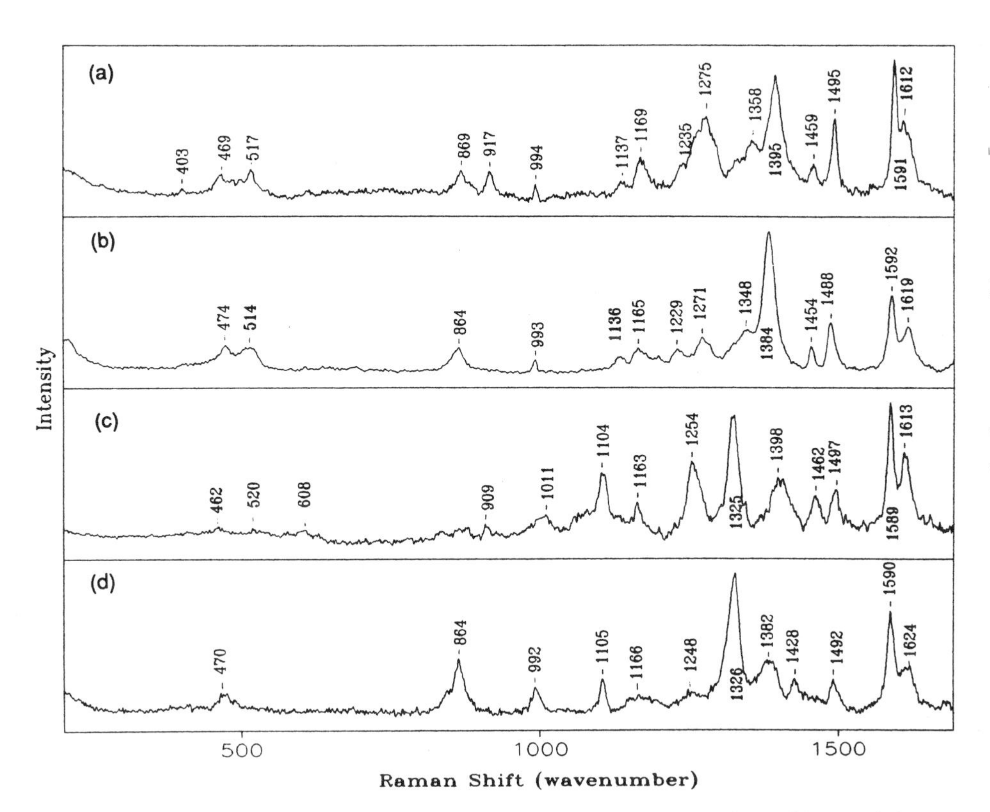

Figure 9 Resonance Raman spectra of labeled bovine plasma amine oxidase and the isolated active-site peptides. (a) The enzyme phenylhydrazone; (b) the phenylhydrazine-labeled peptide; (c) the enzyme *p*-nitrophenylhydrazone; (d) the *p*-nitrophenylhydrazine–labeled peptide. All spectra were collected at room temperature using 457.9-nm excitation. (Reproduced with permission from Ref. 31.)

and purification. In this context it should be noted that the c-DNA–derived sequence for rat lysyl oxidase is not homologous to the yeast sequence and apparently does not contain the conserved Asn-tyr(topa)-asp(glu)-tyr-val sequence [35,36]. The question of possible diversity in the structure of the quinone cofactor among amine oxidases deserves to be vigorously pursued.

Given the difficulty of obtaining resonance Raman spectra from native amine oxidases (owing to the relatively low extinction and high fluorescence), the various hydrazines are attractive labels. It might be possible to monitor active-site perturbations via resonance Raman spectroscopy of the enzyme derivatives. For example, the resonance Raman spectrum of the dinitrophenylhydrazone of copper-depleted bovine plasma amine oxidase displays some shifts and changes in

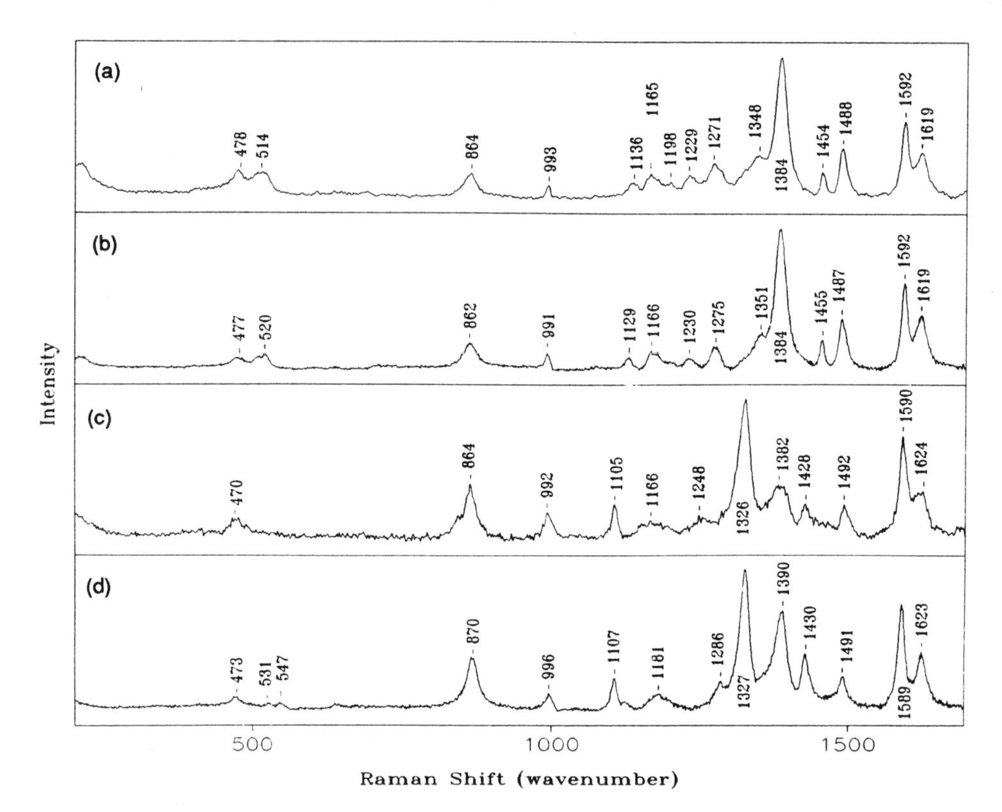

Figure 10 Comparison of the resonance Raman spectra of the labeled peptides from bovine plasma amine oxidase to the corresponding synthetic topa-hydantoin derivatives. (a) The peptide phenylhydrazone; (b) the phenylhydrazone of topa-hydantoin; (c) the peptide *p*-nitrophenylhydrazone; (d) the *p*-nitrophenylhydrazone of topa-hydantoin. Conditions as in Figure 9. (Reproduced with permission from Ref. 31.)

the relative intensities of certain bands compared to the dinitrophenylhydrazone of the native enzyme [37]. Comparable results were obtained when the effects of copper removal from *Arthrobacter* P1 methylamine oxidase were examined [38]. These perturbations may reflect copper-quinone interactions, either arising directly via weak coordination or mediated by the polypeptide chain.

B. Methylamine Dehydrogenases

The molecular properties of these enzymes are described in Chapter 5. The cofactor in methylamine dehydrogenase (MADH) was originally suspected to be a covalently bound derivative of PQQ [39]. A comparison of the 2-hydrazino-

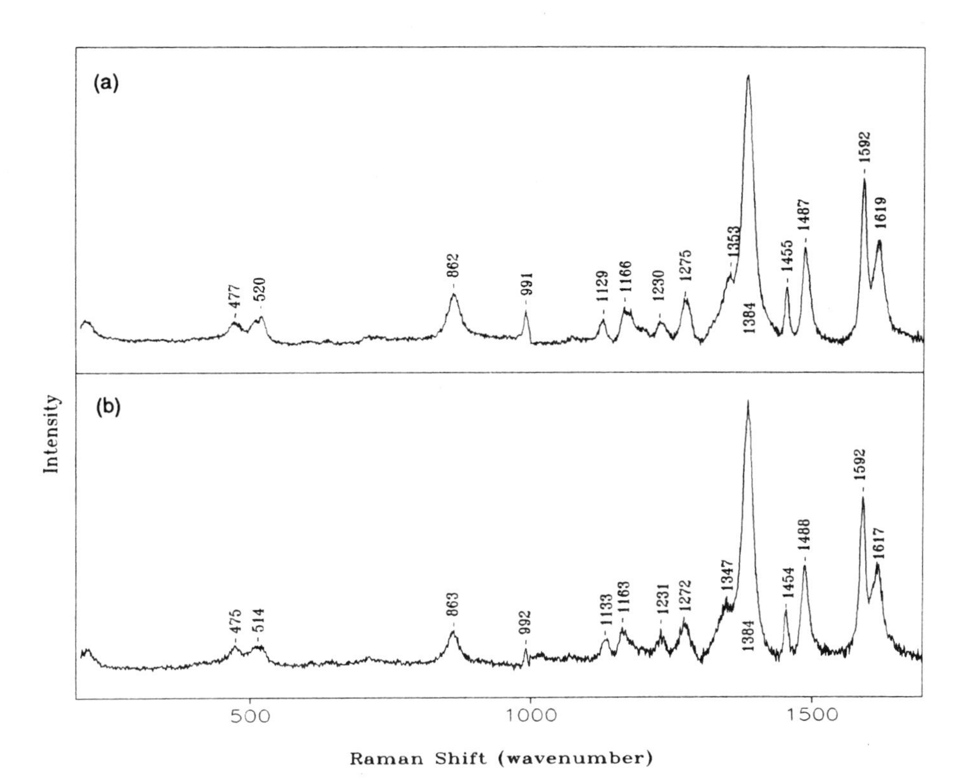

Figure 11 Comparison of the resonance Raman spectra of the labeled peptide from yeast (*Hansenula polymorpha*) amine oxidase to the corresponding synthetic topa-hydantoin derivative. (a) The phenylhydrazone of topa-hydantoin; (b) the peptide phenylhydrazone. Conditions as in Figure 9. (Reproduced with permission from Ref. 32.)

pyridine derivatives of W3A1 MADH, *Arthrobacter* methylamine oxidase, and PQQ is shown in Figure 12. On the basis of these and other resonance Raman data, we concluded that the quinone cofactor in MADH could not be unmodified PQQ or topa quinone [40]. Thus the resonance Raman spectra showed that three different kinds of quinoproteins could be identified. As already mentioned and reviewed elsewhere in this volume, amine oxidases are now recognized to contain topa quinone [30], and the MADH cofactor has been identified as a quinone derived from covalently linked tryptophans and designated TTQ [23].

Resonance Raman spectra of oxidized MADH from bacterium W3A1 and the isolated β subunit of this enzyme, which contains the TTQ moiety, are displayed in Figure 13. The spectra of the MADH enzymes from *Paracoccus denitrificans* (*Pd*) and *Thiobacillus versutus* (*Tv*) are sufficiently similar so that one can say with

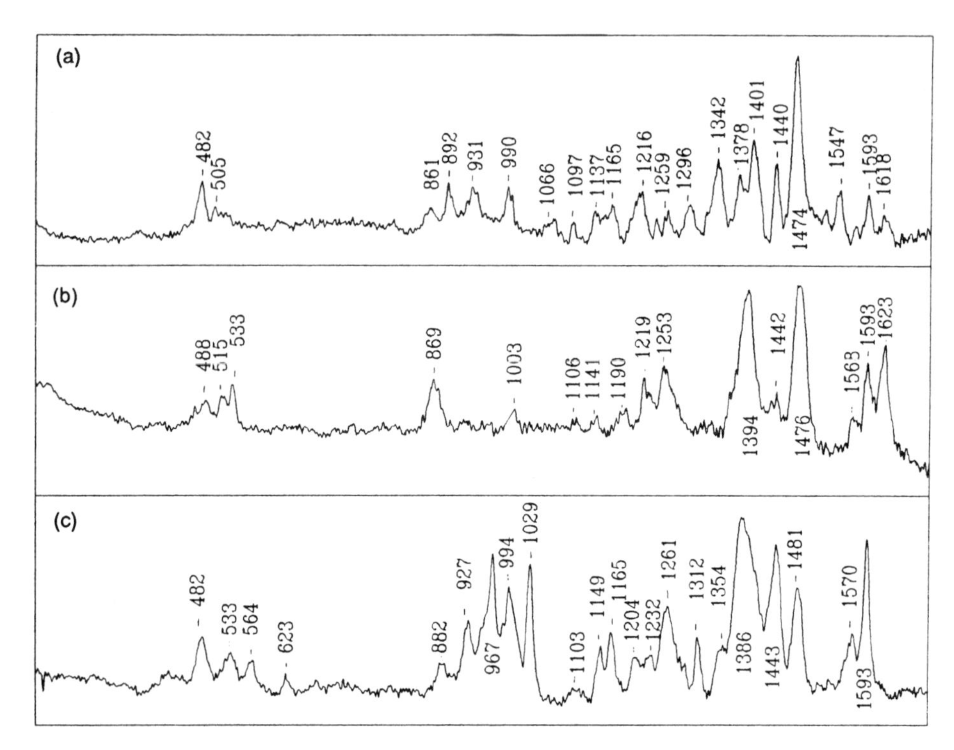

Figure 12 Resonance Raman spectra of 2-hydrazinopyridine derivatives at pH 7.0. Excitation wavelength was 457.9 nm. (a) Oxidized MADH from W3A1; (b) oxidized *Arthrobacter* P1 methylamine oxidase; (c) PQQ. (Part of this figure was reproduced with permission from Ref. 40.)

confidence that the cofactor is identical in all three enzymes [41]. In other words, the observed similarities among the Raman spectra are consistent with the presence of TTQ in all three, but with some variation in the local environment of the quinone. Enhancement profiles for selected resonance Raman peaks of *Pd* MADH are shown in Figure 14 [41]. Most of the vibrational modes are in resonance with the broad, asymmetric absorption band, characteristic of MADHs, which peaks at 440 nm. This band is probably composed of overlapping electronic transitions including a $\pi \rightarrow \pi^*$ transition of the conjugated α,β dicarbonyl moiety of TTQ. Note that the 1452 cm^{-1} band is enhanced at longer wavelengths but is maximally enhanced at shorter wavelengths, compared to the other modes. This indicates, as expected from its breadth and asymmetry, that multiple electronic transitions are represented in the broad 440-nm band of MADH [41].

Significant differences are apparent between the intact W3A1 enzyme and its

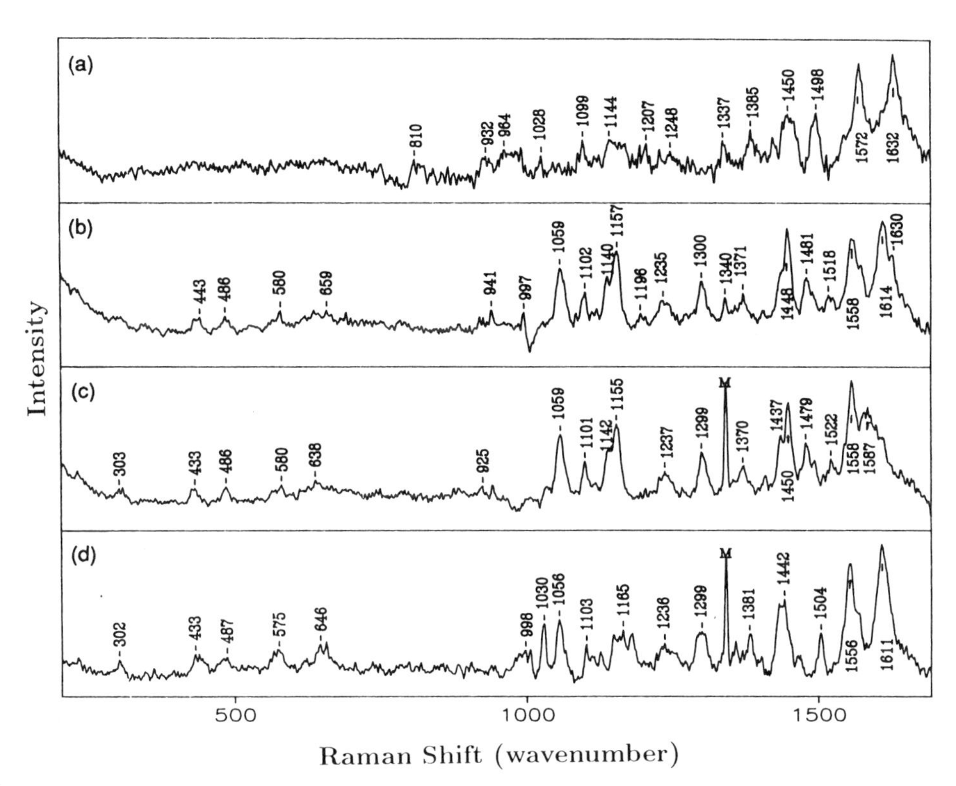

Figure 13 Resonance Raman spectra of underivatized W3A1 MADH measured with 457.9-nm excitation. (a) Isolated β subunit; (b) native oxidized MADH; (c) native oxidized MADH following exchange with ^{18}O; (d) oxidized MADH in deuterated 10 mM potassium phosphate buffer, pD 7.1. The band marked M is a wavelength marker from a second laser. (Reproduced with permission from Ref. 40.)

isolated β subunit (see Fig. 13) [40]. For example, some intense peaks in the native enzyme spectrum shift to higher frequency following subunit dissociation (1481 → 1498 cm^{-1}; 1558 → 1572 cm^{-1}; 1614 → 1632 cm^{-1}). Two prominent peaks of the native enzyme, at 1059 and 1300 cm^{-1}, are not evident in the spectrum of the β subunit. Crystal structures of the *Tv* and *Pd* enzymes place the TTQ cofactor at the αβ subunit interface in the tetrameric $\alpha_2\beta_2$ protein structure [42,43], it is likely that the TTQ moiety in MADH from W3A1 is also located at the subunit interface. This is consistent with the resonance Raman spectra, which suggest that the microenvironment of TTQ is different in the intact enzyme as compared to the isolated β subunit. The resonance Raman spectra of the W3A1 MADH-phenylhydrazone is also altered significantly by dissociation of the subunits [40].

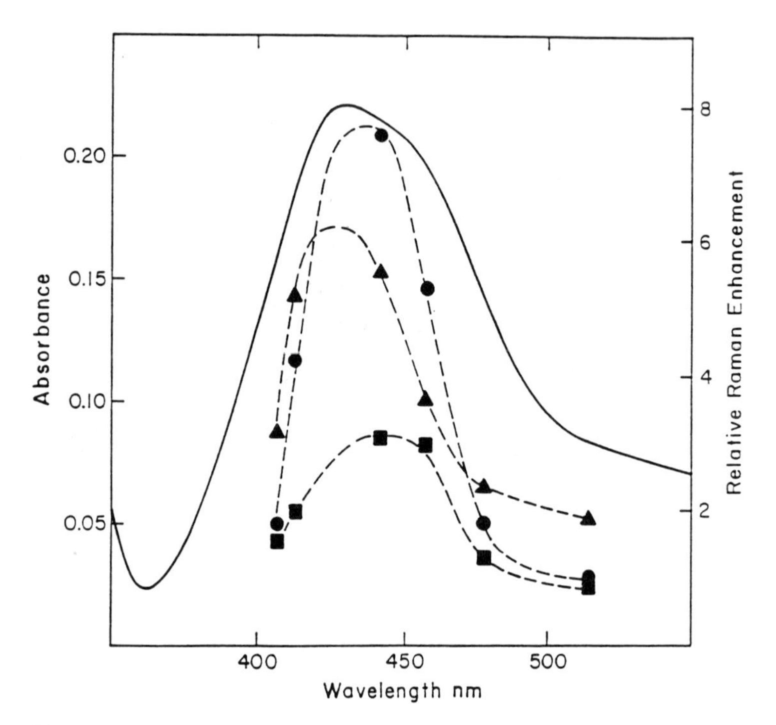

Figure 14 Absorption spectrum of oxidized *Pd* MADH (—) and enhancement profiles (---) for resonance Raman peaks at 1064 (■), 1452 (▲), and 1619 (●) cm^{-1}. The Raman enhancement is calculated from the area of the sample peak relative to the area of the 983 cm^{-1} peak from a sulfate standard. Enhancement profiles represent the best fit to the experimental points from available laser lines. (Reproduced with permission from Ref. 41, which may be consulted for additional details.)

Effects of $^{18}O/^{16}O$ exchange and D/H exchange on the resonance Raman spectrum of W3A1 MADH are also shown in Figure 13; for comparison the results of $^{18}O/^{16}O$ exchange on the *Pd* MADH spectrum are shown in Figure 15. Based on the magnitude of the ^{18}O-induced shift, the bands at 1614 cm^{-1} (W3A1, -27 cm^{-1} shift), 1625 cm^{-1} (*Pd*, -16 cm^{-1} shift), and 1628 cm^{-1} (*Tv*, -22 cm^{-1} shift) may be assigned to the $(C{=}O)_{\text{in phase}}$ stretching mode [40,41]. Each MADH spectrum contains a band or shoulder to higher frequency that also shifts to lower frequency upon ^{18}O substitution: 1630 (-20) cm^{-1}, 1648 (-6) cm^{-1}, and 1634 (-12) cm^{-1}, for W3A1, *Pd*, and *Tv*, respectively. This behavior is remarkably analogous to that observed for *p*-benzoquinone [13–15] and is thus also attributable to Fermi resonance between the $(C{=}O)_{\text{in phase}}$ fundamental and a combination or overtone band. Only one other band in the resonance Raman spectra of MADHs from W3A1 and *Pd* shifted significantly following ^{18}O substitution: W3A1, 443 (-10)

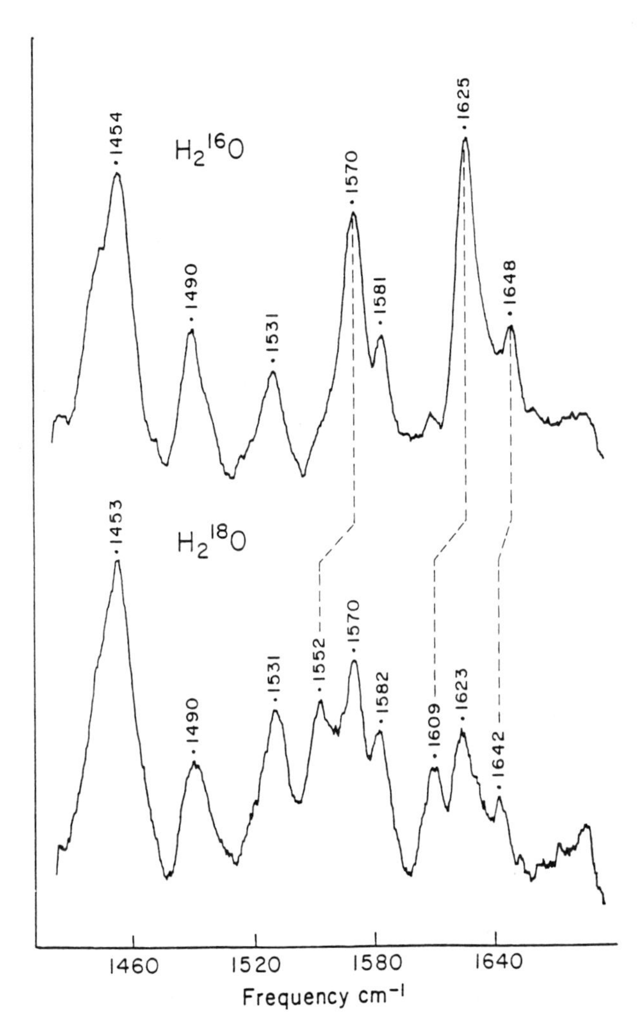

Figure 15 Resonance Raman spectra of *Pd* MADH at 15 K equilibrated with $H_2{}^{16}O$ (top) and $H_2{}^{18}O$ (bottom). Excitation wavelength was 457.9 nm. (Reproduced with permission from Ref. 41.)

cm^{-1}; *Pd*, 486 (−5) cm^{-1}. Again using *p*-benzoquinone as a model (Table 1), these bands may be assigned as the $(C{=}O)_{bending}$ vibration [14]. Ring deformation modes (primarily ring bending) may fall in the 400–500 cm^{-1} region as well [13–15,20], and these could be sensitive to $^{18}O/^{16}O$ exchange, but evidently not in TTQ. Corresponding bands were apparently not resolved in the resonance Raman spectrum of *Tv* MADH [41].

It is evident from Figure 13 that D/H exchange perturbs several frequencies in the resonance Raman spectrum of W3A1 MADH. In marked contrast, D/H exchange was reported to have no effect on the resonance Raman spectra of *Pd* or *Tv* MADH [41]. The Raman spectrum of free tryptophan is sensitive to deuterium substitution at the pyrrole nitrogen (see examples in Table 1). Generally, deuterium substitution produces shifts to lower frequency as observed for the 1518 → 1504 cm^{-1}, 1481 → 1472 cm^{-1}, and 1448 → 1438 cm^{-1} bands of W3A1 MADH. In addition, the 1614 cm^{-1} $(C{=}O)_{in\ phase}$ vibration shifts to 1611 cm^{-1}. The loss of the resolved feature at 1630 cm^{-1} suggests that the Fermi resonance splitting is also altered. Such effects are attributable to a perturbation of hydrogen bonding to the carbonyl group, which may also account for the 443 → 437 cm^{-1} shift. Some bands apparently shift to higher frequency (e.g., 1371 → 1381 cm^{-1}) following D/H exchange, which might reflect the appearance of new overtone or combination bands or additional Fermi resonance coupling. Since incubation in D_2O-containing buffers does not alter the resonance Raman spectra of *Pd* or *Tv* MADH, their TTQ pyrrole NH groups must be inaccessible to solvent. Another indication of quinone inaccessibility to solvent is lack of a deuterium-induced shift in the carbonyl stretching or bending modes, even though the crystal structures of *Pd* and *Tv* MADH indicate that the C-7 carbonyl is hydrogen bonded to an amide NH of the polypeptide chain. Sanders-Loehr and coworkers have suggested that the lower frequency of the $(C{=}O)_{stretching}$ vibration in MADH compared to free quinones reflects hydrogen bonding to the carbonyls [41]. Consequently, MADHs are divisible into at least two classes, with W3A1 MADH in one class and *Tv* and *Pd* MADHs in the other, which can be distinguished on the basis of the accessibility of the TTQ cofactor. These differences in the local structure around the TTQ site might be related to the differences in chemical reactivity towards ammonia and hydroxide previously observed for the two MADH classes [41].

Figure 16 compares the resonance Raman spectra of oxidized *Pd* MADH, the semiquinone form, and the hydroxide and ammonia adducts. Sanders-Loehr and coworkers have ascribed the differences between the spectrum of oxidized MADH and the other forms to the loss of conjugation that occurs upon reduction or addition of hydroxide or ammonia [41]. They suggest that the spectra in Figure 16 are consistent with a similar degree of conjugation in the semiquinone and covalent adducts and propose the structures shown in Figure 17 for these species. Plausible candidates for the $(C{=}C)_{stretching}$ vibration are the bands located at 1530–1580 cm^{-1}, which are present in all MADH resonance Raman spectra. If this is the correct assignment for the strong features at ~1530 cm^{-1} in the spectra of the semiquinone and the adducts, then it suggests that the degree of conjugation in

---

Figure 16 *Pd* MADH resonance Raman spectra at 15 K. (a) Oxidized MADH at pH 7.5; (b) semiquinone form of MADH at pH 7.5; (c) MADH at pH 9.0; (d) MADH plus 0.4 M NH_4Cl (pH 7.6). Arrows indicate spectral features that differ from the quinone. (Reproduced with permission from Ref. 41.)

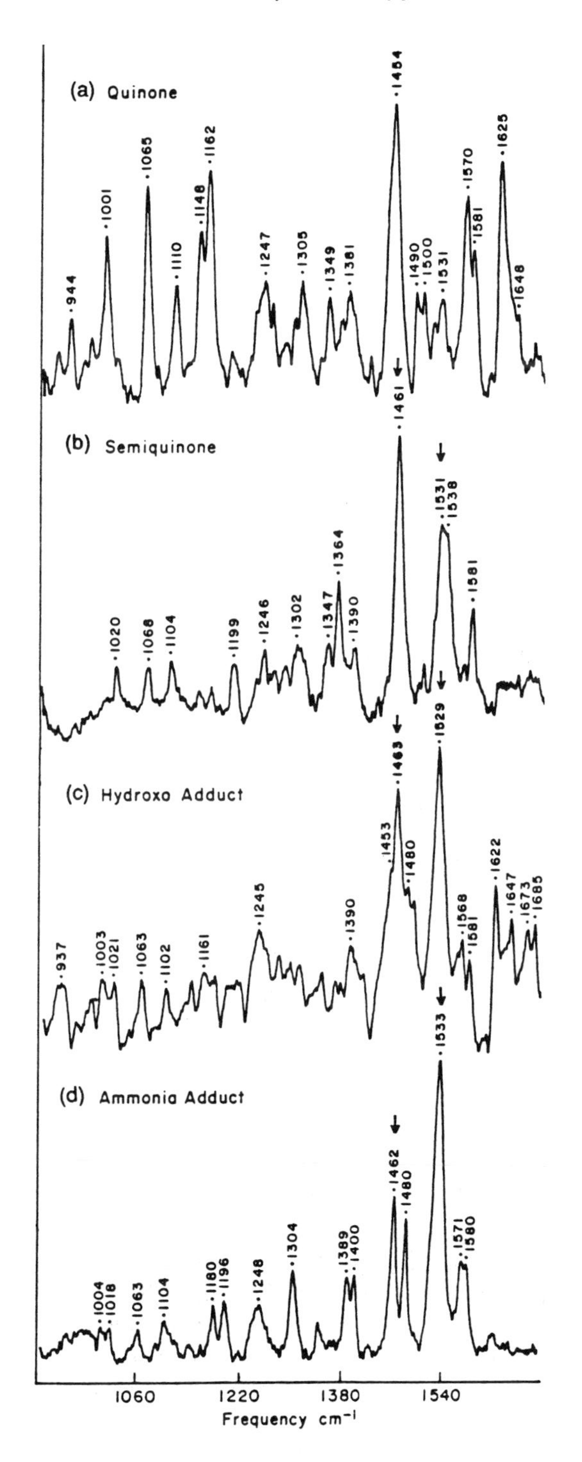
(a) Quinone
944
1001
1065
1110
1148
1162
1247
1305
1349
1381
1454
1490
1500
1531
1570
1581
1625
1648
(b) Semiquinone
1020
1068
1104
1199
1246
1302
1347
1364
1390
1461
1531
1538
1581
(c) Hydroxo Adduct
937
1003
1021
1063
1102
1161
1245
1390
1453
1463
1480
1529
1568
1581
1622
1647
1673
1685
(d) Ammonia Adduct
1004
1018
1063
1104
1180
1196
1248
1304
1389
1400
1462
1480
1533
1571
1580
1060
1220
1380
1540
Frequency cm⁻¹

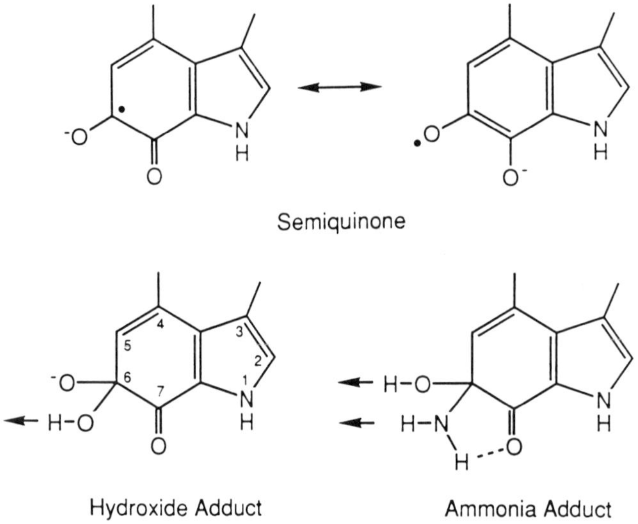

Semiquinone

Hydroxide Adduct

Ammonia Adduct

Figure 17 Structures proposed by Sanders-Loehr and coworkers for the semiquinone (only two of many possible resonance forms shown) and the hydroxide and ammonia adducts of the TTQ cofactor in *Pd* MADH. Arrows depict potential H-bonding interactions. (Reproduced with permission from Ref. 41.)

these species is similar. Note that the $(C{=}C)_{stretching}$ vibration (υ_{8a}) (Table 1) is one of the strongest features in the resonance Raman spectrum of *p*-benzosemiquinone. The absorption spectrum of the W3A1 MADH-ammonia complex is considerably different from that of the ammonia adducts of the other MADHs [44], ammonia activates the W3A1 enzyme [45], but not *Tv* or *Pd* MADH [41]. Unfortunately, the W3A1 MADH-ammonia complex was unstable in the laser beam, rapidly and irreversibly reverting to the oxidized, uncomplexed, form [40]. The structural bases for all the variations in TTQ reactivity among MADHs mentioned above has yet to be elucidated.

C. PQQ-Containing Dehydrogenases

To our knowledge the only PQQ-containing enzymes that have been investigated by resonance Raman spectroscopy are glucose dehydrogenase [26] and methanol dehydrogenase (G. Backes and J. Sanders-Loehr, unpublished results). Figure 18 compares the resonance Raman spectra of PQQ and glucose dehydrogenase from *Acinetobacter calcoaceticus*. The overall similarity of the spectra is evident. Curiously, no band readily attributable to a C=O stretching vibration is observed in either spectrum, perhaps suggesting that there is no distortion in the C=O modes in the excited state produced by 350.7-nm excitation. The resonance

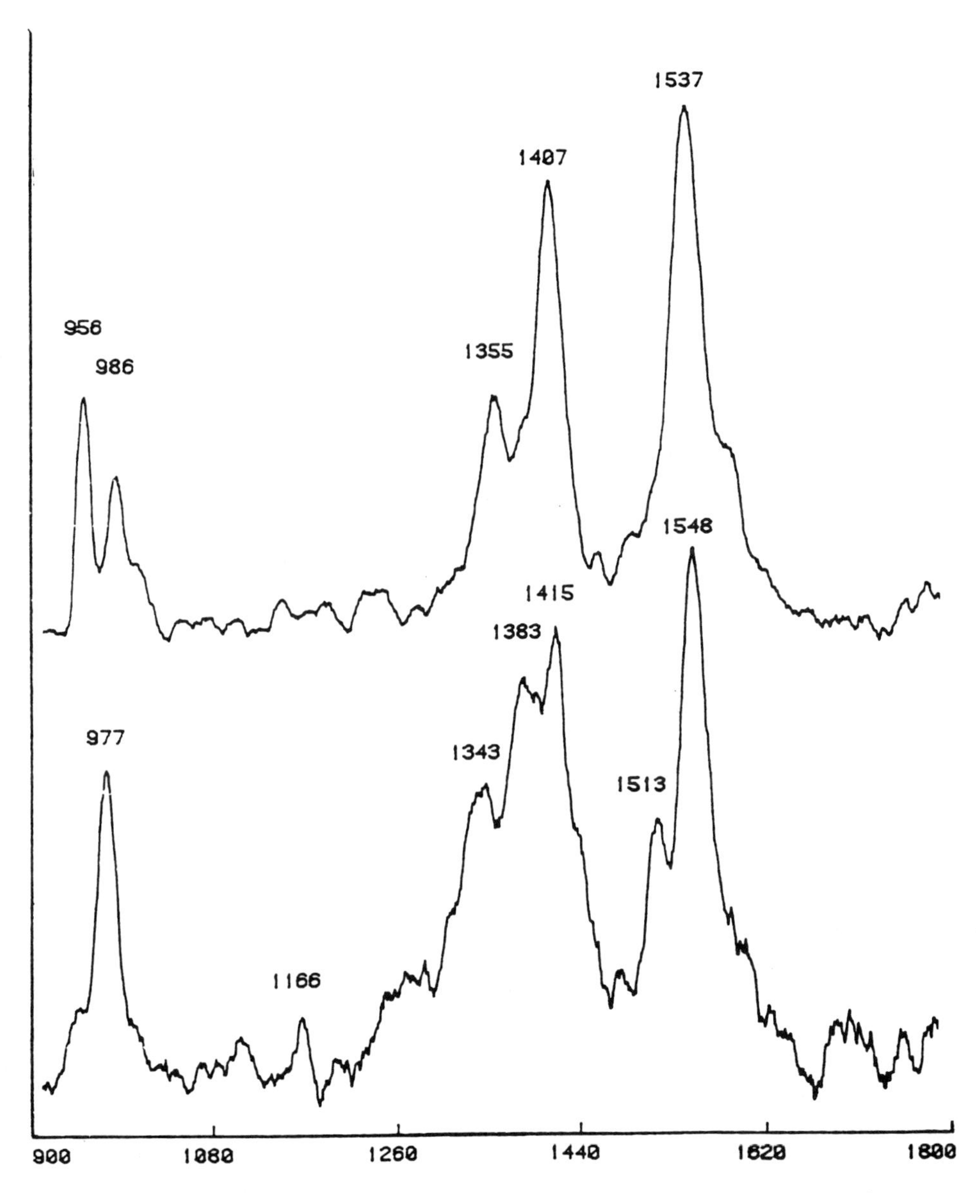

Figure 18 Comparison of the resonance Raman spectra of glucose dehydrogenase from *Acinetobacter calcoaceticus* (upper) at 15 K and PQQ (lower) at 90 K. Both spectra were obtained with 350.7 nm excitation. (Figure courtesy of Dr. J. Sanders-Loehr.)

Raman spectrum of methanol dehydrogenase is distinctly different from the PQQ spectrum and is consistent with a substantial fraction of the semiquinone form in the "as-isolated" enzyme.

IV. CONCLUDING REMARKS

In many respects resonance Raman spectroscopy has already proven to be a valuable technique in the study of quinones and quinoproteins. In other respects the capabilities of the technique have yet to be fully exploited. Refinements in Raman sampling techniques and improvements in the actual preparation of the samples could further increase the analytical sensitivity. Consequently, resonance Raman spectroscopy could become a very powerful method for identifying low levels of quinonoid compounds in biological samples. We anticipate considerable growth in the use of resonance Raman spectroscopy to probe the structure and function of quinoproteins. This growth will come from an increase in the number of quinoproteins investigated and from the desire to probe the active-site structure and reactivity of previously characterized proteins in greater detail and with greater sophistication.

ACKNOWLEDGMENTS

Our research on the resonance Raman spectroscopy of quinoproteins is supported by NIH Grant GM 27659. We are very grateful for numerous helpful discussions with Professor Joann Sanders-Loehr and for her generosity in sharing her data with us in advance of publication.

REFERENCES

1. Spiro, T. G., ed. (1987). *Biological Applications of Resonance Raman Spectroscopy*, Vols. 1–3, Wiley, New York.
2. Carey, P. R. (1982). *Biochemical Applications of Raman and Resonance Raman Spectroscopies*, Academic Press, New York.
3. Carey, P. R. (1988). Raman and resonance Raman spectroscopy. In *Modern Physical Methods in Biochemistry Part B* (A. Neuberger and L. L. M. Van Deenen, eds.), Elsevier, Amsterdam, p. 27.
4. Raman, C. V., and Krishnan, K. S. (1928). A new type of secondary radiation. *Nature 121*: 501.
5. Johnson, B. B., and Peticolas, W. L. (1976). The resonant Raman effect. *Ann. Rev. Phys. Chem. 27*: 465.
6. Strommen, D. P., and Nakamoto, K. (1984). *Laboratory Raman Spectroscopy*, Wiley, New York.
7. Heller, E. J. (1981). The semiclassical way to molecular spectroscopy. *Acc. Chem. Res. 14*: 368.

8. Heller, E. J., Sundberg, R. L., and Tannor, D. (1982). Simple aspects of Raman scattering. *J. Phys. Chem. 86*: 1822.
9. Lee, S.-Y, and Heller, E. J. (1979). Time-dependent theory of Raman scattering. *J. Chem. Phys. 71*: 4777.
10. Hishimura, Y., Hirakawa, A. Y., and Tsuboi, M. (1978). Resonance Raman spectroscopy of nucleic acids. In *Advances in Infrared and Raman Spectroscopy*, Vol. 5 (R. J. H. Clark and R. E. Hester, eds.), Heyden, London, p. 217.
11. Gardiner, D. J., and Graves, P. R., eds. (1989). *Practical Raman Spectroscopy*, Springer-Verlag, Berlin.
12. Berger, S., and Rieker, A. (1974). Identification and determination of quinones. In *The Chemistry of Quinonoid Compounds* (S. Patai, ed.), Wiley, New York, p. 163.
13. Becker, E. D., Charney, E., and Anno, T. (1964). Molecular vibrations of quinones. VI. A vibrational assignment for p-benzoquinone and six isotopic derivatives. Thermodynamic functions of p-benzoquinone. *J. Chem. Phys. 42*: 942.
14. Becker, E. D. (1991). Raman spectra of isotopic derivatives of p-benzoquinone: Revised vibrational assignments. *J. Phys. Chem. 95*: 2818.
15. Stammreich, H., and Sans, T. T. (1964). Molecular vibrations of quinones. IV. Raman spectra of p-benzoquinone and its centrosymmetrically substituted isotopic derivatives and assignment of observed frequencies. *J. Chem. Phys. 42*: 920.
16. Palmo, K., Pietila, L.-O., Mannfors, B., Karonen, A., and Stenman, F. (1983). Raman scattering from p-benzoquinone. *J. Molec. Spectros. 100*: 368.
17. Tripathi, G. N. R. (1981). Resonance Raman scattering of semiquinone radical anions. *J. Chem. Phys. 74*: 6044.
18. Hester, R. E., and Williams, K. P. J. (1982). Resonance Raman studies of p-benzosemiquinone and p,p′-biphenylsemiquinone free-radical anions. *J. Chem. Soc. Faraday Trans. 2 78*: 573.
19. Tripathi, G. N. R., and Schuler, R. H. (1987). Resonance Raman spectra of p-benzosemiquinone radical and hydroquinone radical cation. *J. Phys. Chem. 91*: 5881.
20. Hirakawa, A. Y., Nishimura, Y., Matsumoto, T., Nakanishi, M., and Tsuboi, M. (1978). Characterization of a few Raman lines of tryptophan. *J. Raman. Spectros. 7*: 282.
21. Harada I., Miura, T., and Takeuchi, H. (1986). Origin of the doublet at 1360 and 1340 cm^{-1} in the Raman spectra of tryptophan and related compounds. *Spectrochim. Acta 42A*: 307.
22. Hudson, B. S., and Mayne, L. C. (1987). Peptides and protein side chains. In *Biological Applications of Raman Spectroscopy*, Vol. 2 (T. G. Spiro, ed.), Wiley, New York, pp. 181–209.
23. McIntire, W. S., Wemmer, D. E., Christoserdov, A., and Lidstrom, M. E. (1991). A new cofactor in a prokaryotic enzyme: Tryptophan tryptophanylquinone as the redox prosthetic group in methylamine dehydrogenase. *Science 252*: 817.
24. Herzberg, G. (1945). *Infrared and Raman Spectra*, Van Nostrand Reinhold, New York, pp. 215–220.
25. Dekker, R. H., Duine, J. A., Frank, jzn, J., Verwiel, P. E. J., and Westerling, J. (1982). Covalent addition of H_2O, enzyme substrates and activators to pyrroloquinoline quinone, the coenzyme of quinoproteins. *Eur. J. Biochem. 125*: 69.

26. Backes, G., Sanders-Loehr, J., and Duine, J. A. (1991). (Personal communication)
27. Lobenstein-Verbeek, C. L., Jongejan, J. A., Frank, J., and Duine, J. A. (1984). Bovine serum amine oxidase: A mammalian enzyme having covalently bound PQQ as prosthetic group. *FEBS Lett. 170*: 305.
28. Ameyama, M., Hayashi, M., Matsushita, K., Shinagawa, E., and Adachi, O. (1984). Microbial production of pyrroloquinoline quinone. *Agric. Biol. Chem. 48*: 561.
29. Moog, R. S., McGuirl, M. A., Coté, C. E., and Dooley, D. M. (1986). Evidence for methoxatin (pyrroloquinoline quinone) as the cofactor in bovine plasma amine oxidase from resonance Raman spectroscopy. *Proc. Natl. Acad. Sci. USA 83*: 8435.
30. Janes, S. M., Mu, D., Wemmer, D., Smith, A. J., Kaur, S., Maltby, D., Burlingame, A. L., and Klinman, J. P. (1990). A new redox cofactor in eukaryotic enzymes: 6-hydroxydopa at the active site of bovine serum amine oxidase. *Science 248*: 981.
31. Brown, D. E., McGuirl, M. A., Dooley, D. M., Janes, S. M., Mu, D., and Klinman, J. P. (1991). The organic functional group in copper-containing amine oxidases. Resonance Raman spectra are consistent with the presence of topa quinone (6-hydroxydopa quinone) in the active site. *J. Biol. Chem. 266*: 4049.
32. Mu, D., Janes, S. M., Smith, A. J., Brown, D. E., Dooley, D. M., and Klinman, J. P. (1992). The tyrosine condon corresponds at 6-hydroxydopa at the active site of copper amine oxidases. *J. Biol. Chem. 267*:7979.
33. Knowles, P. F., Pandeya, K. B., Rius, F. X., Spencer, C. M., Moog, R. S., McGuirl, M. A., and Dooley, D. M. (1987). The organic cofactor in plasma amine oxidase: Evidence for pyrroloquinoline quinone and against pyridoxal phosphate. *Biochem. J. 241*: 603.
34. Williamson, P. R., Moog, R. S., Dooley, D. M., and Kagan, H. M. (1986). Evidence for pyrroloquinoline quinone as the carbonyl cofactor in lysyl oxidase by absorption and resonance Raman spectroscopy. *J. Biol. Chem. 261*: 16302.
35. Trackman, P. C., Pratt, A. M. Wolanski, A., Tang, S.-S., Offner, G. D., Troxler, R. F., and Kagan, H. M. (1990). Cloning of rat aorta lysyl oxidase cDNA: Complete codons and predicted amino acid sequence. *Biochemistry 29*: 4863.
36. Trackman, P. C., Pratt, A. M. Wolanski, A., Tang, S.-S., Offner, G. D., Troxler, R. F., and Kagan, H. M. (1990). Cloning of rat aorta lysyl oxidase cDNA: Complete codons and predicted amino acid sequence. *Biochemistry 30*: 8282. [Erratum to ref. 35].
37. Dooley, D. M., Coté, C. E., McGuirl, M. A., Bates, J. L., Perkins, J. B., Moog, R. S., Singh, I., Knowles, P. F., and McIntire, W. S. (1989). Copper-PQQ interactions in amine oxidases. *PQQ and Quinoproteins* (J. A. Jongejan and J. A. Duine, eds.), Kluwer, Dordrecht, p. 307.
38. Dooley, D. M., McIntire, W. S., McGuirl, M. A., Coté, C. E., and Bates, J. L. (1990). Characterization of the active site of *Arthrobacter* P1 methylamine oxidase: evidence for copper-quinone interactions. *J. Am. Chem. Soc. 112*: 2782.
39. McIntire, W. S., and Stults, J. T. (1986). On the structure and linkage of the covalent cofactor of methylamine dehydrogenase from methylotrophic bacterium W3A1. *Biochem. Biophys. Res. Commun. 141*: 562.
40. McIntire, W. S., Bates, J. L., Brown, D. E., and Dooley, D. M. (1991). Resonance Raman spectroscopy of methylamine dehydrogenase from bacterium W3A1. *Biochemistry 30*: 125.

41. Backes, G., Davidson, V. L., Huitema, F., Duine, J. A., and Sanders-Loehr, J. (1991). Characterization of the tryptophan-derived cofactor of methylamine dehydrogenase by resonance Raman spectroscopy. *Biochemistry 30*: 9201.
42. Vellieux, F. M. D., Huitema, F., Groendijk, H., Kalk, K. H., Frank, jzn, J., Jongejan, J. A., Duine, J. A., Petratos, K., Drenth, J., and Hol, W. G. J. (1989). Structure of quinoprotein methylamine dehydrogenase at 2.25 Å resolution. *EMBO J. 8*: 2171.
43. Chen, L., Mathews, F. S., Davidson, V. L., Huizinga, E., Vellieux, F. M. D., Duine, J. A., and Hol, W. G. J. (1991). Crystallographic investigations of the tryptophan-derived cofactor in the quinoprotein methylamine dehydrogenase. *FEBS Lett. 287*: 163.
44. Kenney, W. C., and McIntire, W. (1983). Characterization of the methylamine dehydrogenase from bacterium W3A1. Interaction with reductants and amino-containing compounds. *Biochemistry 22*: 3858.
45. McIntire, W. S. (1987). Steady-state kinetic analysis for the reaction of ammonium and alkylammonium ions with methylamine dehydrogenase from bacterium W3A1. *J. Biol. Chem. 262*: 11012.

IV

Analysis and Properties of PQQ and Quinonoid Cofactors

12

The Chemistry of PQQ and Related Compounds

Yoshiki Ohshiro and Shinobu Itoh

Osaka University, Osaka, Japan

I. INTRODUCTION

Since pyrroloquinoline quinone (PQQ) has been shown to be involved in various biologically important processes [1], much attention has been focused on the chemistry of PQQ and its related compounds. In particular, the redox reactions are most attractive, since PQQ exists as a coenzyme in several important oxidoreductases [1]. Moreover, the pharmaceutical activity of PQQ is now partially attributed to its facile electron transfer ability [2,3]. The adduct formation reactions are also intriguing, not only from the viewpoint of organic chemistry but also in connection with development of the detection methods of PQQ in living systems [4]. The high reactivity of PQQ towards nucleophilic substances makes it difficult to detect the exact levels of PQQ in biological tissues and fluids.

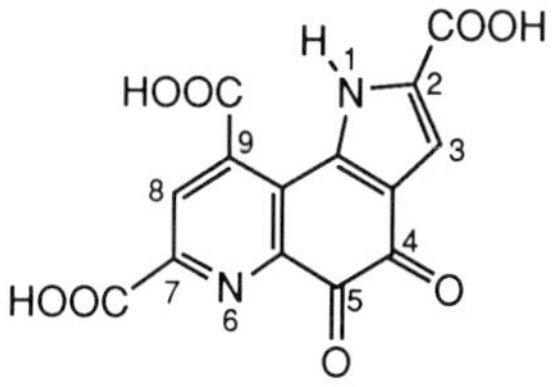

Coenzyme PQQ

PQQ has a unique heterocyclic *o*-quinone structure. The *o*-quinone ring, the active site, is condensed with a pyridine nucleus, which has an electron-withdrawing nature, and also with a pyrrole nucleus, which, on the other hand, has an electron-donating nature. Therefore the electronic nature of the *o*-quinone moiety of PQQ is considered to be relatively different from that of other simple *o*-quinone compounds. Furthermore, the carboxyl groups at the 2-, 7-, and 9-positions must play very important roles not only for binding to the active site of quinoproteins but also for the reactivity of PQQ itself.

In this chapter, we will survey the chemistry of PQQ and its derivatives from the viewpoints of synthetic methods, physical and chemical properties, and redox reactions. The physical and chemical characteristics of PQQ have been summarized elsewhere [5,6].

II. SYNTHESIS OF PQQ AND ITS DERIVATIVES

So far, total synthesis of PQQ has been accomplished in five different ways [7–14]. The outline of each synthetic procedure has been summarized in the literature [15]. Among them, the Corey method [7], which was reported only 2 years after the discovery of PQQ, is mentioned here (Scheme 1), since his synthetic strategy has been widely applied to the syntheses of several PQQ model compounds as

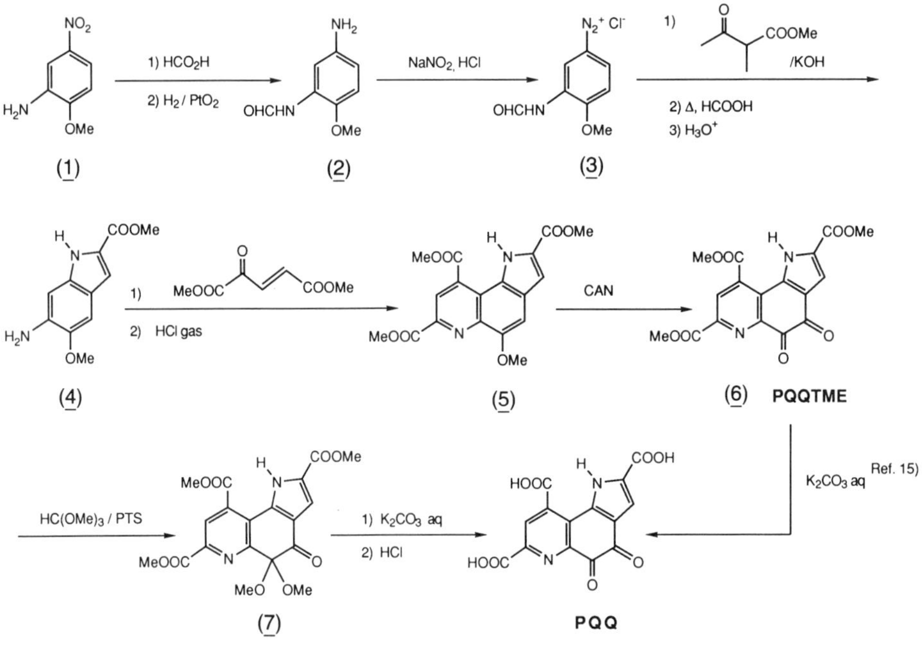

Scheme 1

mentioned below. the Büchi method [14] is also interesting since it is accomplished by using tyrosine and glutamic acid derivatives, the starting materials of PQQ biosynthesis [16–18].

Corey's method follows the traditional synthetic procedures of combining the Japp-Klingemann reaction with Fischer indolization (for the preparation of indole (4) and Doebner–von Miller type annulation (from (4) to (5)). Direct oxidation of (5) by CAN (ceric ammonium nitrate) gives PQQTME (6) trimethyl ester of PQQ), which is converted into PQQ via monoketal protection, ester hydrolysis, and a deprotection sequence. However, it has been found that monoketal protection of the quinone function is not necessary and that simple alkaline hydrolysis of the ester groups in a K_2CO_3 aqueous solution is enough to generate PQQ from (6) directly [19].

Decarboxylated derivatives of PQQ ((8–10), so-called decarboxymethoxatin) have been prepared by Bruice and coworkers using basically similar strategies, although several steps are required in each synthesis [20,21].

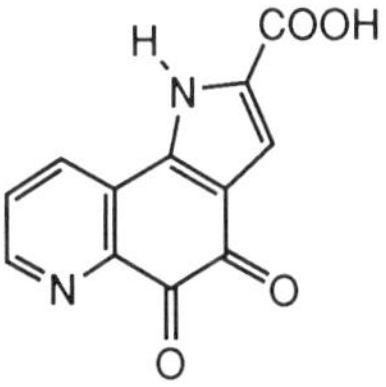

(8) 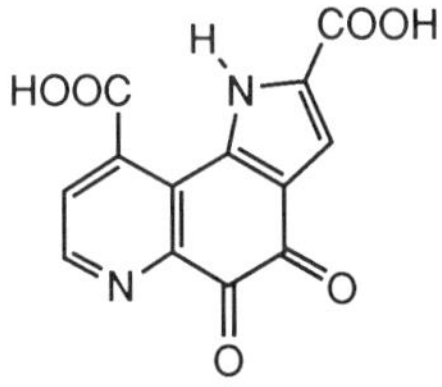 (9) 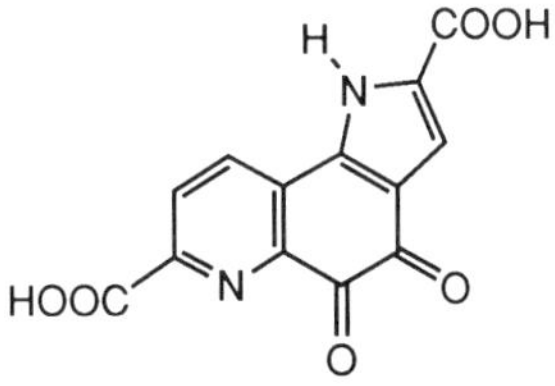 (10)

The decarboxylated models can be also synthesized very easily using (5) as a common starting material [22,23]. Regioselective hydrolysis of the three methyl ester groups of (5) under several conditions gives the tri-, di-, and monocarboxylic acid derivatives (11), (12), (13), and (14), respectively, as shown in Scheme 2. Thermal decarboxylation of the carboxyl groups of (12), (13), and (14) and subsequent oxidation by CAN affords the ester derivatives of (8), (9), and (10), respectively. Regioselective functionalization of PQQ through an amide linkage has been also accomplished from the carboxylic acid derivatives (11–14) [23]. The pyrrole proton was alkylated by applying the phase-transfer system, iodomethane/18-crown-6/potassium *tert*-butoxide, to indole (4). 1-Methylated PQQ (15) was prepared from the 1-methylated indole by Corey's method [22].

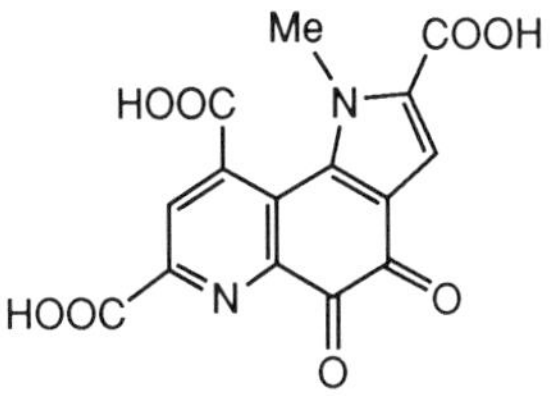

(15)

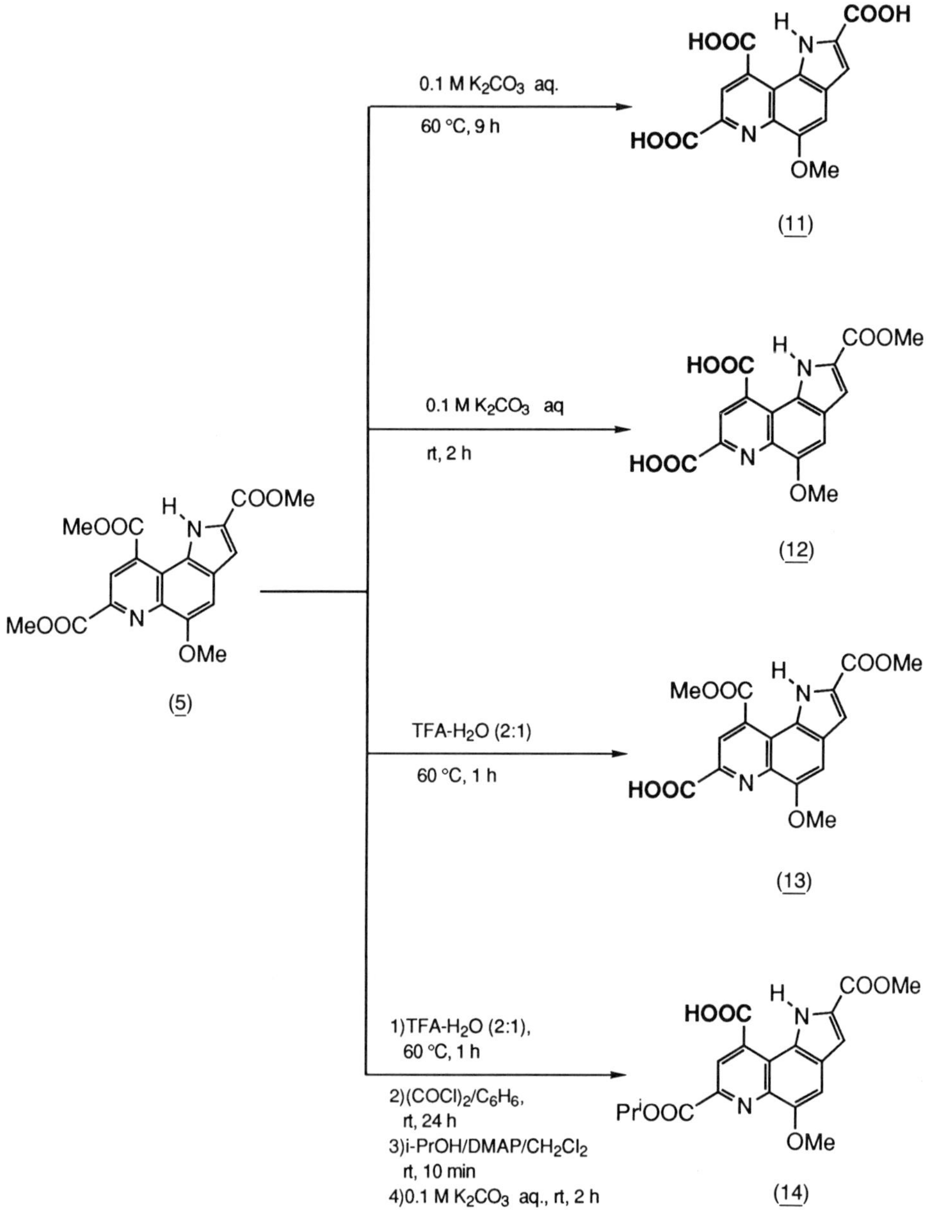
HOOC
COOH
H
N
0.1 M K_2CO_3 aq.
60 °C, 9 h
HOOC
OMe
(11)
COOMe
HOOC
0.1 M K_2CO_3 aq
rt, 2 h
HOOC
OMe
(12)
MeOOC
COOMe
MeOOC
OMe
(5)
TFA-H_2O (2:1)
60 °C, 1 h
MeOOC
COOMe
HOOC
OMe
(13)
1)TFA-H_2O (2:1), 60 °C, 1 h
2)$(COCl)_2/C_6H_6$, rt, 24 h
3)i-PrOH/DMAP/CH_2Cl_2 rt, 10 min
4)0.1 M K_2CO_3 aq., rt, 2 h
HOOC
COOMe
Pr^iOOC
OMe
(14)

Scheme 2

Table 1 Spectral Data of Coenzyme PQQ

^{1}H NMR[a] (δ, ppm) in DMSO-d_6	7.14 (H-3, *d*, *J* = 2.0 Hz), 8.62 (H-8, *s*) 16.1 (H-1, *br s*, exchangeable)
^{13}C NMR[a] (ppm) in DMSO-d_6	113.8 (C-3), 123.3 (C-3a), 126.1 (C-9a), 127.6 (C-2), 130.3 (C-8), 136.7 (C-1a), 142.3 (C-9), 146.4 (C-7), 148.0 (C-5a), 161.2 (C-2′), 165.4 (C-7′), 167.0 (C-9′), 173.3 (C-4), 179.3 (C-5)
UV-vis.[b] (λ_{max}, nm) (ϵ, $M^{-1}cm^{-1}$)	249 (20800), 273 (sholder, 16350), 331 (10050) 475 (very broad, 680)

[a]Recorded on JEOL FT-NMR JNM-GSX-400.
[b]pH = 7.0.

III. PHYSICAL PROPERTIES

Spectroscopic characterization of PQQ has been well documented [6], and some representative data are summarized in Table 1. The absorption spectrum is largely influenced by both pH and temperature. The former effect is due to the acid-base equilibria of the three carboxyl groups, the pyridine nitrogen, and the pyrrole proton (see Scheme 3). The latter effect has been attributed to hydration of the quinone function. It has been proposed that the hydration equilibrium is strongly dependent on temperature; lowering temperature shifts the equilibrium in favor of the hydrated species [24]. An aqueous solution of PQQ shows green fluorescence, which is attributed to the hydrated species [24]. Full assignment of ^{1}H and ^{13}C NMR spectra in DMSO-d_6 has been performed as shown in Table 1 [16,17,25]. Because of the three carboxyl groups, IR and mass spectra are obscure and elemental analysis does not always give satisfactory results. Details of the crystal structural data of PQQ are also available [26]. PQQ is insoluble in most organic

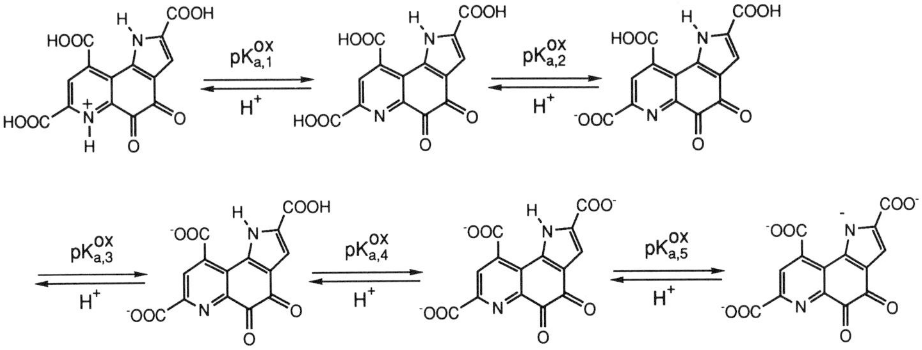

Scheme 3

solvents but soluble in highly polar solvents such as DMSO. Esterification of the carboxyl groups simplifies manipulation and clarifies the spectra.

The two-electron redox potential of the quinone/quinol couple was first determined to be 419 and 90 mV (vs. NHE) at pH 2.0 and 7.0, respectively, by potentiometric titration [25]. These values are relatively higher than those of flavin coenzymes but comparable to those of fused heterocyclic *o*-quinones such as phenanthrolinequinone derivatives [27]. The one-electron redox potentials of the quinone/semiquinone and semiquinone quinol couples were reported to be −218 and −242 mV at pH 13.0, respectively [6]. Recently, the electrochemical behavior of PQQ in aqueous media has been investigated in more detail by cyclic voltammetry [28,29]. The one-electron redox potentials for quinone/semiquinone and semiquinone/quinol couples in a wider pH range (6–12) are calculated by computer simulation. The PQQ-semiquinone is generated by electrochemical reduction under alkaline conditions (pH 12) and is well characterized by ESR. The semiquinone formation constant K is estimated to be 0.02 at pH 6–8 and 0.8 at pH 11–12 [28].

PQQ has at least five pK_as, as shown in Scheme 3. Those values could not be determined accurately by ordinary spectrophotometric titration, but recently, Kano et al. electrochemically determined the pK_a^{ox} values to be 0.3, 1.6, 2.2, 3.3, and 10.3, respectively [28,29]. The acid-base equilibria of the reduced PQQ become more complex since there are dissociations of the phenolic protons in addition to those of the protonated pyridine nitrogen, the carboxyl groups, and the pyrrole proton. They also estimated the pK_a^{red} values as follows; $pK_{a,1}^{red} = 0.9$, $pK_{a,2}^{red} = 1.8$, $pK_{a,3}^{red} = 2.7$, $pK_{a,4}^{red} = 4.5$, and $pK_{a,5}^{red} = 8.5$ (for the phenolic proton).

The pK_a^{ox} for the protonated pyridine nitrogen ($pK_{a,1}^{ox}$) is, however, relatively different from those of 9-decarboxymethoxatin (<u>10</u>) ($pK_{a,1}^{ox} = -2.9$) and 7,9-didecarboxymethoxatin (<u>8</u>) ($pK_{a,1}^{ox} = 1.41$) [20,30]. Judging from those pK_a values, the carboxyl groups on the pyridine nucleus seem to affect the electron density of PQQ considerably. In fact, the absorption spectra of those model compounds are relatively different from that of PQQ itself ((<u>8</u>), $\lambda_{max} = 276$ and 316 nm; (<u>10</u>), $\lambda_{max} = 271$ and 306 nm at pH 7.0). Changes in electron density with substitution must be reflected to some extent in the chemical properties of the quinones as well. However, it is of interest that the redox potentials of PQQ are not altered by removal of both 7- and 9-carboxyl groups [31].

IV. CHEMICAL PROPERTIES

A. Adduct Formation

The quinone carbonyl of PQQ is very reactive against several nucleophilic substances [6]. For example, acetone is easily added to PQQ to form an aldol-type adduct under weakly alkaline conditions. The first X-ray crystallographic analysis of PQQ was performed on this adduct [32]. The addition of H_2O to the quinone function (hydration, Eq. 1) has been studied spectrophotometrically, and the

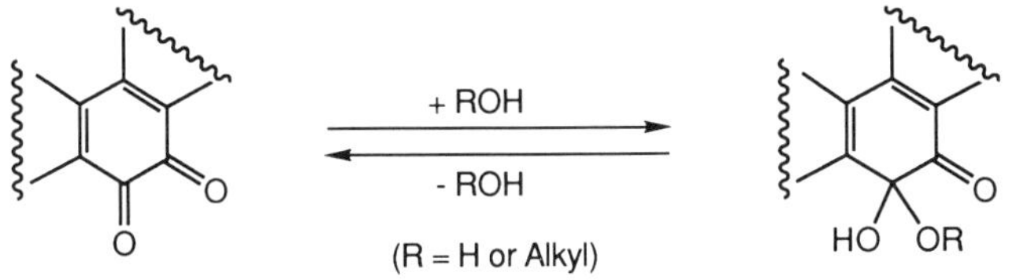

+ ROH

- ROH

(R = H or Alkyl)

HO OR

(1)

equilibrium of the hydration has been demonstrated to be highly dependent on temperature [24]. The pK_{app} values for the hydroxide addition (pseudo base formation) to (8) and (10) are 10.15 ± 0.03 and 10.73 ± 0.3, respectively [20,31]. Alcohols also add to the quinone reversibly to form a hemiketal-type adduct [24]. The addition position of these species (water and alcohols) has always been considered to be C-5, but without any direct evidence. However, our recent study showed that methanol addition occurs at C-4 depending on reaction conditions.

It has been suggested that nucleophilic addition is a key step in the redox reactions of PQQ with biologically important substances such as amines, amino acids, hydrazines, and thiols, and those are discussed in the next section. Nitroalkanes, however, form dioxol derivatives with several *o*-quinones. One plausible mechanism is that electron transfer results in a radical pair between the semiquinone and the substrate followed by radical coupling in the solvent cage. Subsequent intramolecular cyclization affords the dioxol derivatives as shown in Scheme 4 [33]. This novel reaction is interesting from the viewpoint of the enzymatic mechanism of the quinoprotein nitroalkane oxidase [1].

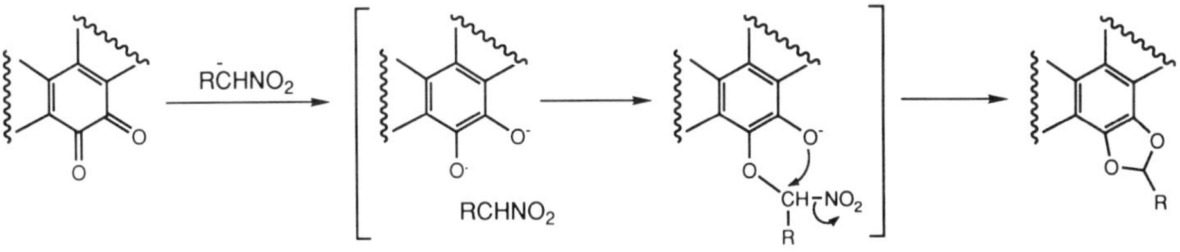

$R\bar{C}HNO_2$

$RCHNO_2$

Scheme 4

B. Redox Reactions

1. Reaction of PQQ with Amines and Related Substrates

The reaction of PQQ and amines has recently received much attention in connection with the role of PQQ in amine metabolism [34]. The nutritional importance of PQQ has been attributed to its functional effect on the cross-linking of collagen and elastin in connective tissue biogenesis and on the regulation of intracellular spermine and spermidine levels [35]. It is also important from the perspective of quinoprotein inhibition and PQQ detection in proteins and biological fluids.

Soon after the new coenzyme was discovered [32], model studies on amine oxidation by PQQ were performed by Bruice et al. using phenanthrolinequinones (16–18) [27,36]. The oxidation of primary amines such as cyclohexylamine and glycine by those quinones proceeds smoothly to provide 6-amino-5-hydroxyphenanthroline, or aminophenol, as the sole isolable reduced product, and the order of reactivity is (16) > (17) > (18) in spite of their almost identical redox potentials.

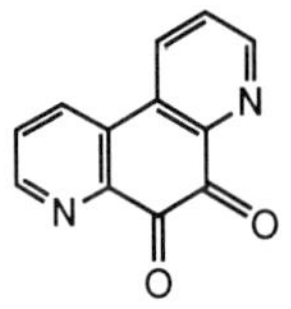

(16) 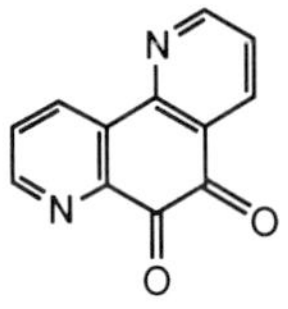 (17) 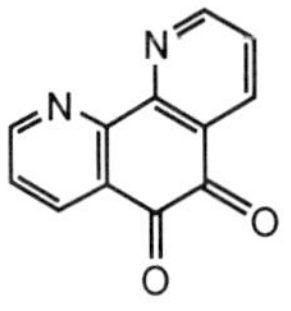 (18)

On the other hand, the reactivities of secondary and tertiary amines are relatively low. These findings, together with observations of general base catalysis by the amine substrate itself and of consecutive first-order kinetics, suggest that the oxidation of primary amines proceeds via a *transamination* mechanism through the imine intermediate [36]. A similar mechanism has been well established in the oxidation of amines by 3,5-di-*tert*-butyl-1,2-benzoquinone [37].

We reported the PQQ-catalyzed oxidation of amines under aerobic conditions, the so-called aerobic autorecycling oxidation of amines, by applying a cationic micellar system [38], where amine substrates are efficiently converted into the corresponding carbonyl compounds (Eq. 2). This was the first example of the

$$RR'CHNH_2 + O_2 + H_2O \xrightarrow[\text{pH 7, 30°C}]{\text{PQQ (1 mol\%), CTAB}} RR'C{=}O + NH_3 + H_2O_2 \qquad (2)$$

PQQ-catalyzed oxidation reaction of amines, suggesting its possible contribution to amine metabolism in living systems. Investigation into the role of micelles using several surfactants and aliphatic amines indicated that the reaction proceeds in the electrostatic layer (Stern layer) of the micelle [39,40].

The model studies were further extended using 7,9-didecarboxymethoxatin (8) to demonstrate that it can oxidize primary amines, but not secondary or tertiary amines [20]. The oxidation of the primary amines by (8) under anaerobic conditions converts the latter to the corresponding quinol with some formation of the aminophenol product. It is assumed that these products arise via two competing covalent addition base-catalyzed elimination mechanisms through the carbinolamine and the imine intermediates, respectively (Scheme 5). Similar results were obtained in the oxidation of amines by PQQ itself in the micellar system [41]. A detailed kinetic analysis using 9-decarboxymethoxatin (10) indicated that the

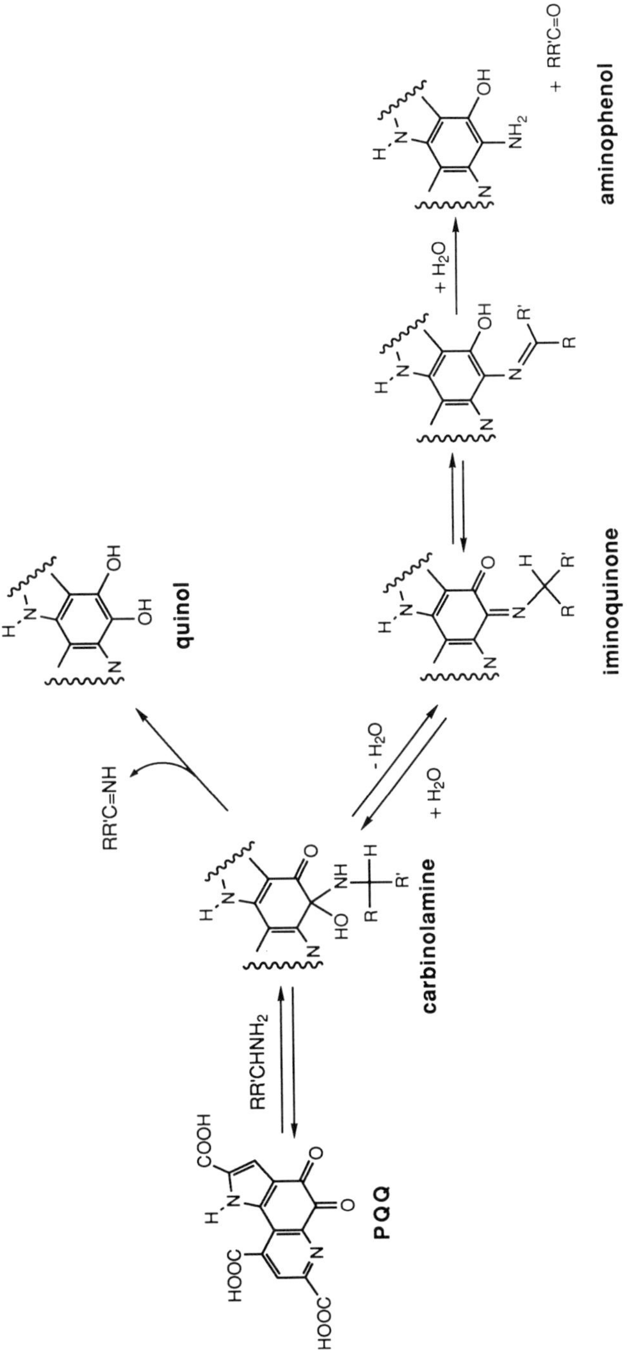

PQQ

$RR'CHNH_2$

carbinolamine

$RR'C{=}NH$

quinol

$- H_2O$

$+ H_2O$

iminoquinone

$+ H_2O$

aminophenol

$+ RR'C{=}O$

Scheme 5

product ratio of the quinol and the aminophenol is controlled by pH conditions of the solution [42]. This ionic mechanism has been further confirmed by examining the reaction using PQQTME (6) in organic media, where the iminoquinone-type intermediate and the quinol and aminophenol products are isolated and well characterized [43,44].

The reaction of PQQ with α,ω-diaminoalkanes provides more evidence for the ionic mechanism [45]. It has been found that the reaction pathway is altered by changing the alkylene chain length of the substrates. If the alkylene chain length is longer than 4, the redox reaction proceeds predominantly to produce reduced PQQ products, indicating that those diamines act as a substrate for the PQQ oxidation. On the other hand, adduct formation mainly occurs to give the pyrazine derivative (19) in the reaction with 1,2-diaminoalkanes (Eq. 3). In this case, the second amino

HOOC, COOH, HOOC, H, N, N, O, O + $H_2N(CH_2)_2NH_2$ —pH 4.6, 50 °C, 1h→ HOOC, COOH, HOOC, H, N, N, N, N (3)

(19) : λ_{max} = 286 nm 86 %

group is close enough in the C-5 adduct intermediate (carbinolamine and/or iminoquinone) to attack the C-4 position, providing the cyclic adduct. These results are in good accordance with the inhibitory action of diamines in enzymatic systems [46].

Both redox and adduct formation are observed in the reactions of hydrazines and aminoguanidine [47,48]. For example, in the reaction of PQQTME (6) with phenylhydrazine or methylhydrazine in methanol, the quinol $6H_2$ is obtained as a major product (Eq. 4), while the introduction of a strong electron acceptor such as a nitro group into the phenyl ring of the substrates alters the reaction product. That is, an almost 1:1 mixture of the quinol and the C-5 hydrazone derivative (20) is obtained in the reaction with 4-nitrophenylhydrazine (Eq. 5), and the C-5 hydrazone (21) becomes the main product in the case of 2,4-dinitrophenylhydrazine (Eq. 6). Examination of the reaction between PQQ and aminoguanidine indicates that the redox reaction predominantly proceeds at pH 10.0 to give $PQQH_2$ (Eq. 7), whereas deactivation of PQQ occurs at pH 6.7 to give the aminotriazine derivative (22) (Eq. 8).

These results can be interpreted by the similar ionic mechanism shown in Scheme 6. The carbinolamine-type adduct is also considered to be a key intermediate from which the redox reaction (quinol formation) proceeds via electron flow from the hydrazino nitrogen to the quinone. If this is not fast enough, dehydration from the intermediate (a) predominates to give the C-5 hydrazone or azo adduct. Electron-withdrawing substituents would retard such electron flow as in 4-nitrophenylhydrazine and 2,4-dinitrophenylhydrazine to give the correspond-

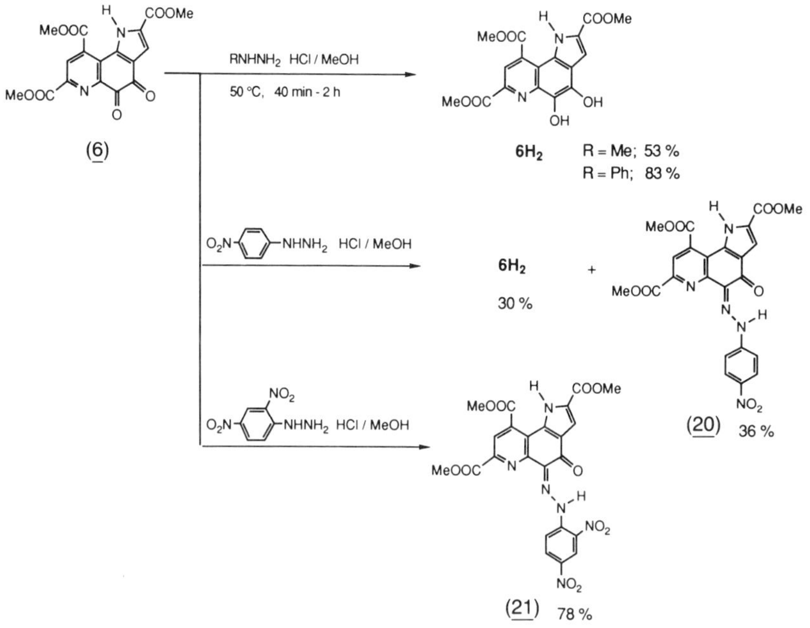

(6)

$RNHNH_2$ HCl / MeOH

50 °C, 40 min - 2 h

$6H_2$ R = Me; 53 %
R = Ph; 83 % (4)

O_2N-C_6H_4-$NHNH_2$ HCl / MeOH

$6H_2$ 30 % + (20) 36 % (5)

O_2N-$C_6H_3(NO_2)$-$NHNH_2$ HCl / MeOH

(21) 78 % (6)

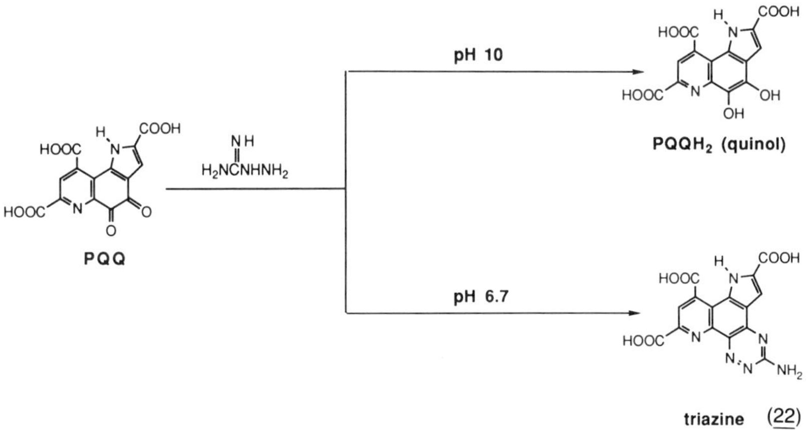

PQQ

$H_2NC(=NH)NHNH_2$

pH 10 → $PQQH_2$ (quinol) (7)

pH 6.7 → triazine (22) (8)

ing adducts, whereas electron-donating substituents such as methyl and phenyl facilitate the reduction of PQQ. Kinetic studies on reactions between PQQ models and hydrazine derivatives also support the ionic mechanism [47].

The pH dependence of the reaction with aminoguanidine can be explained similarly as follows. At lower pH, where almost all the guanidino groups are protonated, the electron flow from the hydrazino nitrogen into the quinone moiety must be suppressed because of the strong electron-withdrawing effect of the protonated guanidino group. Acid-catalyzed dehydration from the intermediate

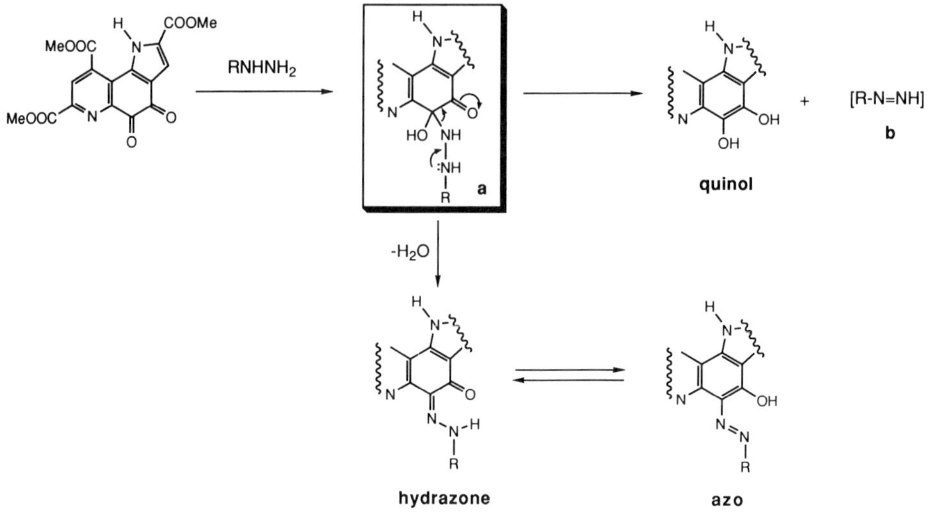

Scheme 6

could be accelerated under such conditions. Consequently, the adduct formation mainly proceeds at pH 6.7. On the contrary, the protonation of the guanidino group and the acid-catalyzed dehydration would be depressed at higher pH (pH 10) allowing the electron flow to yield the quinol. The importance of the electron-withdrawing effect of the substituent attached to the hydrazino group is also found in the reaction between PQQ and semicarbazide and acetohydrazide; the azo adducts are the major products even at higher pH conditions (pH 10). In conclusion, the more electron-withdrawing the nature of the substituent, the more preferable is the adduct formation. These findings are interesting not only from the viewpoint of amine-oxidation mechanisms but also in connection with the inhibitory action of those substrates on quinoproteins [49,50].

As an extension of the model studies, the authors first reported the PQQ-catalyzed *oxidative decarboxylation* of α-amino acids in the cationic micellar system [51]. α-Phenylglycine is efficiently converted into benzaldehyde in the presence of a catalytic amount of PQQ (1 mol%) and CTAB surfactant in a neutral aqueous solution under aerobic conditions (Eq. 9). During the catalytic reactions, PQQ is gradually converted into a redox inactive oxazole derivative (23).

Efficient oxidative C_{α}-C_{β} cleavage (*dealdolation*) occurs in the reactions of β-hydroxy amino acids [52]. For example, β-phenylserine is converted into benzaldehyde in the presence of a catalytic amount of PQQ and CTAB surfactant under alkaline aerobic conditions (Eq. 10). In this reaction, mandelaldehyde, which is a predicted product of the oxidative decarboxylation, and glycine, which is a common product of the dealdolation reaction catalyzed by PLP and metal ions

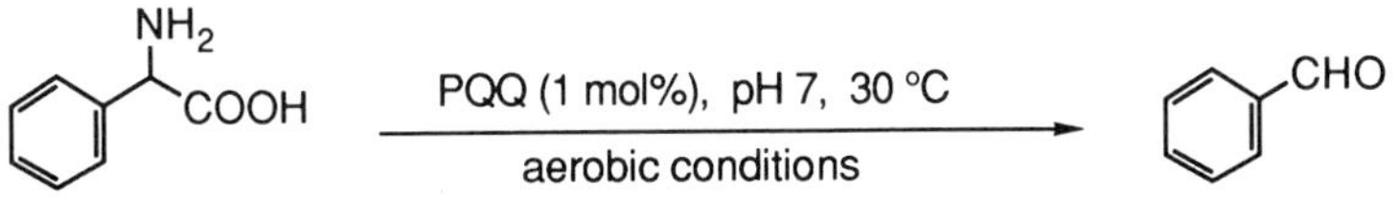

$$\text{PhCH(NH}_2\text{)COOH} \xrightarrow[\text{aerobic conditions}]{\text{PQQ (1 mol\%), pH 7, 30 °C}} \text{PhCHO} \qquad (9)$$

3400 % (based on PQQ)

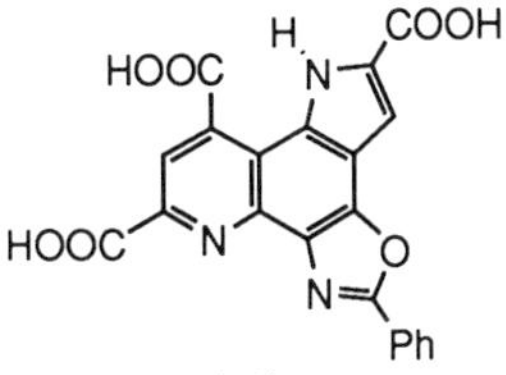

(23)

[53], are not detected. It should be noted that in this case the nonsubstituted oxazole derivative (24) is obtained as a deactivated product of PQQ. The oxazole (24) is also produced in reactions with a series of amino acids such as serine, threonine, tyrosine, and tryptophan, as well as reactions with β-hydroxy amines such as 2-amino-1-phenylethanol. A similar ionic mechanism through the carbinolamine-type intermediate has been proposed for those reactions [52].

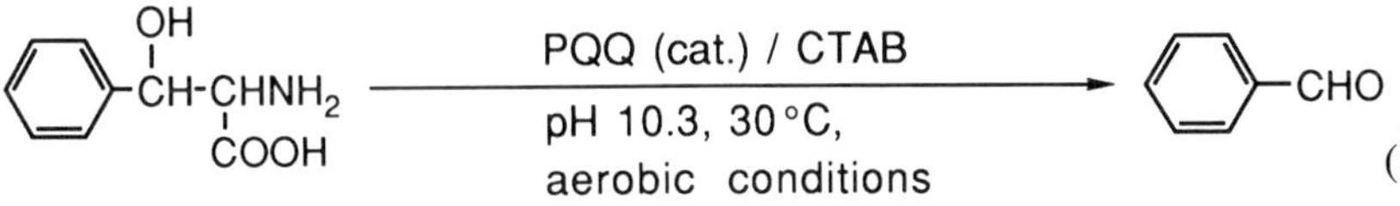

$$\text{PhCH(OH)CH(NH}_2\text{)COOH} \xrightarrow[\text{pH 10.3, 30 °C, aerobic conditions}]{\text{PQQ (cat.) / CTAB}} \text{PhCHO} \qquad (10)$$

[PhCH(OH)CHO and H_2NCH_2COOH are not obtained.]

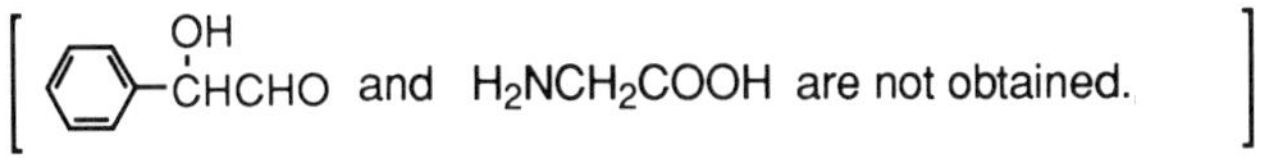

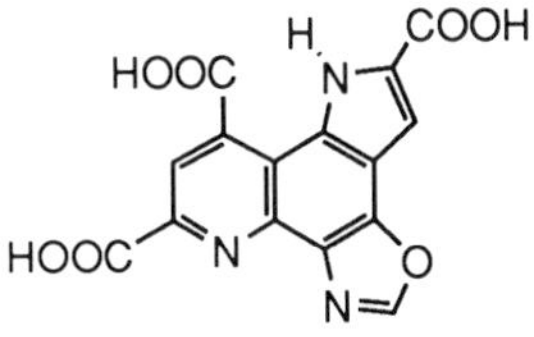

(24)

2. Reactions of PQQ with Other Biologically Important Substrates

Alcohols, one of the most important substrates of quinoproteins, simply add to the quinone moiety of PQQ, but no redox reaction is observed in vitro (see Eq. 1). The hemiketal type adduct thus formed is, however, considered to be an intermediate for enzymatic alcohol oxidation as in amine oxidation [34]. An Oppenauer-type oxidation of alcohols by PQQTME (6) using aluminum salts as a mediator has been reported by the authors (Table 2) [54]. In this reaction, the aluminum salt

Table 2 Oxidation of Alcohols by PQQTME

$$RR'CHOH \xrightarrow{PQQTME/AlCl_3/CH_3CN} RR'C{=}O$$

Substrate	Temp. (°C)	Time (h)	Yield (%)
$PhCH_2OH$	rt	24	76
$PhCH_2OH$	52	9	99
$PhCH(CH_3)OH$	52	9	64
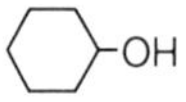 -OH	rt	24	44
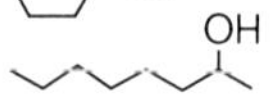 OH	rt	24	72
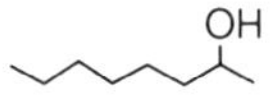 OH	52	9	92

rt = room temperature.

activates both alcohols and the quinone by forming an alkoxide and by coordinating to the quinone carbonyl oxygen as a Lewis acid, respectively. The active site of the enzymes might play the similar role of a concerted general acid-base catalyst.

PQQ is reduced to the quinol by glucose in an alkaline aqueous solution. Kinetic studies on the redox reaction using glucose and related substrates indicate that the active species is 1,2-enediolate and its formation is the rate-determining step (Scheme 7) [55]. PQQ is readily reduced by ascorbic acid and by enediolate-

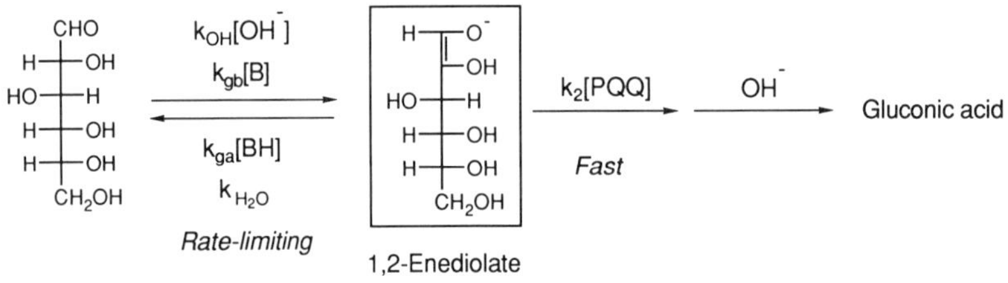

CHO, H–OH, HO–H, H–OH, H–OH, CH_2OH — $k_{OH}[OH^-]$, $k_{gb}[B]$ / $k_{ga}[BH]$, k_{H_2O} (*Rate-limiting*) ⇌ [H–O⁻, –OH, HO–H, H–OH, H–OH, CH_2OH] 1,2-Enediolate — $k_2[PQQ]$ (*Fast*) → OH^- → Gluconic acid

Scheme 7

type adducts of aldehydes and CN^- or thiazolium salts, so-called activated aldehyde, in a cationic micellar system (Eqs. 11 and 12) [56]. These reactions must be useful in understanding the enzymatic mechanisms of quinoprotein glucose dehydrogenase and aldehyde dehydrogenase. The mechanism of enediolate-oxidation by PQQ has not yet been clarified, but an electron transfer mechanism seems to be plausible.

Oxidation of thiols by PQQ to the corresponding disulfides proceeds very

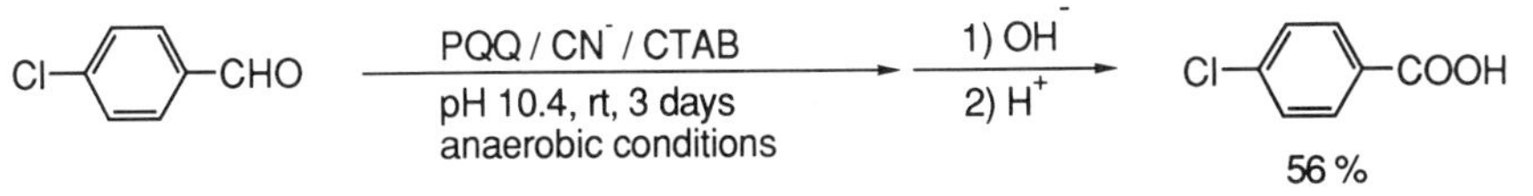

$$\text{Cl-C}_6\text{H}_4\text{-CHO} \xrightarrow[\text{pH 10.4, rt, 3 days, anaerobic conditions}]{\text{PQQ / CN}^- \text{/ CTAB}} \xrightarrow[2)\ \text{H}^+]{1)\ \text{OH}^-} \text{Cl-C}_6\text{H}_4\text{-COOH} \quad (56\%) \tag{11}$$

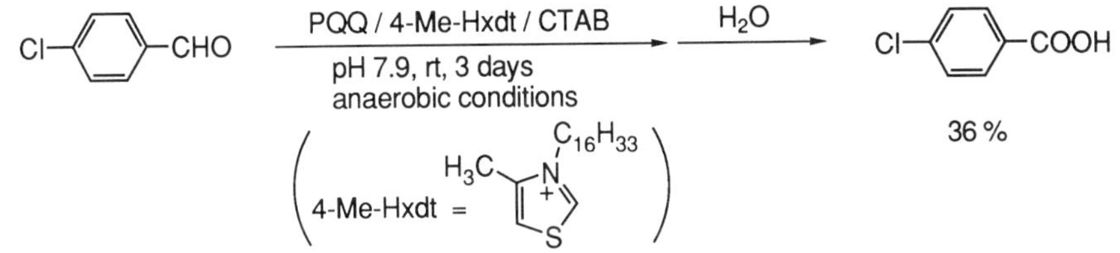

$$\text{Cl-C}_6\text{H}_4\text{-CHO} \xrightarrow[\text{pH 7.9, rt, 3 days, anaerobic conditions}]{\text{PQQ / 4-Me-Hxdt / CTAB}} \xrightarrow{\text{H}_2\text{O}} \text{Cl-C}_6\text{H}_4\text{-COOH} \quad (36\%) \tag{12}$$

(4-Me-Hxdt = 3-hexadecyl-4-methylthiazolium; H_3C, $C_{16}H_{33}$, N^+, S)

efficiently [57,58]. Kinetic studies on the reaction under anaerobic conditions provide a bell-shaped pH-rate profile having a maximum rate at around the pK_a of the thiol. Interestingly, the rate-determining step is different at both sides of the bell-shaped pH-rate profile, indicating the existence of at least one intermediate in the course of the reaction. A covalent adduct is also proposed to be the key intermediate for the thiol oxidation (Scheme 8) [58]. PQQ acts as a very efficient

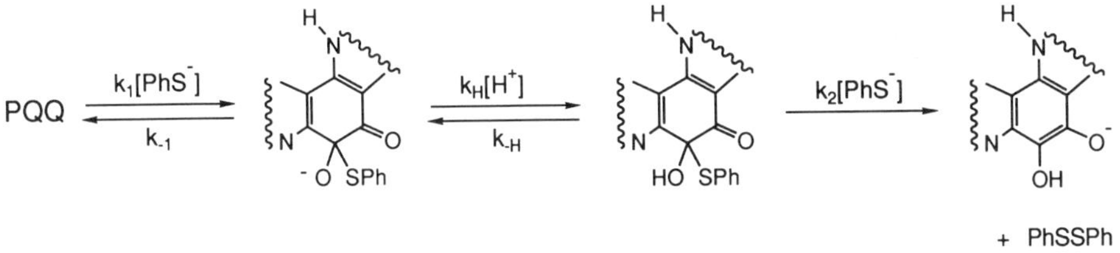

$$\text{PQQ} \underset{k_{-1}}{\overset{k_1[\text{PhS}^-]}{\rightleftharpoons}} \text{adduct}(\text{-O}^-,\ \text{SPh}) \underset{k_{-H}}{\overset{k_H[\text{H}^+]}{\rightleftharpoons}} \text{adduct}(\text{HO},\ \text{SPh}) \xrightarrow{k_2[\text{PhS}^-]} \text{reduced PQQ}(\text{OH},\ \text{O}^-) + \text{PhSSPh}$$

Scheme 8

turnover catalyst in the autoxidation of thiols [57]. PQQ and its related heterocyclic *o*-quinones have been examined as electron transfer catalysts in other systems such as the aerobic regeneration of NAD^+ from NADH [59,60] and some synthetic organic reactions [61–63]. In particular, the redox reaction between PQQ and NADH [59,60] seems to be very important from the viewpoint of energy transportation in living systems. PQQ must interact with NAD(P)H and other redox cofactors in a variety of biologically important processes.

C. Metal Coordination

Metal coordination is another interesting part of the chemistry of PQQ. Most of the research so far reported was performed under the commonly held premise that PQQ is the second organic cofactor of copper-containing amine oxidases. Although recent studies revealed that this is not so [64], the metal coordination chemistry of PQQ is still worthwhile since there should be several interactions between PQQ and metals in enzymatic systems [1]. PQQ itself has some possible metal coordination sites. The ternary copper (II) complexes containing PQQ or $PQQH_2$ (quinol) and bipyridine or terpyridine were first prepared and characterized using UV-visible and EPR spectroscopy by Suzuki et al. [65,66]. It has been proposed that PQQ binds to Cu(II) through the pyridine nitrogen and the carboxylate at the 7-position (Scheme 9), since 1-methylated PQQ (15) gave a similar copper (II) complex [67].

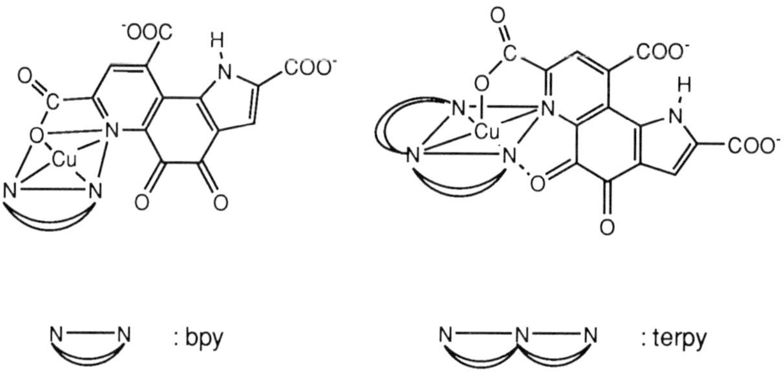

N—N : bpy

N—N—N : terpy

Scheme 9

Bruice et al. investigated metal coordination chemistry using the decarboxymethoxatins (8) and (10) [30]. They pointed out the importance of the carboxylate group at the 7-position for metal coordination. Lack of the carboxyl group at the 7-position drastically diminishes complexing ability. X-ray crystallographic analysis of PQQ crystallized from a sodium phosphate buffer solution clearly indicates that the sodium ion coordinates at the same position between N(6) and COO^{-}(7) [26]. Metal ions may play very important roles in binding PQQ to the active sites of enzymes and for activation of PQQ as a Lewis acid. Electron transfer between PQQ and metals should take part in several biological processes.

V. CONCLUDING REMARKS

In this chapter, the chemistry of PQQ and its related compounds have been reviewed, and several interesting aspects of their chemical functions have been introduced. Although many issues remain to be resolved, interesting characteristics of the PQQ (pyrroloquinoline quinone) molecule have been identified. For example, it is apparent that the pyrroloquinoline quinone structure of PQQ is essential for the efficient amine oxidation in vitro. The pyridine nucleus may facilitate the nucleophilic addition of amines to the C-5 quinone carbon and stabilize the carbinolamine intermediate thus formed through intramolecular hydrogen bonding [36]. The pyrrole proton must also be very important for conducting intramolecular general base catalysis to abstract the α-proton of the amine (see Scheme 5). Mechanistic details and the structure-reactivity relationship of the adduct formation reactions and the redox reactions of PQQ should be addressed in the next investigative stages.

PQQ itself is now commercially available, so interest in PQQ and quinoproteins has been increasing in various research fields during the past few years. In particular, studies by Duine's group in the Netherlands have stimulated many investigators to discover many important findings. A typical example is the discovery of new organic cofactors; TOPA (6-hydroxydopa) in bovine serum amine oxidase [64], the Tyr-Cys cross-linking residue in galactose oxidase [68], and TTQ (tryptophan tryptophylquinone) in methylamine dehydrogenase [69]. Comparison of the chemical characteristics of these cofactors to that of PQQ will undoubtedly provide another interesting avenue of research.

REFERENCES

1. Jongejan, J. A., and Duine, J. A. (1988). PQQ and quinoproteins. In *Proceedings of the First International Symposium on PQQ and Quinoproteins*, Delft, The Netherlands, 1988.
2. Watanabe, A., Hobara, N., and Tsuji, T. (1988). Protective effect of pyrroloquinoline quinone against experimental liver injury in rats. *Curr. Ther. Res. 44*: 896.
3. Nishigori, H., Yasunaga, M., Mizumura, M., Lee, J. W., and Iwatsuru, M. (1989). Preventive effects of pyrroloquinoline quinone on formation of cataract and decline of lenticular and hepatic glutathione of developing chick embryo after glucocorticoid treatment. *Life Sci. 45*: 593.
4. Duine, J. A., van der Meer, R. A., and Groen, B. W. (1990). The cofactor pyrroloquinoline quinone. *Annu. Rev. Nutr. 10*: 297.
5. Ohshiro, Y., and Itoh, S. (1986). New coenzyme PQQ—profiles of organic chemistry. *Kagaku* (Kyoto) *41*: 495.
6. Duine, J. A., Frank, J., and Jongejan, J. A. (1987). Enzymology of quinoproteins. *Adv. Enzymol. 59*: 170.
7. Corey, E. J., and Tramontano, A. (1981). Total synthesis of the quinonoid alcohol dehydrogenase coenzyme (1) of methylotrophic bacteria. *J. Am. Chem. Soc. 103*: 5599.
8. Gainor, J. A., and Weinreb, S. M. (1981). Total synthesis of methoxatin, the coenzyme of methanol dehydrogenase and glucose dehydrogenase, *J. Org. Chem. 46*: 4319.
9. Gainor, J. A., and Weinreb, S. M. (1982). Synthesis of the bacterial coenzyme methoxatin. *J. Org. Chem. 47*: 2833.
10. Hendrickson, J. B., and deVries, J. G. (1982). A convergent total synthesis of methoxatin. *J. Org. Chem. 47*: 1148.
11. Hendrickson, J. B., and deVries, J. G. (1982). Total synthesis of the novel coenzyme methoxatin. *J. Org. Chem. 50*: 1688.

12. MacKenzie, A. R., Moody, C. J., and Rees, C. W. (1983). Synthesis of the bacterial coenzyme methoxatin. *J. Chem. Soc. Chem. Commun.* 1372.
13. MacKenzie, A. R. Moody, C. J., and Rees, C. W. (1986). Synthesis of the bacterial coenzyme methoxatin. *Tetrahedron 42*: 3259.
14. Büchi, G., Botkin, J. H., Lee, G. C. M., and Yakushijin, K. (1985). A synthesis of methoxatin. *J. Am. Chem. Soc. 107*: 5555.
15. Naruta, Y., and Maruyama, K. (1988). *The Chemistry of Quinonoid Compounds*, Vol. II (Patai, S., and Rappoport, Z., eds.), John Wiley & Sons, p. 374.
16. Houck, D. R., Hanners, J. L., and Unkefer, C. J. (1988). Biosynthesis of pyrroloquinoline quinone. 1. Identification of biosynthetic precursors using ^{13}C labeling and NMR spectroscopy. *J. Am. Chem. Soc. 110*: 6920.
17. Houck, D. R., Hanners, J. L., and Unkefer, C. J. (1991). Biosynthesis of pyrroloquinoline quinone. 2. Biosynthetic assembly from glutamate and tyrosine. *J. Am. Chem. Soc. 113*: 3162.
18. van Kleef, M. A. G., and Duine, J. A. (1988). *L*-Tyrosine is the precursor of PQQ biosynthesis in hyphomicrobium X. *FEBS Lett. 237*: 91.
19. Itoh, S., and Ohshiro, Y. (1986). Hydrolysis of ester groups of coenzyme PQQ-trimethyl ester. *Chem. Exp. 1*: 315.
20. Sleath, P. R., Noar, J. B., Eberlein, G. A., and Bruice, T. C. (1985). Synthesis of 7,9-didecarboxymethoxatin (4,5-dihydro-4,5-dioxo-1H-pyrrolo[*2,3-f*]quinoline-2-carboxylic acid) and comparison of its chemical properties with those of methoxatin and analogous *o*-quinones. Model studies directed toward the action of PQQ requiring bacterial oxidoreductases and mammalian plasma amine oxidase. *J. Am. Chem. Soc. 107*: 3328.
21. Noar, J. B., and Bruice, T. C. (1987). Decarboxylated methoxatin analogues. Synthesis of 7- and 9-decarboxymethoxatin. *J. Org. Chem. 52*: 1942.
22. Itoh, S., Kato, J., Inoue, T., Kitamura, Y., Komatsu, M., and Ohshiro, Y. (1987). Syntheses of pyrroloquinoline quinone derivatives: Model compounds of a novel coenzyme PQQ (methoxatin). *Synthesis* 1067.
23. Itoh, S., Inoue, T., Fukui, Y., Huang, X., Komatsu, M., and Ohshiro, Y. (1990). Regioselective transformation of the functional groups of coenzyme PQQ. *Chem. Lett.* : 1675.
24. Dekker, R. H., Duine, J. A., Frank, J., Verwiel, P. E. J., and Westerling, J. (1982). Covalent addition of H_2O, enzyme substrates and activators to pyrroloquinoline quinone, the coenzyme of quinoproteins. *Eur. J. Biochem. 125*: 69.
25. Duine, J. A., Frank, J., and Verwiel, P. E. J. (1981). Characterization of the second prosthetic group in methanol dehydrogenase from hyphomicrobium X. *Eur. J. Biochem. 118*: 395.
26. Ishida, T., Doi, M., Tomita, K., Hayashi, H., Inoue, M., and Urakami, T. (1989). Molecular and crystal structure of PQQ (methoxatin), a novel coenzyme of quinoproteins: Extensive stacking character and metal ion interaction. *J. Am. Chem. Soc. 111*: 6822.
27. Eckert, T. S., Bruice, T. C., Gainor, J. A., and Weinreb, S. M. (1982). Some electrochemical and chemical properties of methoxatin and analogous quinones. *Proc. Natl. Acad. Sci. USA 79*: 2533.

28. Kano, K., Mori, K., Uno, B., Kubota, T., Ikeda, T., and Senda, M. (1990). Voltammetric and spectroscopic studies of pyrroloquinoline quinone coenzyme under neutral and basic conditions. *Bioelectrochem. Bioeng. 23*: 227.
29. Kano, K., Mori, K., Uno, B., Kubota, T., Ikeda, T., and Senda, M. (1990). Voltammetric determination of acid dissociation constants of pyrroloquinoline quinone and its reduced form under acidic conditions. *Bioelectrochem. Bioeng. 24*: 193.
30. Noar, J. B., Rodriguez, E. J., and Bruice, T. C. (1985). Synthesis of 9-decarboxymethoxatin. Metal complexation of methoxatin as a possible requirement for its biological activity. *J. Am. Chem. Soc. 107*: 7198.
31. Rodriguez, E. J., Bruice, T. C., and Edmondson, D. E. (1987). Studies on the radical species of 9-decarboxymethoxatin. *J. Am. Chem. Soc. 109*: 532.
32. Salisbury, S. A. Forrest, H. S., Cruse, W. B. T., and Kennard, O. (1979). A novel coenzyme from bacterial primary alcohol dehydrogenases. *Nature 280*: 843.
33. Itoh, S., Nii, K., Mure, M., and Ohshiro, Y. (1987). Novel addition of nitroalkanes to *o*-quinones. *Tetrahedron Lett. 28*: 3975.
34. Hartmann, C., and Klinman, J. P. (1988). Pyrroloquinoline quinone: A new redox cofactor in eukaryotic enzymes. *BioFactors 1*: 41.
35. Killgore, J., Smidt, C., Duich, L., Romero-Chapman, N., Tinker, D., Reiser, K., Melko, M., Hyde, D., and Rucker, R. B. (1989). Nutritional importance of pyrroloquinoline quinone. *Science 245*: 850.
36. Eckert, T. S., and Bruice, T. C. (1983). Chemical properties of phenanthroline-quinones and the mechanism of amine oxidation by *o*-quinone of medium redox potentials. *J. Am. Chem. Soc. 105*: 4431.
37. Corey, E. J., and Achiwa, K. (1969). A new method for the oxidation of primary amines to ketones. *J. Am. Chem. Soc. 91*: 1429.
38. Ohshiro, Y., Itoh, S., Kurokawa, K., Kato, J., Hirao, T., and Agawa, T. (1983). Micellar enhanced oxidation of amines by coenzyme PQQ. *Tetrahedron Lett. 24*: 3465.
39. Itoh, S., Kitamura, Y., and Ohshiro, Y. (1986). Effects of surfactants on the oxidative deamination of amines by coenzyme PQQ. *J. Jpn. Oil Chem. Soc. (YUKAGAKU) 35*: 91.
40. Itoh, S., Mure, M., and Ohshiro, Y. (1987). Oxidative deamination of aliphatic amines by coenzyme PQQ in the micellar system. *J. Jpn. Oil. Chem. Soc. (YUKAGAKU) 36*: 882.
41. Itoh, S., Kitamura, Y., Ohshiro, Y., and Agawa, T. (1986). Kinetics and mechanisms of the oxidative deamination of amines by coenzyme PQQ. *Bull. Chem. Soc. Jpn. 59*: 1907.
42. Rodriguez, E. J., and Bruice, T. C. (1989). Reaction of methoxatin and 9-decarboxymethoxatin with benzylamine. Dynamics and products. *J. Am. Chem. Soc. 111*: 7947.
43. Mure, M., Itoh, S., and Ohshiro, Y. (1989). Preparation and characterization of iminoquinone and aminophenol derivatives of coenzyme PQQ. *Tetrahedron Lett. 30*: 6875.
44. Itoh, S., Mure, M., Ogino, M., and Ohshiro, Y. (1991). Reaction of trimethyl ester of coenzyme PQQ (PQQTME) and amines in organic media. Products and mechanism. *J. Org. Chem. 56*:6857.

45. Mure, M., Itoh, S., and Ohshiro, Y. (1989). Chemical behavior of coenzyme PQQ toward diamines. *Chem. Lett.* : 1491.
46. Gacheru, S. N., Trackman, P. C., Calaman, S. D., Greenaway, F. T., and Kagan, H. M. (1989). Vicinal diamines as pyrroloquinoline quinone-directed irreversible inhibitors of lysyl oxidase. *J. Biol. Chem. 264*: 12963.
47. Mure, M., Nii, K., Inoue, T., Itoh, S., and Ohshiro, Y. (1990). The reaction of coenzyme PQQ with hydrazines. *J. Chem. Soc., Perkin Trans II*: 315.
48. Mure, M., Nii, K., Itoh, S., and Ohshiro, Y. (1990). Chemical behavior of coenzyme PQQ toward aminoguanidine: Redox reaction and adduct formation. *Bull. Chem. Soc. Jpn. 63*: 417.
49. Petterson, G. (1982). *Structure and Functions of Amine Oxidases* (Mondovi, B., ed.), CRC Press, Boca Raton, FL, pp. 2–50.
50. Bieganski, T., Osinska, Z., and Maslinski, C. (1982). Inhibition of plant and mammalian diamine oxidases by hydrazine and guanidine compounds. *Int. J. Biochem. 14*: 949.
51. Itoh, S., Kato, N., Ohshiro, Y., and Agawa, T. (1984). Oxidative decarboxylation of α-amino acids with coenzyme PQQ. *Tetrahedron Lett. 25*: 4753.
52. Mure, M., Suzuki, A., Itoh, S., and Ohshiro, Y. (1990). Oxidative C_{α}-C_{β} Fission (Dealdolation) of β-Hydroxy Amino Acids by Coenzyme PQQ. *J. Chem. Soc. Chem. Commun.* 1608.
53. Martell, A. E. (1989). Vitamin B_6 catalyzed reactions of α-amino acids and α-keto acids: Model system. *Acc. Chem. Res. 22*: 115.
54. Itoh, S., Mure, M., Ohshiro, Y., and Agawa, T. (1985). Biomimetic oxidation of alcohols by coenzyme PQQ-trimethyl ester. *Tetrohedron Lett. 26*: 4225.
55. Itoh, S., Mure, M., and Ohshiro, Y. (1987). Oxidation of *D*-glucose by coenzyme PQQ: 1,2-enediolates as substrates for PQQ oxidation. *J. Chem. Soc., Chem. Commun.* 1580.
56. Ohshiro, Y., and Itoh, S. (1989). Chemical functions of novel coenzyme PQQ. *J. Synth. Org. Chem. Jpn. 47*: 855.
57. Itoh, S., Kato, N., Ohshiro, Y., and Agawa, T. (1985). Catalytic oxidation of thiols by coenzyme PQQ. *Chem. Lett.*: 135.
58. Itoh, S., Kato, N., Mure, M., Ohshiro, Y. (1987). Kinetic studies on the oxidation of thiols by coenzyme PQQ. *Bull. Chem. Soc. Jpn. 60*: 420.
59. Itoh, S., Kinugawa, M., Mita, N., and Ohshiro, Y. (1989). Efficient NAD^+-regeneration system with heterocyclic *o*-quinones and molecular oxygen. *J. Chem. Soc. Chem. Commun.*: 694.
60. Itoh, S., Mita, N., and Ohshiro, Y. (1990). NAD^+-regeneration system with heterocyclic *o*-quinones and molecular oxygen catalyzed by diaphorase. *Chem. Lett.* : 1949.
61. Hirao, T., Murakami, T., Ohno, M., and Ohshiro, Y. (1989). Redox system of palladium-trimethyl ester of coenzyme PQQ. *Chem. Lett.* : 785.
62. Hirao, T., Ohno, M., and Ohshiro, Y. (1990). Mediation of *o*-quinones in the MnTPPCl-catalyzed epoxidation with hydrogen peroxide. *Tetrahedron Lett. 31*: 6039.
63. Hirao, T., Murakami, T., Ohno, M., and Ohshiro, Y. (1991). Trimethyl ester of coenzyme PQQ in redox reactions with transition metals. *Chem. Lett.* : 299.

64. Janes, S. M., Mu, D., Wemmer, D., Smith, A. J., Kaur, S., Maltby, D., Buringame, A. L., and Klinman, J. P. (1990). A new redox cofactor in eukaryotic enzymes: 6-hydroxydopa at the active site of bovine serum amine oxidase. *Science 248*: 981.
65. Suzuki, S., Sakurai, T., Itoh, S., and Ohshiro, Y. (1988). Preparation and characterization of ternary copper (II) complexes containing coenzyme PQQ and bipyridine or terpyridine. *Inorg. Chem. 27*: 591.
66. Suzuki, S., Sakurai, T., Itoh, S., and Ohshiro, Y. (1988). Characterization of ternary copper (II) complexes containing reduced PQQ ($PQQH_2$) and bipyridine or terpyridine. *Chem. Lett.* : 777.
67. Suzuki, S., Sakurai, T., Itoh, S., and Ohshiro, Y. (1988). Characterization of ternary copper (II) complexes containing 1-methyl derivative of PQQ. *J. Chem. Soc. Jpn. Chem. Ind. Chem.* : 421.
68. Ito, N., Phillips, S. E. V., Stevens, C., Ogel, Z. B., McPherson, M. J., Keen, J. N., Yadav, K. D. S., and Knowles, P. F. (1991). Novel thioether bond revealed by a 1.7 Å crystal structure of galactose oxidase. *Nature 350*: 87.
69. McIntire, W. S., Wemmer, D. E., Chistoserdov, A., and Lidstrom, M. E. (1991). A new cofactor in a prokaryotic enzyme: Tryptophan tryptophylquinone as the redox prosthetic group in methylamine dehydrogenase. *Science 252*: 817.

13

Glycine-Dependent Redox Cycling and Other Methods for PQQ and Quinoprotein Detection

Rudolf Flückiger, Mercedes A. Paz, Edward Henson, and Paul M. Gallop

Children's Hospital Medical Center, Harvard Schools of Medicine and Dental Medicine, Boston, Massachusetts

Peter R. Bergethon

Boston University School of Medicine, Boston, Massachusetts

I. INTRODUCTION

The initial observation that pyrroloquinoline quinone (PQQ) efficiently catalyzes redox cycling [1,2] led to an understanding of the factors involved in this extraordinary property of PQQ. The efficiency of redox cycling is greatly determined by the extent of quinone polymerization. In quinoproteins where the quinone is covalently bound, polymerization is limited and glycine-dependent redox cycling constitutes a sensitive method for the detection of the protein-bound quinones. Furthermore, the characterization of the dialyzable, low molecular weight redox-cycling material present in biological fluids has shown that free PQQ is definitely responsible for this activity in bovine skim milk and is also likely to be the source of such redox-cycling activity in blood cells and in a variety of tissue extracts. Three aspects of our research—(1) quinone catalyzed redox cycling in NBT/glycinate, (2) quinoprotein detection based on this assay, and (3) the characterization of the redox-active PQQ-like material in biological fluids— are discussed below.

II. REDOX CYCLING IN NBT/GLYCINATE

A. Principle of Assay

Quinones, upon aerobic oxidation and reduction, generate superoxide and may also polymerize. In this reaction, the reactive semiquinone radical, formed by a one-electron transfer reduction, is reoxidized by dioxygen with superoxide generation. The quinone and its reduction products may also undergo self-polymerization to redox-inert products. This polymerization limits the extent and rate of redox cycling. In the case of the glycine-dependent redox cycling of PQQ, a Schiff's base adduct is formed, which can hydrolyze to a reactive quinoneimine, or become stabilized as a redox-inert oxazole (Fig. 1).

The glycine-dependent redox-cycling activity is determined by incubating quinones or hydroquinones with 1 M potassium glycinate at pH 10 for 1 hour at 25°C in the dark. Reduction of the yellowish nitro-blue tetrazolium (NBT) to blue formazan products is determined spectrophotometrically at 530 nm. NBT may be reduced both directly by reduced PQQ or indirectly by the superoxide formed in the dioxygen oxidation of reduced PQQ; at pH 10 about 80% is through superoxide and 20% by direct reduction. Because of the limited water solubility of the diformazan, albumin (700 μg/ml) is added to keep the diformazan in solution. If glycine is replaced by valine, PQQ does not reduce NBT. To test for other reducing

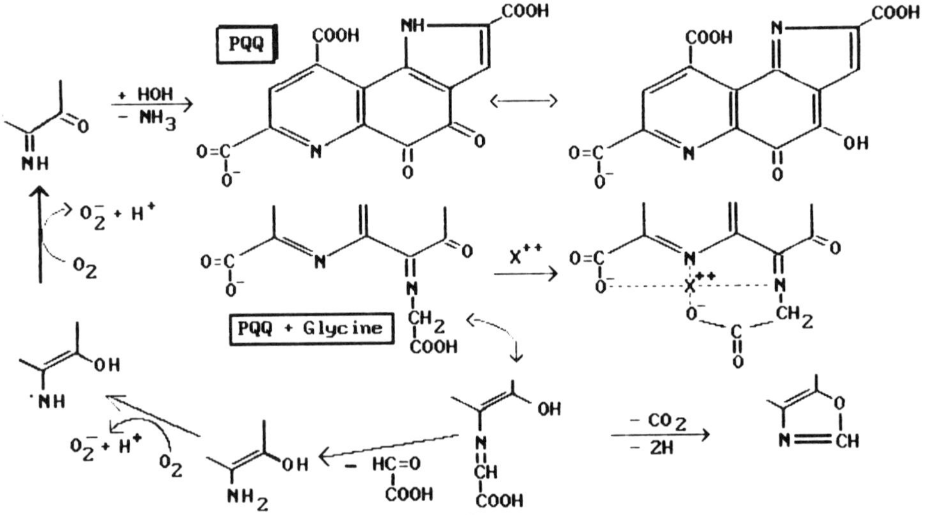

Figure 1 Chemistry of PQQ-catalyzed redox cycling fueled by glycine as reductant and dioxygen as oxidant. Tetrazolium reduction by the superoxide generated is used to follow the reaction.

compounds that, if present in high concentration, may directly reduce NBT, a control reaction is run in 0.5 M potassium valinate buffer, pH 10 (the lower concentration of valine is used because of its limited solubility). Ascorbate, which directly reduces NBT, can be rendered unreactive by borate at a concentration (10 mM) where the redox-cycling of PQQ is only marginally affected [3]. The rate of redox cycling in glycinate-NBT-borate, therefore, measures free PQQ-like material in quinoprotein-free samples.

B. Relative Reactivities of Quinones

Table 1 compares the redox-cycling activity of various quinones and other compounds in NBT/glycinate, pH 10. Of the compounds tested, PQQ redox cycles with the greatest efficiency—it undergoes an estimated 2000 redox cycles/hour under the conditions of the assay. In contrast, the isoelectronic 1,7-phenanthroline-dione (1,7-PD) undergoes about 300 redox cycles/hour. This may reflect a greater propensity for polymerization of the latter compound, compared to PQQ. With both compounds, polymerization is attenuated by the fused ring systems blocking the highly reactive nucleophilic sites adjacent to the quinone carbonyl groups [5]. In PQQ, dimerization is further limited by mutual charge repulsion. Other electronic structures of the aromatic nucleus are one order of magnitude less redox-active than 1,7-PD and show activities comparable to the aromatic triols that form *o*- or *p*-quinones. Simple *o*- or *p*-quinones show very little redox-cycling activity. The influence of polymerization in limiting redox cycling is also evident from the different reactivities of 5- and 3-hydroxyanthranilic acid (HAT). 3-HAT is quite unreactive, because it can form the redox-inert cinnabarinic acid [6]. PQQ reacts with certain amino acids to form redox-inert oxazoles [7,8]. In the presence of the 1 M glycinate used in the redox-cycling assay, oxazole formation could occur, but the alkaline pH will not favor decarboxylation.

From the above discussion, it appears that four factors contribute to the efficient redox-cycling of PQQ: (1) limited dimer formation by mutual charge repulsion by its three carboxylate groups, (2) restricted nucleophilic sites on the aromatic ring, (3) enhancement of the *ortho*-quinone reactivity by quinone-imine formation, and (4) marginal oxazole formation. Because of the high redox-cycling efficiency of PQQ, this compound is detected with considerable specificity and sensitivity (picomole range) by the glycinate/NBT/borate assay.

C. Inhibition of Redox Cycling

1. Metal Inhibition

The glycine-dependent redox cycling of PQQ is inhibited by divalent cations because of a complex formation that probably occurs with the Schiff's base adduct. Such inhibition caused lower than expected redox-cycling efficiencies (10–30%) of PQQ with certain batches of glycine. Inclusion of 0.01 M EDTA in the

Table 1 Redox Cycling with Glycinate-Nitroblue Tetrazolium: Reactivities of Quinones and Related Compounds

Compound	Absorbance/μmol (530 nm)
2,7,9-Tricarboxypyrroloquinoline quinone (PQQ)	20,000
PQQ + borate	18,000
1,7-Phenanthroline-5, 6-dione (1,7-PD)	3,150
4,7-Phenanthroline-5,6-dione (phanquone)	387
9,10-Phenanthrenedione (phenanthrenequinone, PAQ)	382[a]
1,10-Phenanthroline-5,6 dione (1,10-PD)	210
Acenaphthenequinone	107
Triols	
1,2,4-Benzenetriol	550
Trihydroxyphenylalanine (TOPA)	400
o-Quinones (catechols)	
1,2-Naphthoquinone	106
Aminoethyl-1,2-benzenediol (dopamine)	100
3,4-dihdroxyphenylacetic acid (DOPAC)	95
p-Quinones	
1,4-Naphthoquinone	30
2,5-Dihydroxyphenylacetic acid (homogentisic acid)	20
p-Benzoquinone	7
Quinoneamines	
5-Hydroxyanthranilic acid (5-HAT)	100
3-Hydroxyanthranilic acid (3-HAT)	3
Directly reducing compounds	
Ninhydrin	25
Ninhydrin + borate	0.02
Ascorbate	14
Ascorbate + borate	0.05

[a]Differs from value in [4] because of dilutional error.

reaction restored the redox-cycling activity of PQQ. Inhibition of redox cycling by divalent cations also occurs with protein-free milk dialysates. Chelex treatment of the PQQ-spiked milk dialysates yields a quantitative recovery of redox-cycling activity.

2. Inhibition by Organic Compounds

PQQ forms adducts with a variety of compounds such as acetone [9], hydrazines [10], cyclopropanol [11], and guanidino- [12] or 1,2-diamino-compounds [13], which also inhibit glycine-dependent redox cycling. The search for new and more potent inhibitors led to the identification of three interesting compounds shown in Figure 2.

The α-hydrogen in malononitrile is more active than in acetone, and this compound inhibits redox cycling more efficiently than does acetone. The IC_{50} for acetone is 8×10^{-3} M and for malononitrile 3.2×10^{-6} M. Aminomalononitrile, which may form a cyclic adduct with PQQ and cyanide release, is a more potent inhibitor ($IC_{50} = 1.8 \times 10^{-6}$ M) than malononitrile.

From the group of cyclopropanol inhibitors, the antidepressant monoamine oxidase inhibitor tranylcypromine was tested. This compound inhibits redox cycling and forms an unusual adduct with PAQ. Similar adduct formation appears to be the basis for its inhibitory action on PQQ-dependent redox cycling.

With regard to diamines, both aromatic diamines, 2,3-diamino-naphthalene (DAN), and 4,5-dimethyl-1,2-phenylenediamine (DIMPDA), form a phenazine adduct with PQQ [14]. (We have previously referred to this adduct structure as quinoxaline, which pertains to the ring system, without the aromatic substituted

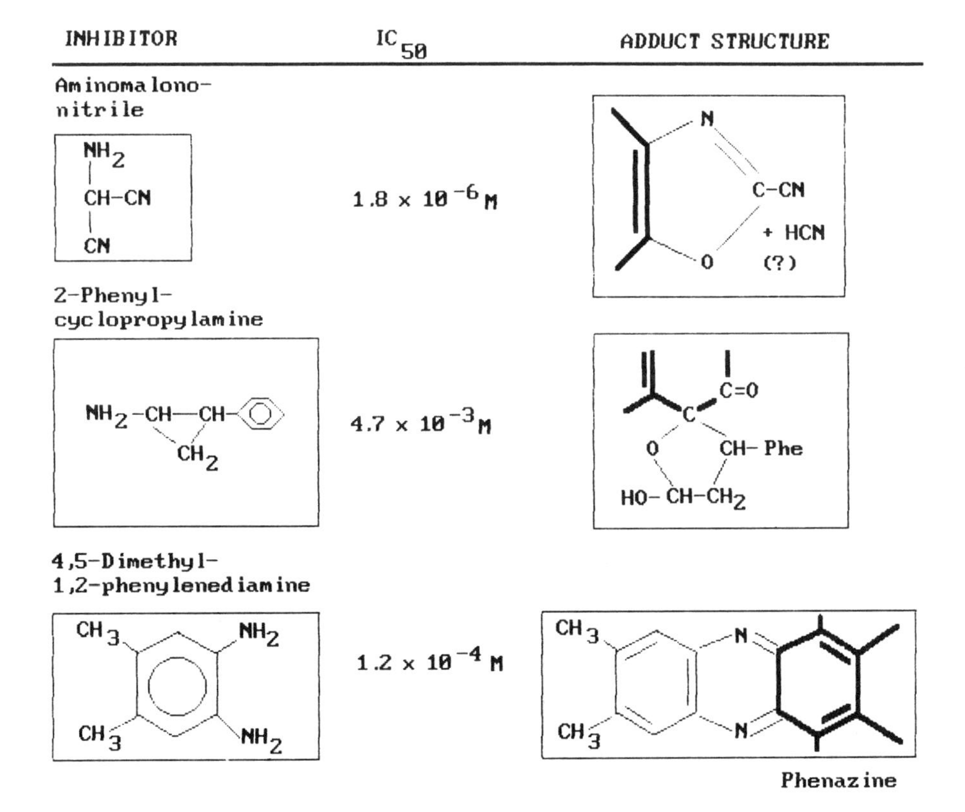

Figure 2 Inhibitors and adducts formed with PQQ (PQQ moiety indicated by dark lines).

benzene ring belonging to PQQ.) The adduct with DIMPDA has proven especially useful because DIMPDA can be tritiated by exposure to tritium gas. The tritiated DIMPDA adduct formed with the PQQ present in bovine milk dialysate can be characterized (see below). Also, the incorporation of radioactivity into quinoproteins exposed to tritiated DIMPDA can be demonstrated by autoradiography after SDS-PAGE separation [15].

III. DETECTION OF QUINOPROTEINS

Glycine-dependent redox cycling is efficient with protein-bound quinones since quinone polymerization is limited. This is exemplified by the dopa-residues in the vitelline proteins [16,17], the catalytic 6-hydroxydopa residue in serum amine oxidase [18], the tryptophan tryptophylquinone prosthetic group in the active site of the bacterial methylamine dehydrogenase [19], and the cysteinyl-dopa moiety in proteins exposed to dopaquinone [20,21] (Fig. 3).

In order to test for the presence of a bound quinone cofactor in a protein, a

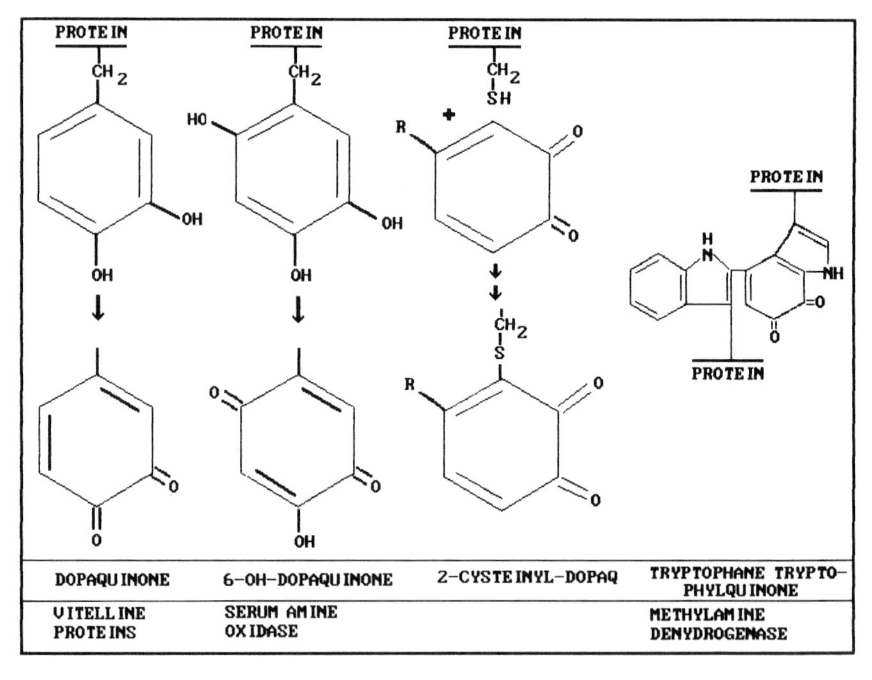

Figure 3 The structures of protein-bound quinones.

purified preparation of this protein is separated from other protein contaminants by SDS-PAGE, electroblotted onto nitrocellulose, and stained with the NBT/glycinate reagent [22]. Quinoproteins are stained blue-purple by the deposition of formazan, while other proteins remain unstained (Fig. 4). After the stained strips are washed in borate buffer, they are counterstained with Ponceau S.

With this methodology, we confirmed the following proteins to be quinoproteins: pig kidney diamine oxidase, rat aorta lysyl oxidase, bovine serum amine oxidase, bacterial methylamine dehydrogenase (the smaller 15-kD or β-subunit), and bacterial methylamine oxidase. Two proteins, soybean lipoxygenase [23] and dopamine β-hydroxylase [22], purported by Duine and colleagues to be PQQ-containing quinoproteins [24], do not react as quinoproteins by redox cycling. We also found bovine serum albumin to be weakly positive and perhaps partially quinated. The dopamine-β-hydroxylase preparation tested was found to contain an unidentified 97-kD quinoprotein.

From these results and others it appears that PQQ does not occur as a covalently bound cofactor. Not only has the identity of the quinone cofactor of the two putative PQQ-containing quinoproteins serum amine oxidase and methylamine dehydrogenase been firmly established and shown not to be PQQ, but the absence of PQQ in dopamine-β-hydroxylase [25], soybean lipoxygenase [25], and bovine serum amine oxidase [27] has been documented by other techniques. Raman

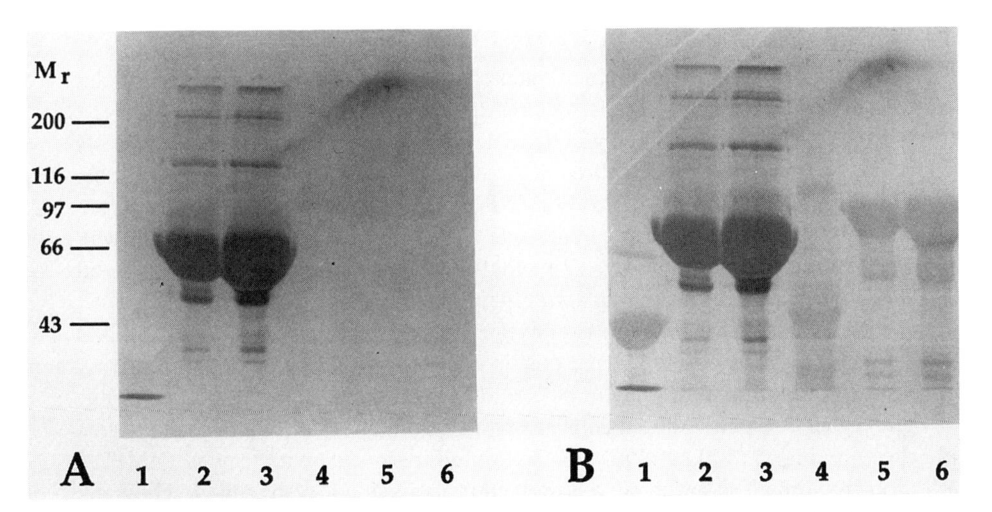

Figure 4 Staining of SDS-PAGE gel with NBT/glycinate (A) and Ponceau S (B). Lane 1: methylamine dehydrogenase (15 kD subunit stains as quinoprotein, not 45 kD subunit). Lanes 2,3: methylamine oxidase. Lane 4: pig kidney diamine oxidase. Lanes 5,6: native lipoxygenase (unstained is not a quinoprotein). (Reprinted with permission from Ref. 23.)

resonance spectra indicate that the organic cofactor of all amine oxidases is very similar [28], rendering it unlikely that covalently bound PQQ is the quinone cofactor in other amine oxidases. Finally, the cofactor in galactose oxidase, which was also believed to be PQQ, has now been shown to be a cross-link derived from a cysteine and a tyrosine residue [29].

IV. IDENTIFICATION OF FREE PQQ IN BOVINE MILK

In view of the protective role of PQQ in mammals (see Chapter 17), it is important to establish the identity of the redox-cycling material in tissues and biological fluids. Most of the work outlined was performed on fresh, unpasteurized bovine milk because of its easy availability (see Ref. 30).

The analytical approach was dictated by the low concentration and high reactivity of the putative PQQ. It was found that the redox-cycling activity of a freshly prepared milk dialysate cochromatographs on reverse-phase chromatography with authentic PQQ, but that with sample storage, even at 4°C, redox-cycling material begins to show an early elution with a decrease in overall activity. This behavior is also reproduced when authentic PQQ is added to biological samples and allowed to age. A fast chromatographic separation of fresh samples with sensitive electrochemical detection was therefore used. Figure 5 shows the separation by HPLC of a protein-free milk dialysate using pH gradient elution and monitoring of the effluent with a 16-channel electrochemical detector. A peak with both a retention time and a 'redox-signature' identical to that of authentic PQQ is apparent with the milk dialysate. With this analytical system elution time is determined by both hydrophobicity and charge properties of the analyzed molecule while the detector responses define the molecule functionally.

Further evidence identifying PQQ as the redox-active material in milk consists of the following: (1) the rechromatographed material collected from multiple analytical runs elutes in the position of PQQ and shows an absorption spectrum which is very similar to that of authentic PQQ; (2) Milk dialysates reacted with DAN or the carbonyl reagent 2-methyl-3-benzothiazolinone hydrazone yield adducts which cochromatograph with the respective PQQ authentic adducts [31]; (3) the radioactivity in a milk dialysate reacted with tritiated DIMPDA and separated by HPLC eluted in the position of authentic PQQ-DIMPDA.

The concentration of PQQ in bovine milk is in the range of 50–500 pmol/ml. The higher concentrations calculate from the quantitative evaluation of the results shown in Fig. 5 (400 pmol/ml) and from the experiments with labeled DIMPDA (500 pmol/ml). The lower value is derived from glycine-driven redox-cycling and may be explained by the presence of inhibitory compounds or dissociable complexes with diminished redox-cycling activity. Until the complex interaction of free PQQ with the components of the analytical matrix is better understood the values given above have to be regarded as estimates.

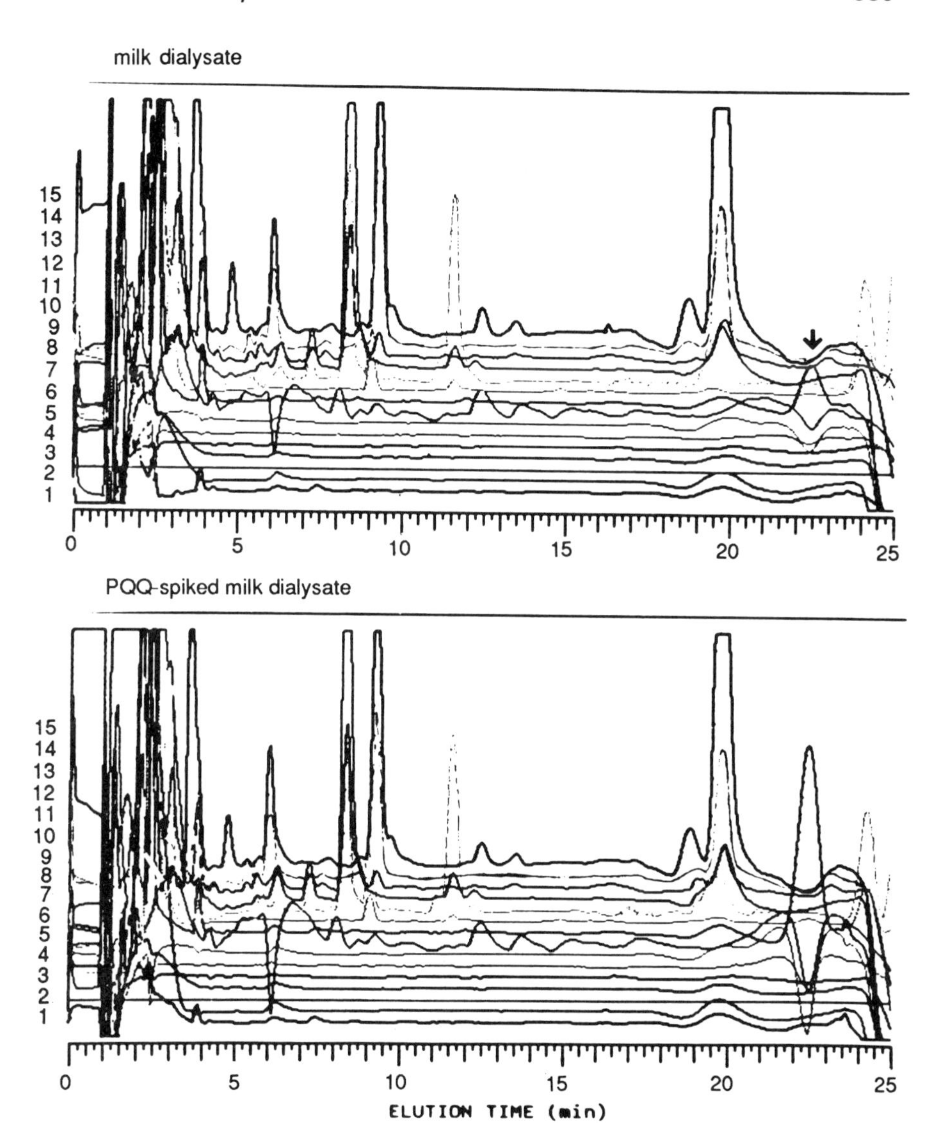

Figure 5 Separation of milk dialysate by HPLC using pH gradient elution and electrochemical detection with a CEAS 5500 16-channel electrochemical detector. Sensor settings: T1 (0 mV), T8 (−630 mV); σE (−90 mV); T9 (0 mv), T16 (910 mV); σE (130 mV). (Arrow indicates elution position of PQQ at 22.46 min.) The signal ratios between the various channels of the peak at 22.4 min. are comparable in both samples, indicating electrochemical identity of putative and authentic PQQ.

The redox-cycling activity, present in other biological samples including plasma, lysates of blood cells, brain homogenates, cerebrospinal fluid, synovial fluid and bile is under further investigation, but it appears likely that it represents free PQQ.

REFERENCES

1. Flückiger, R., Woodtli, T., and Gallop, P. M. (1988). The interaction of aminogroups with pyrroloquinoline quinone as detected by the reduction of nitroblue tetrazolium. *Biochem. Biophys. Res. Commun. 153*: 353.
2. Paz, M. A., Flückiger, R., Henson, E., and Gallop, P. M. (1989). Direct and amplified redox-cycling measurements of PQQ in quinoproteins and biological fluids: PQQ-peptides in pronase digests of DBH and DAO. In *PQQ and Quinoproteins* (Jongejan, J. A., and Duine, J. A., eds.), Kluwer Academic Publishers, Norwell, pp. 131–143.
3. Paz, M. A., Flückiger, R., and Gallop, P. M. (1990). Redox-cycling is a property of PQQ but not of ascorbate. *FEBS Lett. 264*: 283.
4. Gallop, P. M., Paz, M. A., and Flückiger, R. (1990). The biological significance of quinonoid compounds in, on, and out of proteins. *Chemtracts-Biochem. & Mol. Biol. 1*: 357.
5. Pierpoint W. S. (1990). PQQ in plants. *Trends Biochem. Sci. 15*: 299.
6. Ishii, T., Iwahashi, H., Sugata, R., Kido, R., and Fridovich, I. (1990). Superoxide dismutase enhances the rate of autooxidation of 3-hydroxyanthranilic acid. *Arch. Biochem. Biophys. 276*: 248.
7. Adachi, O., Okamoto, K., Shinagawa, E., Matsushita, K., and Ameyama, M. (1988). Adduct formation of pyrroloquinoline quinone and amino acid. *Biofactors 1*: 251.
8. van Kleef, M. A. G., Jongejan, J. A., and Duine, J. A. (1989). Factors relevant in the reaction of pyrroloquinoline quinone with amino acids: Analytical and mechanistic implications. *Eur. J. Biochem. 183*: 41.
9. Duine, J. A., Frank, J., and Jongejan, J. A. (1986). Enzymology of quinoproteins. *Adv. Enzymol. 59*: 169.
10. Mure, M., Nii, K., Inoue, T., Itoh, S., and Ohshiro, Y. (1990). The reaction of coenzyme PQQ with hydrazines. *J. Chem. Soc. Perkin Trans. 2*: 315.
11. Dijkstra, M., Frank, J., Jongejan, J. A., and Duine, J. A. (1984). Inactivation of quinoprotein alcohol dehydrogenases with cyclopropane-derived suicide substrates. *Eur. J. Biochem. 140*: 369.
12. Mure, M., Nii, K., Itoh, S., and Ohshiro, Y. (1990). Chemical behavior of coenzyme PQQ toward aminoguanidine: Redox reaction and adduct formation. *Bull. Chem. Soc. Jpn. 63*: 417.
13. Mure, M., Itoh, S., and Ohshiro, Y. (1989). Chemical behavior of coenzyme PQQ towards diamines. *Chem. Lett.* : 1491.
14. Gallop, P. M., Henson, E., Paz, M. A., Greenspan, S. L., and Flückiger, R. (1989). Acid-promoted tautomeric lactonization and oxidation-reduction of pyrroloquinoline quinone (PQQ). *Biochem. Biophys. Res. Commun. 163*: 755.
15. Paz, M. A., and Apekin, V. (1992). Radiochemical detection of quinoproteins (submitted).

16. Waite, J. H., and Rice-Ficht, A. C. (1987). Presclerotized eggshell protein from the liver fluke *Fasciola hepatica*. *Biochemistry 26*: 7819.
17. Waite, J. H., and Rice-Ficht, A. C. (1989). A histidine-rich protein from the vitellaria of the liver fluke *Fasciola hepatica*. *Biochemistry 28*: 6104.
18. Janes, S. M., Mu, D., Wemmer, D., Smith, A. J., Kaur, S., Maltby, D., Burlingame, A. L., and Klinman, J. P. (1990). A new redox cofactor in eukaryotic enzymes: 6-Hydroxydopa at the active site of bovine serum amine oxidase. *Science 248*: 981.
19. McIntire, W. S., Wemmer, D. E., Christoserdov, A., and Lindstrom, M. E. (1991). A new cofactor in a prokaryotic enzyme: Tryptophane tryptophylquinone as the redox prosthetic group in methylamine dehydrogenase. *Science 252*: 817.
20. Kato, T., Ito, S., and Fujita, K. (1986). Tyrosinase-catalyzed binding of 3,4-dihydroxyphenylalanine with proteins through the sulfhydryl group. *Biochim. Biophys. Acta 881*: 415.
21. Ito, S., Kato, T., and Fujita, K. (1988). Covalent binding of catechols to proteins through the sulfhydryl group. *Biochem. Pharmacol. 37*: 1707.
22. Paz, M. A., Flückiger, R., Boak, A., Kagan, H., and Gallop, P. M. (1991). Specific detection of quinoproteins by redox-cycling staining. *J. Biol. Chem. 266*: 689.
23. Veldink, G. A., Boelens, H., Maccarone, M., van der Lecq, F., Vliegenthart, J. F. G., Paz, M. A., Flückiger, R., and Gallop, P. M. (1990). Soybean lipoxygenase-1 is not a quinoprotein. *FEBS Lett. 270*: 135.
24. Duine, J. A., and Jongejan, J. A. (1989). Quinoproteins, enzymes with pyrroloquinoline quinone as cofactor. *Annu. Rev. Biochem. 58*: 403.
25. Robertson, J. G., Kumar, A., Mancewicz, J. A., and Villafranca, J. J. (1989). Spectral studies of bovine dopamine-β-hydroxylase: absence of covalently bound pyrroloquinoline quinone. *J. Biol. Chem. 264*: 19916.
26. Michaud-Soret, I., Daniel, R., Chopard, C., Mansuy, D., Cucurou, C., Ullrich, V., and Chottard, J. C. (1990). Soybean lipoxygenases-1, -2a, -2b and -2c do not contain PQQ. *Biochem. Biophys. Res. Commun. 172*: 1122.
27. Kumazawa, T., Seno, H, Urakami, T., and Suzuki, O. (1990). Failure to verify the presence of pyrroloquinoline quinone (PQQ) in bovine plasma amine oxidase by gas chromatograpy/mass spectrometry. *Arch. Biochem. Biophys. 283*: 533.
28. McIntire, W. S., Dooley, M. A., McGuirl, M. A., Cote, C. E., and Bates, J. L. (1990). Methylamine oxidase from Arthrobacter P1 as a prototype of eukaryotic plasma amine oxidase and diamine oxidase. *J. Neural. Trans. Suppl. 32*: 315.
29. Itoh, N., Phillips, S. E. V., Stevens, C., Ogel, Z. B., McPherson, M. J., Keen, J. N., Yadav, K. D. S., and Knowles, P. F. (1991). Novel thioether bond revealed by a 1.7 Å crystal structure of galactose oxidase. *Nature 350*: 87.
30. Flückiger, R., Paz, M. A., Bergethon, P. R., Bishop, A., Henson, E., Matson, W. R., and Gallop, P. M. (1992). Identification of free PQQ in bovine milk (submitted).
31. Flückiger, R., Paz, M. A., Bergethon, P. R., Greenspan, S. L., Goodman, S., Henson, E., and Gallop, P. M. (1991). First isolation of methoxatin (PQQ) from bovine skim milk and other mammalian sources. *FASEB J. 5* (Suppl 3):A 1510

14

Biosynthesis of PQQ

Clifford J. Unkefer

Los Alamos National Laboratory, Los Alamos, New Mexico

I. INTRODUCTION

As discussed elsewhere in this book, *o*-quinone prosthetic groups have been positively identified in alcohol dehydrogenases [1,2], glucose dehydrogenase [3], amine oxidases [4], and amine dehydrogenases [5]. These prosthetic groups can be divided into two classes based on the presence or absence of a covalent linkage between cofactor and enzyme. The covalently linked *o*-quinone cofactors of the amine oxidases (TOPA, trihydroxyphenylalanine) [4] and amine dehydrogenases (TTQ, trytophan tryptophylquinone) [5] are apparently the product of posttranslational modification of the side chains of tyrosyl or tryptophanyl residues present in the parent protein. Pyrroloquinoline quinone (PQQ), though tightly bound to methanol dehydrogenase (MDH), is not covalently linked to the enzyme and can be removed by nonhydrolytic enzyme denaturation [6]. As discussed below, mutants of methylotrophic bacteria have been isolated which require exogenous PQQ for the expression of MDH activity [7–9]. In addition, the PQQ-dependent glucose dehydrogenase is produced by wild-type *Escherichia coli* as an apoenzyme, and active enzyme/PQQ complex is produced only when the organism is cultured in the presence of PQQ [10]. Moreover, activity is restored to the apoform of the glucose dehydrogenase in vitro by incubation with PQQ [11] or the structural analog 4,7-phenanthroline-5,6-dione [12]. Clearly, PQQ is not produced by posttranslational modification of amino acyl residues in these bacterial dehydrogenases and must be the product of an independent biosynthetic pathway.

PQQ biosynthesis is complicated by the fact that it is synthesized in the cytoplasm and must then find its way to the periplasm where the active enzyme/cofactor complex is assembled. Interestingly, during this process, methylotrophs release PQQ into the environment. As discussed below, during growth on methanol and methylamine, all methylotrophs tested release relatively large quantities of PQQ into their growth medium [13–15]. On the other hand, *E. coli* have retained the ability to express PQQ-dependent glucose dehydrogenase but have apparently lost the ability to express the pathway for de novo PQQ biosynthesis and must rely on an endogenous supply of PQQ to produce active glucose dehydrogenase. This chapter reviews studies that have been carried out on the biosynthesis of PQQ. The chapter is divided into two major sections based on the general approaches used to probe the biosynthesis of PQQ. First, stable isotope-labeling studies have been used to identify the biosynthetic precursors to PQQ. In addition, classical and molecular genetic approaches have been used to identify the genes required for PQQ biosynthesis.

II. BIOSYNTHETIC PRECURSORS TO PQQ

A. Production of PQQ

Preliminary to biosynthetic studies, Ameyama and coworkers reported that PQQ is produced in the growth medium of methylotrophic bacterium M5 when cultured with methanol as the sole source of carbon [13]. Since then a variety of methylotrophs including *Methylobacterium extorquens* AM1, *Hyphomicrobium* X, and methylotrophic bacterium W3A1 have been shown to excrete PQQ during growth on C-1 compounds including methanol and methylamine [14,15]. These organisms produce up to 30 μmol of PQQ per liter of culture medium. PQQ is easily isolated from the culture filtrate using two chromatography steps [13]. At neutral pH, PQQ binds tightly to DEAE chromatography resins and is eluted at high KCl concentration. Under acidic conditions (pH = 2.0), PQQ binds to a reversed-phase chromatography column and is eluted with methanol. These studies provided a convenient source of PQQ for the biosynthetic studies described below.

B. Labeling Studies in *Methylobacterium* AM1

During growth on methanol, methylotrophs derive essentially all of their carbon from the methanol [16]; therefore, it is impossible to extract information pertinent to the biosynthesis of PQQ from experiments using labeled methanol as the sole carbon and energy source. Because one can determine which carbons in PQQ are derived from C-1 and/or C-2 of ethanol, it is a more useful precursor for labeling studies of biosynthesis. Anthony and coworkers [17,18] demonstrated that the facultative methylotroph *Methylobacterium extorquens* AM1 could grow using ethanol as a sole carbon source and further that, during growth on ethanol, methanol dehydrogenase was expressed and used by the organism for the oxidation of ethanol. Therefore, AM1 biosynthesizes and was shown to excrete PQQ during growth on ethanol [19].

The ^{13}C enrichments in PQQ biosynthesized from [1-^{13}C]ethanol based on analysis of NMR intensities are summarized in Table 1 [19]. C-1 of ethanol labels predominantly the three carboxylates (C-2′,7′, and 9′) and carbons 5, 5a, and 9a. The singlet character of the carboxylates indicates that they are incorporated into positions in which their neighbors arise from C-2 of ethanol. Carbons 5, 5a, and 9a each yield three resonances, which are the combination of a singlet from singly labeled species and a doublet ($^1J_{C\text{-}C}$ = 60 Hz) from species labeled at C-5 and C-5a or C-9a and C-5a. The precursors were identified by comparing the selective ^{13}C-labeling patterns in PQQ with those observed in amino acids also derived from *M. extorquens* AM1 cultured on [1-^{13}C]ethanol. In PQQ, C-1 of ethanol significantly labels C-7′ (59%) and C-9′ (>99%), but not C-9 (<2%); similarly, C-2 of ethanol labels PQQ at C-7 (64%), C-8 (61%), and C-9 (76%), but not C-9′. These

Table 1 Chemical Shift Assignments and ^{13}C Enrichments of PQQ

		^{13}C Enrichments[b] (atom %^{13}C)	
Carbon	δ (ppm)[a]	[1-^{13}C]Ethanol	[2-^{13}C]Ethanol
1a	136.7	16	54
2	127.6	23	68
2′	161.3	82	24
3	113.8	16	65
3a	123.4	16	59
4	173.4	13	59
5	179.2	27 "triplet"	46
5a	148.1	54 "triplet"	27
7	146.5	16	64
7′	165.4	59	24
8	130.3	17	61
9	142.2	n.o.	76
9′	167.2	101	n.o.
9a	126.1	35 "triplet"	37

[a]Chemical shifts of a natural abundance sample (17.2 mg/ml in d_6-DMSO at 25°C) of PQQ obtained from Fluka Chemical Co.
[b]^{13}C NMR spectra were obtained at 25°C using a 45° pulse and with the ^1H decoupler gated off for 10 s to minimize NOE effects. Biosynthetic samples of PQQ contained 10 mg/3.0 ml DMSO.
n.o. = not observed.

labeling patterns are essentially identical to those observed in glutamate. An equivalent result has been presented in detail based on labeling patterns in glutamate and PQQ derived from [2-^{13}C]ethanol [20].

[1-^{13}C]Ethanol also labels the phenol ring of tyrosine at C-3′ and C-4′ yielding a NMR spectrum that exhibits $^{1}J_{C\text{-}C}$ coupling; this labeling pattern is identical to that observed in tyrosine isolated from *E. coli* cultured on [1-^{13}C]lactate [21]. This adjacent labeling of C-3′ and C-4′ of tyrosine arises from the C-1 to C-1 joining of two trioses in gluconeogenesis and is diagnostic of compounds that arise from the shikimate pathway. The incorporation of C-1 of ethanol into C-2′, 5, 5a, and 9 of PQQ is equivalent to its incorporation into C-1, 3′, and 4′ of tyrosine. The adjacent labeling evident from the high degree of ^{13}C coupling at C-3′ and C-4′ in tyrosine is also observed in the orthoquinone-containing ring in PQQ, which has adjacent ^{13}C labeling (doublets) at C-5a and C-5, or C-5a and C-9a. This labeling implies that the orthoquinone-containing ring arises from a symmetric compound (C_2 axis through C-1′ and C-4′). The requirement of C_2 symmetry supports tyrosine or phenylalanine and rules out indole as a precursor for that portion of PQQ containing the orthoquinone and pyrrole rings. The [1-^{13}C]ethanol-labeling experiment coupled with the obvious structural homologies provided a working hypothesis for the biosynthetic origins of PQQ (Fig. 1). It was proposed that glutamate provides N-6 and carbons 7′, 7, 8, 9, and 9′, while the remaining nine carbons and N-1 are donated by an amino acid from the shikimate pathway, most likely tyrosine [18].

C. Direct Evidence for Tyrosine as a Biosynthetic Precursor to PQQ

1. Labeling with [3′,5′-$^{13}C_2$]Tyrosine

Experiments with ^{13}C-labeled ethanols described above indicated that a symmetric product of the shikimate pathway, either tyrosine or phenylalanine, is the precursor to the six-membered ring containing the *o*-quinone and to the attached pyrrole-2-

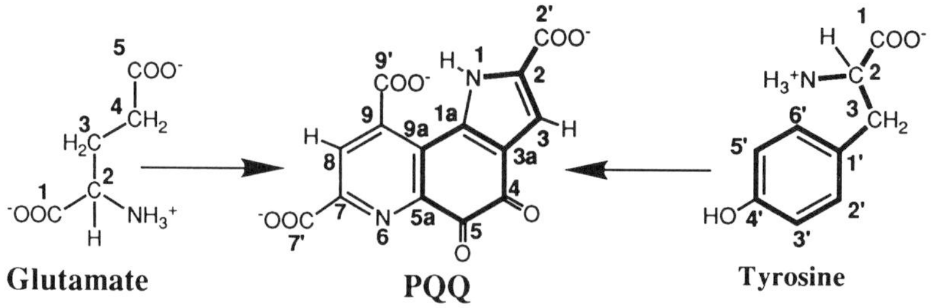

Figure 1 Precursors for PQQ biosynthesis.

carboxylate moiety. This transformation would entail an intramolecular cyclization of the amino acid backbone with C-2 of the phenyl (or phenol) group. Because such a reaction is known to occur in the biosynthesis of melanin [22], tyrosine seemed to be the most likely candidate as a direct precursor to PQQ. This hypothesis was tested by feeding ^{13}C-labeled tyrosine to *M. extorquens* AM1 and analyzing the resulting labeling pattern in PQQ [20,23]. Details of the feeding experiments with tyrosine have been reported [20]; in general the organism was grown in the presence of 0.5 mM ^{13}C-labeled tyrosine. PQQ isolated from cultures that contained L-[3′,5′-$^{13}C_2$]tyrosine was labeled only at C-5 and C-9a. Further, the ^{13}C resonances for C-5 and C-9a are doublets as a result of geminal ^{13}C—^{13}C coupling, demonstrating that the phenol group of tyrosine is incorporated intact into the *o*-quinone–containing ring of PQQ. ^{1}H NMR analysis provided quantitative data on the ^{13}C enrichment: C-9a contained 83 atom % ^{13}C.

2. Labeling with [3′,5′-$^{13}C_2$] and [β-^{13}C]Tyrosine

The results outlined above demonstrate that the phenol ring of tyrosine provides the six carbons of the *o*-quinone–containing ring. To examine the possibility that internal cyclization of the tyrosyl backbone forms the pyrrole-2-carboxylic acid moiety, PQQ was labeled with an equimolar mixture of L-[3′,5′-$^{13}C_2$] and L-[β-^{13}C]tyrosine [20]. Indirect incorporation of tyrosine, involving an intermediate step such as β-elimination of phenol (i.e., phenol ammonia lyase), would dilute the label at C-3 relative to that at C-9a and C-5. However, the equal labeling of C-9a (41%) and C-3 (42%) observed in PQQ derived from the 50:50 mixture of L-[3′,5′-$^{13}C_2$] and L-[β-^{13}C]tyrosine implies intact incorporation of tyrosine into PQQ. These data are consistent with a biosynthetic pathway in which the tyrosyl backbone is cyclized to form the pyrrole-2-carboxylic acid moiety in PQQ, the reactions being similar to those that occur in melanin biosynthesis [22]. In addition, these data rule out more complicated biosynthetic routes to the pyrrole ring such as that used in lincomycin biosynthesis to produce the pyrrolidine ring from tyrosine [24].

3. ^{13}C Labeling Studies in Hyphomicrobium X

The biosynthesis of PQQ has been examined in *Hyphomicrobium* X using a clever isotope dilution approach [23,25]. The direct incorporation of amino acids into PQQ was determined by culturing the organism in a medium that contained [^{13}C]methanol as a growth substrate and was supplemented with unlabeled amino acids. When PQQ was purified from the culture medium of *Hyphomicrobium* X grown in the presence of ^{13}C-methanol plus L-phenylalanine, no significant changes in ^{1}H and ^{13}C NMR spectra were found as compared to the spectra of [U-^{13}C]PQQ, indicating that L-phenylalanine is not incorporated into the PQQ skeleton. However, PQQ isolated from cultures that were supplemented with tyrosine yielded, ^{1}H and ^{13}C NMR spectra that were clearly altered. The ^{1}H NMR resonance from H-3 collapsed to a singlet (7.15 ppm); H-8 (8.64 ppm) was a doublet (168 Hz) as a result of one-bond C-H coupling. Thus, label at C-3 had been diluted to less than 5% with ^{12}C and must have been derived from tyrosine; C-8 was enriched with ^{13}C (94%) and was derived from the labeled methanol.

The tyrosine labeling [20,23] and dilution [23,25] experiments provide evidence that the carbon skeleton of tyrosine is incorporated intact into PQQ; the phenol side chain provides the six carbons of the ring containing the *ortho*-quinone, whereas internal cyclization of the amino acid backbone forms the pyrrole-2-carboxylic acid moiety.

4. *Labeling with [α-^{15}N, 2′,6′-$^{13}C_2$]Tyrosine*

To determine if the pyrrole nitrogen derives from the α-amino group of tyrosine, L-[α-^{15}N, 2′,6′-$^{13}C_2$)tyrosine was fed to *M. extorquens* AM1 cultures [20]. Intramolecular cyclization of this compound would yield PQQ with a ^{15}N (N-1) directly bonded to ^{13}C (C-1a) and would result in a direct ^{13}C-^{15}N spin-spin coupling network. The spectrum of PQQ isolated from this culture was labeled with ^{13}C at C-4 and C-1a; the ^{13}C enrichment, estimated from the ^{1}H NMR of signal of H-3, was 50.7%. From integration of the ^{13}C NMR signal of C-1a, it is estimated that 15.7% of the molecules labeled with ^{13}C also contain ^{15}N ($^{1}J_{C1a\text{-}N}$ = 14.8 Hz). The overall ^{15}N enrichment was determined to be 10.6% from integration of ^{1}H NMR signal of the N-H resonance, which is directly coupled to N-1 ($^{1}J_{N\text{-}H}$ = 97 Hz). As expected, there is significant dilution of the ^{15}N label due to transamination; however, the incorporation of ^{15}N into PQQ was 30-fold greater than that expected if the tyrosine nitrogen equilibrated with the free ammonium pool. In addition, there is a strong correlation between ^{13}C and ^{15}N labeling. That is, of the PQQ molecules labeled with ^{15}N in the pyrrole nitrogen, a greater fraction contains ^{13}C at C-1a (75%) than contains ^{12}C at C-1a (25%). Clearly the α-nitrogen of tyrosine was not randomized in PQQ, demonstrating that tyrosine, and not its requisite α-keto acid, is the precursor for PQQ biosynthesis.

The labeling studies recounted provide direct evidence that carbons 2′,2,3,3a,4,5,5a,9a,1a, and N-1 of PQQ are derived from tyrosine. Less direct ethanol-labeling experiments indicate that the remainder of PQQ is derived from glutamate. Thus far, efforts to label PQQ directly with glutamate have failed. Glutamate added to *M. extorquens* AM1 cultures was removed rapidly from the medium. It is likely that, as is the case with succinate [26], glutamate added to the medium induces the expression of a functional TCA cycle in *M. extorquens* AM1, and labeled glutamate is rapidly oxidized.

III. GENETIC STUDIES OF PQQ BIOSYNTHESIS

PQQ biosynthesis has been probed by molecular and classical genetics by several groups. Goosen and coworkers screened for mutants of *Acinetobacter calcoaceti-*

cus that required exogenous PQQ for expression of PQQ-dependent glucose dehydrogenase [7]. These mutants belonged to four complementation groups. A 5 kb fragment of DNA was isolated that complemented all four classes of *pqq* mutants. Mapping studies carried out on this DNA fragment demonstrated that it carries all four PQQ biosynthesis genes, probably located in three transcriptional units. In addition, insertion mutations between the three transcriptional units do not affect PQQ expression. This 5 kb DNA fragment has been sequenced [27]. It contains reading frames that correspond to four PQQ biosynthesis genes. Three of the genes encode for proteins large enough for enzyme function (Mr = 29,700, 10,800, and 43,600). However, the most probably reading frame for the fourth gene encodes for a peptide of only 24 amino acids. In order to test if expression of these four genes is sufficient to encode for biosynthesis of PQQ, a plasmid was constructed that contained the four PQQ biosynthesis genes under control of the *lac* promoter [27]. The plasmid was transformed to an *E. coli* strain that carries the *pstI* mutation. This strain will not grow using glucose as its sole carbon source unless PQQ is added to the growth medium. The transformants were able to form colonies on minimal medium plates that contained glucose, indicating they biosynthesized PQQ.

The genetic studies of *A. calcoaceticus* indicate these that four genes are both necessary and sufficient for the biosynthesis of PQQ. Only three of these genes encode for proteins large enough for enzymatic activity. The fourth gene product is apparently a 24-amino-acid polypeptide; its role in PQQ biosynthesis has been the subject of some speculation [27]. Because the peptide contains both PQQ precursors glutamate and tyrosine, it has been proposed that the peptide rather than the free amino acids could serve as the precursor for PQQ biosynthesis. Alternatively, this gene could serve in a regulatory role.

These results in *A. calcoaceticus* were perplexing because it is difficult to imagine a pathway from glutamate and tyrosine to PQQ that requires only three unique enzymatic steps. In addition, these results seem to be in partial conflict with mutant studies carried out in *M. extorquens* AM1 by Lidstrom and coworkers [8] and in *Methylobacterium organophilum* DSM 760 by Gasser and coworkers [9]. Rather than using a screening protocol to identify PQQ biosynthesis in *A. calcoaceticus*, these organisms were subjected to a selection technique [28] where cells treated with a chemical mutagen are cultured in the presence of allyl alcohol. Allyl alcohol is oxidized by methanol dehydrogenase to yield acrolein, which is very toxic. Mutants which cannot express active MDH (*mox*) are resistant to allyl alcohol and are able to grow in its presence. When cultured in the presence of exogenous PQQ, one class of the *mox* mutants grow with methanol as their sole carbon source. The mutants are classified phenoltypically as PQQ biosynthesis mutants. The PQQ biosynthesis mutants fall into seven complementation groups in *M. extorquens* AM1 [8] and six groups in *M. organophilum* DSM 760 [9]. This could provide for a more reasonable six- or seven-step pathway for the conversion of glutamate and tyrosine to PQQ. In cross-feeding experiments, extracts from PQQ biosynthesis mutants have been checked for their ability to reconstitute PQQ biosynthesis. None of the PQQ biosynthesis mutants in *M. extorquens* AM1 (M. E. Lidstrom, personal communication) or *A. calcoaceticus* [29] were able to cross-feed each other. It is possible that intermediates in the pathway for PQQ biosynthesis are unstable or, more likely, cannot be transported across the cellular membrane.

The apparent conflict between the results obtained in *A. calcoaceticus* and the methylobacterium species was resolved recently. Glasser and coworkers [30] isolated a spontaneous mutant (EF mutant) of an *E. coli* strain carrying the *pstI* mutation, which apparently biosynthesizes PQQ. The parent strain lacks a functional phosphotransferase system (PTS^-) and cannot grow on a variety of sugars including glucose. Addition of exogenous PQQ to the growth medium of the PST^- strain restores their ability to grow using glucose as a sole source of carbon. The EF mutant strain grew on glucose and expressed active glucose dehydrogenase in the absence of added PQQ. Apparently *E. coli* contain the genetic information required for the biosynthesis of PQQ [30]. This result complicates the interpretation of the *E. coli* transformation/PQQ expression experiments described above [27]. The four *A. calcoaceticus* genes identified by Goosen and coworkers may not be sufficient for PQQ biosynthesis.

IV. BIOSYNTHETIC PATHWAY

In contrast to the biosynthetic pathways leading to other cofactors such as riboflavin or folic acid, the biochemical transformation that leads to PQQ is remarkably efficient in the sense that all the carbons and probably both nitrogens of the precursors are conserved in the product. Genetic studies indicate that the pathway for conversion of glutamate and tyrosine requires only three to seven enzymes. Because the process involves the loss of 12 electrons in the conversion of tyrosine and glutamate to product, PQQ biosynthesis involves primarily oxidative reactions. Oxidation of the phenol side chain must be an early step in the pathway because it is requisite for the formation of the pyrrole ring by cyclization of the tyrosine backbone. Given this data, a possible route for PQQ biosynthesis (Fig. 2) was outlined [20]. In this route, tyrosine or some derivative of tyrosine is oxidized to dopaquinone in a reaction catalyzed by a monophenol monooxygenase–like enzyme (tyrosinase, EC 1.14.18.1). Glutamate could form a Schiff base with dopaquinone. The cyclization of the tyrosine backbone to form the pyrrole ring could occur by a Michael-type addition analogous to the known nonenzymatic cyclization of dopaquinone to form dopachrome [22]. Alternatively, dopachrome has been proposed as an intermediate in the biosynthesis of PQQ and glutamate added in a subsequent reaction [25]. Several groups have failed to detect tyrosinase activity in cell-free extracts. In addition, no sequence homology exists

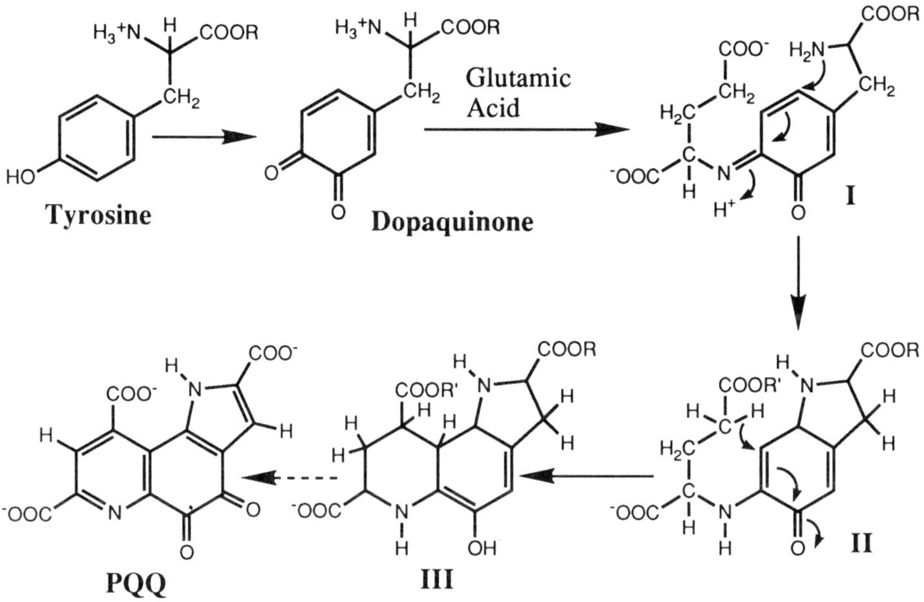

Figure 2 Possible route for PQQ biosynthesis.

between tyrosinase genes from *Streptomyces glaucescens* [31] or *Neurospora crassa* [32] and the PQQ biosynthesis genes examined to date from *Acinetobacter calcoaceticus* [27]. The absence of tyrosinase activity may indicate that a derivatization of tyrosine precedes oxidation of its phenol side chain. One suggestion is that the tyrosine precursor is part of a 24-amino-acid polypeptide encoded by one of the PQQ biosynthesis genes [27]. Alternatively, the enzymatic oxidation of tyrosine may donate electrons to an intermediate redox cofactor rather than directly to O_2.

REFERENCES

1. Salisbury, S. A., Forrest, H. S., Cruse, W. B. T., and Kennard, O. (1979). A novel coenzyme for bacterial primary alcohol dehydrogenases. *Nature* (London) *280*: 843.
2. Duine, J. A., Frank, J. J., and Verwiel, P. E. J. (1981). Structure and activity of the prosthetic groups of methanol dehydrogenase. *Eur. J. Biochem. 118*: 395.
3. Duine, J. A., Frank, J. J., and van Zeeland, K. (1979). Glucose dehydrogenase from *Acinetobacter calcoaceticus*: A quinoprotein. *FEBS Lett. 108*: 443.
4. Janes, S. M., Mu, D., Wemmer, D., Smith, A. J., Kaur, S., Maltby, D., Burlingame, A. L., and Klinman, J. P. (1990). A new redox cofactor in eukaryotic enzymes: 6-Hydroxydopa at the active site of bovine serum amine oxidase. *Science 248*: 981.

5. McIntეer, W. S., Wemmer, D. E., Chistoserov, A., and Lidstrom, M. E. (1991). A new cofactor in prokaryotic enzyme: Trytophan tryptophylquinone as a redox prosthetic group in methylamine dehydrogenase. *Science 252*: 817.
6. Ameyama, M., Nonobe, M., Shinagawa, E., Matsushita, K., Takimoto, K., and Adachi, O. (1986). Purification and characterization of the quinoprotein D-glucosed dehydrogenase apoenzyme from *Escherichia coli*. *Agric. Biol. Chem. 50*: 49.
7. Goosen, N., Vermaas, D. A. M., and Putte, P. van de (1987). Cloning of the genes involved in the synthesis of coenzyme pyrroloquinoline quinone from *Acinetobacter calcoaceticus*. *J. Bacteriol. 169*: 303.
8. Lidstrom, M. E. (1991). Genetics of carbon metabolism in methylotrophic bacteria. *FEMS Microbiol. Rev. 87*: 431.
9. Biville, F., Turlin, E., and Gasser, F. (1989). Cloning and genetic analysis of six pyrroloquinoline quinone biosynthesis genes in *Methylobacterium organophilum* DMS 760. *J. Gen. Micro. 135*: 291.
10. Hommes, R. W. J., Postma, P. W., Neijssel, O. M., Tempest, D. W., Dokter, P., and Duine, J. A. (1984). Evidence of a quinoprotein glucose dehydrogenase apoenzyme in several strains of *Escherichia coli*. *FEMS Microbiol. Lett. 24*: 329.
11. Kilty, C. G., Maruyama, K., and Forrest, H. S. (1982). Reconstitution of glucose dehydrogenases using synthetic methoxatin. *Arch. Biochem. Biophys. 218*: 623.
12. Conlin, M., Forrest, H. S., and Bruice, T. C. (1985). Replacement of methoxatin by 4,7-phenanthroline and the inability of other phenanthrolines, as well as 7,9-didecarboxy methoxatin, to serve as cofactors for the methoxatin-requiring glucose dehydrogenase. *Biochem. Biophys. Res. Commun. 131*: 564.
13. Ameyama, M., Hayashi, M., Matsushita, K., Shinagawa, E., and Adachi, O. (1984). Microbial production of pyrroloquinoline quinone. *Agric. Biol. Chem. 48*: 561.
14. McIntეer, W. S., and Weyler, W. (1987). Factors affecting the production of pyrroloquinoline quinone by the methylotrophic bacterium W3A1. *Appl. Environ. Microbiol. 53*: 2183.
15. van Kleef, M. A. G., and Duine, J. A. (1989). Factors relevant in bacterial pyrroloquinoline quinone production. *Appl. Environ. Microbiol. 55*: 1209.
16. Anthony, C. (1982). *The Biochemistry of Methylotrophs*. Academic Press, Inc., New York, pp. 1–40.
17. Dunstan, P. M., Anthony, C., and Drabble, W. T. (1972). Microbial metabolism of C_1 and C_2 compounds. The involvement of glycollate in the metabolism of ethanol and of acetate by *Pseudomonas* AM1. *Biochem. J. 128*: 99.
18. Dunstan, P. M., Anthony, C., and Drabble, W. T. (1972). Microbial metabolism of C_1 and C_2 compounds. The role of glyoxylate, glycollate and acetate in the growth of *Pseudomonas* AM1 on ethanol and on C_1 compounds. *Biochem. J. 128*: 107.
19. Houck, D. R., Hanners, J. L., and Unkefer, C. J. (1988). Biosynthesis of pyrroloquinoline quinone. Identification of the biosynthetic precursors using ^{13}C labeling and NMR spectroscopy. *J. Am. Chem. Soc. 110*: 6920.
20. Houck, D. R., Hanners, J. L., and Unkefer, C. J. (1991). Biosynthesis of pyrroloquinoline quinone. Biosynthetic assembly from glutamate and tyrosine. *J. Am. Chem. Soc. 113*: 3162.
21. LeMaster, D. M., and Cronan, J. E., Jr. (1982). Biosynthetic production of ^{13}C-labeled amino acids with site-specific enrichment. *J. Biol. Chem. 257*: 1224.

22. Canovas, F. G., Garcia-Carmona, F., Sanchez, J. V., Pastor, J. L. I., and Teruel, J. A. L. (1982). The role of pH in the melanin biosynthesis pathway. *J. Biol. Chem. 257*: 8738.
23. Houck, D. R., Hanners, J. L., Unkefer, C. J., van Kleef, M. A. G., and Duine, J. A. (1989). PQQ biosynthetic studies in *Methylobacterium* AM1 and *Hyphomicrobium* X using specific ^{13}C labeling and NMR. *Antonie van Leeuwenhoek 56*: 93.
24. Brahme, N. M., Gonzales, J. E., Rolls, J. P., Hessler, E. J., Mizsak, S., and Hurley, L. H. (1984). Biosynthesis of the lincomycins. 1. Studies using stable isotopes on the biosynthesis of the propyl- and ethyl-L-hygric acid moieties of lincomycins A and B. *J. Am. Chem. Soc. 106*: 7873.
25. van Kleef, M. A. G., and Duine, J. A. (1981). L-Tyrosine is the precursor of PQQ biosynthesis in *Hyphomicrobium* X. *FEBS Lett. 237*: 91.
26. Taylor, I. J., and Anthony, C. (1976). A biochemical basis for obligate methylotrophy: Properties of mutant of Pseudomonas AM1 lacking 2-oxoglutarate dehydrogenase. *J. Gen. Micro. 93*: 259.
27. Goosen, N., Horsman, H. P. A., Huinen, R. G. M., and Putte, P., van de (1989). *Acinetobacter calcoaceticus* genes involved in biosynthesis of the coenzyme pyrroloquinoline-quinone: Nucleotide sequence and expression in *Escherichia coli*. *J. Bacteriol. 171*: 447.
28. Nunn, D., and Lidstrom, M. E. (1986). Isolation and complementation analysis of 10 methanol oxidation mutant classes and identification of the methanol dehydrogenase structural gene of *Methylobacterium* sp. AM1. *J. Bacteriol. 166*: 581.
29. van Kleef, M. A. G. (1988). A search for intermediates in the biosynthesis of PQQ. In *The Biosynthesis of Cofactor Pyrroloquinoline Quinone*, Krips Repo Meppel, pp. 95–112.
30. Biville, F., Turlin, E., and Gasser, F. (1991). Mutants of *Escherichia coli* producing pyrroloquinoline quinone. *J. Gen. Micro. 137*: 1775.
31. Huber, M., Hintermann, G., and Lerch, K. (1985). Primary structure of tyrosinase from *Streptomyces glaucescens*. *Biochem. 24*: 6038.
32. Lerch, K. (1982). Primary structure of tyrosinase from Neurospora crassa II. complete amino acid sequence and chemical structure of a tripeptide containing an unusual thioether. *J. Biol. Chem. 257*: 6414.

15

PQQ, A Growth-Stimulating Substance for Microorganisms

Kazunobu Matsushita and Osao Adachi

Yamaguchi University, Yamaguchi, Japan

I. INTRODUCTION

Two types of growth-stimulating effects by PQQ on microorganisms have been observed thus far. An example of one type of PQQ effect is seen in a symbiotic polyvinyl alcohol (PVA) degradation, in which one species of *Pseudomonas* excretes PQQ, which in turn enables the other species to grow on PVA [1]. The growth extent of the PQQ-accepting species is completely dependent on the PQQ supplied, and the growth rate as well as the cell yield is enhanced as seen during auxotrophic growth. However, an appreciable reduction of the lag phase of growth is scarcely observed. This type of PQQ effect is illustrated and referred to as Type I in Figure 1. The alternative type of growth stimulation by PQQ is referred to as Type II in Figure 1. It is characterized by a marked reduction of the lag period in the presence of a trace amount of PQQ in the culture medium. In this case, the subsequent growth rate at the exponential phase is not affected, and neither is the total cell yield at the stationary phase. Unlike the Type I effect, PQQ is not always an essential growth factor and normal cell growth (but with a relatively prolonged lag period) can be seen even in the absence of exogenous PQQ or PQQ-adducts. In this chapter, the growth stimulating effects of PQQ and PQQ-adducts on microorganisms are described.

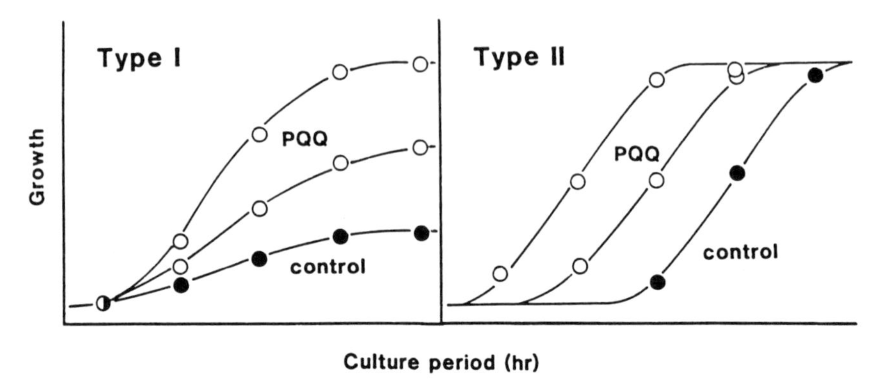

Figure 1 Schematic drawings of the growth-stimulating effect of PQQ.

II. GROWTH STIMULATION BY PQQ

A. Authentic PQQ as an Essential Growth Factor

In a mixed continuous culture of *Pseudomonas* sp. VM15C and *Pseudomonas putida* VM 15A, the former produces a PVA-degrading enzyme and the latter an essential growth factor (trivial name, factor A) for PVA utilization by the former. From the supernatant of this mixed culture, the factor A has been purified and identified as PQQ by Shimao et al. [1]. Spectral evidence and chromatographic analysis showed a fine correlation between the factor A and authentic PQQ. Authentic PQQ is effective not only in causing growth of the PVA-degrading bacterium on PVA, but also in enhancing the growth rate and the cell yield. Such a PQQ effect can be seen even at a PQQ concentration of 0.05 ng PQQ per ml of culture medium. This was the first report that PQQ is recognized as a bacterial growth factor.

B. Growth-Stimulating Factors for Microorganisms in Naturally Occurring Substances that Cause Reduction of the Lag Period for Growth

It is a well-known phenomenon that the growth of microorganisms is stimulated markedly by the addition of extracts of naturally occurring substances to the culture medium. One of the reasonable explanations for this phenomenon is that the culture medium might be supplemented with growth nutrients such as vitamins, nucleic acids, amino acids, sugars, and minerals derived from the added extract. These nutrients may not be present in sufficient quantities in the chemically defined medium. Recently, the existence of a substance that causes reduction of the lag period of microbial growth has been indicated in such naturally occurring substances. Furthermore, several lines of evidence suggest that the biological activity of this substance is the same as that observed with PQQ and PQQ-adducts.

In the course of investigations on the growth of acetic acid bacteria, it was observed that the growth of *Acetobacter* species, which show no appreciable nutritional requirements, was stimulated markedly by the addition of yeast extract to a synthetic medium in which generally known growth factors, vitamins, amino acids, and nucleic acids were fully supplemented [2]. For strains of the genus *Gluconobacter*, on the other hand, some species have been shown to require vitamins or other factors. Addition of yeast extract to a complete synthetic medium for such *Gluconobacter* strains further stimulated the growth of the organisms. A growth-stimulating effect similar to that observed with yeast extract is also found with other naturally occurring substances used for culture ingredients for microorganisms [3]. These include peptone, corn steep liquor, malt extract, koji extract, casamino acids, meat extract, blood powder or blood serum, soybean cake, and rumen juice. It is also generally accepted that the growth stimulation observed with such naturally occurring substances is not restricted to acetic acid bacteria. The growth of other bacteria and eukaryotic microbes including yeast and fungi is also similarly stimulated.

In general, a wild-type strain of *Escherichia coli* or soil *Pseudomonas* exhibits a rapid rate of growth on a chemically defined simple medium such as glucose mineral medium. However, growth stimulation by reduction of the lag period can be seen with such strains if the inoculum suspension is prepared carefully, as discussed later. The growth-stimulating substance in these materials and in the culture broths of *E. coli*, methylotrophic bacteria, and pseudomonads has been isolated and identified to be either PQQ or PQQ-adducts [1,3,4]. Given our present knowledge, the substance is more likely to be PQQ-adducts rather than PQQ itself, as is also discussed later.

It is interesting to note that the *E. coli* broth of glucose mineral medium also showed a growth-stimulating activity. Thereafter, it was clarified that *E. coli* as well as other microorganisms also produce the growth-stimulating substance that is released into the culture medium. The purified substance from *E. coli* broth, yeast extract, casamino acids, rumen juice, and other nutritional supplements has been characterized. All preparations show common optical properties that are identical to PQQ or PQQ-adducts [3]. These substances can function as the coenzyme of an apoquinoprotein glucose dehydrogenase, though the isolated compounds show a reduced activity compared to that of authentic PQQ. The activity is less than expected from their PQQ contents as estimated on the basis of optical density. It should be noted, however, that when the isolation of the PQQ-chromophore is attempted from quinoproteins in which PQQ is tightly bound to enzyme protein, the estimated PQQ content is usually far less than expected [5–9]. On the other hand, despite the apparent low activity in the apoenzyme assay, a considerable growth-stimulating activity for *Acetobacter aceti* is observed with such PQQ-chromophores.

C. Authentic PQQ as a Growth-Stimulating Factor for Microorganisms

When authentic PQQ is fed at the level of picograms to micrograms per ml of a synthetic medium for microorganisms, in which any naturally occurring substance is excluded, growth stimulation or rapid growth of inoculated microorganism is observed, while the control shows no appreciable growth [10,11]. Addition of a trace amount of PQQ or PQQ-adducts to the culture medium causes a marked reduction of the lag phase but does not increase the subsequent growth rate in the exponential phase or the total cell yield in the stationary phase. The length of the lag phase of microorganisms decreases with increasing inoculum size. It has been observed, however, that the effect of the growth-stimulating substance can be seen at any inoculum size. Compared to controls, reduction of the lag phase was observed even with very heavy inoculation [4].

III. PQQ–AMINO ACID ADDUCTS

A. Effect of PQQ–Amino Acid Adducts on Quinoprotein Glucose Dehydrogenase Activity

It can be readily presumed, given the known chemical reactivity of PQQ, that when a small amount of PQQ is exposed to a large excess of amino compounds during procedures such as acid hydrolysis, extraction, concentration, etc., PQQ will be readily converted to PQQ-adducts. These adducts are less active than PQQ as a functional coenzyme when assayed with the apoquinoprotein glucose dehydrogenase [12]. Conversely, PQQ does not lose its growth-stimulating activity after it is converted to a PQQ-adduct via Schiff base formation.

Almost all of the L-amino acids are capable of reducing the coenzyme activity of PQQ to some extent. Glycine, serine, threonine, tyrosine, arginine, and lysine show a marked effect, and the activation of apoenzyme activity by the adducts is less than 10% of that usually observed with authentic free PQQ. These results indicate that a primary amino group of the amino acids is selectively bound to the active carbonyl, probably the C-5 carbonyl of PQQ [13–15], to form a Schiff base. The Schiff base formation with PQQ is similarly observed with other amino compounds such as D-amino acids, amines, amino sugars, and carbonyl reagents, which yield adducts having an absorption maximum at 420 nm.

B. Preparation of PQQ-Serine as a Model Compound of a PQQ-Adduct

A PQQ-adduct formed with serine has been prepared and isolated by DEAE-Sephadex A-25 chromatography [16]. Unreacted serine was excluded from chromatography by washing the column before gradient elution. Fractions containing free PQQ, which is probably derived from unreacted PQQ or PQQ that has dissociated from the adduct, or both, eluted at about 0.8–1.0 M KCl, similar to previous reports [5]. Two peaks containing PQQ-adducts eluted from the column somewhat later in the gradient than free PQQ. PQQ activity as a coenzyme for the apoquinoprotein glucose dehydrogenase was surveyed with the chromatographed fractions. As expected, a marked coenzyme activity was found at the position where authentic PQQ eluted. On the other hand, a very poor coenzyme activity for the enzyme, estimated to be more than 100 times less active than authentic PQQ, was found with both peaks of PQQ-serine when compared with each other on the basis of an equimolar amount of authentic PQQ. Conversely, when assayed for growth-stimulating activity, the two fractions of PQQ-serine equally stimulated the growth of *A. aceti*. This was examined under conditions using a biophotometer as described by Ameyama et al. [11].

The relative levels of formation of the two species of adducts from PQQ and serine can be controlled. As shown in Figure 2, when adduct formation is performed at acidic pH and alkaline pH, a single different PQQ-adduct species, with different retention time in HPLC, is formed under each condition. When incubation of PQQ and serine is carried out at pH 3.5, the adduct is hydroxymethylated and designated as PQQ-ser I (Fig. 3). When serine is mixed with PQQ at alkaline pH, on the other hand, the resulting PQQ-adduct is structurally the same as the oxazole that is obtained from PQQ and glycine. This is designated PQQ-ser II in Figure 3.

The growth-stimulating effect of these PQQ-adducts has been shown to be

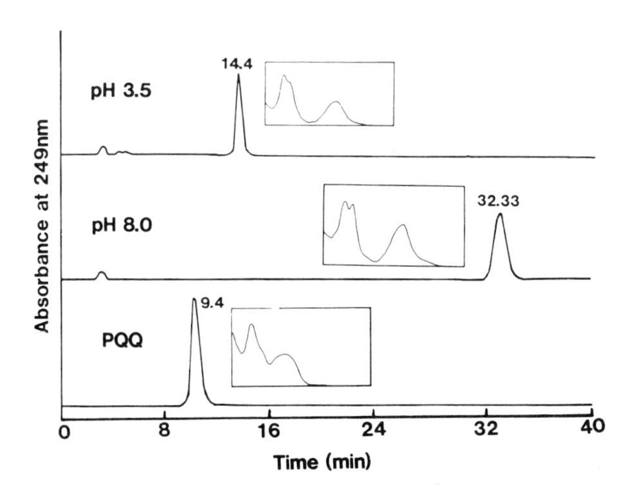

Figure 2 HPLC profiles of PQQ-adducts formed with serine at different values of pH and their absorption spectra.

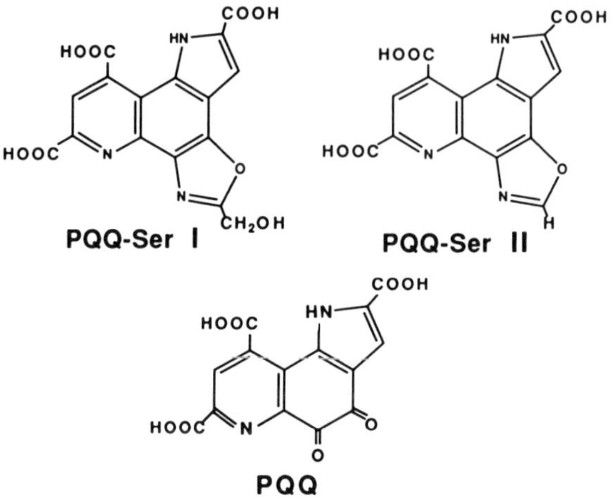

Figure 3 Chemical structures of PQQ-adducts and authentic PQQ.

more effective than that of authentic PQQ (Fig. 4). Furthermore, the presence of PQQ in too high a concentration appears to be inhibitory to the bacterial growth. Thus, although binding of serine to the active carbonyl group of PQQ renders it almost inert as the prosthetic group for apo-glucose dehydrogenase, this PQQ-adduct appears to be a more efficient species than authentic PQQ for the

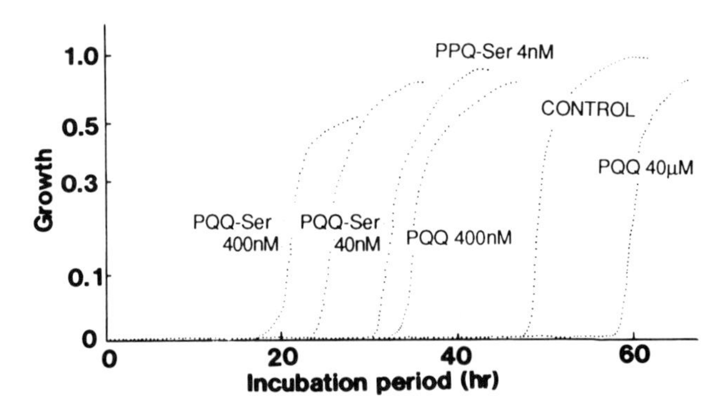

Figure 4 Growth-stimulating effects of PQQ and PQQ-serine for *A. aceti*. Different concentrations of PQQ or PQQ–serine were added to the basal medium as indicated. Inoculum cell suspensions of *A. aceti* (0.4 ml), having an optical density of 0.1–0.2 at 660 nm and which had been rinsed several times to remove trace amounts of PQQ, were added to 9.6 ml of the medium in an L-tube. Control cultures were incubated without test compounds. Incubation was at 30°C for the period indicated, and the bacterial growth was recorded automatically.

incorporation of PQQ, which is required for the initial stages of proliferation of microorganisms.

IV. PROBLEMS IN ASSAYING THE GROWTH-STIMULATING ACTIVITY OF PQQ

In general, there are many problems in performing bioassays. The importance of the preparation of the inoculum cell suspension is emphasized below. Furthermore, the growth of the microorganisms becomes variable with experimental conditions and seems to be affected by the ingredients in the culture medium. One attempt has been made with *A. aceti* to establish an ideal synthetic culture medium for the microorganisms in order to obtain an exact estimation of the growth stimulating effect of PQQ and PQQ-adducts [17]. The most effective bioassay medium for *A. aceti* that has been established includes glycerol, sodium-glutamate, glutathione, vitamins, and minerals. The data shown in Figure 4 were obtained using this basal medium.

At the same time, it is emphasized that the procedure for the preparation of the inoculum suspension is critical for the proper assay of the growth-stimulating activity of PQQ and PQQ-adducts [18]. In particular, repeated rinsing of the cells should be performed until free PQQ is no longer detectable in the inoculum suspension. This can be checked enzymatically using the quinoprotein apo-glucose dehydrogenase assay [12]. The water-rinsed inoculum suspension, in whose supernatant no PQQ can be detected, is found to be a favorable inoculum for assay of the growth-stimulating activity of PQQ. When an inoculum prepared after only a few repeated rinses is used, it is difficult to observe any difference in the length of the lag period in the presence and absence of exogenous PQQ in the culture medium. In this case, the control sample with no exogenous PQQ also shows a relative short lag period compared with another sample, which was prepared with an extensively rinsed inoculum. This means that a significant amount of PQQ, which causes a noticeable reduction of the lag period, is introduced from the insufficiently rinsed inoculum suspension. It can be said that all cells in the unrinsed inoculum suspension carry enough of a trace amount of PQQ to function as a starter at the initial stage of cell proliferation.

V. AN ENZYME CATALYZING THE LIBERATION OF PQQ FROM PQQ-ADDUCTS

It is important to understand how a PQQ-adduct exerts its physiological activity as a growth stimulant after introduction into a culture medium. Recently, the existence of an interesting enzyme has been identified in the cell free extracts of some microorganisms. *Pseudomonas* sp. VM15C, an auxotrophic strain for PQQ occurring symbiotically in PVA degradation [1], shows considerable growth on

PVA in the presence of yeast extract. Since the strain cannot grow independently on PVA in the absence of exogenous PQQ or a PQQ-donor, the growth extent is strictly dependent on the PQQ fed into the culture medium. It can be deduced that, if *Pseudomonas* sp. VM15C could use only the free form of PQQ to grow on PVA, the organism would have an enzyme able to generate PQQ from PQQ-adducts when it is grown on the growth substrate in the presence of PQQ-adducts but not free PQQ. Search for such an enzyme has been made with cell free extracts and PQQ-DNP (dinitrophenylhydrazine). The liberated PQQ is followed by the assay that utilizes apoquinoprotein glucose dehydrogenase [12]. Appearance of glucose dehydrogenase activity indicates formation of holo-glucose dehydrogenase, formed from PQQ that was generated from the PQQ-adduct. As stated earlier, no PQQ-adducts have any appreciable coenzyme activity in this assay. The enzyme activity obtained was small but reasonable, in that PQQ exerts its biochemical activity even at the picomolar level in nature. In addition to PQQ-DNP, PQQ liberation is seen with other PQQ-adducts. The same enzyme activity occurs in a PQQ-deficient strain of *Acinetobacter calcoaceticus* [19].

VI. ADDITIONAL COMMENTS

Before closing this chapter, it is worth making some additional comments about PQQ as a growth stimulant. It is important to understand the functional role of PQQ during the initial stage of cell proliferation. At this moment, it is convincing that PQQ is deeply associated with the mechanism of initiation of cell division. PQQ may exist in vivo free form or be associated with a quinoprotein or a PQQ carrier protein. However, with the exception of PQQ production by methylotrophic bacteria [6,11], it is difficult to isolate PQQ in its free form from other biomaterials. It is reasonable to consider that, in the case of PQQ fermentation by methylotrophs, since more PQQ exists than other amino compounds, PQQ is excreted into the culture medium and isolated as free PQQ. On the other hand, in yeast cells and other natural biological materials, the PQQ content is not nearly so high, and PQQ is probably converted spontaneously to a PQQ-adduct during the manipulations required for its isolation.

A matter of controversy that should also be addressed is the question of whether or not *E. coli* synthesizes PQQ. Recently, two groups from the Pasteur Institute have presented evidence that *E. coli* produces PQQ [20,21]. The purified growth-stimulating substance from the culture broths of *E. coli* and yeast extract, which have been concluded to be PQQ [3,5], is probably a PQQ-adduct. As noted above, with such a PQQ-adduct, PQQ content cannot be estimated exactly with the apoquinoprotein glucose dehydrogenase assay, whereas growth stimulating activity can be estimated in the same manner as free PQQ.

From the observations described above, it is important to note that PQQ activity should be followed by both criteria—coenzyme activity for apoquinoproteins and growth-stimulating activity for microorganisms. This should allow the detection of both free PQQ and PQQ-adducts.

REFERENCES

1. Shiamao, M., Yamamoto, H., Ninomiya, K., Kato, N., Adachi, O., Ameyama, M., and Sakazawa, C. (1984). Pyrroloquinoline quinone as an essential growth factor for a poly(vinyl-alcohol)-degrading symbiont, *Pseudomonas* sp. VM15C. *Agric. Biol. Chem. 48*: 2873.
2. Ameyama, M., and Kondo, K. (1966). Carbohydrate metabolism by the acetic acid bacteria. Part V. On the vitamin requirements for the growth. *Agric. Biol. Chem. 30*: 203.
3. Ameyama, M., Shinagawa, E., Matsushita, K., and Adachi, O. (1985). Growth stimulating activity for microorganisms in naturally occurring substances and partial characterization of the substance for the activity as pyrroloquinoline quinone. *Agric. Biol. Chem. 49*: 699.
4. Ameyama, M., Shinagawa, E., Matsushita, K., and Adachi, O. (1984). Growth stimulating substance for microorganisms produced by *Escherichia coli* causing the reduction of the lag phase in microbial growth and identity of the substance with pyrroloquinoline quinone. *Agric. Biol. Chem. 48*: 3099.
5. Ameyama, M., Hayashi, M., Matsushita, K., Shinagawa, E., and Adachi, O. (1984). Microbial production of pyrroloquinoline quinone. *Agric. Biol. Chem. 48*: 561.
6. Kawai, F., Yamanaka, H., Ameyama, M., Shinagawa, E., Matsushita, K., and Adachi, O. (1985). Identification of the prosthetic group and further characterization of a novel enzyme polyethyleneglycol dehydrogenase. *Agric. Biol. Chem. 49*: 1071.
7. Ameyama, M., Shinagawa, E., Matsushita, K., Takimoto, K., Nakashima, K., and Adachi, O. (1985). Mammalian choline dehydrogenase is a quinoprotein. *Agric. Biol. Chem. 49*: 3623.
8. Nagasawa, T., and Yamada, H. (1987). Nitrile hydratase is a quinoprotein: A possible new function of pyrroloquinoline quinone; activation of H_2O in an enzymatic hydration reaction. *Biochem. Biophys. Res. Commun. 147*: 701.
9. Shinagawa, E., Matsushita, K., Nakashima, K., Adachi, O., and Ameyama, M. (1988). Crystallization and properties of amine dehydrogenase from *Pseudomonas* sp. *Agric. Biol. Chem. 52*: 2255.
10. Ameyama, M., Shinagawa, E., Matsushita, K., and Adachi, O. (1984). Growth stimulation of microorganisms by pyrroloquinoline quinone. *Agric. Biol. Chem. 48*: 2909.
11. Ameyama, M., Matsushita, K., Shinagawa, E., and Adachi, O. (1988). Pyrroloquinoline quinone: Excretion by methylotrophs and growth stimulation for microorganisms. *BioFactors 1*: 51.
12. Ameyama, M., Nonobe, M., Shinagawa, E., Matsushita, K., and Adachi, O. (1985). Methods of enzymatic determination of pyrroloquinoline quinone. *Anal. Biochem. 151*: 263.
13. Lobenstein-Verbeek, Jongejan, J. A., Frank, J., and Duine, J. A. (1984). Bovine serum amine oxidase: A mammalian enzyme having covalently bound PQQ as prosthetic group. *FEBS Lett. 170*: 305.

14. Itoh, S., Kato, N., Ohshiro, Y., and Agawa, T. (1984). Oxidative decarboxylation of α-amino acids with coenzyme PQQ. *Tetrahedron Lett. 25*: 4753.
15. van der Meer, R. A., and Duine, J. A. (1986). Covalently bound pyrroloquinoline quinone is the organic prosthetic group in human placental lysyl oxidase. *Biochem. J. 239*: 789.
16. Adachi, O., Okamoto, K., Shinagawa, E., Matsushita, K., and Ameyama, M. (1988). Adduct formation of pyrroloquinoline quinone and amino acid. *BioFactors 1*: 251.
17. Adachi, O., Okamoto, K., Matsushita, K., Shinagawa, E., and Ameyama, M. (1990). An ideal basal medium for assaying growth stimulating activity of pyrroloquinoline quinone with *Acetobacter aceti* IFO 3284. *Agric. Biol. Chem. 54*: 2751.
18. Ameyama, M., Shinagawa, E., Matsushita, K., and Adachi, O. (1985). How many times should the inoculum cells be rinsed before inoculation in the assay for growth stimulating activity of pyrroloquinoline quinone? *Agric. Biol. Chem. 49*: 853.
19. Adachi, O., Okamoto, K., Matsushita, K., Shinagawa, E., and Ameyama, M. (1990). An enzyme catalyzing liberation of pyrroloquinoline quinone (PQQ) from PQQ-adducts. *Agric. Biol. Chem. 54*: 2481.
20. Bouvet, O. M. M., Pascal, L., and Grimont, P. A. D. (1989). Taxonomic diversity of the D-glucose oxidation pathway in the *Enterobacteriaceae*. *Intl. J. Syst. Bacteriol. 39*: 61.
21. Biville, F., Turlin, E., and Gasser, F. (1991). Mutants of *Escherichia coli* producing pyrroloquinoline quinone. *J. Gen. Microbiol. 137*:1775.

V

PQQ and Quinoproteins in Medicine and Biotechnology

16

PQQ: Effect on Growth, Reproduction, Immune Function, and Extracellular Matrix Maturation in Mice

Francene M. Steinberg, Carsten Smidt, Jack Kilgore, Nadia Romero-Chapman, David Tran, David Bui, and Robert B. Rucker

University of California, Davis, California

I. INTRODUCTION

Recent advances made in the detection of pyrroloquinoline quinone (PQQ) and the appreciation that a number of bacterial oxidoreductases utilize PQQ have encouraged researchers to examine eukaryotic systems with respect to possible functions for PQQ in plants and animals. It has been shown that when PQQ is omitted from a chemically defined diet containing antibiotics, growth impairment is observed in mice and rats, particularly if the diet is introduced prior to neonatal development [1]. A normal growth response is observed when PQQ is added to the deficient diet or whenever PQQ or PQQ-like compounds are detected in fecal samples or as a diet or water contaminant. Moreover, attempts to breed young female mice previously fed a PQQ-deficient diet for 8–9 weeks often result in either no litters or small litters. Pups are sometimes cannibalized at birth, and those surviving to weaning are stunted. In the pups that do survive, additional signs of PQQ deprivation include friable skin, general unfitness, and a hunched posture. In contrast, mice fed PQQ are normal [1].

Most intriguing is that the positive responses to PQQ supplementation, i.e., normal growth, reproduction, immune function, and extracellular matrix maturation, occur at pico- to nanomolar levels of dietary PQQ. This level is equivalent to the requirement for those nutrients, which are needed at the lowest of quantities in experimental diets, e.g., biotin, vitamin B_{12}, folic acid, or vitamin K [2–5].

II. NUTRITIONAL IMPORTANCE OF PQQ

A. Effect on Growth, Development, and Reproductive Performance

Many of the descriptions that follow are taken from several recent papers on the characteristics of PQQ deprivation [1,6–8]. For the experiments summarized, female BALB/c or male Swiss-Webster mice (Charles River Laboratories, Wilmington, MA) were used: Swiss-Webster mice in experiments to estimate PQQ absorption, and BALB/c in studies to assess growth and reproductive performance.

To minimizc bactcrial contamination in thc growth and rcproductivc studics, individual mice were housed in steel wire cages in a BioClean incubator (Duo-Flo, BioClean Lab Products, Inc., Maywood, NJ), and the water supply was filtered through an activated carbon cartridge (Carbon Capsule 12011, Gelman Sciences, Inc. Ann Arbor, MI) and a 0.2-μm bacterial filter (Mini Capsule 12122, Gelman Sciences, Inc. Ann Arbor, MI). All mice were fed a chemically defined, amino acid–based diet (Table 1). The diet was administered in feeders designed to avoid contamination and spillage [9]. All glassware, feeders, and containers were autoclaved or acid-washed to reduce potential sources of PQQ contamination. Often, 1–2% succinyl sulfathiazole was added to diets to retard bacterial growth. Other nonabsorbed antibiotics were also substituted (nystatin, 0.01%; neomycin, 0.05%; or bacitracin, 0.05%) whenever microflora were easily detected in routine screening of fecal material and appeared resistant to the effects of succinyl sulfathiozole.

Quality control of foodstuffs and the water supply was assessed by estimating PQQ with respect to PQQ contamination. Methods for PQQ detection and quantitation were adapted from procedures published by Gallop, Paz, Flückinger, and associates (see Ref. 10 and references cited therein). For routine analysis of multiple samples, we modified their assay, which is based on the reduction of nitroblue tetrazolium (NBT) by PQQ to formazan. For example, it is possible to carry out multiple assays if the samples are mixed and incubated in microelisa plates (Dynatech Laboratories, Chantilly, VA). Formazan formation is then measured using a commercial ELISA reader. Further, by adding sodium borate buffer and chelating agents to the assay, it is also possible to reduce interfering side reactions, such as the ascorbate catalyzed reduction of NBT [7,8,10].

When PQQ in complex samples was assayed, chromatographic procedures were also used to separate PQQ from other compounds with redox cycling potential. For example, tissue extracts or samples were chromatographed on columns (16 × 100 mm) of DEAE-Sephadex A-25 using linear gradients of KCl 0.002 M potassium phosphate (pH 7.0) to elute PQQ (Fig. 1). In this regard, recovery of PQQ added to the synthetic diets and other materials was usually excellent (>80%), but only if samples were extracted and assayed immediately.

Table 1 Diet Composition

Amino acid mix	16.0
Sucrose[a]	30.0
Corn starch	26.0
Corn oil	15.0
Cellulose	2.5
Vitamin mix	0.5
Mineral mix	10.0
Amino acid mix (g/kg diet)	
L-Arginine	12.00
L-Histidine·HCl·H_2O	6.00
L-Isoleucine	8.00
L-Leucine	12.00
L-Lysine·HCl	14.00
L-Methionine	6.00
L-Phenylalanine and L-tyrosine[b]	10.00
L-Threonine	8.00
L-Tryptophan	2.00
L-Valine	8.00
L-Alanine	12.00
L-Asparagine·H_2O	11.50
L-Cystine	4.00
L-Glutamate	12.50
Glycine	12.00
L-Proline	10.00
L-Serine	12.00
Total amino acids	160.00
Vitamin mix (per kg diet)	
Choline·HCl	2.00 g
Thiamin·HCl	100.00 mg
Niacin	50.00 mg
Riboflavin	15.00 mg
Calcium pantothenate	30.00 mg
Vitamin B_{12}	0.20 mg
Pyridoxine·HCl	30.00 mg
Biotin	3.00 mg
Folic acid	20.00 mg
myo-Inositol	100.00 mg
Ascorbic acid	250.00 mg
Retinyl acetate	5200 IU
Cholecalciferol	900 IU
DL-α-Tocopherol acetate	400.00 mg
Phylloquinone	10.00 mg
Glucose	to 5.00 g

Table 1 Continued

Mineral mix (per kg diet)	
Calcium carbonate	3.00 g
Calcium phosphate, tribasic	28.00 g
Potassium phosphate, dibasic	9.00 g
Sodium chloride	8.80 g
Sodium bicarbonate	10.00 g
Magnesium sulfate	1.85 g
Manganous sulfate·H_2O	650.00 mg
Ferric citrate	500.00 mg
Zinc carbonate	100.00 mg
Copper sulfate·$5H_2O$	50.00 mg
Boric acid	5.00 mg
Sodium molybdate·$2H_2O$	5.00 mg
Sodium fluoride	5.00 mg
Nickel (II) chloride·$6H_2O$	5.00 mg
Potassium iodide	1.00 mg
Cobalt (II) sulfate·$7H_2O$	1.00 mg
Chromium (III) chloride·$6H_2O$	1.00 mg
Sodium selenite	0.25 mg
Corn starch	to 100.00 g

[a]Antibiotics (most often succinyl-sultathiazole) were added at the expense of sucrose.
[b]L-tyrosine was eliminated from some diets, since it is a putative PQQ precursor. However, for most of the experiments described L-tyrosine was added at 5 g/kg and L-phenylalanine at 5 g/kg.

Recovery of PQQ from solutions containing free amino acids was very poor (<20%), particularly if solutions were allowed to stand more than a few hours (see Ref. 11).

The data given in Tables 2 and 3 and Figure 2 confirm our previously published data [1] that PQQ deficiency causes poor growth and reproductive failure. Only half of P_0-deficient female mice produce litters. Often, only half of the normal number of pups are born per litter, and about half of these pups survive to weanling (Table 2). Subsequent growth depression is also observed (Table 3). Although there were often no differences between groups by week 8, when the F_1 generation mice were continued on diets containing no PQQ, complete reproductive failure was noted when attempts were made to produce litters from such mice [1].

As a part of these nutritional studies, we also examined the extent to which PQQ is absorbed and its ability to be synthesized by gut microflora. With assistance from D. R. Houck and C. J. Unkefer (Los Alamos National Laborato-

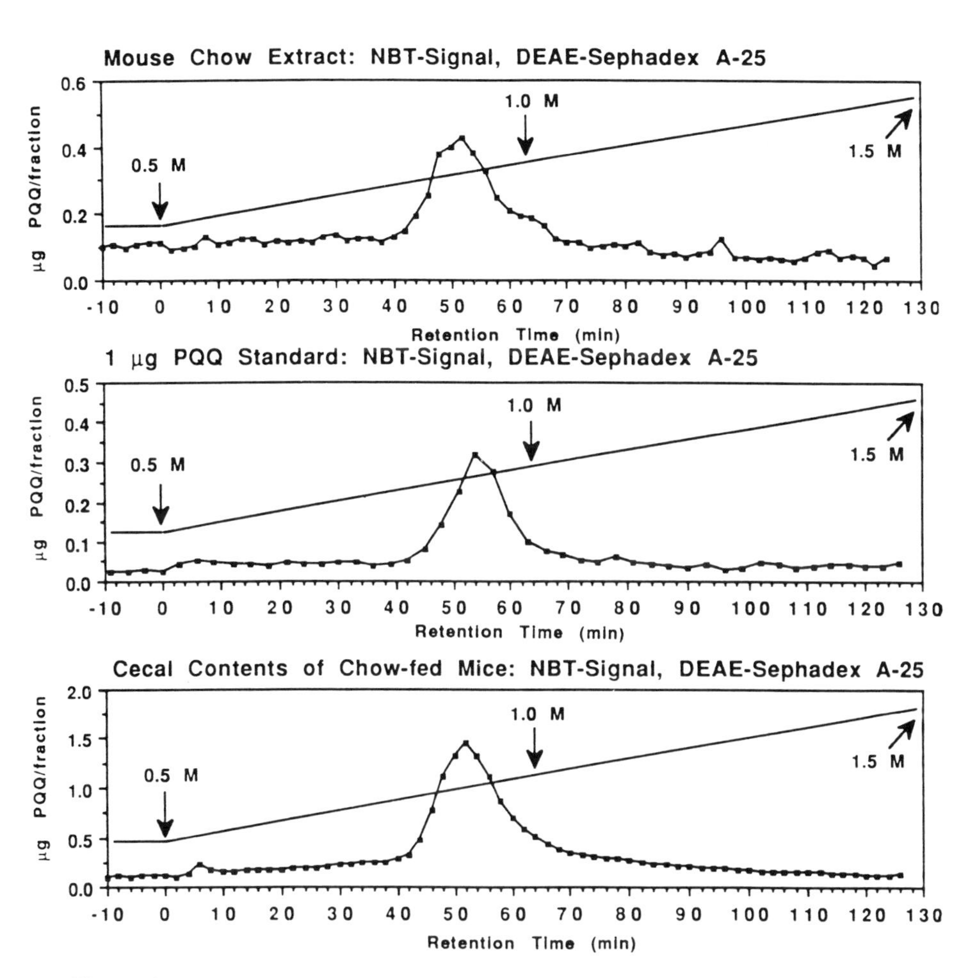

Figure 1 Chromatographic elution of PQQ extracted from commercial mouse chow utilizing a column of Sephadex A-25 (see text and Ref. 8 for details).

ries, Los Alamos, NM), ^{14}C-PQQ was prepared utilizing *Methylobacterium extorquens* AM1 (Pseudomonas AM1, ATCC 14718) and [^{14}C]-tyrosine as a radiochemically labeled precursor (see Refs. 7 and 12). The ^{14}C-PQQ was ultimately purified and used to assess intestinal absorption.

For the administration of the ^{14}C-PQQ, Swiss-Webster mice weighing 18–20 g were fed the chemically defined diet for 7 days prior to oral administration. The specific activity of PQQ was 0.42 μCi/μmol, and each animal received about 0.1 μmol of PQQ (about 80,000 dpm, corresponding to 28 μg PQQ). The isotopic PQQ was dissolved in saline buffered with 5 mM potassium phosphate, pH 7.0,

Table 2 Effect of PQQ Deprivation on Reproductive Performance in P_0 BALB/c Mice

Parameters	PQQ+ (1000 or 5000 ng/g)	PQQ− (no added PQQ)
Consequence to dam:		
Infertile females	0/20[a]	10/20[b]
Consequence to offspring:		
Pups/litter (mean)	9[a]	5[b]
Cannibalized pups	15%[a]	25%[a]
Abandoned pups	7%[a]	36%[b]
Pups surviving to week 4	78%[a]	38%[b]

Values with differing superscripts are significant at $p < 0.05$ for the group of mice designated PQQ+ versus the group of mice designated PQQ−.

and administered per os using a gavage needle designed for small rodents and mice (Perfectum, 24 gauge, 1 inch, Popper & Sons, Inc., New York, NY). Animals were then placed immediately into metabolic cages (100 × 200 cm) equipped for collection of excreta and expired CO_2, which was trapped in ethanolamine. To reduce spillage during ^{14}C study, the diet was provided as agarose gel cubes prepared by dissolving agarose at 1% (w/v) into a mixture of the chemically defined diet and H_2O (1:4, w/v). After 6 and 24 hours, five animals were killed by a carbon dioxide overdose, and blood and tissues were collected for ^{14}C determinations.

After 6 hours, 3.3% of the label was absorbed, compared to 62.0% (range: 19–89%) after 24 hours. PQQ concentrations estimated chemically as remaining in the intestinal lumen agreed within 10–20% with the dose calculated not to be absorbed. Since a large single dose of PQQ was administered (relative to that fed in diets), it was inferred that absorption was relatively efficient. Moreover, studies

Table 3 Growth of BALB/c Mice Fed Varying Amounts of PQQ

PQQ level (ng/g)	Growth (g)			
	Birth	4 weeks	8 weeks	12 weeks
0	1.3 ± 0.2[a]	6.8 ± 0.7[a]	14.4 ± 0.7[a]	18.1 ± 1.1[a]
100	1.2 ± 0.2[a]	6.9 ± 0.2[a]	13.3 ± 1.3[a]	16.9 ± 3.0[a]
200	1.2 ± 0.1[a]	9.1 ± 1.2[b]	16.3 ± 2.4[b]	18.8 ± 1.7[a]
300	1.5 ± 0.1[a]	12.7 ± 1.4[b]	18.2 ± 1.8[b]	20.5 ± 1.5[a]
1000	1.4 ± 0.1[a]	10.8 ± 1.7[b]	18.4 ± 2.0[b]	20.5 ± 2.7[a]
5000	1.4 ± 0.2[a]	10.7 ± 1.5[b]	18.4 ± 2.0[b]	20.9 ± 2.4[a]

Within a column, values with differing superscript are significant at $p < 0.05$.

Figure 2 PQQ-deficient (left) and supplemented female BALB/c mice (right). Note the rougher coat and subtle differences in facial cranial features. These features are similar to those previously reported in Swiss-Webster male mice (see Ref. 1).

were also conducted to assess the extent to which the intestinal microflora incorporate label from ^{14}C-L-tyrosine into PQQ [8]. When the medium of *Methylobacterium extorquens* M.AM1 (after incubation with ^{14}C-tyrosine) was passed over a DEAE Sephadex A-25 anion exchange column, about 0.3% of the labeled material coeluted with authentic PQQ. However, when cecal material of mice (200 mg/ml of labeled medium) was cultured (50 μCi of ^{14}C-L-tyrosine/ml) and the medium analyzed, no radioactivity was detected as PQQ.

That a large oral dose of PQQ was absorbed fulfills an obvious and important requirement, if PQQ-like compounds are to be considered metabolically or nutritionally important in animals. Renal excretion appeared to be the major route of excretion of absorbed PQQ; 62.1% of the dose was excreted at 6 hours, and 81.4% of the absorbed PQQ was excreted after 24 hours.

B. Effect on Extracellular Matrix Maturation

The relative solubility of tissue collagen is an index of its degree of maturation and cross-linking. Since it was speculated that lysyl oxidase, the enzyme responsible for cross-linking [13], was PQQ dependent [1], we measured skin collagen solubility, an excellent source of fibrillar collagen [14]. The skin was extracted sequentially with 1 M NaCl in 0.05 M disodium phosphate buffer (pH 7.5), 0.5 M acetic acid, and 4 M urea in 0.05 disodium phosphate buffer (pH 7.5), each for 24 hours. Aliquots were taken for estimation of soluble collagen as described by Marotto and Martino [15]. Collagen-derived cross-linking amino acids, such as hydroxylysinorleucine (HLNL) and dihydroxylysinorleucine (DHLNL), were determined as described by Reiser and Last [16]. Lysyl oxidase accumulation was measured in 4 M urea extracts of skin samples using an enzyme-linked immunoassay [17].

Figure 3a,b shows values for collagen solubility and skin lysyl oxidase. The values are from mice (Swiss-Webster) fed diets containing 0 or 1000 ng/g PQQ (see Ref. 1). The amounts of lysyl oxidase were significantly reduced, and the solubility of collagen was markedly increased in the skin from mice exposed to the PQQ-deficient diet throughout gestation and neonatal development. Likewise, when weanling Swiss-Webster mice (4 weeks of age) were fed the same diets for 8 weeks, the differences in collagen solubility persisted (30% increase in deficient mice), although no differences in weight were observed. Moreover, differences in collagen solubility were observed in 12-week-old BALB/c female mice fed diets varying in PQQ content since weaning (4 weeks of age). Expressed as mg collagen/g skin, values for the BALB/c mice were: for no added PQQ, 73 mg/g; for 100 ng added PQQ/g of diet, 75 mg/g; for 200 ng added PQQ/g of diet, 73 mg/g; for 300 ng added PQQ/g of diet, 66 mg/g; and for 1000 ng added PQQ/g of diet, 63 mg/g. The values represent the total collagen extracted from skin using 1 M NaCl followed by 0.1 M acetic acid as extractants.

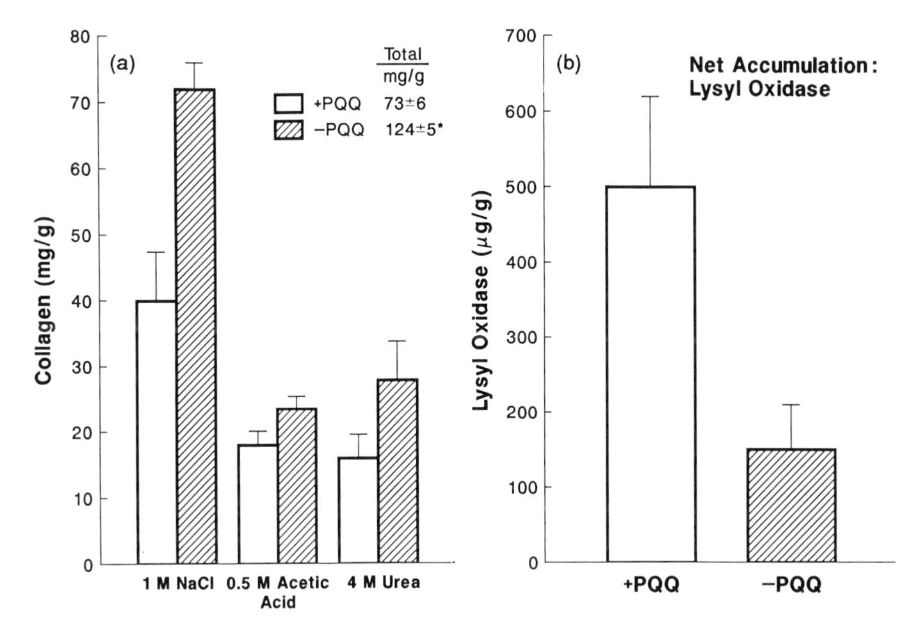

Figure 3 (a) The content of lysyl oxidase in skin of mice fed diets deficient or sufficient in PQQ. (b) The solubility of collagen from skin of mice fed diets deficient or sufficient in PQQ.

C. Effect on Immune Function

Spleen mitogen response was assayed as previously described [18]. The single-cell suspensions were cultured in RPM1 1640 tissue culture medium supplemented with L-glutamate, sodium pyruvate, penicillin, and streptomycin. Following two washes, the cells were resuspended in complete RPM1 1640 media supplemented with 6% fetal calf serum. An aliquot of cells was next counted, and cell suspensions containing 5×10^6 cells/ml were prepared. Viability was >90%, as determined by trypan blue exclusion.

Concanavalin A (Con A, Sigma Chemical Co., St. Louis, MO), a T-cell mitogen, and lipopolysaccharide from *Escherichia coli* K235 (LPS, List Biological Laboratories Inc., Campbell, CA), a B-cell mitogen, were diluted to concentrations of 5 μg/ml and 25 μg/ml, respectively, in RPM1 1640 media. The cells were cultured for a total of 48 hours and pulsed with 1 μCi/well of ^{3}H-thymidine 4–6 hours prior to harvest onto glass fiber filters, which were analyzed by liquid scintillation counting. Stimulation indices were calculated by dividing cpm in the presence of mitogen by cpm in the absence of mitogen.

Table 4 Effect of Dietary PQQ on Mitogen Reactivity of Splenocytes from 20-Week-old BALB/c Mice

Mitogen	PQQ (ng/g diet)					
	0	100	200	300	1000	5000
None (dpm $\times 10^3$)	$0.94 \pm .4^{a}$	$1.91 \pm .5^{a}$	5.49 ± 1.0^{b}	$0.74 \pm .3^{a}$	$1.17 \pm .3^{a}$	$1.79 \pm .5^{a}$
Concanavalin A (fold-stimulation)	6.4 ± 1.8^{a}	9.1 ± 3.7^{a}	43.5 ± 6.7^{b}	41.4 ± 21.7^{b}	51.0 ± 24.2^{b}	37.8 ± 10^{b}
Lipopolysaccharide (fold-stimulation)	12.2 ± 2.9^{a}	23.2 ± 5.8^{a}	39.9 ± 4.9^{b}	28.7 ± 8.2^{b}	26.0 ± 6.3^{b}	41.9 ± 8^{b}

Within a row, values with differing superscripts are significant at $p < 0.05$.

Values for mitogen-stimulated reactivity of splenocytes from mice at 20 weeks of age are given in Table 4. At levels below 200 ng of PQQ per g of diet, mitogen reactivity was clearly depressed with respect to stimulation by concanavalin A and lipopolysaccharide.

III. DISCUSSION AND PERSPECTIVE

Although there is clearly a need to reassess and reinterpret many of the roles originally proposed for PQQ as a cofactor, the nutritional and physiological observations pertaining to PQQ remain intriguing and indicate unusual functions for PQQ, or a compound very similar to it.

If PQQ is omitted from a chemically defined diet and that diet is introduced prior to neonatal development, growth impairment is observed. Normal growth is observed when PQQ is added in the pico- to nanomolar range in diets. Further, attempts to breed P_0- or F_1-generation young female mice fed a PQQ-deficient diet throughout gestation result in either no litters or in pups that are stunted.

Studies using ^{14}C-labeled PQQ also show that PQQ is readily absorbed, chiefly from the large intestine, but may not be made by intestinal microflora. We therefore conclude that PQQ plays some type of role that has yet to be clearly defined, or alternatively that PQQ is substituting for a compound(s) with similar functional properties. Support for the latter rests in part on the fact that deficiency signs are not observed when healthy weanling mice are used as experimental subjects and that mice eventually recover in some aspect from their deficiency, e.g., changes in the rate of growth and maturation of connective tissue are not apparent after week 12.

Immunological defects, however, as measured in mitogen-stimulated splenocyte assays persist up to 20 weeks. One possibility is that the requirement for PQQ (or a related substance) to maintain a normal index of immune system activity is greater than that needed to maintain an optimal early rate of growth. The deficiency may also result in decreased production of cytokines necessary for cellular signaling and proliferation, in particular interleukin-2. Another possibility might be a defect in maturation and processing of the T cells in the thymus prior to its involution. The mature T cells continue to circulate over the lifetime of the animal and could account for the persistance of the depressed stimulatory response that is noted. More research is needed on the physiological and nutritional effects of PQQ before the answer will be known. The immunological response, however, is not unlike that observed in other vitamin and mineral deficiencies [19–23].

We are currently testing the possibility that a PQQ-like substance may be synthesized by mammalian cells and that the putative neonatal requirement for PQQ is conditional, much like the conditional neonatal requirements for choline, inositol, or taurine (see Ref. 6).

In summary, it is possible to prepare chemically defined diets sufficiently deficient in PQQ-like compounds to negatively affect neonatal development. As little as 500 pmol to 1 nmol per gram of diet reverses the negative or abnormal responses that relate to growth and connective tissue maturation. At the very least, physiological studies utilizing PQQ should provide novel models for examining the functions of water-soluble quinones. We feel our evidence strongly suggests a nutritionally important role for PQQ or compounds with its chemical properties.

ACKNOWLEDGMENTS

This work was supported in part by NIH grants HL 15956 and DK 35747 and USDA grant 89-37200-4429.

REFERENCES

1. Killgore, J., Smidt, C., Duich, L., Romero-Chapman, N., Tinker, D., Reiser, K., Melko, M., Hyde, D., and Rucker, R. B. (1989) Nutritional importance of pyrroloquinoline quinone. *Science 245*: 850.
2. Suttie, J. W. (1991). Vitamin K. In *Handbook of Vitamins* (L. J. Machlin, ed.), Marcel Dekker, New York, p. 145.
3. Bonjour, J.-P. (1991). Biotin. In *Handbook of Vitamins* (L. J. Machlin, ed.), Marcel Dekker, New York, p. 393.
4. Brody, T. (1991). Folic acid. In *Handbook of Vitamins* (L. J. Machlin, ed.), Marcel Dekker, New York, p. 453.
5. Ellenbogen, L., and Cooper, B. A. (1991). Vitamin B_{12}. In *Handbook of Vitamins* (L. J. Machlin, ed.), Marcel Dekker, New York, p. 491.
6. Smidt, C. R., Myers-Steinberg, F., and Rucker, R. B. (1991). Physiologic importance of pyrroloquinoline quinone. *Proc. Soc. Expl. Biol. Med. 197*: 19.
7. Smidt, C. R., Unkefer, C. J., Houck, D. R., and Rucker, R. B. (1991). Intestinal absorption and tissue distribution of [^{14}C]-pyrroloquinoline quinone in mice. *Proc. Soc. Expl. Biol. Med. 197*: 27.
8. Smidt, C. R., Bean-Knudsen, D., Kirsch, D. G., and Rucker, R. B. (1991). Does the intestinal microflora synthesize pyrroloquinoline quinone? *BioFactors 3*: 53.
9. Miller, D. L. (1990). A new feeder for powdered diets. *Proc. Soc. Expl. Biol. Med. 194*: 81.
10. Paz, M. A., Fluckinger, R., Bouk, A., Kagen, H. M., and Gallop, P. (1991). Specific Detection of Quinoproteins by redox-cycling staining. *J. Biol. Chem. 266*: 689.
11. van Kleef, M. A. G., Jongejan, J. A., and Duine, J. A. (1989). Factors relevant in the reaction of PQQ with amino acids. Analytical and mechanistic implications. In *PQQ and Quinoproteins* (J. A. Jongejan and J. A. Duine, eds.), Kluwer Academic Publishers, Boston, p. 217.
12. Houck, D. R., Hanners, J. L., and Unkefer, C. C. (1988). Biosynthesis of pyrroloquinoline quinone, I. Identification of biosynthetic precursers, using ^{13}C labelling and NMR spectroscopy. *J. Am. Chem. Soc. 110*: 6020.

13. Kagan, H. M. (1986). Characterization and regulation of lysyl oxidase. In *Biology of Extracellular Matrix: A Series. Regulation of Matrix Accumulation* (R. P. Mechan, ed.), Academic Press, Orlando, FL, p. 321.
14. Tinker, D., and Rucker, R. B. (1985). Role of selected nutrients in synthesis, accumulation, and chemical modification of connective tissue proteins. *Physiol. Revs. 65*: 607.
15. Marotta, M., and Martino, M. (1985). Sensitive spectrophotometric method for the quantitative estimation of collagen. *Anal. Biochem. 150*: 86.
16. Reiser, K. M., and Last, J. A. (1986). Biosynthesis of collagen crosslines: In vivo labelling of neonatal skin, tendon, and bone in rats. *Connect. Tissue Res. 14*: 293.
17. Romero-Chapman, N., Lee, J., Tinker, D., riu-Hare, J. Y., Keen, C. L., and Rucker, R. B. (1991). Lysyl oxidase: Purification, properties and influence of dietary copper on accumulation and functional activity in rat skin. *Biochem. J. 174*: 297.
18. Gery, I., Gershon, R. K., and Waksman, B. H. (1972). Potentiation of the T-lymphocyte response to mitogens. I. The responding cell. *J. Exp. Med. 136*: 128.
19. Wade, S., Lemonnier, D., Bleiberg, F., and Delorme, J. (1983). Early nutritional experiments: Effects on the humoral and cellular immune responses in mice. *J. Nutr. 113*: 1131.
20. Boissonneault, G. A., and Johnston, P. V. (1983). Essential fatty acid deficiency, prostaglandin synthesis and humoral immunity in Lewis rats. *J. Nutr. 113*: 1187.
21. Bendich, A., D'Apolito, P., Gabriel, E., and Machlin, L. J. (1983). Interaction of dietary vitamin C and vitamin E on guinea pig immune responses to mitogens. *J. Nutr. 113*: 1187.
22. Boissonneault, G. A., and Johnson, P. V. (1984). Humoral immunity in essential fatty acid-deficient rats and mice: effect of route of injection of antigen. *J. Nutr. 114*: 89.
23. Fraker, P. J., Hildebrandt, K., and Luecke, R. W. (1984). Alteration of antibody-mediated responses of suckling mice to T-cell-dependent and independent antigens by maternal marginal zinc deficiency: Restoration of responsivity by nutritional repletion. *J. Nutr. 114*: 170.

17

The Biomedical Significance of PQQ

Mercedes A. Paz, Rudolf Flückiger, and Paul M. Gallop

Children's Hospital Medical Center, Harvard Schools of Medicine and Dental Medicine, Boston, Massachusetts

PQQ was discovered as a noncovalently bound bacterial cofactor involved in the oxidative metabolism of alcohols. It was first isolated from cultures of methylotrophic bacteria as an acetone adduct and its structure determined by X-ray diffraction [1] and confirmed by organic synthesis [2]. Claims were made later that PQQ was covalently bound to certain mammalian enzymes designated as "quinoproteins" [3]. The claim for covalently protein-bound PQQ has not been substantiated. There is, however, convincing evidence for the covalent attachment of other types of quinones in certain enzymes. These quinones are not related to PQQ and arise, either by the posttranslational modification of a tyrosine residue to a 6-hydroxydopa residue, as shown with bovine plasma monamine oxidase [4], or by the posttranslational cross-linking of two sequence-separated tryptophan residues to form a quinonoid compound called tryptophan tryptophylquinone (TTQ) as shown in bacterial methylamine dehydrogenase [5]. The term quinoprotein, initially applied to designate a protein with covalently bound PQQ, has been retained to identify now a protein that contains any type of covalently bound quinonoid compound.

What is most significant, as we have shown [6–8], is that animal cells, tissues, and biological fluids contain free PQQ, which has important physiological functions. Earlier hints that PQQ was important to mammals derive from nutritional experiments in Rucker's laboratory, which indicate that PQQ might be an essential nutrient for mouse pups [9–12]. In our laboratory, we noticed that animal cells, tissues, and fluids contained a dialyzable anionic quinonoid compound that catalyzed efficient redox cycling [6] and was PQQ-like in nature. We have now proven that this substance is, indeed, free PQQ, present in its quinone, hydroquinone, and semiquinone redox states. Chapter 13 deals with the isolation and identification of free PQQ from mammalian sources. The present chapter presents an overview of the biomedical significance of PQQ. Many of the concepts expressed are based on recent experiments, not yet complete. However, the biological cofactor role of PQQ as a catalytic reversible shuttler of electrons promises to be so important that an early preview and speculative overview are warranted.

I. COFACTOR PQQ AS A MAJOR CATALYST IN DIOXYGEN-SUPEROXIDE INTERCONVERSION

We have found that PQQ catalyzes and establishes a dynamic equilibrium between dioxygen, an accessible electron, a proton, and superoxide:

$$O_2 + e^- + H^+ \leftarrow [PQQ \leftrightarrow PQQH\cdot] \leftarrow O_2- + H^+$$

PQQ will capture electrons from primary amines, catechols, catecholamines, ascorbic acid, and NAD(P)H to form PQQ-semiquinone and/or PQQ-hydroquinone. These reduced forms of PQQ can be reoxidized by dioxygen in one-electron steps to generate superoxide anion. The equilibrium is driven to the right by the presence of (1) a high concentration of an appropriate reducing substrate, (2) a high level and accessibility of dioxygen, (3) superoxide oxidizing reagents like ferricytochrome *c* or tetrazolium salts, and, especially, (4) the enzyme superoxide dismutase (SOD). It is noteworthy that PQQ, in combination with SOD, forms a broad-spectrum, hydrogen peroxide–generating oxidase system with SOD acting as an apoenzyme and PQQ acting as a non-covalently bound redox coenzyme. This system may be very important in the degradative metabolism of catecholamines.

Collaborative experiments with Dr. Herbert Kagan's laboratory also indicate that free PQQ, in the presence of copper ions and oxygen, acts as a potent amine-oxidizing agent and is able to oxidize lysine residues in elastin and collagen to aldehydes [13] with some similarities to lysyl oxidase. Moreover, experiments by Drs. Ohshiro, Itoh, and their colleagues [14–19] have shown that PQQ has the capacity to catalyze the oxidation of a wide variety of organic compounds including catechols, amines, amino acids, and related compounds. They have also examined the role of copper in nonenzymatic PQQ-catalyzed oxidations [20].

We have shown that the dioxygen-superoxide equilibrium catalyzed by PQQ, as mentioned above, can be driven to the left by a high concentration of superoxide, usually generated by xanthine oxidase (XO) and xanthine oxidase substrates like xanthine, hypoxanthine, or acetaldehyde with an electron acceptor like NAD^+.

This statement implies that PQQ transforms xanthine oxidase (XO) into xanthine dehydrogenase (XD). XD promotes a reaction in which electrons, taken from a xanthine oxidase substrate (represented as H—X—OH), end up in NADH without the net release or consumption of dioxygen. In the xanthine dehydrogenase system, XO is the apopenzyme and PQQ serves as the noncovalently bound redox coenzyme:

$$\text{H—X—OH} + \text{HOH} + 2\text{O}_2 \xrightarrow{\text{XO}} \text{HO—X—OH} + 2\text{O}_2^- + 2\text{H}^+$$

$$2\text{O}_2^- + 2\text{H}^+ + 2\text{PQQ} \xleftrightarrow{\text{XO}} 2\text{O}_2 + 2\text{PQQH}\cdot$$

$$2\text{PQQH}\cdot + \text{NAD}+ \xleftrightarrow{\text{XO}} \text{NADH} + \text{H}^+ + 2\text{PQQ}$$

$$\text{H—X—OH} + \text{HOH} + \text{NAD}^+ \xrightarrow{\text{XO + PQQ}} \text{HO—X—OH} + \text{NADH} + \text{H}^+$$

Here the combination of XO + PQQ generates XD.

In a 1982 study of milk XO and XD, it was apparent that the addition of a quinone like benzoquinone to purified but undegraded XO in the presence of NAD^+ and sulfhydryl compounds regenerated XD activity [21]. More recent studies [22] suggest that an anionic component present in XD may be missing in XO. Definitive proof for PQQ, loosely bound to XD, may require X-ray diffraction structural studies on XD crystals. However, at present it appears that PQQ may be the actual anionic quinonoid cofactor present in intact and enzymatically active, XD. One should note that raw fresh milk, a rich source of both free PQQ and xanthine oxidase, is also an excellent source of XD. When one attempts to purify XD activity from milk, the dialyzable PQQ, not known to be present in milk until recently, is lost during purification, and XO is obtained. Free PQQ has been recently isolated from raw milk by our laboratory [7,8].

According to the preceding arguments, superoxide may be generated from dioxygen and reduced PQQ or removed by its oxidation by PQQ; the direction of the process depends on the chemical properties, concentrations, and accessibility of other redox reactants present. The rate of the reaction, left or right, depends on the concentrations of PQQ and its redox products. If a PQQ-sequestering agent is added (for example, 4,5-dimethylphenylene-1,2-diamine, DIMPDA), it reacts with PQQ to form a stable phenazine adduct which prevents the PQQ-dependent catalysis of redox-cycling and the formation of superoxide.

Recent evidence from our laboratory indicates that PQQ is present in neutrophils and may be involved in the respiratory burst that follows neutrophil activation. This is illustrated below using phorbol myristate acetate–activated guinea pig peritoneal exudate neutrophils, which generate and release superoxide in the complex NADPH oxidase reaction in respiratory burst. The generation of superoxide is inhibited by DIMPDA, which enters the cells and sequesters PQQ and perhaps any other *ortho*-quinones present.

II. PQQ AND THE RELEASE OF SUPEROXIDE FROM ACTIVATED NEUTROPHILS

The respiratory burst is a characteristic property of neutrophils, eosinophils, monocytes, and macrophages; these cells are collectively called phagocytes because they engulf particular material. Phagocytes have a common function, the defense of the organism against invading microbes; this function is related to their special ability to generate and release large amounts of superoxide and other oxidants when they recognize and phagocytose their prey [23]. In phagocytic cells, the presence of certain stimuli like bacterial or yeast membranes, special peptides, phorbol myristate acetate (PMA), and lymphokines like interferon-gamma [24] stimulate the respiratory burst. First, the increase in oxygen uptake and glucose oxidation through the hexose monophosphate pathway leads to the augmented generation of cellular NADPH. NADPH is then oxidized stepwise through several reactions that use flavins, cytochromes, and other redox compounds, and perhaps *ortho*-quinones like PQQ. Ultimately and rapidly, superoxide is released from the plasma membrane of the activated phagocyte. In a collaborative study with Dr. Manfred Karnovsky, an expert in respiratory burst biochemistry [25], we have recently shown that superoxide may be derived from a reaction involving the dioxygen oxidation of a reduced *ortho*-quinone, probably PQQ, which is present in PMA-activated guinea pig peritoneal exudate neutrophils. Addition to neutrophils of either DIMPDA, or other lipoidal *ortho*-quinone–sequestering agents to which the cells are permeable, inhibits the release of superoxide. By using tritiated DIMPDA, we have isolated from neutrophils a labeled PQQ-DIMPDA phenazine adduct. We view the reaction sequences as follows:

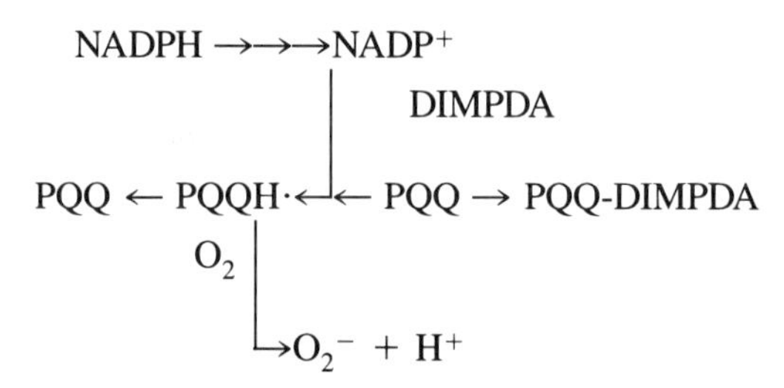

PQQH· (PQQ semiquinone) is formed directly or indirectly from NADPH in a step of the NADPH oxidase electron chain. The oxidation by dioxygen of appropriately located PQQH· generates superoxide at a plasma membrane site and is released from the cells. If cellular PQQ has been sequestered by added DIMPDA, there is decreased formation of the PQQH· needed for the one-electron reduction of dioxygen and the formation of superoxide. Our results indicate that an *ortho*-quinone, likely to be PQQ, plays a major role in the generation and release of superoxide in the respiratory burst unique to activated phagocytic cells.

It is interesting to note that there is little release of intermediate superoxide during the nonenzymatic spontaneous oxidation of NAD(P)H catalyzed by PQQ in the presence of ferricytochrome *c* at physiological pH [26]. In this homogeneous reaction, PQQ oxidizes NAD(P)H directly. The reduced forms of PQQ react more rapidly with ferricytochrome *c* than with dioxygen so that little superoxide is generated and SOD has a minor effect on the extent of ferricytochrome *c* reduction. To a major extent, the reduction of ferricytochrome to ferrocytochrome is by the reduced PQQ not by superoxide. The aerobic oxidation of NADH catalyzed by PQQ has also been studied in the presence of diaphorase and catalase, which markedly increase the efficiency of NAD^+ regeneration directed by PQQ catalysis [14].

In the intact activated phagocytic cells with a normal respiratory burst, reduced PQQ is probably placed at a membrane site accessible to dioxygen, but not accessible to the extracellular ferricytochrome *c* indicator. Superoxide is released from the activated cells into the ferricytochrome *c*–containing media, and its detection by the cytochrome *c* indicator is therefore very sensitive to SOD added to the extracellular milieu.

Important recent results on the released products of respiratory burst indicate that the hormone, endothelium-derived relaxing factor (EDRF), now known to be nitric oxide [27], is secreted from activated macrophages and other phagocytic cells [28,29]. Nitric oxide is a radical derived by the nitric oxide synthetase reaction from the five-electron oxidation of a terminal guanidino nitrogen atom of L-arginine to form citrulline [30,31]. Released nitric oxide, being lipophilic, permeates the plasma membrane of target cells and forms a complex with the heme group of soluble guanylate cyclase. The activated cyclase raises the level of cyclic-GMP, a second messenger in the target cells. For example, platelet aggregation is inhibited by a nitric oxide–like factor released from human neutrophils in vitro [29,32].

Nitric oxide is destroyed by superoxide; consequently, SOD potentiates the hormonal action of nitric oxide. Nitric oxide release and hormonal activity might be controlled by the coupled or timed release of superoxide anion. The human neutrophil has been widely implicated in both the host defense and tissue destruction found in inflammatory diseases. Our recent demonstration [25] that PQQ may be involved in the formation of SOD by activated neutrophils suggests that PQQ could have an important role in the mediation of inflammatory events. The recent reports on the anti-inflammatory and antioxidant actions of PQQ discussed below support this concept. One would expect from the chemistry of PQQ and from its direct reactions with ascorbic acid, NAD(P)H, and superoxide that PQQ could directly oxidize and destroy nitric oxide or convert it to other physiologically significant products. The reduced forms of PQQ may also reductively remove or modify nitric oxide. At present, the most likely extracellular role of PQQ involves its antioxidant–anti-inflammatory actions. We suggest that these actions may be mediated through the removal of both superoxide and nitric oxide or by the formation of superoxide, which reacts with nitric oxide.

A recent report describes another action of nitric oxide not mediated by cyclic GMP, namely, the inhibition of osteoclastic bone resorption [33]. The osteoclast is the only cell capable of resorbing bone, and its enhanced activity plays an important role in osteoporosis, Paget's disease of bone, rheumatoid arthritis, and bone metastasis. PQQ, by its catalysis of superoxide formation or removal, may regulate the levels of nitric oxide and play a role in the etiology or prevention of osteoporosis. The list of actions ascribed to the nitric oxide continues to grow. They include intercellular communication in neutrophils, brain tissue, renal epithelial cells, mast cells, and autonomic nerves, in addition to its originally recognized function as a vasodilator produced by vascular endothelium.

For the purposes of a more thorough discussion of the important subject of nitric oxide, superoxide, and the role of PQQ in radical reactions, we will examine a mechanism through which cellular PQQ could be involved in the terminal step of nitric oxide synthesis and with a coupled or uncoupled synthesis of superoxide. In the nitric oxide synthetase reaction, L-arginine is oxidized to L-citrulline in two stages (four electrons) with one of the guanidino amino groups being oxidized stepwise in a process requiring mixed-function redox reactions involving NADH or tetrahydrobiopterin:

1. $R{-}NH{-}C({=}NH){-}NH_2 + O_2 + 2H \rightarrow\rightarrow R{-}NH{-}C({=}NH)N{-}O + 2H + HOOH$
2. $R{-}NH{-}C({=}NH)N{=}O + HOH \rightarrow$
   $[H{-}N{=}O] + R{-}NH{-}C({=}NH){-}OH$
   (Citrulline)
3. $[H{-}N{=}O] + PQQ \rightarrow PQQH\cdot + \cdot N{=}O$
   $PQQH\cdot + O_2 \rightarrow PQQ + O_2^- + H^+$

Here we assume that after a net four-electron oxidation (reactions 1 and 2), a hypothetical precursor of nitric oxide, [H—N=O], is generated along with citrulline. The hypothetical one-electron oxidation of [H—N=O] by PQQ (reaction 3) generates nitric oxide and PQQ-semiquinone. Depending on how the semi-quinone is handled by the cell, superoxide anion may or may not be produced almost simultaneously with nitric oxide. However, if superoxide is formed and SOD is present, hydrogen peroxide rather than superoxide could be released with nitric oxide. Since nitric oxide also reacts with superoxide and is removed as peroxynitrite anion ($O_2^- + \cdot N{=}O \rightarrow O{=}N{-}O{-}O^-$) [34], the prior removal of superoxide with SOD enhances the physiological actions of nitric oxide. Peroxynitrite anion is a potent oxidizer of sulfhydryl groups, and it may exert cytotoxic effects by oxidizing tissue sulfhydryls. Extracellular PQQ is converted by ascorbic acid to its hydroquinone, PQQ(2H), which may be involved in the scavenging of nitric oxide and/or peroxynitrite anion and this may also account for some of the antioxidant and anti-inflammatory actions of PQQ.

III. PQQ AND THE OXIDATIVE REMOVAL OF SUPEROXIDE IN RED CELLS

We have found that PQQ can be isolated from ultrafiltrates of human red cell hemolysates. Red cells are believed to generate superoxide during both the reoxygenation of hemoglobin to oxyhemoglobin and during the release of dioxygen from oxyhemoglobin. From a classical perspective, the unwanted superoxide produced by red cells is rapidly dismutated by the erythrocyte SOD; the hydrogen peroxide formed is removed by catalase and/or glutathione peroxidase. Every molecule of superoxide formed in the red cell uses an electron, and dismutation of superoxide by SOD leads to the net loss of one (catalase) or two electrons (GSH peroxidase and GSSG reductase):

$$2O_2 + 2e^- \rightarrow 2O_2^- \xrightarrow{SOD} O_2 + HOOH \xrightarrow{Cat} HOH + \tfrac{1}{2}O_2$$
$$2O_2 + 2e^- \rightarrow 2O_2^- \xrightarrow{SOD} O_2 + HOOH$$
$$HOOH + 2GSH \xrightarrow{\text{GSH peroxidase}} 2HOH + GSSG$$
$$GSSG + 2e^- + 2H^+ \xrightarrow{\text{GSSG reductase}} 2GSH$$

We propose an alternate scheme, likely to be less costly in electron losses. PQQ may be used to oxidize superoxide, and the reduced PQQ may then return some of the electrons to the cell in a useful form, directly as NADH, or indirectly as glutathione:

$$2O_2 + 2e^- \rightarrow 2O_2^-;$$
$$2O_2^- + NAD^+ + H^+ \xrightarrow{PQQ} NADH + 2O_2$$
$$NADH + H^+ + GSSG \xrightarrow{\text{GS-SG reductase}} 2GSH + NAD^+$$

The use of PQQ to remove superoxide by oxidation coupled with the recovery of electrons may be very important and could explain some of the antioxidant and anti-inflammatory actions of PQQ reported recently in other systems [35–37]. Here we emphasize that PQQ is likely to be crucial for red cell viability and survival and that an acquired or genetic deficiency of PQQ in erythrocytes is likely to lead to the formation of methemoglobin and red cell hemolysis. In blood, PQQ is present in plasma, red cells, and buffy coat, and the cellular levels of PQQ greatly exceed the extracellular levels. It appears likely that blood contains an active mechanism for transporting and concentrating PQQ from plasma into cells. One can speculate that neonates, and especially preterm neonates, may have an immature PQQ-handling system, which might lead to biochemical and functional abnormalities of their polymorphonuclear leukocytes. The human neonate is uniquely susceptible to serious and overwhelming bacterial and fungal infections. There may be developmental defects in oxygen metabolism and in intracellular antioxidant mechanisms [39].

Smidt et al. have found that labeled PQQ administered per os to mice was readily absorbed in the lower intestine and excreted mostly by the kidneys in 24 hours [11]. In 6 hours, most of the labeled PQQ in blood was associated with red cells, rather than plasma. It will be interesting to determine if pharmacological amounts of PQQ will help to protect red cells that have heritable or acquired propensity to form methemoglobin and to undergo hemolysis. (see note added in proof, page 390.)

IV. THE PROTECTIVE ACTIONS OF PQQ: PQQ AS ANTIOXIDANT AND ANTI-INFLAMMATANT

In animal models of oxidative stress, injected PQQ is protective. Rats given a variety of hepatotoxins capable of destroying the liver through the induction of oxidative stress injury are protected by PQQ in a dose-dependent manner [35]. Chick embryos oxidatively stressed by excessive administration of glucocorticoids developed massive cataracts and are protected against cataracts and glutathione depletion by injected PQQ in a dose-dependent manner [36]. Rats injected in their paws with carregheen develop an inflammatory reaction with rat paw edema. This can be prevented by i.v. PQQ injection, showing that PQQ has a potent anti-inflammatory activity [37]. Neurons in culture are excited by the glutamate-receptor agonist *N*-methyl-D-aspartic acid (NMDA). PQQ added to these cultures protects the neurons from the excitotoxicity of NMDA [38]. PQQ also protects animals from the toxic actions of adriamycin, a highly toxic antibiotic, which generates toxic oxidants and also intercalates into DNA (H. Nishigori et al., unpublished).

In the future, it will be important to ascertain if PQQ protects animals against other toxic agents that target specific organs and tissues and generate superoxide and other toxic oxidants through redox-cycling. Will PQQ protect rats against the induction of diabetes mellitus by alloxan and streptozotocin, agents that destroy pancreatic β cells? Will PQQ protect animal lungs against the toxic actions of viologens like diquat and paraquat, and red cells against the methemoglobin-inducing hemolytic action of agents like divicine, dialuric acid, isouramil, and a variety of pharmaceutical agents [40,41]?

1,2,3,6-Tetrahydro-1-methyl-4-phenylpyridine (MPTP) is a piperidine derivative that causes irreversible symptoms of parkinsonism in humans and monkeys. In animal models of Parkinson's disease, MPTP crosses the blood brain barrier. This agent is chemically related to physiologically active catecholamines and to drugs like amphetamines, which are selectively transported within the brain to dopaminergic sites. Interaction of MPTP with amine oxidases may convert this compound to an *N*-methyl-4-phenylpyridinium compound capable of redox-cycling with the generation of superoxide and depletion of reducing equivalents. A modified lipophilic form of PQQ that may efficiently cross the blood-brain barrier and release free PQQ may be a useful agent in preventing the toxicity of MPTP. Indeed, such a compound may be useful as an antiparkinsonism drug. In a preliminary study, our laboratory [8] has found that the CSF concentration of PQQ in Parkinson's patients (PD) was significantly lower than in age-matched controls. The serum levels were higher in PD than in controls. While these results derive from a limited number of patients, they suggest a decreased active transport of PQQ across the blood-brain barrier, causing the neurons in PD patients to be less protected against oxidative stress. Preliminary experiments with aging rats suggest that the CSF level of PQQ may decrease with age.

Another example of how PQQ protects tissues against noxious agents is provided by the work of Hobara et al. [42]. Using ethanol loaded rats, they showed that PQQ accelerated the oxidation of acetaldehyde but did not accelerate the oxidation of ethanol. Acetaldehyde is the major noxious agent in alcohol toxicity. Their results indicate, based on our proposal, that PQQ is the coenzyme for xanthine dehydrogenase (XD). XD without PQQ is XO, an enzyme that has a broad substrate specificity and generates superoxide. Acetaldehyde is an excellent substrate for XO and for XD. In alcoholic rats, the oxidation of acetaldehyde by XO will generate large amounts of superoxide with oxidative stress damage likely to result. However, if sufficient PQQ and NAD^+ is present, acetaldehyde can be oxidized by XD (XO + PQQ) without superoxide generation and with the formation of NADH and indirect regeneration of GSH. The accelerated removal of acetaldehyde with PQQ may result from more favorable kinetic parameters with XD than with XO. This remains to be established by a kinetic analysis of XO and XD with acetaldehyde as substrate. Suffice it to say that PQQ through the formation of XD may be a prooxidant for acetaldehyde and for other XO substrates, preventing their toxic effects.

We have described the isolation of PQQ from mammals [7,8]. It has become apparent that recent reports concerning the presence of as yet unidentified antioxidants in colostrum and milk [43,44] is now explained by the presence of PQQ in milk. PQQ in milk and colostrum, along with ascorbic acid, constitutes the major source of the antioxidant activity present.

V. SUMMARY AND PROSPECTUS: THE IMPORTANCE OF PQQ AND REMAINING QUESTIONS

Since the first isolation of free PQQ from bacteria in the late 1970s and its isolation from milk and other mammalian sources in the early 1990s, great progress has been made in spite of severe technical problems inherent in working with such a reactive and elusive compound. PQQ may be an essential nutrient for vertebrate animals and perhaps a vitamin for growing mammals. It still remains to be established if there is any de novo synthesis of PQQ in vertebrate animals. This is a crucial question, which should be solved in a short time. PQQ has unique special redox-cycling properties, which are related to its cofactor functions. PQQ catalyzes dioxygen and superoxide interconversion and converts xanthine oxidase into xanthine dehydrogenase. It may also serve as a cofactor and/or redox mediator in a variety of other redox systems. PQQ's interactions with ascorbate and other reducing agents like NAD(P)H are important, and it appears that the high extracellular concentrations of ascorbate may serve to keep extracellular PQQ in its hydroquinone form. In this form it can be carried in biological fluids without promoting unwanted oxidations or the formation of unwanted adducts with reactive nucleophilic metabolites. PQQ, with and without ascorbate, may be found to replace or substitute for flavins (FMN or FAD) and flavoproteins in several electron transport systems (diaphorases) that also require NAD(P)H. A recent report by Itoh et al. [14] suggests this possibility.

PQQ is a potent antioxidant and anti-inflammatory agent, which could be used pharmaceutically. Whether or not PQQ will be useful in the therapy of inflammatory joint and bowel diseases awaits additional testing. Prophylactic use of PQQ in the preterm neonates may help to prevent some of their unique clinical problems, especially those associated with oxidative stress injury. In the elderly, PQQ and prodrugs that generate PQQ in the brain may have value in the prevention and therapy of neuromotor and other neurological disorders. Drugs developed from agents like aromatic diamines that sequester PQQ may also be useful in clinical medicine. These agents may also induce localized PQQ deficiencies and produce interesting new animal models of human disease. That PQQ may be involved, not only in superoxide metabolism but in nitric oxide metabolism as well, could have medical implications, which should be explored in the next few years. A great stimulus to PQQ research would be the discovery of a heritable or acquired human or animal disorder that involves defects in either the active transport, metabolism, or biosynthesis of PQQ.

NOTE ADDED IN PROOF

Recently, a NADPH-dependent reductase, isolated from the cytosolic fraction of bovine erythrocytes, has been shown to catalyze the transfer of reducing equivalents from NADPH to PQQ (Hultquist, D. E. et al., in press, 1992). This 26,000 Dalton protein has also been detected in the cytosolic fraction of liver and other cells and has been shown to be equivalent to the human erythrocyte enzyme *NADPH-dependent methemoglobin reductase* or *flavin reductase*. Using a coupled reaction in which PQQ reduction is linked to ferrous cytochrome c reduction, the apparent Km of the reductase for PQQ at pH 8.1 was found to be approximately 2 μM.

REFERENCES

1. Salisbury, S. A., Forrest, H. S., and Cruse, W. B. T. (1979). A novel coenzyme from bacterial primary alcohol dehydrogenases. *Nature 280*: 843.
2. Corey, E. J., and Tramontano, A. (1981). Total synthesis of the quinonoid alcohol dehydrogenase coenzyme (1) of methylotrophic bacteria. *J. Am. Chem. Soc. 103*: 5599.
3. Duine, J. A., and Jongejan, J. A. (1989). Quinoproteins, enzymes with pyrrolo-quinoline quinone as cofactor. *Ann. Rev. Biochem. 58*: 403.
4. Janes, S. M., Mu, D., Wemmer, D., Smith, D. J., Kaur, S., Maltby, D., Burlingame, A., and Klinman, J. P. (1990). A new redox cofactor in eukaryotic enzymes: 6-Hydroxydopa at the active site of bovine serum amine oxidase. *Science 248*: 981.
5. McIntire, W. S., Wemmer, D. E., Chistoserbov, A., and Lidstrom, M. E. (1991). A new cofactor in a prokaryotic enzyme: tryptophan tryptophylquinone as the redox prosthetic group in methylamine dehydrogenase. *Science 252*: 817.
6. Gallop, P. M., Paz, M. A., and Flückiger, R. (1990). The biological significance of quinonoid compounds in, on and out of proteins. *CHEMTRACTS-Biochem. and Mol. Biol. I*: 357.
7. Flückiger, R., Paz, M. A., Bergethon, P., Bishop, A., Henson, E., Matson, W., and Gallop, P. M. (1992). Isolation of PQQ from milk (submitted).
8. Flückiger, R., Paz, M. A., Bergethon, P., Greenspan, S., Goodman, S., Henson, E., and Gallop, P. M. (1991). First isolation of methoxatin (PQQ) from bovine skim milk and other mammalian sources. *FASEB J. 5* (Suppl 3), Part III A1510 (#6609).
9. Killgore, J., Smidt, C., Duich, L., Romero-Chapman, N., Tinker, D., Reiser, K., Melko, M., Hyde,D., and Rucker, R. B. (1989). Nutritional importance of pyrrolo-quinoline quinone. *Science 245*: 850.
10. Smidt, C. R., Steinberg, F. M., and Rucker, R. B. (1991). Physiologic importance of pyrroloquinoline quinone. *Proc. Soc. Exp. Biol. Med. 197*: 19.
11. Smidt, C. R., Unkefer, C. J., Houck, D. R., and Rucker, R. B. (1991). Intestinal absorption and tissue distribution of [^{14}C]-pyrroloquinoline quinone in mice. *Proc. Soc. Exp. Biol. Med. 197*: 27.
12. Smidt, C. R., Bean-Knudsen, D., Kirsch, D. G., and Rucker, R. B. (1991). Does the intestinal microflora synthesize pyrroloquinoline quinone? *Biofactors 3*: 53.
13. Shah, M., Bergethon, P., Boak, A., Gallop, P. M., and Kagan, H. (1992). Oxidation of lysine in elastin and collagen by PQQ: Copper complexes and lysyl oxidase. (in preparation).
14. Itoh, S., Mita, N., and Ohshiro, Y. (1990). Efficient NAD^+-regeneration system with heterocyclic o-quinones and molecular oxygen catalyzed by diaphorase. *Chem. Lett.* 1949.
15. Mure, M., Itoh, S., and Ohshiro, Y. (1989). Preparation and characterization of iminoquinone and aminophenol derivatives of coenzyme PQQ. *Tetrahedron Lett. 30*: 6875.
16. Mure, M., Itoh, S., and Ohshiro, Y. (1989). Chemical behavior of coenzyme PQQ toward diamines. *Chem. Lett.*: 1491.

17. Mure, M., Nii, K., Inoue, T., Itoh, S., and Ohshiro, Y. (1990). The reaction of coenzyme PQQ with hydrazines. *J. Chem. Soc. Perkin Trans. 2*: 315.
18. Mure, M., Suzuki, A., Itoh, S., and Ohshiro, Y. (1990). Oxidative C_α-C_β fission (dealdolation) of β-hydroxy amino acids by coenzyme PQQ. *Chem. Commun. 22*: 1608.
19. Itoh, S., Kitamura, Y., Ohshiro, Y., and Agawa, T. (1986). Kinetics and mechanism of oxidative deamination of amines by coenzyme PQQ. *Bull. Chem. Soc. Jpn. 59*: 1979.
20. Suzuki, S., Sakurai, T., Itoh, S., and Ohshiro, Y. (1988). Preparation and characterization of ternary copper(II)complexes containing coenzyme PQQ and bipyridine or terpyridine. *Inorg. Chem. 27*: 591.
21. Nakamura, M., and Yamazaki, I. (1982). Preparation of bovine milk xanthine oxidase as a dehydrogenase form. *J. Biochem. 92*: 1279.
22. Massey, V., Schopfer, L. M., Nishino, T., and Nishino, T. (1989). Differences in protein structure of xanthine dehydrogenase and xanthine oxidase revealed by reconstitution with flavin active site probes. *J. Biol. Chem. 264*: 10567.
23. Baggiolini, M., and Wyman, M. P. (1990). Turning on the respiratory burst. *Trends Biochem. Sci. 15*: 69.
24. Hill, H. R., Augustine, N. H., and Jaffe, N. H. (1991). Human recombinant interferon gamma enhances neonatal polymorphonuclear leukocyte activation and movement, and increases free intracellular calcium. *J. Exp. Med. 173*: 767.
25. Karnovsky, M. L., Paz, M. A., and Gallop, P. M. (1992). 2,7,9-Tri-carboxypyrroloquinoline quinone (PQQ, methoxatin) may be essential for the respiratory burst of activated macrophages (in preparation).
26. Sugioka, K., Nakano, M., Naito, I., Tero-Kubota, S., and Ikegami, Y. (1988). Properties of a coenzyme, pyrroloquinoline quinone: Generation of an active oxygen species during a reduction-oxidation cycle in the presence of NAD(P)H and O_2. *Biochim. Biophys. Acta 964*: 175.
27. Palmer, R. M. J., Ferrige, A. G., and Moncada, S. (1987). Nitric oxide release accounts for the biological activity of endothelium-derived relaxing factor. *Nature 327*: 324.
28. Hibbs, J. B., Taintor, R. R., and Vavrin, S. (1987). Macrophage cytotoxicity: Role for L-arginine deaminase and imino nitrogen oxidation to nitrite. *Science 235*: 473.
29. Faint, R. W., Mackie, I. J., and Machin, S. J. (1991). Platelet aggregation is inhibited by a nitric oxide-like factor released from human neutrophils in vitro. *Br. J. Haemat. 77*: 539.
30. Sakuma, I., Stuehr, D. J., Gross, S. S., Nathan, C., and Levi, R. (1988). Identification of arginine as a precursor of endothelium-derived relaxing factor. *Proc. Natl. Acad. Sci. USA 85*: 6328.
31. Ignarro, L. J. (1990). Biosynthesis and metabolism of endothelium-derived nitric oxide. *Annu. Rev. Pharmacol. Toxicol. 30*: 535.
32. McCall, T. B., Boughton-Smith, N. K., Palmer, R. M. J., Whittle, B. J. R., and Moncada, S. (1989). Synthesis of nitric oxide from L-arginine by neutrophils. *Biochem. J. 261*: 293.
33. MacIntyre, I., Zaidi, M., Towhidul, A., Datta, H. K., Moonga, B. S., Lidbury, P. S., Hecker, M., and Vane, J. R. (1991). Osteoclastic inhibition: An action of nitric oxide not mediated by cyclic GMP. *Proc. Natl. Acad. Sci. USA 88*: 2936.

34. Radi, R., Beckman, J. S., Bush, K. M., and Freeman, B. A. (1991). Peroxynitrite oxidation of sulfhydryls: the cytotoxic potential of superoxide and nitric oxide. *J. Biol. Chem. 266*: 4244.
35. Watanabe, A., Hobara, N., and Tsuji, T. (1988). Protective effect of pyrroloquinoline quinone against experimental liver injury in rats. *Curr. Ther. Res. 44*: 896.
36. Nishigori, H., Yasanaga, M., Mizumura, M., Lee, J. W., and Iwatsuru, M. (1989). Preventive effect of pyrroloquinoline quinone on formation of cataract and decline of lenticular and hepatic glutathione of developing chick embryo after glucocorticoid treatment. *Life Sci. 44*: 593.
37. Hamagishi, Y., Murata, S., Kamei, H., Oki, T., Adachi, O., and Ameyama, M. (1990). New biological properties of pyrroloquinoline quinone and its related compounds: Inhibition of chemiluminescence, lipid peroxidation and rat paw edema. *J. Pharmacol. Exp. Ther. 255*: 980.
38. Aizenman, E., Harnett, K. A., Gallop, P. M., and Rosenberg, P. (1992). Interaction of the putative essential nutrient pyrroloquinoline quinone with the NMDA receptor redox site. *Neurosci. 12*:2362.
39. Hill, H. R. (1987). Biochemical, structural and functional abnormalities of polymorphonuclear leukocytes in the neonate. *Pediatr. Res. 22*: 375.
40. Winterbourn, C. C., Cowden, W. B., and Sutton, H. C. (1989). Autooxidation of dialuric acid, divicine and isouramil. Superoxide dependent and independent mechanisms. *Biochem. Pharmacol. 38*: 611.
41. Winerbourn, C. C., and Munday, R. (1990). Concerted action of reduced glutathione and superoxide dismutase in preventing redox cycling of dihydroxypyridines, and their role in antioxidant defense. *Free Radical Res. Commun. 8*: 287.
42. Hobara, N., Watanabe, A., Kobayshi, M., and Tsuji, T. (1989). Quinone derivatives lower blood and liver acetaldehyde but not ethanol concentrations following ethanol loading to rats. *Pharmacology 37*: 264.
43. Buescher, E. S., and McIheran, S. M. (1988). Antioxidant properties of human colostrum. *Pediatr. Res. 24*: 14.
44. Buescher, E. S., McIlheran, S. M., and Frenck, R. W. (1989). Further characterization of human colostral antioxidants: Identification of an ascorbate-like element as an antioxidant component and demonstration of antioxidant heterogeneity. *Pediatr. Res. 25*: 266.

18

Pharmacological Application of PQQ

Akiharu Watanabe and Toshihiro Tsuchida

Toyama Medical and Pharmaceutical University, Toyama, Japan

Hideo Nishigori

Teikyo University, Kanagawa, Japan

Teizi Urakami

Mitsubishi Gas Chemical Company, Tokyo, Japan

Norio Hobara

Okayama University Medical School, Okayama, Japan

I. INTRODUCTION

Pyrroloquinoline quinone (4,5-dihydro-4,5-dioxo-IH-pyrrolo-[2,3-*f*]-quinoline-2,7,9-tricarboxylic acid; PQQ), a newly discovered coenzyme, functions as the prosthetic group of several dehydrogenases and oxidases in microorganisms [1]. Unique effects of PQQ on microbial growth have been reported [2], but the metabolism and pharmacological functions of PQQ in the higher animals have not as yet been elucidated. However, recent studies in mammals revealed the occurrence of quinoproteins such as amine oxidase [3] and also suggested the nutritional importance of PQQ [4]. Because of its interesting properties and actions, the development of this agent as a new drug is probable. Thus, the pharmacological effects of PQQ should be investigated in experimental animals, and toxicological studies on PQQ should be performed for its general use.

The present report summarizes our previous experimental data on pharmacological effects of PQQ specifically with respect to acetaldehyde metabolism following ethanol loading to rats [5] and the protective effects of PQQ against hepatotoxin-induced liver injury in rats [6] and against hydrocortisone-induced cataract formation in chick embryos [7]. Furthermore, the possibility of a toxic effect of PQQ on the kidney was also tested in rats and mice [8].

II. SUMMARY OF EXPERIMENTAL RESULTS

Pharmacological application of PQQ to the animals was performed, particularly with respect to the acceleration of alcohol metabolism and prevention of experimental liver injury and cataract formation. Toxic effects of PQQ on the kidney were also demonstrated. The following results were obtained.

1. A rise in blood and liver acetaldehyde concentrations following an intragastric administration of ethanol to rats was significantly inhibited when PQQ was injected intraperitoneally prior to the ethanol load. When acetaldehyde was incubated in vitro with PQQ at 40°C, the concentration of the former slowly decreased with incubation time.
2. PQQ protected rats from experimental liver injury, which was induced by hepatotoxins such as carbon tetrachloride (CCl_4) and D-galactosamine. The effect was observed after an intraperitoneal injection of PQQ was given either before or within 3–4 hours following the administration of hepatotoxins. Oral administration of PQQ was less effective in CCl_4-induced liver injury.
3. PQQ administration to 15-day-old fertile eggs, which had been previously treated with hydrocortisone succinate, caused a preventive effect against hydrocortisone-induced cataract formation. PQQ also prevented the decline of reduced glutathione (GSH) content in the cataractous lens.
4. When a large dose of PQQ was intraperitoneally injected into rats daily for 4 days, functional and morphological changes of the kidneys were clearly observed; necrotic and degenerative changes of the proximal tubular epithelium, as well as hematuria and an elevation of serum creatinine. High mortality of PQQ-treated mice was observed after a single intraperitoneal dose of more than 40 mg/kg body weight. Therefore, new derivatives of PQQ, which have no toxic effect on the kidney and are easily absorbed from the gut, should be developed for the further application of PQQ to human diseases.

III. PHARMACOLOGICAL STUDIES

A. Experimental Protocols

Male Sprague-Dawley rats (Clea Japan, Inc., Tokyo), weighing 230–350 grams each, and male Wistar rats (Kyudo Co., Ltd., Kumamoto), weighing 360–430 grams each, were used in the present study. The rats were kept at 24 ± 2°C and 55 ± 10% relative humidity under an alternating light-dark cycle of 12 hours each. The rats were kept on a basal diet (CE-2, Clea Japan, Inc., Tokyo) for one week prior to the experiments. Male SPF-ICR mice (Charles River Japan, Inc., Tokyo), weighing 26–32 grams each, were kept at 23 ± 2°C and 55 ± 6% relative humidity and on a basal diet (MF, Oriental Yeast Co., Ltd., Tokyo). One-day-old fertile White Leghorn eggs were purchased from a local hatchery and incubated at 37.5°C and 68% relative humidity.

PQQ disodium (Ube Industries, Ltd., Ube) dissolved in 2% (w/v) $NaHCO_3$ solution (115 mg/dl) was administered intraperitoneally to rats at a dose of 11.5 mg/kg body weight (10 ml/kg). Control rats were similarly injected with a 2% $NaHCO_3$ solution alone. For toxicological studies, mice were treated with an intraperitoneal injection of a solution of PQQ (1 mg/ml in 2% $NaHCO_3$) at a dose of 20–160 mg/kg body weight (1.2–4.8 ml/mouse). For cataract production, PQQ dipotassium (Kanto Co., Ltd., Tokyo) was dissolved in 25 mM potassium acetate, and each egg was treated with 0.25–2.50 μmol of PQQ in a volume of 0.2 ml/egg. Control eggs were similarly treated by 25 mM potassium acetate alone.

All of the data are expressed as means ± SE, unless otherwise stated. Statistical differences between the mean values were determined by Student's *t*-test after analysis of variance.

B. Experiment 1. Effect of PQQ on Blood and Liver Acetaldehyde Levels After Administration of Ethanol

1. *Methods*

A 15% ethanol solution was administered intragastrically to Sprague-Dawley rats at a dose of 1.5 g/kg body weight. PQQ was administered at a dose of 11.5 mg/kg body weight twice, at 1 hour and 10 minutes, prior to the ethanol administration. The rats were killed by exsanguination from the carotid artery 3 hours after the ethanol load. The liver was immediately removed and frozen in liquid nitrogen, and the frozen samples were pulverized in mortars cooled with liquid nitrogen. Liver and blood ethanol and acetaldehyde contents were determined by a head-space gas chromatographic (Hitachi, type 163) procedure [9].

The nonenzymatic reaction of PQQ (1.4–4.9 mM) with acetaldehyde (final concentration of 160 μM in 20 mM sodium phosphate buffer, pH 8.8) was examined in vitro at 0 and 40°C for 3 hours in test tubes capped with rubber stoppers. The incubation was terminated by addition of a 1.06 M perchloric acid solution containing 20 mM thiourea and 1.3 mM *n*-propanol, and the acetaldehyde concentrations were immediately determined as described above.

2. *Results*

Ethanol contents in the blood and liver 3 hours following ethanol loading to rats were not different, whether PQQ was given or not (Table 1). However, blood and liver acetaldehyde levels were much lower in the PQQ-treated rats than in

Table 1 Effect of PQQ on Ethanol and Acetaldehyde Levels Following Ethanol Loading in Rats

	Acetaldehyde		Ethanol	
	Blood (μM)	Liver (nmol/g)	Blood (mM)	Liver (μmol/g)
Control	3.7 ± 0.5	6.9 ± 0.9	7.8 ± 2.8	6.4 ± 1.4
PQQ	2.6 ± 0.2*	2.5 ± 0.5*	6.7 ± 2.1	5.7 ± 2.5

*$p < 0.05$.
Source: Ref. 5.

untreated rats. Similar results (data not shown) were also observed in rats treated with other quinone derivatives such as coenzyme Q_{10} (10 mg/kg body weight) and idebenone (30 mg/kg body weight).

A nonenzymatic reaction of PQQ with acetaldehyde was observed in vitro (Fig. 1). Acetaldehyde levels diminished with time in the presence of PQQ at 40°C but not 0°C. Added acetaldehyde was recovered at the levels of 43.4, 18.5, and 12.6% at times of 0.5, 1, and 3 hours, respectively, after the start of incubation with 4.9

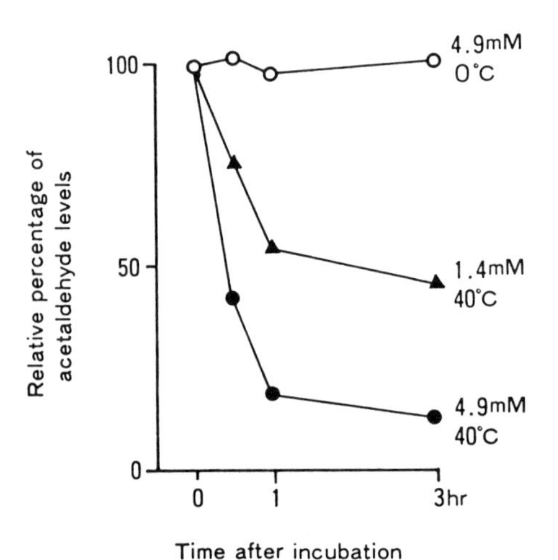

Figure 1 In vitro reaction of acetaldehyde with PQQ at 0 and 40°C. PQQ (1.4 and 4.9 mM) was incubated with 160 μM acetaldehyde at 0 and 40°C for 3 hours in a final volume of 0.5 ml of 20 mM sodium phosphate, pH 8.8. Before and during the incubation, acetaldehyde levels were serially determined at the time indicated. Each value represents the mean of two experiments. (From Ref. 5.)

mM PQQ. The similar experiment of the effect of PQQ on ethanol oxidation was carried out, but PQQ did not oxidize ethanol to acetaldehyde in vitro. However, 1,4-benzoquinone (13 mM) oxidized ethanol to acetaldehyde to a small extent at 40°C (data not shown).

C. Experiment 2. Protection by PQQ of Experimentally Induced Liver Injury

1. Methods

Carbon tetrachloride (CCl_4) was given intragastrically to Sprague-Dawley rats under ether anesthesia at a dose of 8 ml/kg body weight of a 20% CCl_4 solution of liquid paraffin, unless otherwise stated. D-Galactosamine, thioacetamide, and allyl formate were intraperitoneally injected at doses of 1000, 200, and 40 mg per kg of body weight, respectively. The PQQ solution was injected intraperitoneally at a dose of 11.5 mg/kg twice, at 1 hour and 10 minutes, before the hepatotoxins were given. The treated rats were starved but offered water ad libitum during the experiment and killed by a blow on the head 24 hours after hepatotoxin administration.

2. Results

CCl_4- and galactosamine-induced liver injury were effectively prevented by an injection of PQQ prior to the hepatotoxins. This can be seen from the findings that serum bilirubin and glutamic-pyruvate transaminase (GPT) levels significantly decreased in PQQ-treated rats (Table 2). The levels of lipid peroxidation in CCl_4-injured liver were significantly lower in PQQ-treated rats than in untreated rats.

Table 2 Effect of PQQ Injection on CCl_4 and Galactosamine-Induced Liver Injury in Rats

	Liver weight (%)	Serum		Liver	
		Bilirubin (mg/dl)	GPT (KU/l)	Triglyceride (mg/g)	Lipid peroxide (μmol/g)
CCl_4 (5)					
−PQQ	3.8 ± 0.2	0.9 ± 0.1	4015 ± 1393	20 ± 1	4.9 ± 0.4
+PQQ	3.9 ± 0.1	0.3 ± 0.1**	103 ± 18**	14 ± 1	3.5 ± 0.2**
Galactosamine (4)					
−PQQ	3.3 ± 0.2	1.1 ± 0.3	2520 ± 635	15 ± 1	1.7 ± 0.6
+PQQ	3.3 ± 0.3	0.4 ± 0.1*	423 ± 68*	8 ± 1	1.5 ± 0.1

$^*p < 0.02$, $^{**}p < 0.01$. ( ) = No. of rats.
−PQQ: control rats treated with 2% $NaHCO_3$ solution.
Source: Ref. 6.

Table 3 Effect of PQQ Administration on Thioacetamide– and Allyl Formate–Induced Liver Injury in Rats

	Liver weight (%)	Serum bilirubin (mg/dl)	Serum GPT (KU/l)
Thioacetamide			
−PQQ (5)	3.6 ± 0.1	0.74 ± 0.07	1363 ± 484
+PQQ (4)	3.6 ± 0.1	0.60 ± 0.14	282 ± 81*
Allyl formate			
−PQQ (6)	2.9 ± 0.1	0.40 ± 0.04	733 ± 121
+PQQ (5)	3.1 ± 0.3	0.22 ± 0.01**	230 ± 64*

*$p < 0.05$, **$p < 0.01$.
( ) = No. of rats. −PQQ, control rats.
Source: Ref. 6.

Liver triglyceride contents decreased in the PQQ-treated rats that had been injured by CCl_4 and galactosamine, although the changes were not significant. Similar findings of PQQ protection against liver injury were observed in thioacetamide- and allyl formate–induced liver injury (Table 3). The relative liver weights were not changed by PQQ treatments in any of the different types of liver injury.

The protective effect of an intraperitoneal PQQ injection against CCl_4-induced liver injury was also observed with a single dose of PQQ given 10 minutes prior to CCl_4 injection (Table 4). The protective effect was observed even when PQQ was injected twice, either 1 and 2 hours or 3 and 4 hours, following the hepatotoxin injection. Although these effects were slight, they were significant. The effect was not observed, however, when PQQ was injected twice, at 5 and 6 hours, after CCl_4 was administered.

The minimum effective dose of PQQ was determined by administration of a single dose of PQQ given 10 minutes prior to CCl_4 administration. Five mg of PQQ/kg body weight was the lowest dose effective in the case of CCl_4-induced liver injury (Table 4). The oral dose of PQQ (11.5 mg/kg, 10 minutes prior to CCl_4 treatment) showed little preventive effect of PQQ on liver injury as judged from serum GPT activity (data not shown).

D. Experiment 3. Prevention by PQQ of Hydrocortisone-Induced Cataract Formation

1. Methods

For the production of cataracts, hydrocortisone succinate sodium was administered at a dose of 0.2 ml of 51.5 mg/dl, in sterile water, to 15-day-old embryos through a small hole in the eggshell over the air sack [10]. Two-tenths ml of PQQ

Table 4 Effect of PQQ Injection on CCL_4 Hepatotoxicity in Rats

	Serum GPT (KU/l)
Time and number of PQQ injections	
None	2465 ± 195 (5)
One (−10 min)	192 ± 52 (4)**
Two (+1 h, +2 h)	1478 ± 298 (5)*
(+3 h, +4 h)	1543 ± 224 (6)*
(+5 h, +6 h)	1806 ± 354 (5)
Dose of PQQ (−10 min)	
None	1870 ± 585 (5)
1 mg/kg	1017 ± 509 (5)
2 mg/kg	915 ± 248 (5)
5 mg/kg	312 ± 59 (5)*

$*p < 0.05$, $**p < 0.01$.
( ) = No. of rats.
Source: Ref. 6.

solution at a dose of 0.1–1.0 mg was administered 3 times, at 3, 10, and 20 hours following the steroid treatment. The lenses were removed from chick embryos 48 hours following hydrocortisone administration and visually classified as described previously [10] except that the IV and V stages were not separated. Determination of GSH contents in lens and liver was performed by the method using Ellmann's reagent [11].

2. Results

When chick embryos were treated with hydrocortisone, all lenses (95%) were classified as stage IV-V of cataract 48 hours following the treatment (Table 5). A dose-dependent protection by PQQ was observed against hydrocortisone-induced cataract formation. A triple application of PQQ (0.5 mg/egg) at 3, 10, and 20 hours showed the strongest protective effect against cataract formation.

The previous paper [10] showed that the levels of GSH in the hydrocortisone-induced cataractous lenses began to decline 20 hours following hydrocortisone treatment and became about one-half of the control levels after 48 hours (Table 6). The stage I lenses in hydrocortisone and PQQ-treated embryos showed 88% of the control GSH levels, and the stage III and stage IV-V lenses were approximately 76% and 64% of the control GSH levels, respectively. Hepatic GSH levels became about 66% of the control levels 24 hours following hydrocortisone treatment. This decline of liver GSH levels was observed earlier than that of lens GSH contents and the appearance of lens opacification. Triple application of PQQ, however, com-

Table 5 Protective Effects of PQQ on Cataract Formation in Hydrocortisone-Treated Chick Embryos

Treatment	Number of eggs at stage: I	II	III	IV–V
Control	20	0	0	0
Hydrocortisone[a]	0	0	0	20
Hydrocortisone + PQQ				
(0.1)[b]	0	0	0	20
(0.5)[c]	5	3	5	7
(0.5)	9	5	6	0
(1.0)	6	1	3	0

[a]Potassium acetate instead of PQQ was injected.
[b]Numbers in parentheses are mg/egg.
[c]PQQ was injected twice 3 and 10 h following hydrocortisone treatment.
Source: Ref. 7.

pletely protected the decline of liver GSH contents in hydrocortisone-treated embryos (data not shown). PQQ showed some elevation of liver GSH levels (136%) in the hydrocortisone-untreated rats.

E. Experiment 4. Toxic Effects of PQQ

1. Methods

A PQQ solution was administered intraperitoneally to Wistar rats for 4 consecutive days at a daily dose of 11.5 mg/kg body weight. The rats were fed ad libitum during the experimental period and killed on the 5th day by venipuncture of the vena cava inferior. The tissues were processed for light microscopic studies (hematoxylin and eosin). Urine was tested for pH, sugar, protein, ketone body, bilirubin, and occult blood using Multi-Sticks III (Miles-Sankyo Co., Ltd., Tokyo). Serum biochemical tests were routinely examined with a Gilford Autoanalyzer (Impact 400). Urine was collected by applying pressure on the inguinal region.

A single dose of PQQ, which ranged from 20 to 160 mg/kg body weight, was injected intraperitoneally into male SPF-ICR mice. Thereafter, several of the mice died. Body weight and urinalysis were serially checked every day until the 14th day of the experiment. The blood biochemistry was also examined 48 hours following the single PQQ dose.

2. Results

No significant difference of changes in the body weight between PQQ-treated (daily dose of 11.5 mg/kg body weight for 4 consecutive days) and untreated rats

Table 6 Effects of PQQ on GSH Contents in Lenses of Chick Embryos Treated with Hydrocortisone

	GSH (nmol/lens)	% of control
Control	14.2	100
Hydrocortisone	8.4	59.2
Hydrocortisone + PQQ		
Stage I cataract	12.5	88.0
Stage III cataract	10.8	76.1
Stage IV-V cataract	9.1	64.1

PQQ (0.5 mg/egg) was administered at 3, 10, and 20 h after hydrocortisone administration. GSH contents in 8-pooled lenses from 8 embryos were determined, and averages of two samples are shown.
Source: Ref. 7.

was observed during the experiment. Examination of the urine in PQQ-treated rats revealed an increased excretion of protein, glucose, ketone body, and occult blood. Blood urea nitrogen (BUN) and serum creatinine levels were markedly higher in PQQ-treated rats, and serum triglyceride levels were lower than untreated controls (Table 7). Swelling of the kidneys was observed macroscopically at autopsy, and the weight of the kidneys (3.43 ± 0.23 g versus control 2.72 ± 0.25 g) as well as percent kidney weight of body weight (0.91 ± 0.07% versus control 0.65 ± 0.04%) also increased, although only the latter was significant ($p < 0.001$).

In the PQQ-treated rats, vacuolar degeneration, atrophy, and necrosis of the

Table 7 Blood Biochemistry in PQQ-Treated Rats

	BUN	Cr	ALP	LDH	Glc	TG
Group	(mg/dl)		(IU/l)		(mg/dl)	
Control						
Mean	23.7	0.53	118	83	203	137
SD	2.3	0.07	24	40	27	38
PQQ						
Mean	62.5*	1.48*	131	87	176	48*
SD	19.5	0.46	37	26	16	12

Cr, creatinine; Glc, glucose, TG, triglyceride. There were 5 rats in each group.
*$p < 0.05$.
Source: Ref. 8.

Table 8 Histological Findings of Kidneys in PQQ-Treated Rats

	Rat 6	Rat 7	Rat 8	Rat 9	Rat 10
Glomerulus	−	−	−	−	−
Tubular (proximal) epithelium					
Dilatation	−	+++	−	−	+++
Necrosis	++	++	++	+++	+++
Vacuolar degeneration	+	++	+	−	+
Regeneration	−	++	++	−	−
Basophilic atrophy	+	+	+	++	++
Hyaline cast	+	+	+	++	++

−: Negative; +: slight; ++: moderate; +++: severe changes.
Source: Ref. 8.

proximal tubular epithelium in the renal cortex were observed microscopically (Table 8). Dilatation and regeneration of the tubules were also observed in two of the PQQ-treated rats. However, few changes of the glomerulus were observed in any of the rats treated with PQQ. A slight decrease in glycogen deposition and an increase in the mitotic process were observed also in the liver, but pathological findings of the spleen were not observed in the PQQ-treated rats. The nephrotoxicity of PQQ was observed only with a daily dose larger than 2.9 mg/kg body weight under similar conditions.

No mouse died from a single intraperitoneal PQQ dose of 20 and 40 mg/kg body weight, but for the doses of 80 and 160 mg/kg, mortality rates were 63 and 100%, respectively (Table 9). Most of the mice died 2–3 days after PQQ injection. The body weight in PQQ (20–80 mg/kg)-treated mice decreased up to the second

Table 9 Effect of Intraperitoneal PQQ Injection on Mortality of Male Mice

	No. of mice	Number of deaths following PQQ injection						Mortality (%)
		Day 1	Day 2	Day 3	Day 6	Day 10	Day 14	
Control	5	0	0	0	0	0	0	0
PQQ[a]								
20	8	0	0	0	0	0	0	0
40	8	0	0	0	nd	nd	nd	0
80	8	0	2	2	1	0	0	63
160	8	0	4	4	—	—	—	100

nd: Not examined.
[a]mg/kg body weight.

week and then tended to increase gradually (data not shown). Urinary sugar was severely excreted up to the fourth day of the experiment in the case of mice that were given doses of 40 and 80 mg PQQ/kg body weight. In PQQ-injected mice, blood glucose levels decreased depending upon PQQ dose: for 40 mg PQQ/kg body weight blood glucose = 70 mg/dl; 80 mg PQQ/kg, 57 mg/dl; 160 mg PQQ/kg, 35 mg/dl. The large dosing of PQQ probably induced severe hypoglycemia and led to death. In PQQ-treated mice, BUN and serum creatinine levels also increased markedly: for 80 mg PQQ/kg body weight, BUN = 130 mg/dl; creatinine = 3.4 mg/dl.

IV. DISCUSSION

The mechanism of acetaldehyde-induced hepatotoxity has been the focus of recent investigations [12], and toxic effects of acetaldehyde on hepatic functions have also been reported [13]. The present study shows the selective reduction by a prior administration of PQQ of blood acetaldehyde concentrations following ethanol loading. A potent in vivo inhibitor of aldehyde dehydrogenase, disulfiram (Antabuse), has been used clinically for treating alcohol addiction, but there are few reports of an agent for treatment of acetaldehyde toxicity by accelerating its metabolism. The mechanism of the reaction of PQQ with acetaldehyde in vivo is not known. PQQ may react nonenzymatically with acetaldehyde by reduction of the *o*-quinone group to produce the *o*-quinol group. However, acetaldehyde oxidation to acetate was not confirmed in the present study. Another possibility is that aldose condensation converts the acetaldehyde to a dimeric semialdehyde (no acetate formation). It has also been reported that PQQ becomes $PQQH_2$ by reacting nonenzymatically with GSH to form oxidized glutathione [14] and by reacting with NAD(P)H to form NAD(P) [15]. However, it has not yet been shown if these reactions occur in vivo.

The protective action of PQQ was commonly observed against various types of hepatotoxin-induced liver injury whose mechanisms differ greatly. Even after liver injury had already occurred, PQQ was effective in preventing ongoing necrosis of the hepatocytes, suggesting that PQQ is a potent antinecrotic agent. CCl_4 is metabolized by an NADPH-dependent cytochrome P-450 in the microsome to form a trichloromethyl radical. However, it is not known whether direct action of radical and/or lipid peroxidation causes necrosis of hepatocytes [16]. CCl_4 produced a marked decline of GSH in the liver, which might be caused by enhanced GSH catabolism not due to gamma-glutamyl transpeptidase (gamma-GTP) [17]. In any case, oxidative stress injury will be possible in CCl_4-induced liver injury, since superoxide dismutase can protect against CCl_4-induced liver injury [18]. PQQ may increase the synthesis of GSH and/or prevent the utilization of GSH in the injured liver, although the mechanisms are obscure.

The results in the present study indicate that some fundamental process of liver injury leading to hepatocyte necrosis may be blocked by PQQ. Lipid peroxide formation as produced in CCl_4-induced liver injury is not common with other hepatotoxins such as galactosamine (Table 2). The weight of the injured livers and fat contents did not change much in any type of hepatotoxin-induced liver injury, indicating that PQQ mainly inhibits hepatocyte necrosis. Many theories for liver injury including the accelerated entry of calcium into hepatocytes [19] have since been proposed, but currently no agreement on a single theory exists. Thus, no clear explanation of the protective effect of PQQ against hepatocyte necrosis can be possible from only the present study.

During the formation of hydrocortisone-induced cataract formation, the decline of hepatic GSH contents and elevation of hepatic and blood lipid peroxidation began sharply, 20 hours after hydrocortisone administration [20]. Following this period, the decline of GSH contents in the lens, the elevation of lipid peroxidation in the lens, and finally the opacification of the lens seemed to start [21]. Therefore, oxidative stress to the lenses may be strongly involved in the initiating events of the cataract formation. The decline of GSH contents in the liver represents an elevation of oxidative state or the production of oxidative stress, which influences the lens through the blood and aqueous humor. The enhancing effects of PQQ on GSH levels in the liver indicates its potential as a type of anticataract agent. PQQ treatment was observed to prevent the decline of hepatic GSH contents in hydrocortisone-treated chick embryos and to increase the haptic GSH levels in control chick embryos.

The dose of PQQ used in the present study was based on the clinical dose of coenzyme Q_{10}, a clinically available quinone derivative. The optimum dose of PQQ for mammals has not yet been determined. In the present study, a tentative dose of 11.5 mg/kg of PQQ was employed in Experiments 1 and 2. No rats died during the study, but spontaneous movements of rats decreased following the administration of this PQQ dose, suggesting that the present dose of PQQ might be large for rats. In mice experiments, the PQQ dose of 40 mg/kg body weight may be the maximum, and the larger doses induce severe hypoglycemia leading to death (urinary sugar loss due to tubular damage). Further studies on the long-term toxicity and teratogenic effects of PQQ should be performed in the future.

Although the mechanism of PQQ nephrotoxicity is not clear, we noticed that green-colored PQQ metabolites were excreted within 1 hour of an intraperitoneal injection of PQQ. The high concentrations of PQQ metabolites in the kidney may cause injury to the proximal tubular components of the renal cortex. Further studies on this green compound are now under study in our laboratory.

REFERENCES

1. Ameyama, M., Hayashi, M., Matsushita, K., Shinagawa, E., and Adachi, O. (1984). Microbial production of pyrroloquinoline quinone. *Agric. Biol. Chem. 48*: 561.

2. Ameyama, M., Shinagawa, E., Matsushita, K., and Adachi, O. (1985). Growth stimulating activity for microorganisms in naturally occurring substances and partial characterization of the substance for the activity as pyrroloquinoline quinone. *Agric. Biol. Chem. 49*: 699.
3. Lobenstein-Verbeek, C. L., Jongejan, J. A., Frank, J., and Duine, J. A. (1984). Bovine serum oxidase: A mammalian enzyme having covalently bound PQQ as prosthetic group. *FEBS Lett. 170*: 305.
4. Killgore, J., Smidt, C., Duich, L., Romero-Chapman, N., Tinker, D., Reiser, K., Melko, M., Hyde, D., and Rucker, R. B. (1989). Nutritional importance of pyrroloquinoline quinone. *Science 245*: 850.
5. Hobara, Y., Watanabe, A., Kobayashi, M., Tsuji, T., Gomita, Y., and Araki, Y. (1988). Quinone derivatives lower blood and liver acetaldehyde but not ethanol concentrations following ethanol loading to rats. *Pharmacology 37*: 264.
6. Watanabe, A., Hobara, N., and Tsuji, T. (1988). Protective effect of pyrroloquinoline quinone against experimental liver injury in rats. *Curr. Ther. Res. 44*: 896.
7. Nishigori, H., Yasunaga, M., Mizumura, M., Lee, J. W., and Iwatsuru, M. (1989). Preventive effects of pyrroloquinoline quinone on formation of cataract and decline of lenticular and hapatic glutathione of developing chick embryo after glucocorticoid treatment. *Life. Sci. 45*: 593.
8. Watanabe, A., Hobara, N., Ohsawa, T., Higashi, T., and Tsuji, T. (1989). Nephrotoxicity of pyrroloquinoline quinone in rats. *Hiroshima J. Med. Sci. 38*: 49.
9. Hobara, N., Watanabe, A., Kobayashi, M., Nakatsukasa, H., Nagashima, H., Fukuda, T., and Araki, Y. (1985). Tissue distribution of acetaldehyde in rats following acetaldehyde inhalation and intragastric ethanol administration. *Bull. Environ. Contam. Toxicol. 35*: 393.
10. Nishigori, H., Lee, J. W., and Iwatsuru, M. (1984). Effect of MPG on glucocorticoid-induced cataract formation in developing chick embryo. *Invest. Ophthalmol. 25*: 1051.
11. Sedlak, J., and Lindsay, R. H. (1968). Estimation of total, protein-bound, and non-proteins sulfhydryl groups in tissue with Ellman's reagent. *Anal. Biochem. 25*: 192.
12. Hasumura, Y., Tesche, R., and Lieber, C. S. (1975). Acetaldehyde oxidation by hepatic mitochondria: Decrease after chronic ethanol consumption. *Science 189*: 727.
13. Cederbaum, A., Lieber, C. S., and Rubin, E. (1974). The effect of acetaldehyde on mitochondrial fraction. *Arch. Biochem. Biophys. 161*: 26.
14. Itoh, S., Kato, N., Mure, M., and Ohshiro, Y. (1987). Kinetic studies on the oxidation of thiols by coenzyme PQQ. *Bull. Chem. Soc. Jpn. 60*: 420.
15. Nakano, M., Sugioka, K., Tero-Kubota, S., and Ikegami, Y. (1988). Non-enzymatic oxidation of NAD(P)H by pyrroloquinoline quinone (PQQ): Generation of H_2O_2 and O_2^-. In *The Role of Oxygen in Chemistry and Biochemistry* (W. Ando, and Y. Moro-Oka, eds.), Elsevier, Amsterdam, p. 443.
16. Tribble, D. L., Aw, T. Y., and Jones, D. P. (1987). The pathophysiological significance of lipid peroxidation in oxidative cell injury. *Hepatology 7*: 377.
17. Watanabe, A., and Nagashima, H. (1983). Glutathione metabolism and glucose 6-phosphate dehydrogenase activity in experimental liver injury. *Acta Med. Okayama 37*: 463.

18. Yasuyama, T., Inoue, K., Kuwabara, Y., Tsuchida, T., Nakayama, Y, and Sasaki, H. (1990). Localization of Cu, Zn superoxide dismutase (SOD) in rat liver following carbon tetrachloride intoxication. *Acta Hepat. Jpn. 31*: 69.
19. Schanne, F. A. X., Kane, A. B., Yound, E. E., and Farber, J. L. (1979). Calcium dependence on toxic cell death: A final common pathway. *Science 206*: 700.
20. Nishigori, H., Lee, J. W., and Iwatsuru, M. (1988). Preventive effect of isocitrate on glucocorticoid-induced cataract formation of developing chick embryo. *Proc. Int. Soc. Eye Res. 5*: 127.
21. Nishigori, H., Lee, J. W., and Iwatsuru, M. (1983). An animal model for cataract research; cataract formation in developing chick embryo by glucocorticoid. *Exp. Eye Res. 36*: 617.

19

Quinoproteins as Biocatalysts

A. Netrusov

Moscow University, Moscow, Russia

I. BIOCATALYSIS BY QUINOPROTEIN ENZYMES (INTRODUCTION TO CATALYTIC PROCESSES BY PQQ-DEPENDENT ENZYMES)

In the past few years it has been established that, in addition to the two well-characterized groups of dehydrogenases, NAD^+/$NADP^+$-dependent and FAD/FMN-dependent, there is a third class of dehydrogenases, the quinoproteins [1]. These proteins contain pyrroloquinoline quinone (PQQ) as their prosthetic group. Although the novel coenzyme was first reported in the early 1960s, the structure was not elucidated until about 15 years later in two independent laboratories on both sides of the Atlantic [2,3]. PQQ is a noncovalently bound quinone redox cofactor first described for methanol dehydrogenase, and its systematic name is 2,7,9-tricarboxy-1H-pyrrolo[2,3-*f*]-quinoline-4,5-dione, or, more commonly, 2,7,9-tricarboxy-PQQ. This quinone has been shown to be the redox cofactor of several other bacterial enzymes, as well as of methanol dehydrogenases.

Quinoprotein dehydrogenases play a definite role in the oxidation of C_1 compounds, but it is not an exclusive one since flavoprotein oxidases and NAD^+-dependent methanol dehydrogenases were discovered [4–6]. The high redox potential of the PQQ_{red}/PQQ_{ox} couple and the specific mode of substrate oxidation (covalent catalysis) might be important factors in preference for a periplasmic location of several of the known quinoprotein dehydrogenases. Like flavins, PQQ is well suited to its function as an electron relay system, transferring two electrons from the substrate, one at a time, to the one-electron acceptors of the electron transport chain [7].

A number of enzymes have been proposed to contain covalently bound PQQ, but in view of the recent publication on a new prosthetic group of bacterial methylamine dehydrogenase, all of these enzymes should be reexamined [8]. Several microbial enzymes thus far described can potentially be used as catalysts in oxidation/transformation processes (Table 1).

As can be seen, the PQQ-containing enzymes catalyze many diverse reactions, including dehydrogenations, oxidations, and decarboxylations. It will not be surprising to find the PQQ-containing/dependent enzymes in all six major enzyme classes, but according to the redox properties of the PQQ molecule, it can be predicted that it will primarily be associated with redox reactions and oxidoreductases.

Table 1 Microbial PQQ-Containing Enzymes with Potential Use as a Catalytic Source

Enzyme	Organism	Ref.
Alcohol dehydrogenase	Gram-negative methylotrophic bacteria	4
	Clostridium thermoautotrophicum	9
	Nocardia sp. 239	10
	Pseudomonas aeruginosa	11, 12
	Pseudomonas putida	12
	Rhodopseudomonas acidophila	13
	Acetobacter sp.	14
Quinohemoprotein alcohol dehydrogenase	*Pseudomonas testosteroni*	15
Aldehyde dehydrogenase	*Acetobacter* sp.	16
	Hyphomicrobium sp.	17
	Rhodopseudomonas acidophila	13
Glucose dehydrogenase (membrane-bound, soluble)	Gram-negative bacteria	18, 19
	Acinetobacter calcoaceticus	19–21
	Zymomonas mobilis	22
Quinate dehydrogenase	Gram-negative bacteria	23
Glycerol dehydrogenase	*Gluconobacter industris*	24
Polyvinylalcohol dehydrogenase	*Pseudomonas* sp.	22
Polyethyleneglycol dehydrogenase	*Flavobacterium* sp.	25
Lactate dehydrogenase	*Propionibacterium pentosaceum*	26
Methylamine oxidase	*Arthrobacter* sp. P1	27
Galactose oxidase	Fungi	7
Lysyl oxidase	*Pichia pastoris*	28
Nitroalkane oxidase	*Fusarium oxysporum*	29
Glutamate decarboxylase	*Escherichia coli*	7

Catalysis with PQQ-dependent enzymes and enzyme systems might find useful applications in biotechnology, primarily because of the broad substrate specificity. Up to now, only a few processes, which utilize a PQQ-dependent enzyme in catalysis are described in the literature. However, we expect more in the future for different catalytic processes.

II. LACTATE PRODUCTION FROM 1,2-PROPANEDIOL BY FREE CELL SUSPENSIONS OF AN OBLIGATE METHYLOTROPH—BIOCATALYSIS BY PQQ-DEPENDENT METHANOL DEHYDROGENASE

Oxidation of 1,2-propanediol by methylotrophic bacteria is related to the broad substrate specificity of the PQQ-containing, cytochrome *c*-dependent methanol dehydrogenase [4]. For a long time it was thought that the purified methanol dehydrogenase of methylotrophs could not oxidize 1,2-propanediol. It was shown recently that two methylotrophic bacteria, the facultative *Methylobacterium extorquens* AM1 [30,31] and the obligate *Methylophilus methylotrophus* [32], oxidize 1,2-propanediol by the PQQ-containing methanol dehydrogenase in the presence of a modifying protein, which can increase the affinity of methanol dehydrogenase for 1,2-propanediol. In these bacteria, an aldehyde dehydrogenase further oxidizes the resulting lactaldehyde to lactic acid. In this way, the methylotrophs are attractive as a means for lactic acid production from the cheap and reliable product of petrochemistry, 1,2-propanediol.

The main intention of this work has to investigate the possibility and conditions for lactic acid accumulation by nongrowing cell suspensions of the obligate methylotrophic bacterium *Methylobacillus flagellatum* KT immobilized in various gels.

A. Methods

1. Cultures

The cultures of *M. flagellatum* KT were grown in a mineral medium with methanol or methylamine as sole carbon and energy source. *M. flagellatum* KT was isolated from activated sludge of a waste water treatment station in Moscow [33].

The cultivation medium contained (g/l): KH_2PO_4, 10.0; NaCl, 1.0; $MgSO_4 \cdot 7H_2O$, 0.2; $(NH_4)_2SO_4$, 4.0; pH was 7.4 before autoclaving. Trace elements solution was added to the medium in final concentrations (mg/l): $FeSO_4 \cdot 7H_2O$, 4.0, $CaCl_2 \cdot 2H_2O$, 0.2; $CoCl_2 \cdot 6H_2O$, 0.04; $CuSO_4 \cdot 5H_2O$, 0.04; $H_2\ BO_3$, 0.03. The filter-sterilized methanol was added to the medium after autoclaving at 1% (v/v) concentration, and sterile methylamine was added at 0.5% (w/v) concentration. When methylamine served as carbon, energy, and nitrogen source, the ammonium sulfate was excluded from the medium.

The culture was grown in an orbital shaker (New Brunswick Sci. Co.) at 37°C in 250-ml flasks with 100 ml medium. The cells were harvested in exponential growth phase by centrifugation, washed three times in 50 mM K-PO_4 (pH 7.0) buffer, and suspended in appropriate buffers. The required amounts of 1,2-propanediol were added to the cell suspensions, and they were incubated at 37°C in the orbital shaker. Samples were withdrawn at appropriate times, cells were centrifuged, and the supernatant fluid was used for lactic acid determination.

2. *Lactic Acid Determination*

The amount of lactic acid formed was measured enzymatically with lactate dehydrogenase following the reduction of NAD^+ at 340 nm in a Hitachi 20-200 spectrophotometer. The reaction mixture contained, in 3 ml: 100 ml glycine-NaOH buffer (pH 9.0), 2.7 ml; 1 mM solution of NAD^+, 0.1 ml; lactate dehydrogenase (1 unit), 0.1 ml; supernatant fluid, 0.1 ml.

3. *Respiration Measurements*

The cell's respiration was measured with a platinum–sliver-chloride electrode (Clark-type) at 37°C. The substrates were added at 10^{-3} M concentrations (unless otherwise indicated).

4. *Protein Determination*

The protein content of the cells was determined by the method of Lowry [34], using demoisturized bovine serum albumin (Sigma Chemical Co.) as a standard.

5. *Kinetic Measurements*

Michaelis-Menten kinetics of substrate oxidation were followed by respiratory measurements during the first several minutes after substrate addition when respiration rates were linear. The K_m and V_{max} of the processes of 1,2-propanediol, lactate, and methanol oxidations were determined from the data expressed in the dual reciprocal coordinates.

B. Kinetics of 1,2-Propanediol and Lactic Oxidation

The suspensions of nongrowing cells of *M. flagellatum* KT oxidize 1,2-propanediol at various rates, depending on the pH of the buffer and substrate in the medium in which they were grown (Fig. 1). Methylamine-grown cells oxidize 1,2-propanediol eight to nine times more slowly than methanol-grown cells due to the activation of methanol dehydrogenase during the cell's growth in methanol-containing medium.

It was also found that the suspensions of the resting cells of methanol- and methylamine-grown *M. flagellatum* KT oxidize lactate at various rates depending on the pH of the buffer or substrate in the medium in which they were grown (Fig. 2). Methylamine-grown cells oxidize lactate with a maximum rate about two times faster than methanol-grown cells. Both batches of cells have a pH optimum around 8.0 for lactate oxidation.

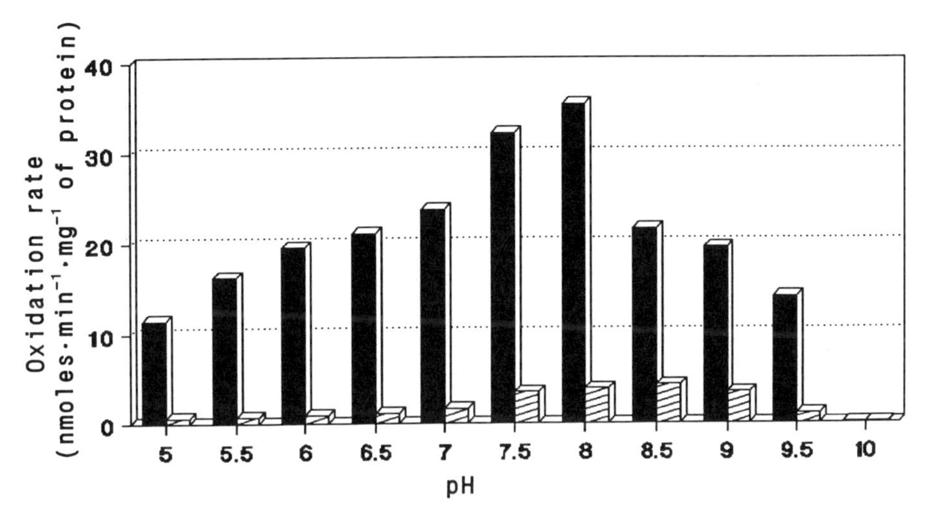

Figure 1 Oxidation of 1,2-propanediol by methanol and methylamine-grown cells of *M. flagellatum* KT depending on pH in the incubation medium. *Conditions*: pH 5–7—0.2M K-PO_4 buffer; pH 8–9—0.2 M Tris-HCl buffer; pH 9.5–10—0.2 M glycine-NaOH buffer. (■), Methanol-grown cells. (▨), Methylamine-grown cells.

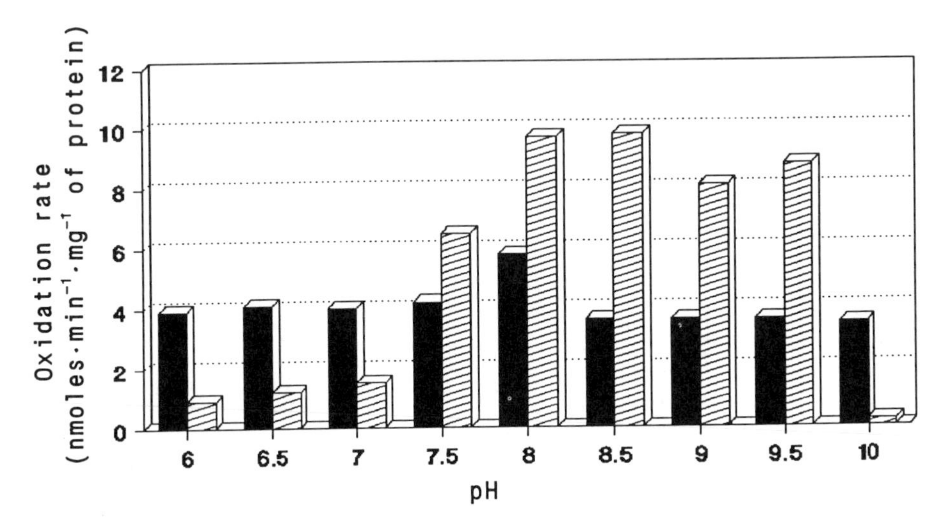

Figure 2 L-lactate oxidation by whole cell suspensions of methanol- and methylamine-grown *M. flagellatum* KT depending on pH in the incubation medium. *Conditions*: pH 5–7—0.2 M K-PO_4 buffer; pH 8–9—0.2 M Tris-HCl buffer; pH 9.5–10—0.2 M glycine-NaOH buffer. (■), Methanol-grown cells. (▨), Methylamine-grown cells.

C. Kinetics of L-Lactate Accumulation

Lactate accumulation rates were measured for the resting cell suspensions of *M. flagellatum* KT grown in methanol- or methylamine-containing medium (Figs. 3 and 4). As shown, lactate accumulates in both cases, but with methanol-grown cells, the lactate output was about 2.5 times higher after 6 hours and about 2 times higher after 24 hours of incubation due to lactate oxidation by the cells (Fig. 2).

In the short-term experiments, the highest rates of lactate accumulation by *M. flagellatum* KT were 2.7 $mM \cdot h^{-1} \cdot g^{-1}$ of protein for methanol-grown cells, and 1.2 $mM \cdot h^{-1} \cdot g^{-1}$ of protein for the cells grown in methylamine-containing medium (Fig. 5).

The reason for the decreasing rate of lactate accumulation is probably that the cells start actively oxidizing lactate after its concentration exceeds the K_M level. Unfortunately, our strain of *M. flagellatum* KT has an active lactate-oxidizing system ($K_M = 2 \times 10^{-3}$ M, $V_{max} = 100$ $nmol \cdot min^{-1} \cdot mg^{-1}$ of protein) (Fig. 6). Other obligate methylotrophs cannot oxidize lactate and can be used for the lactate production process with better efficiency [32].

Our strain of the obligate methylotrophic bacterium *M. flagellatum* KT has a K_M for 1,2-propanediol oxidation of 22×10^{-3} M and V_{max} of 187 $nmol \cdot min^{-1} \cdot mg^{-1}$ of protein (Fig. 7). The K_M for methanol oxidation by whole cells of

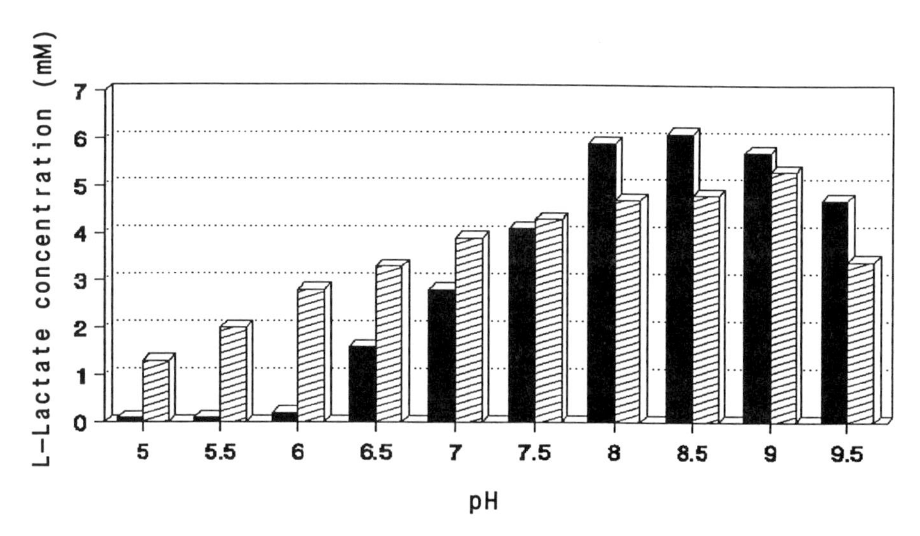

Figure 3 L-lactate accumulation by the resting free cell suspensions of methanol-grown *M. flagellatum* KT depending on the medium pH after 6 and 24 hours of incubation with 100 mM of 1,2-propanediol. (▨), 6 hours of incubation. (■), 24 hours of incubation.

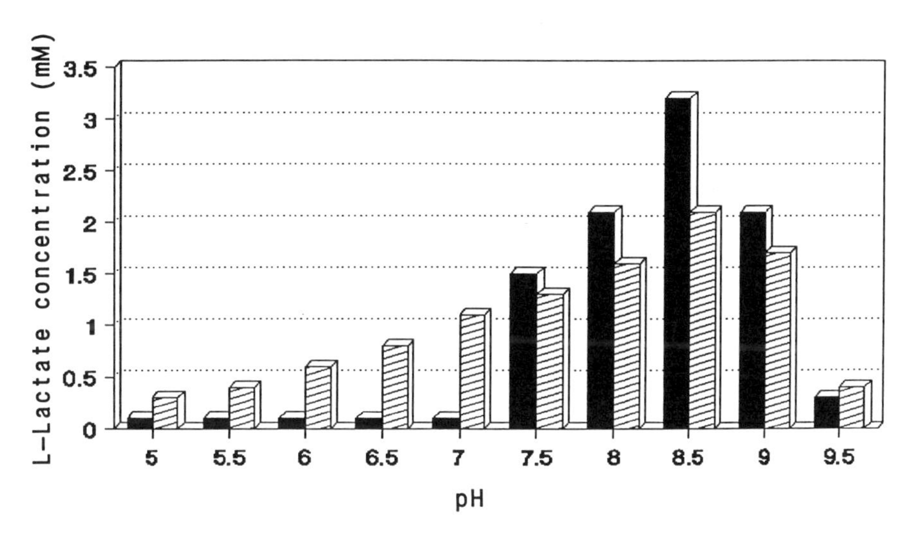

Figure 4 L-lactate accumulation by the resting free cell suspensions of methylamine-grown *M. flagellatum* KT depending on the pH in the medium after 6 and 24 hours of incubation with 100 mM of 1,2-propanediol. (▨), 6 hours of incubation. (■), 24 hours of incubation.

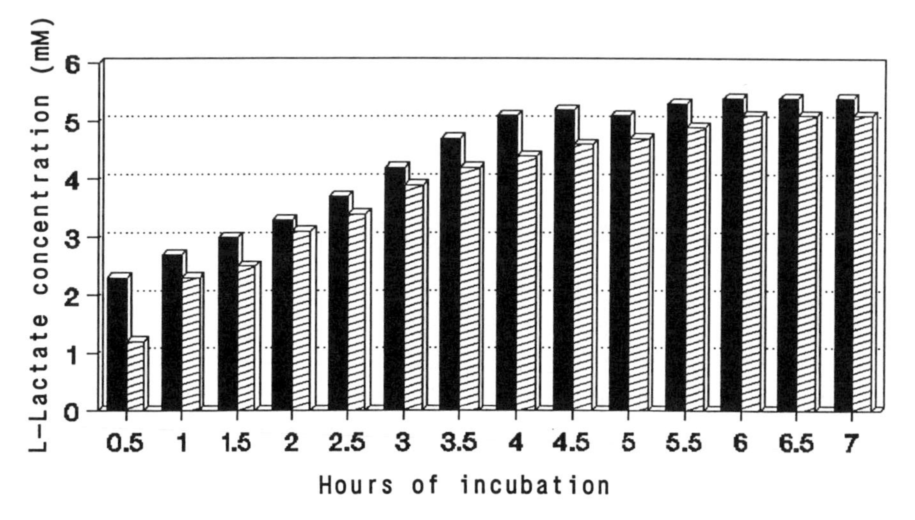

Figure 5 Short-course experiments of 1,2-propanediol oxidation and L-lactate accumulation by whole free cell suspensions of methanol- and methylamine-grown cells of *M. flagellatum* KT. (■), Methanol-grown cells; $V_{lact} = 2.7\ mM \cdot h^{-1} \cdot g^{-1}$ of protein. (▨), Methylamine-grown cells; $V_{lact} = 1.2\ mM \cdot h^{-1} \cdot g^{-1}$ of protein.

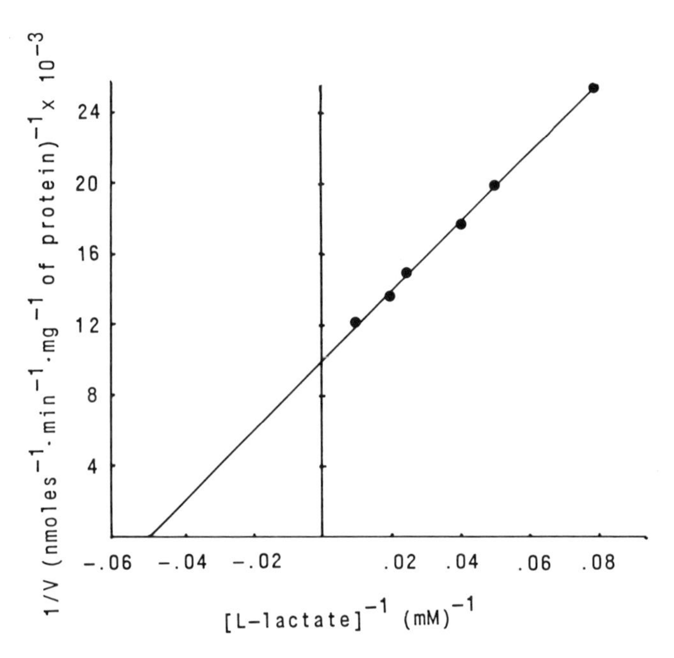

Figure 6 Kinetics of L-lactate oxidation by methanol-grown *M. flagellatum* KT. *Conditions*: Cells were suspended in 0.2 M Tris-HCl buffer, pH 8.5; $K_M = 2 \times 10^{-3}$ M, V_{max} = 100 nmol·min^{-1}·mg^{-1} of protein.

methanol-grown *M. flagellatum* KT was 8.3×10^{-4}, and V_{max} was 111 nmol·min^{-1}·mg^{-1} of protein (Fig. 8).

It was also observed that high concentrations of lactate (2 mM) added at the beginning of experiments with 1,2-propanediol oxidation slowed down the process by approximately twofold. This could be another reason for the rate of lactate accumulation decreasing with time. The temperature optimum for cell growth and for L-lactate accumulation by both free and immobilized cell suspensions was 45°C. Compared to lactate production by the batch-grown free cell suspensions of *Arthrobacter oxydans* that oxidize 1,2-propanediol in a jar fermenter, our strain of obligate methylotroph had activity about 2.5 times higher for the free cell suspensions and about 3.3 times higher for the cells immobilized in κ-carrageenan [35].

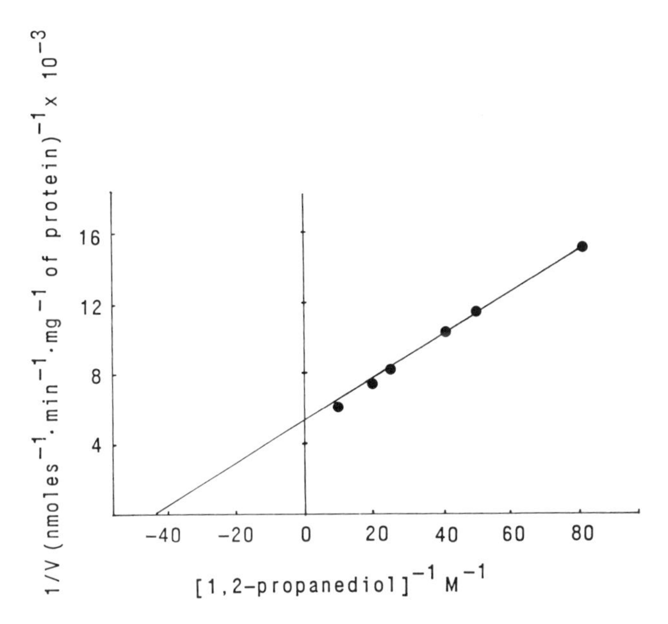

Figure 7 Kinetics of 1,2-propanediol oxidation by methanol-grown *M. flagellatum* KT. *Conditions*: Cells were suspended in 0.2 M Tris-HCl buffer, pH 8.5; $K_M = 22 \times 10^{-3}$ M, V_{max} = 181 nmol·min^{-1}·mg^{-1} of protein.

III. LACTATE ACCUMULATION DUE TO 1,2-PROPANEDIOL OXIDATION BY IMMOBILIZED CELLS OF THE OBLIGATE METHYLOTROPH *METHYLOBACTERIUM FLAGELLATUM* KT

A. Immobilization of Cells of the Obligate Methylotroph in Polyacrylamide Gel, Agarose, and κ-Carrageenan

1. *Immobilization of Cells by Polyacrylamide Gel Entrapment*

About 10 g of fresh methanol-grown wet cells were suspended in 20 ml of 0.2 M Tris-HCl buffer (pH 8.0) and chilled in ice while bubbling with oxygen-free nitrogen gas. Oxygen-free nitrogen gas was prepared by passing commercially available nitrogen gas (99.95% purity) over copper wire heated to 300°C to remove all traces of oxygen. The resulting nitrogen gas was passed through sterile distilled water in a gas-washing flask to be saturated with water vapors. Oxygen removal from the polymerization mixture of polyacrylamide gels is essential to protect the viability of the microbial cells during the process of polymerization.

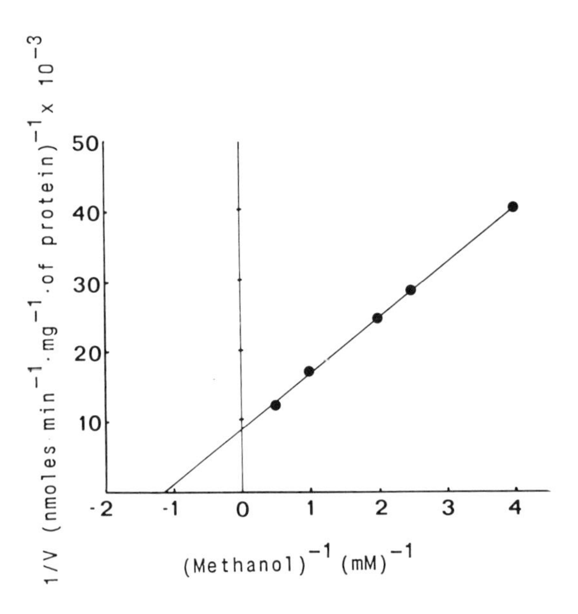

Figure 8 Kinetics of methanol oxidation by methanol-grown *M. flagellatum* KT. *Conditions*: Cells were suspended in 0.2 M Tris HCl buffer, pH 8.5; $K_M = 8.3 \times 10^{-4}$ M, V_{max} = 111 nmol·min^{-1}·mg^{-1} of protein.

5.7 g of acrylamide, 0.3 g of N,N′-methylene-bis-acrylamide, 20 mg of ammonium persulfate, and 0.2 ml of TEMED (N,N,N′,N′-tertamethylethylenediamine) (Sigma Chemical Co.) were dissolved and mixed in a total of 20 ml of ice-cooled 0.2 M Tris-HCl buffer (pH 8.0) which had been gassed with oxygen-free nitrogen. After immediately mixing with the cell suspension, the buffered polymerization mixture was poured into five plastic petri dishes, covered, and chilled on ice. After about an hour of polymerization, the resulting gels were passed through a coarse nylon sieve, which allowed generation of sieved gel particles about 0.5 mm in diameter. The sieved gel particles were then suspended in 200 ml of 0.2 M Tris-HCl (pH 8.0) buffer and washed three times by centrifugation (3000 g, 10 min, 2°C) to remove any impurities and free cells. The resulting gel particles were stored in the refrigerator on ice at 0°C under 0.2 M Tris-HCl (pH 8.0) buffer until required. To maintain the stability of lactate accumulation, the immobilized cells should be washed with growth medium, supplemented with methanol, at least every 2 days during experiments.

2. Immobilization of Cells by Agarose-Gel Entrapment

Agarose (Sigma Chemical Co.) has been used for the entrapment of cells in the form of blocks, membranes, and beads of 0.5 mm in diameter; the latter was found most convenient. A solution of the immobilizing medium (2% of agarose) was heated at a temperature and time sufficient to liquefy the agarose. The cells were then added after the solution was cooled to about 50°C, mixed, and allowed to form the gel. The gel was passed through the coarse nylon sieve to obtain particles of agarose beads of about 0.5 mm in diameter. The particles were washed two times by centrifugation (3000 *g*, 10 min, 2°C), suspended in the growth medium (1:5, v/v), and stored in a refrigerator on ice until required. Agarose beads were washed once with appropriate buffer before each experiment was started.

3. Immobilization of Cells by Entrapment in κ-Carrageenan

Carrageenans are heterogenous polysaccharides containing predominantly α-D-galactopyranosyl sulfate esters. κ-Carrageenan is the insoluble fraction obtained on addition of potassium ions to an aqueous extract of carrageenan.

For the gel preparation, 4% (w/v) of κ-carrageenan (Sigma Chemical Co.) in 0.9% (w/v) NaCl was melted by heating at 60°C to dissolve the polysaccharide, and then the solution was degassed and maintained at 45°C in a water bath. Removal of air bubbles is essential, since otherwise they will be trapped in the beads, causing them to float, which creates problems.

A cell slurry was prepared [1 g of wet cells/10 ml of 2% (w/v) KCl] and kept at 45°C. The warm carrageenan solution was mixed with the cell slurry at a ratio of 9:1, and while still at 45°C the mixture was added to the KCl solution (2% w/v) dropwise using a syringe. The mixture was left to cool and harden in the KCl solution for 20 minutes at 10°C. The formed beads were passed through a coarse nylon sieve to generate particles about 0.5 mm in diameter, washed by centrifugation (3000 *g*, 10 min, 2°C), and stored in the refrigerator on ice until required in the basic growth medium, which included 75 mM KCl. Before each experiment, the immobilized κ-carrageenan cells were washed once with appropriate buffer.

B. Kinetics of 1,2-Propanediol Oxidation by *M. flagellatum* KT Cells Immobilized in Various Gels

The kinetics of 1,2-propanediol oxidation by methanol-grown cells of *M. flagellatum* KT entrapped in polyacrylamide, agarose, and κ-carrageenan gels were followed using an oxygen electrode technique. It was shown that the K_M and V_{max} for 1,2-propanediol oxidation was the same for all cell suspensions immobilized in the different gels, and these values were close to the K_M and V_{max} of 1,2-propanediol oxidation of methanol-grown cells of *M. flagellatum* KT.

C. Kinetics of L-Lactate Accumulation by Immobilized Cells of the Obligate Methylotroph *M. flagellatum* KT

The L-lactate formation activity of *M. flagellatum* KT cells grown in methanol-containing medium and immobilized into three different gels (agarose, κ-carra-

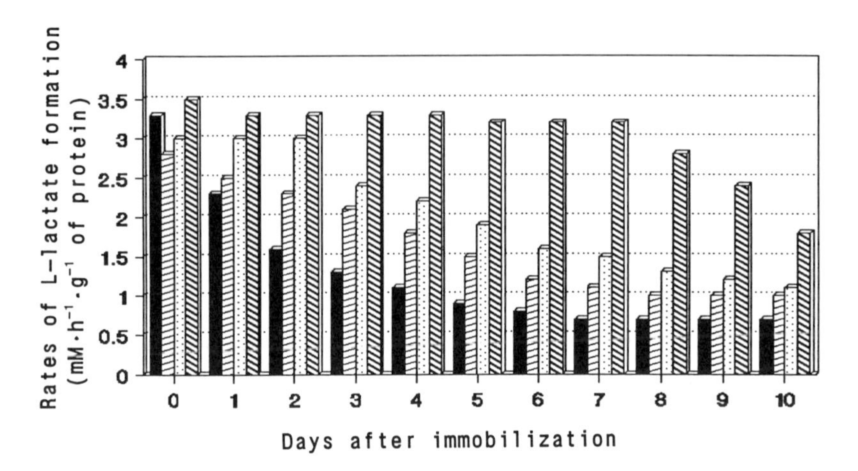

Figure 9 Time-dependent stability of L-lactate formation activity by free and immobilized cells of methanol-grown *M. flagellatum* KT. *Conditions*: Cells were immobilized in various gels and were stored in the refrigerator at 4°C between experiments. The cells and the free cell suspensions and immobilized beads were used for 1,2-propanediol oxidation and L-lactate accumulation experiments in 0.2 M Tris-HCl, pH 8.5 buffer. Before each experiment free and immobilized cells were washed twice in 0.2 M Tris-HCl buffer, pH 8.5. (■), Free cell suspensions. (▨), Polyacrylamide-immobilized cell suspensions. (▦, Agarose-immobilized cell suspensions. (▧), κ-Carrageenan–immobilized cell suspensions.

geenan, and polyacrylamide) was checked every day after the gels were prepared. The gel beads were washed twice by centrifugation and then suspended in a small amount of the 200-mM Tris-HCl buffer at pH 8.5 with 100 mM of 1,2-propanediol. The kinetics of L-lactate accumulation were followed, and the rate was estimated for each gel each day. The results are plotted in Figure 9.

Cell suspensions of *M. flagellatum* KT grown on methanol and immobilized into different gels show an increase of the half-life of the L-lactate accumulation reaction. It is clearly shown that the immobilization of the cells of *M. flagellatum* KT into κ-carrageenan gel gives the best results with the half-life increasing up to six times that of the free cell suspensions.

IV. PQQ-DEPENDENT OXIDATIONS OF GLUCOSE TO GLUCONIC ACID BY VARIOUS BACTERIA

Several processes have been described by which glucose is oxidized to gluconic acid by cells of *Acinetobacter*, *Klebsiella*, *Salmonella*, *Rhizobium*, *Agrobacterium*, *Gluconobacter*, *Escherichia*, and *Pseudomonas* genera. The glucose can be mobilized by cells via two general processes: a high affinity glucose phosphorylating uptake system (glucose phosphotransferase mechanism) and a low affinity glucose dehydrogenase, which was found in several gram-negative bacteria [36,37]. In the first system glucose is taken into the bacteria cell as glucose-6-phosphate for further oxidation. In the second system glucose can be oxidized to gluconic acid by PQQ-dependent, periplasmic enzymes, which then allow gluconate to be transported into the cell or to be excreted into the medium. Using the second system, it is possible to produce gluconic acid using a high concentration of glucose and appropriate bacterial cells.

It was shown that the rate of gluconate formation by washed cell suspensions of potassium-limited *Klebsiella pneumoniae* might be as much as 580 $nmol \cdot min^{-1} \cdot mg^{-1}$ of protein at low pH. At high pH, the production of gluconate is completely replaced by acetic acid [38]. Gluconate can also be produced by washed suspensions of *E. coli* grown in phosphate-limited conditions after PQQ additions (0.1 μM) with a rate of up to 360 $nmol \cdot min^{-1} \cdot mg^{-1}$ of dry cells at pH 5.5 [37].

Cells of *Rhizobium leguminosarum* [39] were shown to produce gluconic acid from glucose (up to 14.4 mM from 10 g of glucose) but only in the presence of added PQQ (4 μM). Two species of *Agrobacterium*, *A. radiobacter* and *A. tumefaciens*, show the same properties: they accumulate 23.5 and 15.3 mM of gluconic acid from glucose, respectively, under the same conditions after PQQ addition. Strain *Pseudomonas* sp. shows the ability to accumulate up to 21 mM of gluconic acid under these conditions [39].

Two strains of *Acinetobacter* genus, *A. calcoaceticus* and *A. lwoffi*, also express the ability to oxidize glucose and produce gluconic acid, but only in the presence of the added PQQ (200 nM). The maximum rate of glucose oxidation/gluconate production for the former strain was 239 $nmol \cdot min^{-1} \cdot mg^{-1}$ of dry cells and, for the latter, 234 $nmol \cdot min^{-1} \cdot mg^{-1}$ of dry cells. The maximum accumulated amount has not been reported [39].

An original method was described recently for the production of gluconic and galactonic acid from whey, the prehydrolyzed byproduct of the milk industry [40,41]. It was shown that the acetic acid bacterium *Gluconobacter oxydans* can produce gluconic and galactonic acids during the oxidation of lactose in whey that had been hydrolyzed by immobilized cells of *Kluyveromyces bulgaricus*.

Cells of *G. oxydans* were also used to oxidize glucose and galactose to gluconic and galactonic acids in a batch fermentation mode. The process allowed recovery of up to 4% of acids from as much as 4% of total sugars (glucose and galactose) but, unfortunately, actual concentrations of gluconic and galactonic acid accumulated were not calculated [41].

The accumulation of gluconic acid by cells of *Salmonella typhimurium* was reported, and it seems to be the same enzyme system that is found in *E. coli* [37]. Whether galactose dehydrogenase activity is catalyzed by the PQQ-dependent glucose dehydrogenase from enteric bacteria or whether there are separate enzymes is still unknown because some purified glucose dehydrogenases also oxidize galactose [42].

V. BIOTECHNOLOGICAL ASPECTS OF COMMERCIAL EXPLOITATION OF PQQ-DEPENDENT ENZYMES AND ENZYME SYSTEMS

For the further possible industrial application of the L-lactate production process from 1,2-propanediol using methanol-grown cells of the obligate methylotroph *M. flagellatum* KT, immobilized in κ-carrageenan gel, it was necessary to create a design for a laboratory-scale continuous flow fermenter which would allow continuous production of L-lactate. The initial requirements for this type of laboratory-scale fermenter should meet the following restrictions.

1. The fermenter vessel should hold the immobilized beads in the fermenter and let the medium flow with any required rate.
2. The fermenter should oxygenate the immobilized cell beads without causing them to float.
3. The fermenter design should allow separation of the beads from the medium after the reaction with 1,2-propanediol, so as to allow removal of the reaction product (L-lactate).
4. The medium should be able to be pumped back to the reactor vessel to continue the closed cycle after separation of the reaction product.
5. The reactor mixture should be kept highly aerobic with the minimum possible force applied to the fermenter mixer to protect the immobilized cells inside the gel beads from mechanical disruption by the stirrer blades.

Keeping in mind all of these restrictions in the fermenter unit design, we were able to construct a membrane-based reaction vessel that allowed us to perform the required reaction and to avoid the inconveniences related to mixing the contents with conventional stirrers (Fig. 10).

Immobilized cells of methanol-grown *M. flagellatum* KT in κ-carrageenan beads were put in the insulated temperature-controlled reaction vessel at 45°C at a concentration of 10% (v/v).The oxidation of 1,2-propanediol was started and L-lactate accumulated in the buffered medium (all conditions were the same as in the experiments with cells immobilized in various gels, Fig. 9). The liquid flow was maintained at a desired level by the operation of speed-regulated peristaltic pumps. Oxygen concentration was maintained at a level not less than 10% during the experiment using an oxygen electrode (Clark-type) and membrane oxygen-exchange regeneration column (Fig. 10, step 8). This column allows oxygen exchange between the system for 1,2-propanediol oxidation/L-lactate accumulation and an external oxygen-exchange device. The oxygen exchange device allows one to maintain the dissolved oxygen concentration up to saturation at the flow

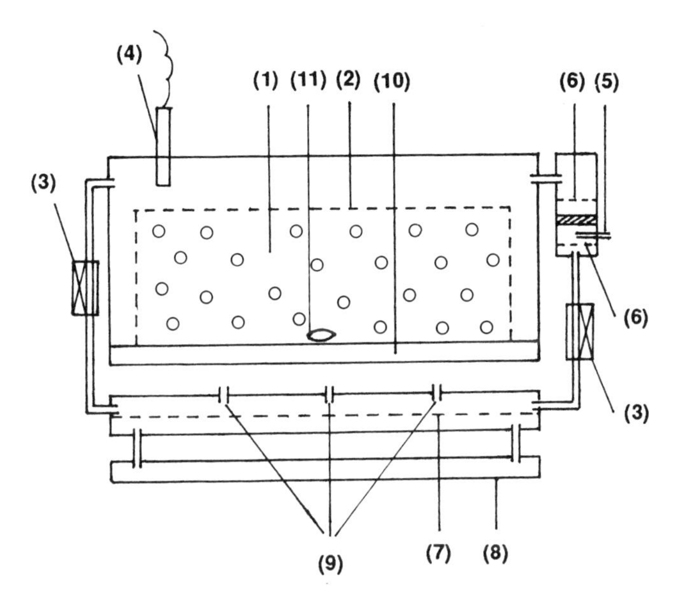

Figure 10 Schematic design of the reaction vessel for the continuous L-lactate production from 1,2-propanediol by methanol-grown cells of *M. flagellatum* KT immobilized into κ-carrageenan gel. (1) Insulated temperature controlled reaction vessel with immobilized cells. (2) Semipermeable membrane. (3) Pumps. (4) Out-reactor space with oxygen electrode probe. (5) Ca-lactate removal device. (6) Nylon sieve. (7) Semipermeable oxygen-exchange membrane. (8) Oxygen regeneration column with injection ports. (9) Injection ports. (10) Magnetic stirrer plate. (11) Stirrer bar.

speed used, and it can be regulated by the rate of oxygen-exchange liquid flow. The semi-permeable oxygen-exchange membrane was not a "bottle-neck point" in the oxygen transfer to the reaction vessel, and the volume of transferred oxygen can be easily increased by the enlargement of the surface of the membrane.

The calcium lactate removal mechanism (Fig. 10, step 5) includes the semipermeable membrane that separates the flow of the reaction mixture from the vessel where L-lactate accumulated as a Ca-salt passing through a precipitated calcium carbonate cartridge. Using the temperature difference, calcium lactate in the form of pentahydrate crystals was removed from the mixture and, after filtration, the clear liquid was pumped into the regeneration column for the saturation with oxygen and 1,2-propanediol additions. The concentration of 1,2-propanediol was maintained at about 100 mM, and it was measured with a probe, with methanol-grown *M. flagellatum* KT cells used to construct calibration curve. The reaction vessel could be run for several weeks until the activity of the immobilized cells reached a rate that was no longer economical. It was also found that washing the immobilized cells with whole growth sterile medium with 0.1% methanol could restore the activity by 12–15%. This type of fermenter vessel was run for 4 weeks continuously with rates of L-lactate production from 3.5 to 0.5 $mmol \cdot h^{-1} \cdot g^{-1}$ of protein. It is possible to operate the unit longer by replacing the beads with immobilized cells, however, this led to problems of sterility, which we are now trying to solve.

In conclusion, we would like to stress that using an obligate methylotroph as the source of L-lactate production shows that it can be used in a commercial process for obtaining pure Ca-lactate.

VI. CONCLUSION

This short review of biocatalytic processes that use PQQ-based enzymatic reactions shows that many processes are potentially available for commercial biotechnological applications in the future. At present it seems that two processes that exploit PQQ-based enzymatic system are more advanced in terms of commercial availability for use as a source of organic acids: glucose oxidation to gluconic acid by enterobacteria [37] and oxidation of 1,2-propanediol to L-lactate by immobilized cells of obligate methylotroph *M. flagellatum* KT [43]. The latter process is ready for biotechnological exploitation, and the necessary steps are underway toward the patenting of the reaction vessel together with the process of obtaining L-lactate in the form of a calcium lactate precipitate from the immobilized cells of obligate methylotrophic bacteirum *M. flagellatum* KT.

REFERENCES

1. Duine, J. A., and Frank, J. (1987). Quinoproteins: A novel class of dehydrogenases. *Trends Biochem. Sci. 6*: 278.
2. Salisbury, S. A., Forrest, H. S., Cruse, W. B. T., and Kennard, O. (1979). A novel coenzyme from bacterial primary alcohol dehydrogenases. *Nature 280*: 843.
3. Duine, J. A., Frank, J., and Verweil, P. E. J. (1980). Structure and activity of the prosthetic group of methanol dehydrogenase. *Eur. J. Biochem. 108*: 187.
4. Anthony, C. (1982). *The Biochemistry of Methylotrophs*, Academic Press, London.
5. Netrusov, A., Guettler, M., and Hanson, R. S. (1988). Methanol oxidizing system of thermophilic *Bacillus*. In *Proceedings Intl. Symp. Microorganisms in Extreme Environments*, Troja, Portugal, #13.
6. Dijkhuizen, L., Arfman, N., Attwood, M. M., Brooke, A. G., Harder, W., and Watling, E. M. (1988). Isolation and initial characterization of thermotolerant methylotrophic *Bacillus* strains. *FEMS Microbiol. Lett. 52*: 209.
7. Duine, J. A. (1989). PQQ and quinoproteins: An important novel field in enzymology. In *PQQ and Quinoproteins* (J. A. Jongejan and J. A. Duine, eds.), Kluwer Academic Publishers, Dordrecht, The Netherlands, p. 357.
8. McIntire, W. S., Wemmer, D. E., Chistoserdov, A., and Lidstrom, M. E. (1991). A

new cofactor in a procaryotic enzyme: Tryptophan tryptophyl-quinone as the redox prosthetic group in methylamine dehydrogenase. *Science 252*: 817.

9. Winters, D., and Ljungdahl, L. G. (1989). PQQ-dependent methanol dehydrogenase from *Clostridium thermoautotrophicum*. In *PQQ and Quinoproteins* (J. A. Jongejan and J. A. Duine, eds.), Kluwer Academic Publishers, Dordrecht, The Netherlands, p. 35.
10. Duine, J. A., Frank, J., and Berkhout, M. P. J. (1984). NAD-dependent, PQQ-containing methanol dehydrogenase: A bacterial dehydrogenase in a multienzyme complex. *FEBS Lett. 168*: 217.
11. Groen, B., Frank, J., and Duine, J. A. (1984). Quinoprotein alcohol dehydrogenase from ethanol-grown *Pseudomonas aeruginosa*. *Biochem. J. 223*: 921.
12. Goerisch, H., and Rupp, M. (1989). Quinoprotein ethanol dehydrogenase from *Pseudomonas*. In *PQQ and Quinoproteins* (J. A. Jongejan and J. A. Duine, eds.), Kluwer Academic Publishers, Dordrecht, The Netherlands, p. 23.
13. Yamanaka, K. (1989). New PQQ-enzyme: Aromatic alcohol and aldehyde dehydrogenases in *Rhodopseudomonas acidophila*. In *PQQ and Quinoproteins* (J. A. Jongejan and J. A. Duine, eds.), Kluwer Academic Publishers, Dordrecht, The Netherlands, p. 40.
14. Adachi, O., Shinigawa, E., Matsushita, K., and Ameyama, M. (1982). Crystallization of membrane-bound alcohol dehydrogenase from acetic acid bacteria. *Agric. Biol. Chem. 46*: 2859.
15. Groen, B. W., Frank, J., and Duine, J. A. (1984). Quinoprotein alcohol dehydrogenase from ethanol-grown *Pseudomonas testosteroni*. *Biochem. J. 223*: 921.
16. Ameyama, M., and Adachi, O. (1982). Aldehyde dehydrogenase from acetic acid bacteria, membrane-bound. *Methods Enzymol. 89*: 491.
17. Kesseler, F. P., Badnus, I., and Schwartz, A. C. (1989). Properties of dye-linked formaldehyde dehydrogenase from *Hyphomicrobium* sp. ZV580 grown on methylamine. In *PQQ and Quinoproteins* (J. A. Jongejan and J. A. Duine, eds.), Kluwer Academic Publishers, Dordrecht, The Netherlands, p. 54.
18. Cleton-Jansen, A.-M., Goosen, N., Vink, K., and van de Putte, P. (1989). Cloning of the genes encoding the two different glucose dehydrogenases from *Acinetobacter calcoaceticus*. In *PQQ and Quinoproteins* (J. A. Jongejan and J. A. Duine, eds.), Kluwer Academic Publishers, Dordrecht, The Netherlands, p. 79.
19. Matsushita, K., Shinagawa, E., Adachi, O., and Ameyama, M. (1989). Quinoprotein D-glucose dehydrogenases in *Acinetobacter calcoaceticus* LMD 79.41: Purification and characterization of the membrane-bound enzyme distinct from the soluble enzymes. In *PQQ and Quinoproteins* (J. A. Jongejan and J. A. Duine, eds.), Kluwer Academic Publishers, Dordrecht, The Netherlands, p. 69.
20. Dokter, P., Frank, J., and Duine, J. A. (1986). Purification and characterization of quinoprotein glucose dehydrogenase from *Acinetobacter calcoaceticus* LMD 79.41. *Biochem. J. 239*: 163.
21. Geiger, O., and Goerisch, H. (1986). Crystalline quinoprotein glucose dehydrogenase from *Acetobacter calcoaceticus*. *Biochemistry 25*: 6043.
22. Strohdeicher, M., Bringer-Meyer, S., Neusz, B., van der Meer, R. A., Duine, J. A., and Sahm, H. (1989). Glucose dehydrogenase from *Zymomonas mobilis*: Evidence for

a quinoprotein. In *PQQ and Quinoproteins* (J. A. Jongejan and J. A. Duine, eds.), Kluwer Academic Publishers, Dordrecht, The Netherlands, p. 103.
23. Van Kleef, M. A. G., and Duine, J. A. (1988). Bacterial NAD(P)-independent quinate dehydrogenase is a quinoprotein. *Arch. Microbiol. 150*: 32.
24. Ameyama, M., Shinigawa, E., Matsushita, K., and Adachi, O. (1985). Solubilization, purification and properties of membrane-bound glycerol dehydrogenase from *Gluconobacter industries*. *Agric. Biol. Chem. 49*: 1001.
25. Kawal, F., Yamanaka, H., Ameyama, M., Shinigawa, E., Matsushita, K., and Adachi, O. (1985). Identification of the prosthetic group and further characterization of a novel enzyme, polyethylene glycol dehydrogenase. *Agric. Biol. Chem. 49*: 1071.
26. Duine, J. A., and Frank, J. (1981). Quinoproteins: A novel class of dehydrogenases. *Trends Biochem. Sci. 6*: 278.
27. Van Iersel, J., van der Meer, R. A., and Duine, J. A. (1986). Methylamine oxidase from *Arthrobacter* P1. A bacterial copper-containing amine oxidase. *Eur. J. Biochem. 161*: 415.
28. Tur, S. S., Royce, P. M., and Lerch, K. (1989). Lysil oxidase from the yeast *Pichia pastoris*. In *PQQ and Quinoproteins* (J. A. Jongejan and J. A. Duine, eds.), Kluwer Academic Publishers, Dordrecht, The Netherlands, p. 327.
29. Tanizawa, K., Moriya, T., Kido, T., Tanaka, H., and Soda, K. (1988). Structural studies on the PQQ-like cofactor of nitroalkane oxidase from *Fusarium oxysporum*. In *PQQ and Quinoproteins* (J. A. Jongejan and J. A. Duine, eds.), Kluwer Academic Publishers, Dordrecht, The Netherlands, p. 43.
30. Bolbot, J. A., and Anthony, C. (1980). The metabolism of 1,2-propanediol by the facultative methylotroph *Pseudomonas* AM1. *J. Gen. Microbiol. 120*: 254.
31. Ford, S., Page, M. D., and Anthony, C. (1985). The role of methanol dehydrogenase in the growth of *Pseudomonas* AM1 on 1,2-propanediol. *J. Gen. Microbiol. 131*: 2173.
32. Page, M. D., and Anthony, C. (1986). Regulation of formaldehyde oxidation by the methanol dehydrogenase and modifier protein of *Methylophilus methylotrophus* and *Pseudomonas* AM1. *J. Gen. Microbiol. 132*: 1553.
33. Govoruchina, N. I., Kletsova, L. V., Tsygankov, Y. D., Trotsenko, Y. A., and Netrusov, A. I. (1987). Characteristics of a new obligate methylotroph. *Microbiology* (USSR) *56*: 859.
34. Lowry, O. H., Rosebrough, N. J., Farr, A. L., and Randal, R. J. (1951). Protein determination with Folin phenol reagent. *J. Biol. Chem. 193*: 265.
35. Yagi, O., and Minoda, Y. (1979). Lactic acid production from 1,2-propanediol by jar culture and resting cells of *Arthrobacter oxydans*. *Agric. Biol. Chem. 43*: 571.
36. Kundig, W., Ghosh, S., and Roseman, S. (1964). Phosphate bound to histidine in a protein as an intermediate in a novel phosphotransferase system. *Proc. Natl. Acad. Sci. USA 52*: 1067.
37. Hommes, R. W. J. (1988). The role of the PQQ-linked glucose dehydrogenase in the physiology of *Klebsiella aerogenes* and *Escherichia coli*. Ph.D. thesis, University of Amsterdam.
38. Hommes, R. W. J., Van Hell, B., Postma, P. W., Neijssel, O. M., and Tempest, D. W. (1985). The functional significance of glucose dehydrogenase in *Klebsiella aerogenes*. *Arch. Microbiol. 143*: 163.

39. Van Schie, B. J., DeMooy, O. H., Linton, J. D., Van Dijken, J. P., and Kuenen, J. G. (1987). PQQ-dependent production of gluconic acid by *Acinetobacter*, *Agrobacterium* and *Rhizobium* species. *J. Gen. Microbiol. 133*: 867.
40. Van Huynh, N., Decleire, M., Voets, M., Motte, J. C., and Monseur, X. (1986). Production of gluconic acid from whey hydrolysate. *Proc. Biochem. 21*: 31.
41. Van Huynh, N., Decleire, M., Motte, J. C., and Monseur, X. (1989). Production of gluconic and galactonic acids from whey. In *PQQ and Quinoproteins* (J. A. Jongejan and J. A. Duine, eds.), Kluwer Academic Publishers, Dordrecht, The Netherlands, p. 97.
42. Ameyama, M., Nonobe, M., Shinagawa, E., Matsushita, K., Takimoto, K., and Adachi, O. (1986). Purification and characterization of the quinoprotein D-glucose dehydrogenase apoenzyme from *Escherichia coli*. *Agric. Biol. Chem. 50*: 49.
43. Dinarieva, T., and Netrusov, A. (1991). Lactic acid formation by free and immobilized cells of an obligate methylotroph. *Biotechnol. Progr. 7*: 234.

20

Quinoproteins as Biosensors

Isao Karube and Kenji Yokoyama

University of Tokyo, Tokyo, Japan

Yasushi Kitagawa

Asahi Breweries, Ltd., Tokyo, Japan

I. INTRODUCTION

Many biosensors using biocatalysts, such as enzymes and microorganisms, as molecular recognition elements have been developed in recent years [1]. These are composed of both transducers, e.g., electrochemical and optical devices, and biocatalysts, and these devices are highly specific for compounds of biochemical relevance. Therefore, a specific compound in a biological fluid containing many kinds of substances can be conveniently assayed using such sensors. These sensors offer many advantages over more traditional methods of assay.

For example, spectrophotometric or spectrofluorometric determinations of the products of an enzyme reaction have been used for some time as a means of an enzyme-based assay of biological substances. However, these methods employ complicated procedures and require much time, and automation of such measurements is difficult to achieve. On the contrary, biosensors can be easily prepared and can quickly analyze samples.

A. Conventional Biosensors

Biosensors, composed of biofunctional molecules and a transducer, have been developed and applied to the fields of analytical chemistry and clinical analysis, as well as the food industry and environmental studies. Immobilized enzymes, microorganisms and antibodies have been used as molecular recognition materials. An oxygen electrode or hydrogen peroxide electrode has often been used as a transducer. An enzyme electrode is composed of an enzyme-immobilized membrane and an electrode. The principle of the enzyme electrode is based on the detection of an electroactive compound produced or consumed by the enzyme reaction. For example, glucose oxidase (GOD) oxidizes glucose with consumption of oxygen and produces gluconolactone and hydrogen peroxide. By measuring either consumption of oxygen with an oxygen electrode or production of hydrogen peroxide with a hydrogen peroxide electrode, glucose concentration can be determined. This type of glucose sensor has been commercialized and used in the diagnosis of diabetes. There are many kinds of biosensors that use the same principle and that are being developed and used in the field of clinical analysis and analysis of foodstuffs.

Microorganisms have been also utilized as molecular recognition elements. A microbial sensor consists of a microorganism-immobilized membrane and an electrode. Various kinds of microbial sensors have been developed and applied to the measurement of organic compounds. The principle of a microbial sensor is based on either the change in the rate of respiration or the amount of produced metabolites that result from assimilation of specific substrates by the microorganism. Furthermore, by using auxotrophic mutants, many kinds of substrates can be selectively determined. For example, a vitamin B_{12} sensor was constructed by using immobilized *Escherichia coli* 215, which requires vitamin B_{12} for growth. A linear relationship was obtained in the range 5×10^{-9} to 2.5×10^{-8} g·ml^{-1} [2]. After 25 days the decrease in the response of the sensor was about 8%.

Recently, microbial sensors using thermophilic bacteria have been developed. The advantages of using thermophilic bacteria are both to reduce contamination by other microorganisms through use at high temperature and to increase the long term stability. For example, a biological oxygen demand (BOD) sensor [3] and CO_2 sensor [4] were constructed by using thermophilic bacteria, which were isolated from a hot spring. A good linear correlation was observed between the BOD sensor response and BOD values in the range of 1–10 mg·ml^{-1} (JIS, Japanese Industrial Standard Committee) at 50°C. The sensor signal was stable and reproducible for more than 40 days. For the CO_2 sensor, a linear relationship was obtained in $NaHCO_3$ concentrations between 1 and 8 mM at 50°C, and the response time was 5–10 minutes. A linear relationship was also observed in CO_2 concentration from 3 to 8%.

Much attention in the field of biosensors is also currently being focused on miniaturization and integration. A miniaturized microbiosensor is advantageous in that it is implantable into the human body and is suitable for in vivo measurement. In addition, many such microbiosensors can be integrated on one chip, and can be used to measure various substrates in a small amount of sample solution simultaneously. Since semiconductor fabrication technology has been applied to the construction of such microbiosensors, it will be possible to develop disposable transducers for biosensors by mass production.

B. Mediated Biosensors

Glucose determination in blood is extremely important in clinical analysis, and, as such, many attempts to develop glucose sensors have been reported in recent years. Most of these sensors used either an oxygen or hydrogen peroxide electrode as a transducer and GOD as the recognition component. GOD is an enzyme catalyzing glucose oxidation. GOD oxidizes glucose to gluconolactone and is itself reduced. It is then normally reoxidized by oxygen (Fig. 1). If, however, the dissolved oxygen concentration is not sufficiently high to rapidly oxidize the reduced form of GOD, then the response of the sensor will be limited by oxygen and not dependent on glucose concentration. For this reason, oxygen- and hydrogen peroxide–based glucose sensors cannot be applied to the determination of blood glucose concentration, which is approximately 5 mM for a hungry, healthy human and reaches 20 mM for a diabetic. A glucose sensor in which oxygen is an oxidant of GOD will also be affected by fluctuations of dissolved oxygen concentration. Another disadvantage of a GOD-based sensor is that a relatively high potential is needed to detect hydrogen peroxide, and a hydrogen peroxide–based glucose sensor is readily interfered with by other electroactive species. Normally, the electrode potential is controlled at around 700 mV versus an Ag/AgCl electrode for the detection of hydrogen peroxide. However, easily oxidizable compounds in blood, such as ascorbic acid and uric acid, are oxidized at such a potential. That is why much attention has been focused on the development of glucose sensors that use electron transfer mediators to oxidize the reduced form of GOD instead of oxygen, and which are then easily reoxidized at an electrode (Fig. 2).

Turner et al. reported several mediated biosensors using ferrocene derivatives [5–10]. The mediated glucose sensor was composed of GOD, 1,1′-dimethylferrocene (DMFc) and a graphite electrode [5]. Covalent immobilization of GOD on the electrode was achieved by using carbodiimide. The measurement of glucose was carried out potentiometrically at 160 mV vs. a saturated calomel electrode. A linear current response was observed to be proportional to the glucose concentration over a range commonly found in diabetic blood samples. The effect of oxygen was found to be small under aerobic conditions. In this research, a number of

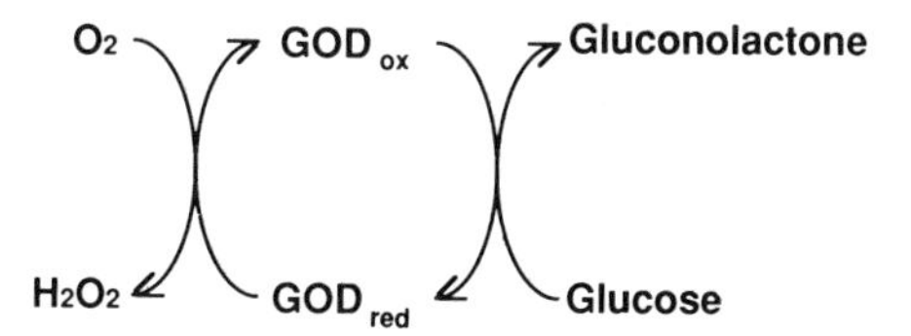

Figure 1 Reaction scheme of glucose oxidase.

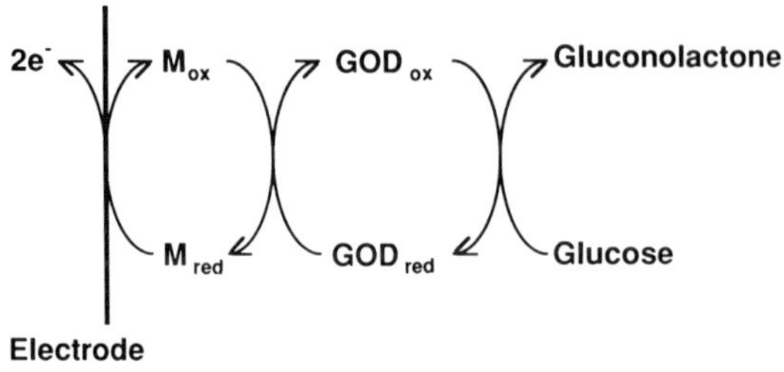

2e⁻ M_{ox} GOD_{ox} Gluconolactone

Electrode M_{red} GOD_{red} Glucose

Figure 2 Reaction scheme of a mediated glucose sensor.

ferrocene derivatives (Table 1) were also investigated by cyclic voltammetry as possible oxidants for GOD. Figure 3a shows a voltammogram of ferrocene monocarboxylic acid in the presence of glucose alone. Upon addition of GOD to the solution, a striking change in the voltammogram occurred (Fig. 3b). No peaks were observed, and a large catalytic current flowed at oxidation potentials. This behavior was particularly apparent at the lower scan rate and was indicative of the regeneration of ferrocene from ferricinium ion, the oxidized form of ferrocene, by the reduced form of GOD. GOD was maintained in this reduced state by the presence of substrate. The theory developed by Nicholson and Shain [11] was used to analyze the reaction. The second-order rate constant for the reaction between ferricinium carboxylic acid and GOD was obtained to be $2.01 \times 10^5\ l \cdot mol^{-1} \cdot s^{-1}$. Kinetic data obtained by using this technique, presented in Table 1, show that all of the ferrocene derivatives investigated acted as rapid oxidants of GOD.

It was known that 1,4-benzoquinone [12–16], ferricyanide [17], and tetrathiafulvalene (TTF) [18] acted as mediators. Senda et al. reported a glucose sensor composed of GOD and a carbon paste electrode that contained benzoquinone as a mediator [13]. This sensor detected glucose in the concentration range of 10–150

Table 1 Rates of Glucose Oxidation Measured at pH 7 and 25°C

Ferrocene derivative	E_0/mV vs. SCE	$10^{-5}\ k_s\ (mol^{-1}\,s^{-1})$
1,1′-Dimethyl	100	0.77
Ferrocene	165	0.26
Vinyl	250	0.30
Carboxy	275	2.01
1,1′-Dicarboxy	285	0.26
(Dimethylamino)methyl	400	5.25

Source: Ref. 5.

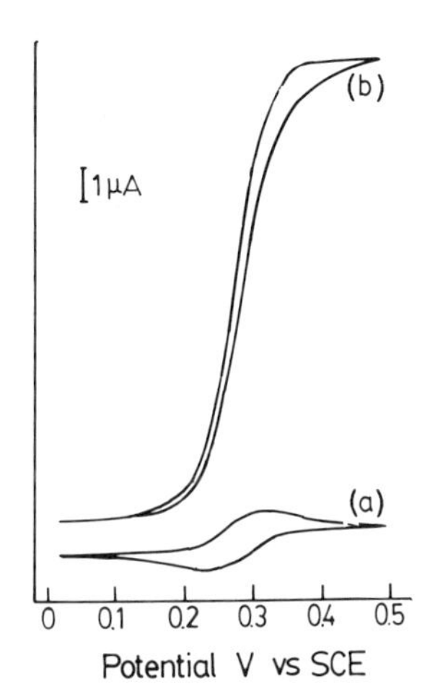

Figure 3 (a) DC cyclic voltammogram of ferrocene monocarboxylic acid (0.5 mM) at pH 7 and 25°C, in the presence of glucose (50 mM) at a scan rate of 1 $mV \cdot s^{-1}$. (b) As for (a), but with the addition of glucose oxidase (10.9 µM). (From Ref. 5.)

mM. Electron transfer also takes place between the TTF-tetracyanoquinodimethane conductive salt and GOD [19–22].

It is normally advantageous to have such mediators highly concentrated in the surface vicinity of the electrode. However, fast electron transfer is known to occur from a GOD molecule neither to the mediator, which is covalently modified on an electrode, nor to the redox polymer, such as polyvinylferrocene. This is because a GOD molecule is very large, 4.3 nm (Stokes radius) [23], and the redox center, FAD, is surrounded by a polypeptide shell. Therefore, it is preferable that the mediator be loaded freely on an electrode surface.

Electron transfer between GOD and oxygen also takes place in the presence of dissolved oxygen. Thus, the reoxidation of GOD in the mediated glucose sensor should be competitive between the mediator and oxygen. The current response of the mediated glucose sensor will be affected by oxygen either at an insufficient mediator concentration or at a high oxygen concentration. To resolve this problem, Yokoyama et al. developed an integrated micro glucose sensor, which consisted of two types of glucose sensors, a mediated glucose sensor and an O_2-based glucose sensor, which were formed on the same glass substrate [24]. The responses affected by dissolved oxygen were compensated for in order to determine the precise glucose concentration. GOD was immobilized on two working electrodes using bovine serum albumin and glutaraldehyde. DMFc was incorporated by dipping the GOD-immobilized electrodes into a DMFc dissolved acetone solution. An Ag/AgCl electrode was used as a reference electrode. The same counter and reference electrodes were used for both the mediated and the oxygen-based glucose sensors. The electrode potentials are maintained at +0.1 V and −0.3 V vs. Ag/AgCl, respectively. The characteristics of the mediated and oxygen-based glucose sensors were evaluated simultaneously in 0.1 M phosphate buffer. The response time of these glucose sensors was about 1 minute when 50 μl of 100 $g \cdot l^{-1}$ glucose sample solution was injected into 30 ml of phosphate buffer. Both glucose sensors responded to glucose in the presence of dissolved oxygen. However, the responses of both sensors to glucose depended on the dissolved oxygen concentration, so that the glucose concentration could be determined from the calibration curves for both sensors at various concentrations of dissolved oxygen.

The disadvantage of the mediated biosensors using DMFc has been indicated in recent years. Even if insoluble mediators such as DMFc are used, the oxidized form of DMFc is soluble in water, and DMFc on the electrode gradually is released with time. As described previously, a redox polymer can often be unavailable for GOD oxidation on account of inflexibility in the mediator. Foulds and Lowe reported GOD immobilization in a ferrocene-modified pyrrole polymer on a platinum electrode and use of this electrode as a glucose sensor [25]. In their former work, they showed that GOD was able to be immobilized when a polypyrrole-modified electrode was prepared by electrooxidation of pyrrole [26]. Ferrocene-modified polypyrrole was deposited by cycling the electrode potential between 0 and +1.0 V in an aqueous perchlorate solution containing pyrrole and [(Ferrocenyl)amidopropyl]pyrrole or ([(Ferrocenyl)amidopentyl]amidopropyl)-pyrrole. GOD was entrapped in the redox copolymer by cycling the working electrode potential between 0 and +0.8 V. Glucose was calibrated in the range of 1–100 mM under nitrogen condition. Hale et al. reported a siloxane polymer bearing a ferrocene group as an electron transfer mediator and fabricated a mediated glucose sensor using a carbon paste electrode [27]. Gregg and Heller synthesized a polymer incorporating an osmium-pyrridinium complex as a mediator [28]. These glucose sensors using redox polymers were affected by the dissolved oxygen concentration because the rate of the electron transfer from GOD to the redox site of the polymer was slow in comparison to oxygen.

II. GLUCOSE SENSORS BASED ON PQQ-DEPENDENT GLUCOSE DEHYDROGENASE

Many studies of GOD-based glucose sensors have been reported. However, these approaches are subject to the inherent limitations associated with the particular enzyme employed. As stated above, the principal disadvantage of GOD is susceptibility to oxygen interference. Alternatively, some researchers have worked on glucose sensors using an NAD^+-dependent glucose dehydrogenase (NAD^+-GDH) [29]. However, the requirement of the enzyme for NAD^+ considerably complicates the construction of practical sensors. The ideal enzyme for an amperometric glucose sensor would have no cofactor requirement and act only on a single substrate. Quinoprotein glucose dehydrogenase (PQQ-GDH) fulfills such requirements for incorporation into practical glucose sensors.

An amperometric glucose sensor based on PQQ-GDH was first reported by D'Costa and coworkers [30]. They purified PQQ-GDH from *Acinetobacter calcoaceticus* by HPLC and fabricated a quinoprotein-based glucose sensor. The second-order rate constants (k_s) for the reaction of GDH with several mediators were determined using a method based on the DC cyclic voltammetry technique, the theory of which was discussed previously [11]. The second-order charge transfer rate constant between PQQ-GDH and the mediators, *N,N,N',N'*-tetramethyl-*p*-phenylenediamine and ferrocene carboxylic acid, at 20°C were calculated to be 7.8×10^6 and 9.4×10^6 $l \cdot mol^{-1} \cdot s^{-1}$, respectively. Although these electroactive compounds were unrelated, the experimentally reported values of k_s were very close. When compared to the rate constants of other oxidoreductases, the value of PQQ-GDH was found to be double that of flavocytochrome b_2 and at least one order magnitude greater than galactose oxidase, xanthine oxidase, and pyruvate oxidase [31]. Furthermore, this result was in accordance with the early work of Hauge [32], which demonstrated that PQQ-GDH was up to 30 times more active than GOD. Such a high rate constant for PQQ-GDH shows that it can be used to modify an electrode for the detection and measurement of glucose by electrochemical means. This result also provides vital information about the type of mediators that can be used to couple PQQ-GDH to an electrode in an amperometric configuration.

Construction of a GDH-immobilized electrode fabricated by D'Costa is shown in Figure 4. DMFc/toluene solution was dropped on a porous graphite electrode, and the purified PQQ-GDH was immobilized onto the graphite using carbodiimide or glutaraldehyde. Figure 5 shows a response of the carbodiimide-treated GDH-immobilized electrode to glucose. The current density of this GDH electrode produced more than twice that of the graphite based-GOD electrode at 4 mM glucose concentration. The calibration curve showed that this sensor could detect glucose in the concentration range of 0.5–4 mM. When this sensor was treated with more PQQ-GDH in the presence of glutaraldehyde, the range of response was increased up to 15 mM. However, this treatment reduced the current response. The effect of oxygen was also examined. Since oxygen does not take part in the enzymatic reaction of PQQ-GDH, the response to glucose was unaffected by oxygen concentration. The response of this electrode to glucose was also unaffected by changes between anerobic and oxygen-saturated environments.

In clinical analysis, simultaneous measurement of several kinds of biological substrates is becoming more important. An integrated multibiosensor for simultaneous measurement of glucose and galactose concentration based on PQQ-GDH

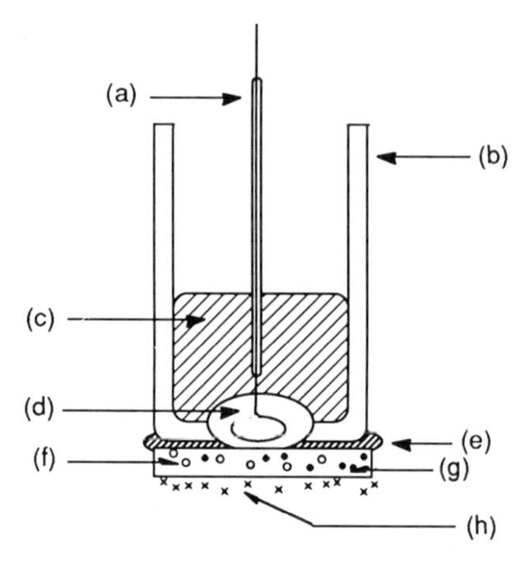

Figure 4 Construction of a GDH-based amperometric electrode: (a) connecting wire, (b) glass tubing, (c) insulating resin, (d) conducting resin, (e) araldite, (f) porous graphite-foil, (g) 1,1′-dimethylferrocene, (h) immobilized GDH. (From Ref. 30.)

and galactose oxidase (GAO) was reported [33]. Because no electrons can transfer from PQQ to oxygen, this glucose sensor is unaffected by the concentration of dissolved oxygen. GAO catalyzes galactose oxidation as follows:

$$\text{galactose} + \text{GAO(ox)} \rightarrow \text{galactohexoaldose} + \text{GAO(red)} \quad (1)$$

$$O_2 + \text{GAO(red)} \rightarrow H_2O_2 + \text{GAO(ox)} \quad (2)$$

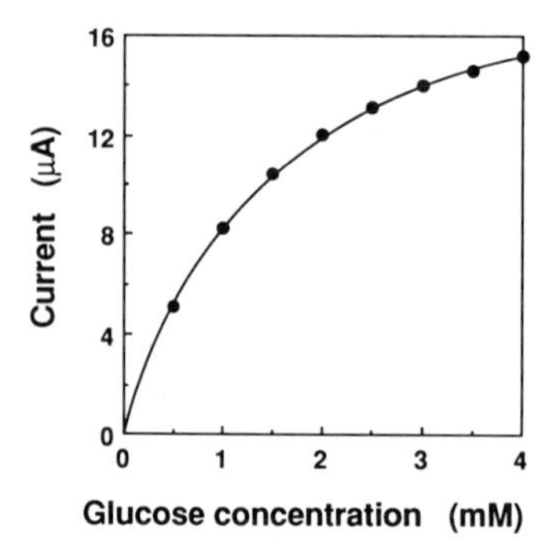

Figure 5 Response of a GDH-based biosensor to glucose concentration. Steady-state currents were measured in stirred 10 mM potassium phosphate buffer, pH 7.0, at 30°C. (From Ref. 30.)

Therefore, it is possible to measure the glucose and galactose concentration independently in the same solution even if PQQ-GDH and GAO are immobilized on two neighboring electrodes on the same glass surface. Glucose can be detected by the current increase resulting from the oxidation of the mediator, and galactose can be measured by the current decrease from the oxygen reduction. Four gold working electrodes and a counterelectrode were formed on a Corning 7059 glass substrate by vapor deposition, as shown in Figure 6. The area of the working electrode is 0.2 mm^2. Two of the working electrodes are used as the GDH and GAO immobilized electrodes. PQQ-GDH and GAO were immobilized on the electrodes as follows. Lyophilized bacterial membrane from *Pseudomonas fluorescens* which contained PQQ-GDH, GAO, and bovine serum albumin were dissolved in 10 mM HEPES buffer solution (pH 7.9) containing 10 mM $MgSO_4$ and 3 mM PQQ. The enzyme solution was spread on the electrode, and subsequently it was exposed to glutaraldehyde vapor at 30°C for 30 minutes. The assay procedure was described as follows. The enzyme electrodes and the reference electrode were immersed in 10 mM HEPES buffer solution (pH 7.4) containing ferrocene monocarboxylic acid with stirring at 30°C. The electrode potentials for the mediator-oxidizing and oxygen electrodes were maintained at +350 and −300 mV, respectively. After the current output stabilized, glucose or galactose solution was injected into the buffer solution and the current changes were measured. The measurement was carried out under air-saturated conditions. Figure 7 shows calibration curves of both electrodes to galactose. Such data implied that the reduced form of GAO mainly reacted with oxygen in the low concentration of galactose and reacted with both oxygen and mediators in the high concentration of galactose since the concentration of dissolved oxygen was not very high. Figure 8 shows the calibration curves of the mediated electrode for glucose in the presence of several concentrations of galactose. The response of the mediated sensor to glucose decreased with the increase in galactose concentration. Because GAO

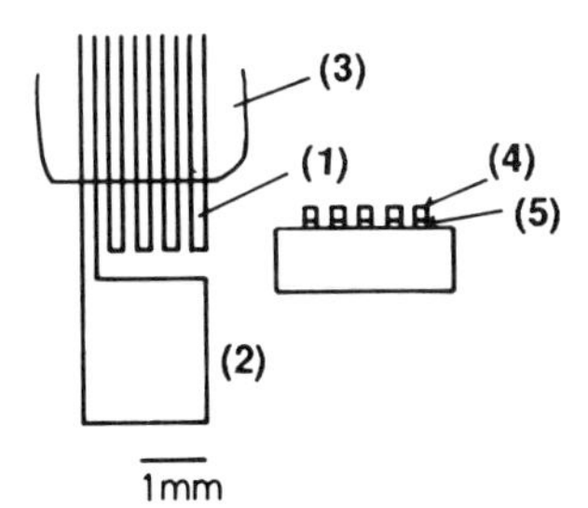

Figure 6 Structure of the integrated microelectrodes: (1) working electrode, (2) auxiliary electrode, (3) epoxy resin, (4) gold, (5) chromium. (From Ref. 33.)

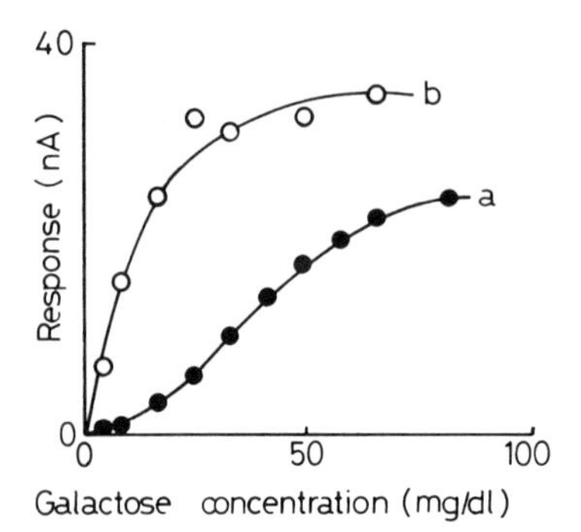

Figure 7 Calibration graph for (a) the oxidizing mediator electrode and (b) the oxygen electrode for galactose. The measurements were carried out under air using HEPES buffer solution (10 mM, pH 7.4) containing ferrocene monocarboxylic acid (2.17 mM). (From Ref. 33.)

reacts with both oxygen and mediators such as ferrocene carboxylic acid, the response of the mediator-oxidizing electrode to glucose depended on the galactose concentration. Therefore, the simultaneous determination of both glucose and galactose can be performed as follows. Galactose can be initially determined on the basis of the response of the oxygen electrode to galactose, and then the glucose concentration can also be determined by choosing an appropriate calibration curve for glucose in the presence of galactose.

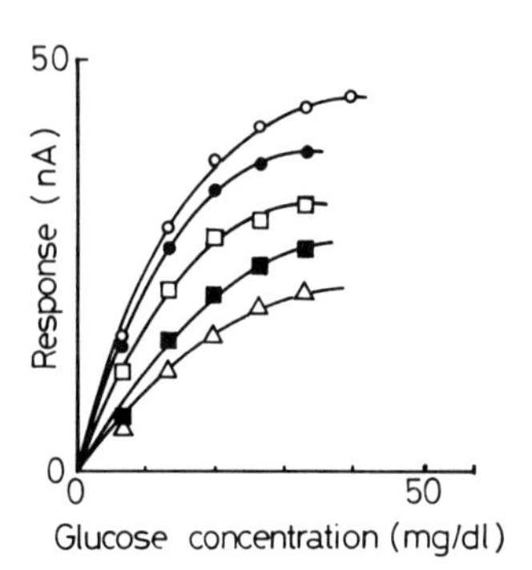

Figure 8 Response of the oxidizing mediators electrode to glucose at several galactose concentrations: (○) 0, (●) 16.6, (□) 33.1, (■) 49.5, (△) 65.8 mg·dl^{-1} (From Ref. 33.)

III. ALCOHOL SENSORS BASED ON PQQ-DEPENDENT ALCOHOL DEHYDROGENASE

The determination of alcohol concentration is probably one of the most important and routine tests for process control in the chemical industries, e.g., fermentation processes. There have been many reports of the determination of alcohol by spectrophotometry, gas chromatography (GC), and enzymatic methods. These methods, however, may involve complicated and delicate procedures. Resulting from the marked progress in techniques for immobilization of enzymes and microorganisms, various types of biosensors have been developed, and many studies have been reported on the application of biosensors using alcohol oxidase (AOD) for the determination of ethanol concentration [34–36]. These kinds of sensor, however, respond not only to ethanol but also to other alcohols and several organic acids. Furthermore, many of these sensors are stable for only 1–2 weeks.

Recently, Ameyama et al. reported the purification and the catalytic and molecular properties of a membrane-bound quinoprotein alcohol dehydrogenase (membrane-bound ADH) from *Gluconobacter suboxydans* [37] and *Acetobacter aceti* [38]. This enzyme is useful for the enzymatic determination of ethanol concentration because of its high substrate specificity and its unique reaction mechanisms [39]. The membrane-bound ADH rapidly catalyzed the oxidation of C_2–C_6 primary aliphatic alcohols, but not secondary or tertiary alcohols, phenols, or methanol. The membrane-bound ADH is part of a complex enzyme system in the cell membrane, which also includes an aldehyde dehydrogenase (ALDH) and an electron transfer system (Fig. 9).

We present here descriptions of a few alcohol sensors that use either the cell

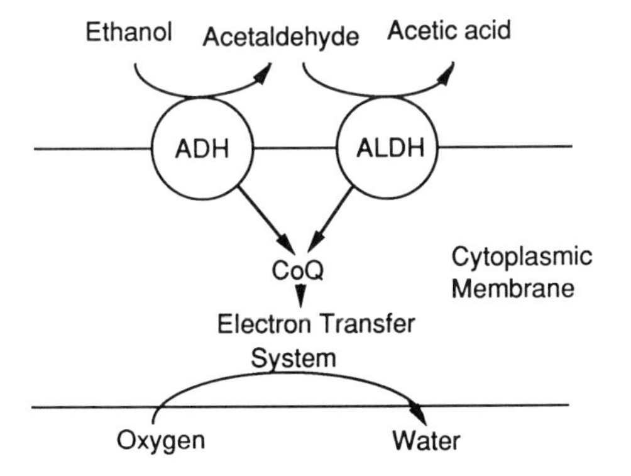

Figure 9 Ethanol oxidation pathway of *Gluconobacter suboxydans* ADH: alcohol dehydrogenase; ALDH: aldehyde dehydrogenase; CoQ: coenzyme Q (From Ref. 40.)

membrane that includes the complex enzyme system or the purified membrane-bound ADH.

A. Alcohol Sensor Using the Cell Membrane

As shown in Figure 9, ethanol is oxidized to acetic acid via acetaldehyde, and simultaneously oxygen is reduced to water in the cell membrane of acetic acid bacteria. By the use of this phenomenon, a simple amperometric alcohol sensor could be constructed using the cell membrane of an acetic acid bacterium (*Gluconobacter suboxydans* IFO 12528), a gas permeable membrane, and an oxygen electrode [40]. The sensor system is schematically illustrated in Figure 10. The cell membrane was adsorbed on a nitrocellulose filter and attached to the Teflon membrane of the oxygen electrode, and then these membranes were covered with a gas-permeable Teflon membrane. When the ethanol solution was

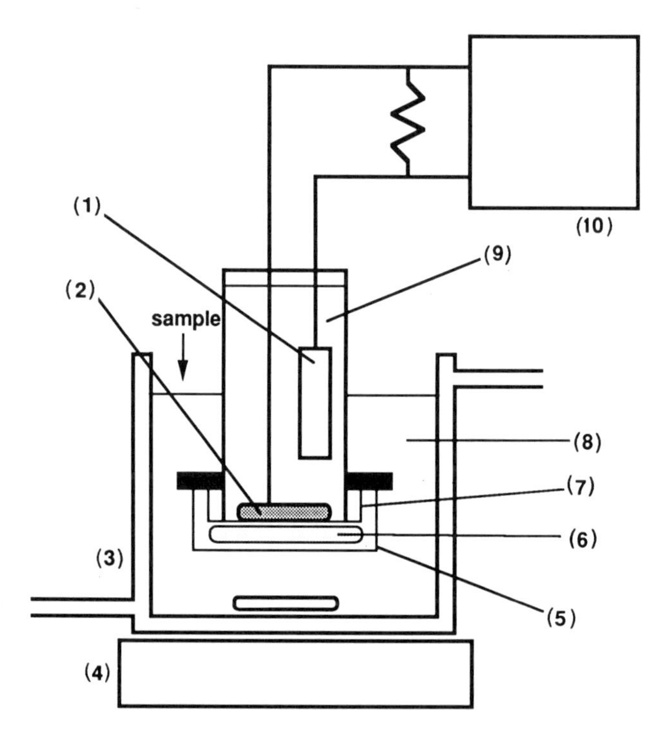

Figure 10 Schematic diagram of the alcohol sensor system. (1) Pb anode, (2) Pt cathode, (3) water jacket, (4) magnetic stirrer, (5) gas permeable membrane, (6) immobilized cell membrane, (7) porous Teflon membrane, (8) distilled water, (9) electrolyte (KOH), (10) recorder. (From Ref. 40.)

injected into the system, ethanol passed through the gas-permeable membrane and was converted to acetic acid by the complex enzyme system in the cell membrane. As a result, the current began to decrease because of the oxygen consumption caused by the action of the complex enzyme system and reached a steady state after approximately 3 minutes.

A linear relationship was observed between the current decrease and the ethanol concentration up to 25 $mg \cdot l^{-1}$. The current decrease was reproducible with a relative standard deviation of 5.2%. The sensor responded to ethanol, 1-propanol, and 1-butanol, but did not respond to methanol, secondary and tertiary alcohols, several organic acids, and glucose. In contrast, conventional enzyme electrode using AOD responded not only to various alcohols, e.g., methanol, ethanol, allyl alcohol, but also to various organic acids, e.g., acetate, formate, lactate. The quinoprotein-based sensor immobilizing the cell membrane, therefore, was more selective than the AOD-based alcohol sensor.

The stability of the sensor was also studied. Although the current response decrease diminished to approximately 70% of the initial response after 15 days, a sufficient response was obtained during this period. The sensor was stored at room temperature during the test period.

The ethanol level in serum was measured to demonstrate a typical application of this sensor. A comparison between serum ethanol levels determined by the sensor, y, and those by gas chromatography, x, gave a linear regression equation $y = 0.87x + 0.12$ ($r = 0.99$, $n = 9$). Although the sensor gave values of 87% of those obtained by GC, there was a good correlation between the two methods. The current response was reproducible with the relative standard deviation of 6.2% in nine experiments.

A microbiosensor for alcohol measurement was developed using an Ion Sensitive Field Effect Transistor (ISFET) and either the cell membrane of whole cells of *Gluconobacter suboxydans* IFO 12528 [41]. The ISFET was used as a micro pH sensitive device. When the cell membrane was immobilized on the ISFET using calcium alginate gel, it was able to detect the pH change in the gel including the cell membrane. The microbiosensor was applied to determination of the ethanol concentration in human serum, and a good correlation was observed between the results obtained by the microbiosensor and those by GC.

B. Alcohol Sensor Using Purified Membrane-Bound ADH

An alcohol sensor system using the purified membrane-bound quinoprotein ADH of *Gluconobacter suboxydans* IFO 12528 was developed to measure the ethanol concentration in a fermentation process [42]. Figure 11 represents the principle of the reaction system of the sensor. Ethanol is oxidized to acetaldehyde in the presence of the enzyme and subsequently the oxidized electron acceptor, ferri-

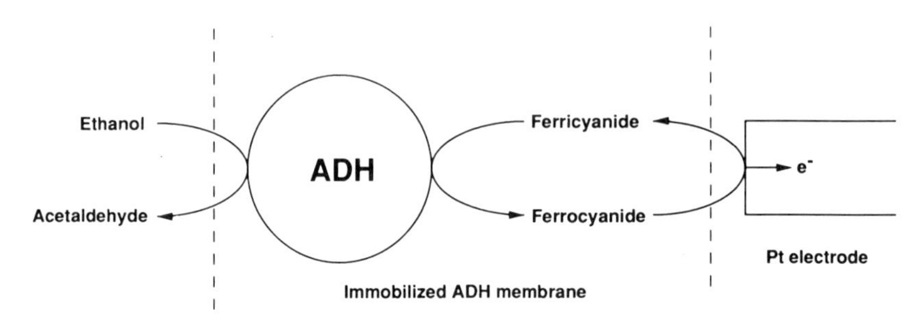

Figure 11 Schematic diagram of reaction system on the immobilized electrode. (From Ref. 42.)

cyanide, is reduced. When the electrode potential was poised at the potential to oxidize the reduced form of ferrocyanide, the reduced electron acceptor was reoxidized and electron transfer to the electrode occurred. Therefore, the current response was proportional to the ethanol concentration.

The enzyme was covalently immobilized on a membrane prepared with 1,8-diamino-4-aminomethyloctane, glutaraldehyde, and cellulose triacetate as materials. Figure 12 shows the experimental setup for the sensor. The immobilized enzyme was attached to a platinum disk electrode (3 mm in diameter) and covered with a dialysis membrane. To correct for the influence of the interfering substances, this alcohol sensor is compensated for a control electrode that has no immobilized enzyme. The potential of these Pt electrodes was maintained at +350 mV vs. Ag/AgCl. These electrodes were placed into the working buffer (McIlvaine buffer, pH 5.5, containing 10 mM potassium ferricyanide and 0.1% Triton-X 100) and maintained at 30°C. After an aliquot of the sample was injected into the working buffer, the current of the enzyme electrode immediately increased and reached a steady state value after approximately 5 minutes. On the other hand, the current of the control electrode was observed to show no change with the ethanol addition. The relationship between the current increase and the ethanol concentration is linear in the range of 0.1–5 mM ethanol. The current increase had a relative standard deviation of 1.9% ($n = 6$).

The optimum pH of the sensor was in the range of 4.5–5.5, and the sensitivity decreased markedly below pH 4. This pH profile is similar to that of the native enzyme reported by Adachi et al. [37]. The maximum sensitivity was attained at 45°C, and decreased at 50°C. The decrease in sensitivity at 50°C was caused by heat denaturation of the membrane-bound ADH. Similar denaturation above 45°C was observed with the native membrane-bound ADH. The characteristics of the native enzyme were reflected in the profile of this sensor, and the immobilization did not affect the pH and temperature profile of the enzyme. The response of the

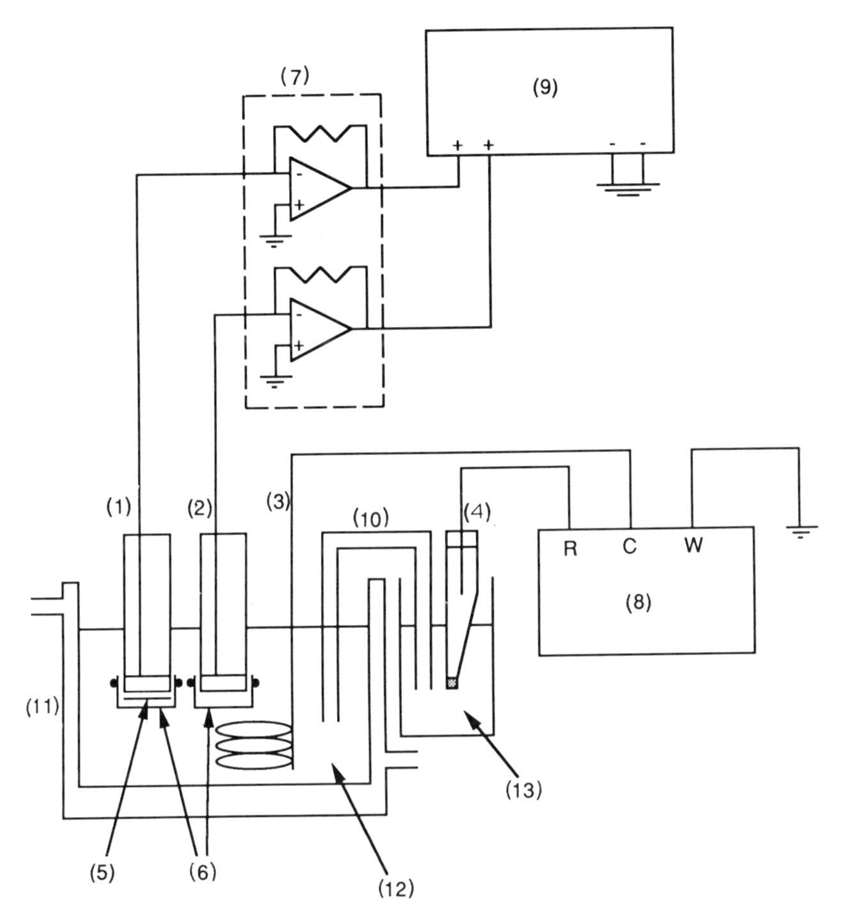

Figure 12 Schematic diagram of alcohol sensor system using membrane-bound ADH. (1) Enzyme electrode, (2) control electrode, (3) counter electrode, (4)Ag/AgCl electrode, (5) immobilized ADH membrane, (6) dialysis membrane, (7) current-voltage converter, (8) potentiostat, (9) recorder, (10) salt bridge, (11) water jacket, (12) reaction solution, (13) 1 M KCl. (From Ref. 42.)

sensor was decreased to 80% of the initial value after one month. In this period, the sensor was stored at 30°C in the working buffer. The sensor responded to ethanol, 1-propanol, and 1-butanol, but no responses were observed to methanol, 2-propanol, 2-butanol, glucose, acetic acid, and L-ascorbic acid. This sensor was more selective and stable than the enzyme sensor based on AOD because the membrane-bound ADH has a more stringent substrate specificity and the sensor system was compensated for by the control electrode to correct for the influence of the interfering substances. The sensor was applied to the determination of ethanol in alcohol beverages. The samples analyzed were beer, wine, sake, and syochu. Samples of high alcoholic content were assayed after appropriate dilution. A comparison of the ethanol concentrations determined by the sensor, Cs, with those determined by GC, Cg, gave a linear correlation: $Cs = 0.987Cg + 0.212$ ($n = 13$, $r = 0.999$). This sensor, therefore, should be applicable to the monitoring of ethanol in fermentation processes Furthermore, when the sensor is combined with an automatic sampling and diluting system, it seems to be very suitable for the control of the fermentation process because it is simple, inexpensive, and highly selective in comparison with other ethanol sensing systems using GC or enzymatic analysis.

REFERENCES

1. Turner, A. P. F., Karube, I., and Wilson, G. S., eds. (1987). *Biosensors—Fundamentals and Applications*, Oxford University Press, Oxford, United Kingdom.
2. Karube, I., Wang, Y., Tamiya, E., and Kawarai, M. (1987). Microbial sensor for vitamin B_{12}. *Anal. Chim. Acta 199*: 93.
3. Karube, I., Yokoyama, K., Sode, K., and Tamiya, E. (1989). Microbial BOD sensor utilizing thermophilic bacteria. *Anal. Lett. 22*: 791.
4. Suzuki, H., Tamiya, E., Karube, I., and Oshima, T. (1988). Carbon dioxide sensor using thermophilic bacteria. *Anal. Lett. 21*: 1323.
5. Cass, A. E. G., Davis, G., Francis, G. D., Hill, H. A. O., Anson, W. J., Higgins, I. J., Plotkin, E. V., Scott, L. D. L., and Turner, A. P. F (1984). Ferrocene-mediated enzyme electrode for amperometric determination of glucose. *Anal. Chem. 56*: 667.
6. Turner, A. P. F., Anson, W. J., Higgins, I. J., Bell, J. M., Colby, J., Davis, G., and Hill, H. A. O. (1984). Carbon monoxide: Acceptor oxidoreductase from *Pseudomonas thermocarboxy-dovorans* strain C2 and its use in a carbon monoxide sensor. *Anal. Chim. Acta 163*: 161.
7. Dicks, J. M., Anson, W. J., Davis, G., and Turner, A. P. F (1986). Mediated amperometric biosensors for D-galactose, glycelate and L-amino acids based on a ferrocene-modified carbon paste electrode. *Anal. Chim. Acta 182*: 103.
8. Brooks, S. L., Ashby, R. E., Turner, A. P. F., Calder, M. R., and Clarke, D. J. (1987). Development of an on-line glucose sensor for fermentation monitoring. *Biosensors 3*: 45.
9. Hall, G. F., Best, D. J., and Turner, A. P. F (1988). Amperometric enzyme electrode for the determination of phenols in chloroform. *Enzyme Microb. Technol. 10*: 543.
10. Bradley, J., Kidd, A. J., Anderson, P. A., Dear, A. M., Ashby, R. E., and Turner, A. P. F (1989). Rapid determination of the glucose content of molasses using a biosensor. *Analyst 114*: 375.
11. Nicholson, R. S., and Shain, I. (1964). Theory of stationary electrode polarography. *Anal. Chem. 36*: 706.
12. Ikeda, T., Katasho, I., Kamei, M., and Senda, M. (1984). Electrocatalysis with a glucose-oxidase-immobilized graphite electrode. *Agric. Biol. Chem. 48*: 1969.

13. Ikeda, T., Katasho, I., and Senda M. (1985). Glucose oxidase-immobilized benzoquinone-mixed carbon paste electrode with pre-minigrid. *Anal. Sci. 1*: 455.
14. Ikeda, T., Hamada, H., Miki, K., and Senda, M. (1985). Glucose oxidase-immobilized benzoquinone-carbon paste electrode as a glucose sensor. *Agric. Biol. Chem. 49*: 541.
15. Ikeda, T., Hamada, H., and Senda, M. (1986). Electrocatalytic oxidation of glucose at a glucose oxidase-immobilized benzoquinone-mixed carbon paste electrode. *Agric. Biol. Chem. 50*: 883.
16. Senda, M., Ikeda, T., Hiasa, H., and Katasho, I. (1987). Biocatalyst electrodes and their use in electrochemical syntheses. *Nihon Kagaku Kaishi (Japanese)* 1987: 358.
17. Schlapfer, P., Mindt, W., and Racine, D. (1974). Electrochemical measurement of glucose using various electron acceptors. *Clin. Chim. Acta. 57*: 283.
18. Gunasingham, H., Tan, C. H., and Ng, H. M. (1990). Pulsed amperometric detection of glucose using a mediated enzyme electrode. *J. Electroanal. Chem. 287*: 349.
19. Albery, W. J. Bartlett, P. N., and Craston, D. H. (1985). Amperometric enzyme electrodes Part II. Conducting salts as electrode materials for the oxidation of glucose oxidase. *J. Electroanal. Chem. 194*: 223.
20. Hill, B. S., and Wilson, G. S. (1988). Enzyme electrocatalysis at organic salt electrodes. *J. Electroanal. Chem. 252*: 125.
21. Hale, P. D., and Wightman, R. M. (1988). Enzyme-modified tetrathiafulvalene tetracyanoquinodimethane microelectrodes: Direct amperometric detection of acetylcholine and choline. *Mol. Cryst. Liq. Cryst. 160*: 269.
22. Gunasingham, H., and Tan, C. H. (1990). Conducting organic salt amperometric glucose sensor in continuous-flow monitoring using a wall-jet cell. *Anal. Chim. Acta 229*: 83.
23. Degani, Y., and Heller, A. (1987). Direct electrical communication between chemically modified enzymes and metal electrodes 1. Electron transfer from glucose oxidase to metal electrodes via electron relays, bound covalently to the enzyme. *J. Phys. Chem. 91*: 1285.
24. Yokoyama, K., Tamiya, E., and Karube, I. (1989). Performance of an integrated biosensor composed of a mediated and an oxygen-based glucose sensors under unknown oxygen tension. *Anal. Lett. 22*: 2949.
25. Foulds, N. C., and Lowe, C. R. (1988). Immobilization of glucose oxidase in ferrocene-modified pyrrole polymers. *Anal. Chem. 60*: 2473.
26. Foulds, N. C., and Lowe, C. R. (1986). Enzyme entrapment in electrically conducting polymers. *J. Chem. Soc., Faraday Trans., I 82*: 1259.
27. Hale, P. D., Inagaki, T., Karan, H. I., Okamoto, Y., and Skotheim, T. A. (1989). A new class of amperometric biosensor incorporating a polymeric electron-transfer mediator. *J. Am. Chem. Soc. 111*: 3482.
28. Gregg, B. A., and Heller, A. (1990). Cross-linked redox gels containing glucose oxidase for amperometric biosensor applications. *Anal. Chem. 62*: 258.
29. György, M. V., Appelqvist, R., and Gorton, L. (1986). A glucose sensor based on glucose dehydrogenase adsorbed on a modified carbon electrode. *Anal. Chim. Acta 179*: 371.
30. D'Costa, E. J., Higgins, I. J., and Turner, A. P. F. (1986). Quinoprotein glucose dehydrogenase and its application in an amperometric glucose sensor. *Biosensors 2*: 71.

31. Davis, G. (1985). Electrochemical techniques for the development of amperometric biosensors. *Biosensors 1*: 161.
32. Hauge, J. G., and Hallberg, P. A. (1964). Solubilization and properties of the structurally-bound glucose dehydrogenase of Bacterium antitratum. *Biochim. Biophys. Acta 45*: 263.
33. Yokoyama, K., Sode, K., Tamiya, E., and Karube, I. (1989). Integrated biosensor for glucose and galactose. *Anal. Chim. Acta 218*: 137.
34. Nanjo, M., and Guilbault, G. G. (1975). Amperometric determination of alcohol and carboxylic acids with an immobilized alcohol oxidase enzyme electrode. *Anal. Chim. Acta 75*: 169.
35. Gulberg, E. L., and Christian, G. D. (1981). The use of immobilized alcohol oxidase in the continuous flow determination of ethanol with an oxygen electrode. *Anal. Chim. Acta 123*: 125.
36. Mason, M. (1963). Ethanol determination in wine with an immobilized enzyme electrode. *Am. J. Enol. Vitic. 34*: 173.
37. Adachi, O., Tayama, K., Shinagawa, E., Matsushita, K., and Ameyama, M. (1978). Purification and characterization of particulate alcohol dehydrogenase from *Gluconobacter suboxydans*. *Agric. Biol. Chem. 42*: 2045.
38. Adachi, O., Miyagawa, E., Shinagawa, E., Matsushita, K., and Ameyama, M. (1978). Purification and characterization of particulate alcohol dehydrogenase from *Acetobacter aceti*. *Agric. Biol. Chem. 42*: 2331.
39. Ameyama, M., Tayama, K., Miyagawa, E., Shinagawa, E., Matsushita, K., and Adachi, O. (1978). A new enzymatic microde-termination procedure for ethanol with particular alcohol dehydrogenase from acetic acid bacteria. *Agric. Biol. Chem. 42*: 2063.
40. Kitagawa, Y., Ameyama, M., Nakashima, K., Tamiya, E., and Karube, I. (1987). Amperometric alcohol sensor based on an immobilized bacteria cell membrane. *Analyst 112*: 1747.
41. Tamiya, E., Karube, I., Kitagawa, Y., Ameyama, M., and Nakashima, K. (1988). Alcohol-FET sensor based on a complex cell membrane enzyme system. *Anal. Chim. Acta 207*: 77.
42. Kitagawa, Y., Kitabatake, K., Kubo, I., Tamiya, E., and Karube, I. (1989). Alcohol sensor based on membrane-bound alcohol dehydrogenase. *Anal. Chim. Acta 218*: 61.

Index